IET MATERIALS, CIRCUITS AND DEVICES SERIES 66

Frontiers in Hardware Security and Trust

Other volumes in this series:

Volume 2	**Analogue IC Design: The current-mode approach** C. Toumazou, F.J. Lidgey and D.G. Haigh (Editors)
Volume 3	**Analogue–Digital ASICs: Circuit techniques, design tools and applications** R.S. Soin, F. Maloberti and J. France (Editors)
Volume 4	**Algorithmic and Knowledge-Based CAD for VLSI** G.E. Taylor and G. Russell (Editors)
Volume 5	**Switched Currents: An analogue technique for digital technology** C. Toumazou, J.B.C. Hughes and N.C. Battersby (Editors)
Volume 6	**High-Frequency Circuit Engineering** F. Nibler *et al.*
Volume 8	**Low-Power High-Frequency Microelectronics: A unified approach** G. Machado (Editor)
Volume 9	**VLSI Testing: Digital and mixed analogue/digital techniques** S.L. Hurst
Volume 10	**Distributed Feedback Semiconductor Lasers** J.E. Carroll, J.E.A. Whiteaway and R.G.S. Plumb
Volume 11	**Selected Topics in Advanced Solid State and Fibre Optic Sensors** S.M. Vaezi-Nejad (Editor)
Volume 12	**Strained Silicon Heterostructures: Materials and devices** C.K. Maiti, N.B. Chakrabarti and S.K. Ray
Volume 13	**RFIC and MMIC Design and Technology** I.D. Robertson and S. Lucyzyn (Editors)
Volume 14	**Design of High Frequency Integrated Analogue Filters** Y. Sun (Editor)
Volume 15	**Foundations of Digital Signal Processing: Theory, algorithms and hardware design** P. Gaydecki
Volume 16	**Wireless Communications Circuits and Systems** Y. Sun (Editor)
Volume 17	**The Switching Function: Analysis of power electronic circuits** C. Marouchos
Volume 18	**System on Chip: Next generation electronics** B. Al-Hashimi (Editor)
Volume 19	**Test and Diagnosis of Analogue, Mixed-Signal and RF Integrated Circuits: The system on chip approach** Y. Sun (Editor)
Volume 20	**Low Power and Low Voltage Circuit Design with the FGMOS Transistor** E. Rodriguez-Villegas
Volume 21	**Technology Computer Aided Design for Si, SiGe and GaAs Integrated Circuits** C.K. Maiti and G.A. Armstrong
Volume 22	**Nanotechnologies** M. Wautelet *et al.*
Volume 23	**Understandable Electric Circuits** M. Wang
Volume 24	**Fundamentals of Electromagnetic Levitation: Engineering sustainability through efficiency** A.J. Sangster
Volume 25	**Optical MEMS for Chemical Analysis and Biomedicine** H. Jiang (Editor)
Volume 26	**High Speed Data Converters** A.M.A. Ali
Volume 27	**Nano-Scaled Semiconductor Devices** E.A. Gutiérrez-D (Editor)
Volume 28	**Security and Privacy for Big Data, Cloud Computing and Applications** L. Wang, W. Ren, K.R. Choo and F. Xhafa (Editors)
Volume 29	**Nano-CMOS and Post-CMOS Electronics: Devices and modelling** S.P. Mohanty and A. Srivastava
Volume 30	**Nano-CMOS and Post-CMOS Electronics: Circuits and design** S.P. Mohanty and A. Srivastava
Volume 32	**Oscillator Circuits: Frontiers in design, analysis and applications** Y. Nishio (Editor)
Volume 33	**High Frequency MOSFET Gate Drivers** Z. Zhang and Y. Liu
Volume 34	**RF and Microwave Module Level Design and Integration** M. Almalkawi
Volume 35	**Design of Terahertz CMOS Integrated Circuits for High-Speed Wireless Communication** M. Fujishima and S. Amakawa
Volume 38	**System Design with Memristor Technologies** L. Guckert and E.E. Swartzlander Jr.
Volume 39	**Functionality-Enhanced Devices: An alternative to Moore's law** P.-E. Gaillardon (Editor)
Volume 40	**Digitally Enhanced Mixed Signal Systems** C. Jabbour, P. Desgreys and D. Dallett (Editors)

Volume 43 **Negative Group Delay Devices: From concepts to applications** B. Ravelo (Editor)
Volume 45 **Characterisation and Control of Defects in Semiconductors** F. Tuomisto (Editor)
Volume 47 **Understandable Electric Circuits: Key concepts, 2nd Edition** M. Wang
Volume 48 **Gyrators, Simulated Inductors and Related Immittances: Realizations and applications** R. Senani, D.R. Bhaskar, V.K. Singh and A.K. Singh
Volume 49 **Advanced Technologies for Next Generation integrated Circuits** A. Srivastava and S. Mohanty (Editors)
Volume 51 **Modelling Methodologies in Analogue Integrated Circuit Design** G. Dundar and M.B. Yelten (Editors)
Volume 53 **VLSI Architectures for Future Video Coding** M. Martina (Editor)
Volume 54 **Advances in High-Power Fiber and Diode Laser Engineering** I. Divliansky (Editor)
Volume 55 **Hardware Architectures for Deep Learning** M. Daneshtalab and M. Modarressi
Volume 57 **Cross-Layer Reliability of Computing Systems** G. Di Natale, A. Bosio, R. Canal, S. Di Carlo and D. Gizopoulos (Editors)
Volume 58 **Magnetorheological Materials and Their Applications** S. Choi and W. Li (Editors)
Volume 59 **Analysis and Design of CMOS Clocking Circuits for Low Phase Noise** W. Bae and D.K. Jeong
Volume 60 **IP Core Protection and Hardware-Assisted Security for Consumer Electronics** A. Sengupta and S. Mohanty
Volume 64 **Phase-Locked Frequency Generation and Clocking: Architectures and circuits for modem wireless and wireline systems** W. Rhee (Editor)
Volume 65 **MEMS Resonator Filters** R.M. Patrikar (Editor)
Volume 67 **Frontiers in Securing IP Cores; Forensic detective control and obfuscation techniques** A. Sengupta
Volume 68 **High Quality Liquid Crystal Displays and Smart Devices: Vol. 1 and Vol. 2** S. Ishihara, S. Kobayashi and Y. Ukai (Editors)
Volume 69 **Fibre Bragg Gratings in Harsh and Space Environments: Principles and applications** B. Aïssa, E.I. Haddad, R.V. Kruzelecky and W.R. Jamroz
Volume 70 **Self-Healing Materials: From fundamental concepts to advanced space and electronics applications, 2nd Edition** B. Aïssa, E.I. Haddad, R.V. Kruzelecky and W.R. Jamroz
Volume 71 **Radio Frequency and Microwave Power Amplifiers: Vol. 1 and Vol. 2** A. Grebennikov (Editor)
Volume 72 **Tensorial Analysis of Networks (TAN) Modelling for PCB Signal Integrity and EMC Analysis** B. Ravelo and Z. Xu (Editors)
Volume 73 **VLSI and Post-CMOS Electronics Volume 1: VLSI and post-CMOS electronics and Volume 2: Materials, devices and interconnects** R. Dhiman and R. Chandel (Editors)

Frontiers in Hardware Security and Trust

Theory, design and practice

Edited by
Chip Hong Chang and Yuan Cao

The Institution of Engineering and Technology

Published by The Institution of Engineering and Technology, London, United Kingdom

The Institution of Engineering and Technology is registered as a Charity in England & Wales (no. 211014) and Scotland (no. SC038698).

First published 2020

The Institution of Engineering and Technology
Michael Faraday House
Six Hills Way, Stevenage
Herts, SG1 2AY, United Kingdom

www.theiet.org

British Library Cataloguing in Publication Data
A catalogue record for this product is available from the British Library

ISBN 978-1-78561-927-4 (hardback)
ISBN 978-1-78561-928-1 (PDF)

Typeset in India by MPS Limited

Contents

About the editors

Chip Hong Chang is a tenured associate professor at the School of Electrical and Electronic Engineering (EEE), Nanyang Technological University, Singapore. He has edited and coedited five books and published 13 chapters, over 100 international journal papers and more than 180 refereed international conference papers. His research interests are hardware security, residue and unconventional number systems, low-power arithmetic circuits, digital filter design and digital image processing. He is a fellow of the IEEE and the IET.

Yuan Cao is a professor at the College of Internet of Things Engineering, Hohai University, China. His research interests include hardware security, silicon physical unclonable function and analog/mixed-signal VLSI circuits and systems.

Preface

With the rollout of gigabit speed communications of 5G becoming a reality, the electronics industry is now bombarded with intensely ubiquitous new, versatile and hype applications, such as augmented reality, virtual reality, artificial intelligence, self-drive cars, Internet of Things (IoT) and intelligent vehicular network. Hardware acceleration and miniaturization powered by advanced semiconductor manufacturing technologies play increasingly important roles to support the significant increase in speed and bandwidth, reliability challenge and energy-cost optimization from centralized cloud computing to distributed edge computing systems. The era of emerging technologies has become a hotbed for new attack vectors on hardware vulnerabilities as security remains an afterthought or has been actively neglected in the hardware design process to avoid missing the golden opportunity of launching new products or introducing new product features ahead of competitors. Recent years have witnessed the release of many hardware vulnerabilities from household names, integrated circuit (IC) design and hardware manufacturing companies that put the whole world of consumer electronics in alert. Meltdown and Spectre are two most well-known hardware flaws that cost a consortium effort of CPU makers, device manufacturers and operating system vendors to patch. These security weaknesses stemmed from the speculative execution to accelerate performance in modern processors. Another classic security hardware exploit is Row hammer on dynamic random-access memory (DRAM). Specially crafted memory access patterns can be used to trigger the unintended coupling of neighboring bits by repeatedly accessing the same memory rows. Row hammer can be used not only for privileged escalation computer exploits, but also through fast network connection to leak or change the memory content of the victim computers. Other equally alarming incidents are the hijacking of Jeep Cherokee in the highway by exploiting Uconnect, an internet-connected computer feature used in thousands of vehicles for in-car entertainment, navigation, phone calls and Wi-Fi access, as well as the Hot Lotto fraud scandal that exploits the compromised random number generator. All in all, the open nature of IoT devices, lack of hardware resources in the edge devices and even the globalization and vertical disintegration of hardware supply chain make the hardware security problems uniquely challenging.

Thanks to the intelligent industry practitioners and academic researchers in the hardware security community, various powerful and effective countermeasures have been proposed over the years to thwart advanced hardware attacks. These technologies span a wide range of electronic devices and products throughout their life cycles and all level of design abstractions, from HDL, synthesis, layout, testing and verification to chip assembly, PCB and system integration, etc. They also tackle impending applications such as post-quantum cryptography, homomorphic encryption, reversible computing, biochips, internet of vehicles, etc. New cryptographic primitives with low power, small footprint and high speed are also developed to enhance security solutions in resource-constrained application scenarios. At this time point, it is important and desirable to comprehensively and effectively review and capture the state-of-the-art development in hardware-based attacks and countermeasures to shed light on potential trend and new direction of this exciting area of research. With this purpose in mind, we embarked this book project dedicated to the very forefront topics in hardware security and trust, which are contributed by the well-established and active researchers in their respective domains.

The book is organized as follows. The first part includes three chapters on new threat models in hardware. In Chapter 1, the modern synthesis approaches on reversible circuits are comprehensively reviewed. New intellectual property (IP) and IC attacks can be launched by exploiting the telltale sign of these synthesis approaches. Countermeasures are also proposed toward the end of this chapter. Chapter 2 presents a new persistent fault attack. The method is simple and yet efficient in breaking some typical AES-128 implementations. In Chapter 3, deployment of electromagnetic compatibility techniques in the design of an IC chip with security functionality is summarized. The side channel information leakage is exploited through the electromagnetic interference of an IC chip.

In the second part, nine state-of-the-art hardware security countermeasures are introduced. In Chapter 4, various hardware obfuscation technologies are investigated for IP protection at different supply chain stages of the semiconductor life cycle. The challenges and new research trend on obfuscation are given. In Chapter 5, a scalable system-on-chip (SoC) bus verification framework is presented. It verifies the security properties of SoC bus implementation where the bus protocol plays the role of a golden reference. Chapter 6 presents a few modern silicon-based true random number generators. The taxonomy is given based on the entropy source. Modern post-processing techniques to improve the randomness and security are also reviewed. Chapter 7 discusses micro-architectural attacks and the countermeasures on public-key implementations. It also develops the first practical software triggered fault analysis of a 2,048-bit RSA implementation by controlling the DRAM vulnerability widely known as the "row hammer" effect. Chapter 8 studies possible solutions to defend a new type of stealthy rootkit attack called CacheKit. It exploits cache incoherence and cache locking to evade detection used by introspection tool against rootkit, making it possible for malicious software to be run undetected in the cache. Chapter 9 exposes the dark side of deep neural networks (DNNs). It reviews existing popular deep learning models and points out the potential information leakage and backdoors of their mapping into hardware that can lead to exploitation for misclassification attacks. It also suggests some defensive mechanisms and relates the taxonomy of current methodologies to achieve transparency with the explainable DNNs. In Chapter 10, two use cases that utilize the feature of non-digital components to improve the security of a given system are studied. One case is the Face Flashing by using light reflections to secure liveness detections. The other case is the secure mobile payment that capitalizes on the imperfection of LCD screens. Chapter 11 discusses the most recent constructions of homomorphic encryption, as well as the algorithms and architectures for the efficient implementation of some performance bottleneck operations in various hardware platforms. Motivated by the effectiveness in mitigating fault injection attacks by embedding security-oriented codes in hardware, Chapter 12 examines whether the use of robust codes with a deterministic encoder can degrade security. The analysis in this chapter shows that given a bound for acceptable information leakage, the designer can easily choose the number of redundant bits required to detect the attack before this bound is reached.

The last part of this book presents three hardware security topics in the communication system. Chapter 13 focuses on the issue of radio resource allocation in overlaying cognitive radio systems, while the primary system employs physical-layer security (PLS) techniques. It describes a radio resource allocation framework to optimize both the energy efficiency of a cognitive communication system while preserving the confidentiality of the primary system. In Chapter 14, typical physical-layer threats in 5G wireless networks are reviewed, including eavesdropping, contamination, spoofing and jamming attacks. It surveys the state-of-the-art research in PLS techniques to alleviate the physical-layer threats in massive multiple input multiple output (MIMO) and millimeter wave (mmWave) 5G networks. PLS schemes based on mmWave and massive MIMO in other 5G communication scenarios are also discussed in this chapter. Finally, Chapter 15 provides an in-depth review of the security of in-vehicle network (IVN). It elaborates on how to manipulate a vehicle in practice without physical access. Toward the end, future research directions on IVN are proposed.

Hardware security is a fashionable research area in both industry and academy. Its scope is growing perpetually with increasing awareness. This book tries to convey the most up-to-date topics and development in the area. We sincerely hope that it is not only a valuable reference for the target readers and researchers in the hardware security community, but the topics discussed in it will also inspire and encourage other researchers and graduate students to jointly participate in the effort to uncover new vulnerabilities and more robust solutions toward improving the trustworthiness of future electronics and computing systems.

Part I

Hardware security threats

Chapter 1

IP/IC piracy threats of reversible circuits

Samah Mohamed Saeed[1]

In recent years, there has been growing interest in reversible computing due to its emerging applications. While designing reversible circuits is of a primary concern, their security implications receive less attention. A reversible synthesis approach generates a reversible circuit for a given target function. This process typically results in additional inputs and outputs to support reversibility. Thus, a reversible circuit realizes not only the target function but also many other functions by varying the value of the additional inputs. This chapter provides a detailed analysis to assess the difficulty in extracting the target function of the reversible circuit. The ultimate goal is to leverage the redundancy of reversible circuits toward security. We first describe reversible circuits and state-of-the-art synthesis approaches to generate these circuits. In particular, we review the telltale signs of these synthesis approaches. We then show how an attacker can exploit these telltale signs to launch IP/IC (intellectual property/integrated circuit) piracy attacks on reversible circuits. Finally, we describe potential countermeasures to thwart these attacks.

1.1 Introduction

We have witnessed progress in reversible computing applications over the past few years. Efforts are underway to design reversible circuits for energy-efficient computation [1–6] and quantum computing [7,8]. The design flow of reversible circuits depends on their applications. While reversible circuits in some of these applications follow the same design flow as conventional circuits (such as low-power design), other applications enforce a different design flow (such as quantum computing). Despite the variation of the reversible circuits' design flow and their underlying technologies, their core concept relies on reversibility.

The widening scope of the reversible circuit has prompted researchers to think of its security vulnerabilities. Reversible circuits are exposed to different attacks that can be applied to conventional ICs under different threat models depending on

[1]Electrical Engineering Department, City College of the City University of New York, New York, USA

their applications.* The distributed design flow of ICs opens backdoor for attackers to launch several attacks, including the insertion of malicious gates (Trojan), IP/IC piracy, reverse engineering, and side channel analysis [9–14]. Similarly, reversible circuits are exposed to different security risks. For instance, an attacker can insert malicious reversible gates (Trojans) into reversible circuits. The difficulty of detecting Trojans in reversible circuits that alter their functionality was indeed analyzed recently, which shows that pre-defined test patterns can trigger and, thus, detect Trojans of small size only [15,16]. This analysis motivates the need to hide the functionality of the reversible circuit, which makes Trojans of different sizes easily detectable by test patterns.

A classical function, referred to as a *target* function, is embedded into a reversible circuit, which may potentially result in additional inputs and outputs, referred to as *ancillary inputs* and *garbage outputs*. Synthesis approaches are used to automate the process of generating reversible circuits. The target function is activated when the ancillary inputs are initialized to the desired value. Therefore, there are exponential number of functions embedded into the reversible circuits depending on the location and the value of the ancillary inputs and the location of the garbage outputs. This naturally raises the question of how difficult it is to recover the target function of the reversible circuit. This chapter answers the previous question. While identifying the target function of a given reversible circuit appears to be hard, it primarily depends on the synthesis approach of the reversible circuits.

In this chapter, we revise the IP/IC piracy attacks on reversible circuits, which were introduced in [17–21]. The attacker first identifies the synthesis approach that generates the reversible circuit. The knowledge of the synthesis approach is vital to determine the location (if unknown) of the ancillary inputs and the garbage outputs and the value of the ancillary inputs to recover the target function. We show the feasibility of these attacks under different threat models. Next, we illustrate how to leverage the redundancy of reversible circuits to enhance the security. We analyze potential countermeasures to hide the target function [20]. While the discussion in this chapter is generic and applicable to different synthesis approaches, we focus on four state-of-the-art synthesis approaches (exclusive sum of products (ESOP), binary decision diagrams (BDD), quantum multivalued decision diagrams (QMDDs), and transformation-based synthesis (TBS)) as case studies.

The remainder of this chapter is organized as follows: Section 1.2 provides the background on reversible logic, their synthesis, and optimization techniques. Section 1.3 illustrates the motivation and the threat model. Section 1.4 provides a detailed description of the IC/IP piracy attacks on reversible circuits. Section 1.5 describes potential countermeasures to hide the target function of a reversible circuit. Finally, we conclude the chapter in Section 1.6.

*Reversible circuits follow different computation paradigm, rely on new technologies, and serve different applications.

1.2 Reversible logic

1.2.1 Reversible circuits

A function $f : B^n \rightarrow B^m$ is *reversible*, if and only if $n = m$, and each input combination maps to a unique output combination. The reversible function can be computed in two directions; the input uniquely determines the output and the output uniquely maps to the input.

Example 1.1. *A function $f : (x, y) \rightarrow (\overline{x}, y)$ is* reversible *since the number of inputs and the number of outputs are equal, and each input combination maps to a unique output combination.*

A reversible logic gate is the building block of the reversible circuit. We focus on *Toffoli gate $TOF(C, t)$*, which is the most commonly used reversible gate [22]. It consists of a set of positive or negative control lines and a single target line. If the number of control lines is zero, the Toffoli gate represents a NOT gate. A Toffoli gate with k control lines behaves as $f : (c_1, c_2, \ldots, c_k, t) \rightarrow (c_1, c_2, \ldots, c_k, ((c_1 c_2 \cdots c_k) \oplus t))$, where all the control lines remain the same, while the target line t is inverted if all the positive (negative) control lines are set to 1 (0). While the number of reversible gates provides an insight on the cost of the reversible circuits, it omits the variation in the number of control lines of different reversible gates. Thus, we define the reversible circuit cost as a function of the weighted sum of the reversible gates [23], for simplicity, while advanced metrics can also be used to measure the reversible circuit cost [24,25]. The cost of a Toffoli gate composed of C positive or negative control lines is computed as $2^{C+1} - 3$ with the exception of a Toffoli gate with negative control lines only, which costs $2^{C+1} - 1$.

Example 1.2. *Figure 1.1 shows a reversible circuit composed of three circuit lines and three Toffoli gates. The positive and negative control lines and the target line of a Toffoli gate are visualized using symbols ●, ○, and ⊕, respectively. Furthermore, the circuit is labeled with the values on the circuit lines for input $x_1x_2x_3 = 001$ before and after each gate. The first gate $g_1 = TOF(\{x_3\}, x_1)$ inverts the value of the target line x_1 since the positive control line x_3 is initialized to 1. Because of the same reason (control lines are initialized accordingly), the second gate $g_2 = TOF(\{x_1, \overline{x}_2\}, x_3)$ inverts the value of the target line x_3. In contrast, the third gate $g_3 = TOF(\{x_3\}, x_2)$ keeps the value of the target line x_2, because the positive control line x_3 is set to 0. The cost of the reversible circuit is 7.*

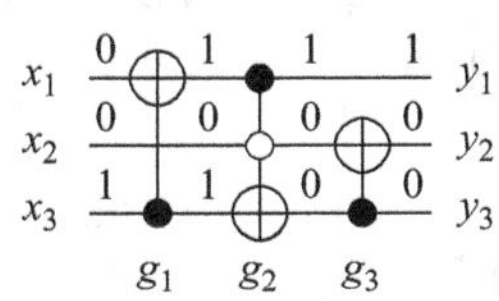

Figure 1.1 A reversible circuit with three Toffoli gates

1.2.2 Reversible synthesis

A reversible synthesis generates a reversible circuit that describes a given Boolean function. Synthesis approaches target minimizing the cost of the reversible circuit, which could be either by reducing the depth or the number of lines of a reversible circuit. As most of the classical Boolean functions are nonreversible, an embedding procedure is required, which can be supported by the synthesis, to generate a reversible function. The embedding procedure may involve adding ancillary inputs and garbage outputs. An ancillary input of a reversible circuit is set to a fixed value (either 0 or 1) to activate the target function. A garbage output of a reversible circuit is a nonfunctional output added to support reversibility.

Example 1.3. *Table 1.1(a) shows the truth table of the AND gate, in which x_1 and x_2 are the inputs and y_1 is the output. An AND gate is not reversible since* (1) *the number of inputs differs from the number of outputs, and* (2) *there is no unique input–output mapping. Clearly, adding a single output to the AND function does not satisfy the one-to-one mapping. Instead, at least one ancillary input and two garbage outputs should be added to guarantee reversibility. The ancillary input and the garbage outputs assignments are determined by the synthesis or the embedding procedure. Possible assignments to the ancillary input and the garbage outputs are shown in Table 1.1(b). Here, the AND gate is obtained if the ancillary input x_3 is set to 0. y_2 and y_3 are garbage outputs.*

Several approaches have been proposed to automatically synthesize nonreversible and reversible functions to reversible circuits. Reversible synthesis approaches can be classified as a structural synthesis, which uses different data structures to implicitly embed the target function into a reversible one, or a functional synthesis, which is preceded by an embedding procedure to convert a nonreversible function into reversible one. We consider reversible circuits generated using ESOP [26] and BDD-based [27] synthesis as examples of structural synthesis, and transformation [28] and QMDD-based [29,30] synthesis as examples of functional synthesis. In both categories,

Table 1.1 Truth tables for (a) a simple AND gate and (b) a reversible function for the AND gate

(a)

x_1	x_2	y_1
0	0	0
0	1	0
1	0	0
1	1	1

(b)

x_1	x_2	x_3	y_1	y_2	y_3
0	**0**	0	**0**	0	0
0	**1**	0	**0**	0	1
1	**0**	0	**0**	1	0
1	**1**	0	**1**	1	1
0	0	1	1	0	0
0	1	1	1	0	1
1	0	1	1	1	0
1	1	1	0	1	1

Note: The bold numbers indicate the truth table of the AND gate.

synthesis approaches leave telltale signs on the generated reversible circuits. In the following subsections, we revise different synthesis approaches and their telltale signs.

1.2.2.1 BDD-based synthesis

BDD-based synthesis approach [27] implicitly embeds the target function into a reversible one to generate the corresponding reversible circuit. The target function f is represented using a BDD [31] that expresses f based on Shannon decomposition as

$$f = \bar{x}_i \cdot f_{x_i=0} + x_i \cdot f_{x_i=1}$$

The function $f_{x_i=0}$ ($f_{x_i=1}$) is the negative (positive) sub-function of f obtained by assigning x_i to 0 (1). Each BDD node (sub-function) is mapped to a predefined reversible sub-circuit. This process may require adding ancillary inputs to realize nonreversible sub-functions. Most of the predefined sub-circuit types are connected to a unique ancillary input value.

- **Telltale sign I:** BDD-based reversible circuits consist of predefined sub-circuits, which indicate the associated ancillary input value.
- **Telltale sign II:** The structure of the BDD implies that the reversible circuit lines are partitioned into control-only and control or target lines.

Example 1.4. *Figure 1.2 shows a BDD of function* $f = x_1x_2\bar{x}_3x_4 + x_1x_2x_3$ *and its corresponding reversible circuit. Shannon decomposition generates four sub-functions; one of them is the identity function* (f_1)*, while the remaining sub-functions are mapped to predefined reversible sub-circuits.*

1.2.2.2 QMDD-based synthesis

QMDD-based synthesis [29] generates a reversible circuit by adding reversible gates, which convert a reversible function to the identity. QMDD [32] describes the permutation matrix of a reversible function. The synthesis traverses each node of the QMDD

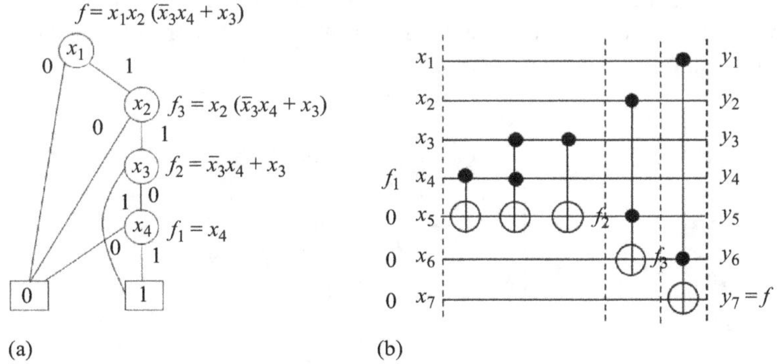

Figure 1.2 (a) BDD of $f = x_1x_2\bar{x}_3x_4 + x_1x_2x_3$*; (b) the corresponding reversible circuit obtained by BDD-based synthesis*

from the top to the bottom. Each node represents a partition according to a variable x_i. To generate a reversible circuit, the reversible function is transformed to the identity by considering one variable at a time; for a given variable x_i, the synthesis approach provides a one-to-one mapping between input x_i and its corresponding output by swapping columns of the permutation matrix using Toffoli gates. A QMDD-based reversible circuit that has n variables is divided into at most n regions. In each region, one variable is transformed to the identity.

- **Telltale sign I:** A unique variable should be used as a control or target line of each Toffoli gate within a region.

QMDD synthesis is preceded by an explicit embedding of the target function into a reversible function. We consider an efficient embedding approach that scales to larger designs and yields a minimum number of ancillary inputs and garbage outputs. The end result is an optimized version of the QMDD-based synthesis [30]. To minimize the number of reversible gates, swap operations should target columns with functional input assignment to the ancillary inputs, which activate the target function. This observation leads to the following telltale sign of QMDD-based synthesis in the presence of the efficient embedding:

- **Telltale sign II:** The functional input assignment to the ancillary inputs activates the largest number of reversible gates.

Example 1.5. *Figure 1.3 provides an example of QMDD-based reversible circuit, which is divided into regions 1 and 2 that transform variable x_5 and x_1, respectively, to the identity. In region 1, there are three Toffoli gates in which x_5 is used as either control or target line. Applying the functional assignment (0) to the ancillary input (x_5) activates the largest number of Toffoli gates.*

1.2.2.3 ESOP-based synthesis

ESOP-based synthesis approach [26] realizes a Boolean function as an ESOP. For a given function f with n primary inputs and m primary outputs, a reversible circuit with $n + m$ circuit lines is generated using ESOP-based synthesis. For each product, a Toffoli gate is added to the reversible circuit in which the control lines are primary inputs and the target line is a primary output.

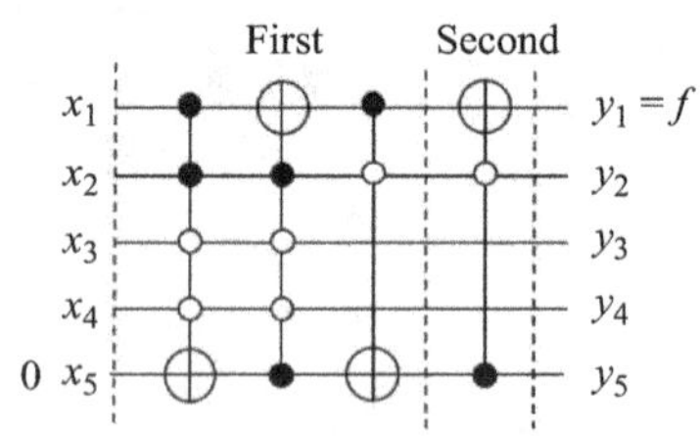

Figure 1.3 *Reversible circuit of $f = x_1x_2\bar{x}_3x_4 + x_1x_2x_3$ obtained by QMDD-based synthesis*

- **Telltale sign I:** The reversible circuit lines are divided into control-only and target-only lines.
- **Telltale sign II:** Ancillary inputs are initialized to zero and considered as the target lines for all the reversible (Toffoli) gates.

Example 1.6. *An ESOP-based reversible circuit is illustrated in Figure 1.4, where control-only lines are connected to primary inputs (x_1, x_2, x_3, x_4) and the target-only line is connected to an ancillary input.*

1.2.2.4 Transformation-based synthesis

Given the truth table description of a given reversible function, a TBS [28] converts the function to the identity by adding Toffoli gates. Unlike QMDD-based synthesis that maps one input line (variable) at a time to the identity output line, TBS traverses the truth table row-by-row to map each output combination to its corresponding input combination. Toffoli gates should be added to transform each output combination, while preserving previously traversed output combinations. The further we proceed with the synthesis algorithm, the more the number of output combinations we should preserve. Since Toffoli gates are added from the output to the input-side of the circuit, the gates toward the input-side of the circuit typically have more control lines, which leads to the following telltale sign of TBS:

- **Telltale sign I:** A Toffoli gate at position i from the input-side of the circuit is likely to have a less number of control lines than a preceding gate at position j, where $i>j$.

Example 1.7. *Figure 1.5 shows a reversible circuit generated using TBS, where the number of control lines of Toffoli gates decreases from the input-side to the output-side of the circuit.*

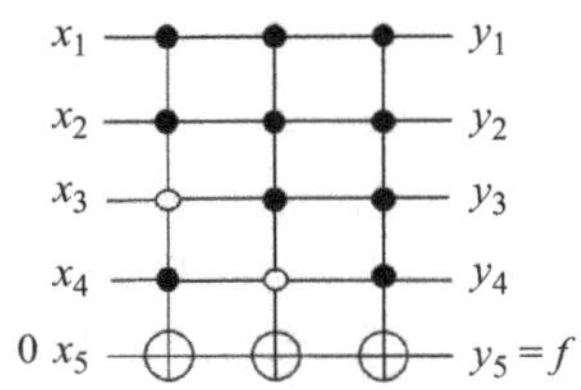

Figure 1.4 Reversible circuit of $f = x_1x_2\overline{x}_3x_4 + x_1x_2x_3$ obtained by ESOP-based synthesis

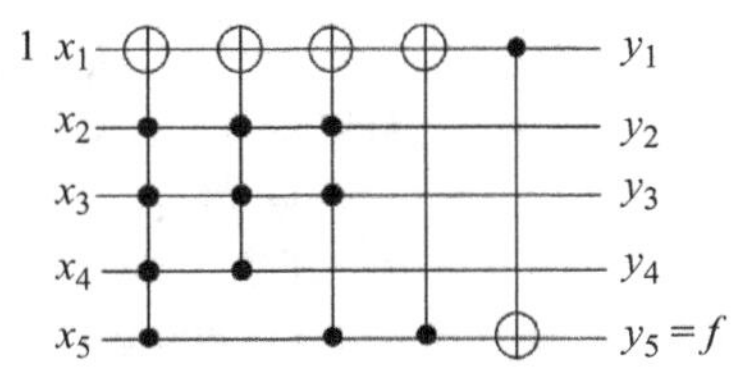

Figure 1.5 Reversible circuit of $f = x_1x_2\overline{x}_3x_4 + x_1x_2x_3$ obtained by TBS

1.2.3 Post-synthesis optimization

As the synthesis approaches are heuristic, they are often suboptimal. Post-synthesis optimization provides further improvement in reversible circuit cost through local optimization. There are different post-synthesis optimization categories, including reducing gate count [33–35], gate cost [28,36–39], reducing circuit depth [23,40], improving locality [41], and resynthesizing part of the design. Many of these post-synthesis optimization approaches provide moderate improvements in the reversible circuit cost at the expense of a very large search space, and thus, high computation complexity. In addition, some of the optimization approaches are not applicable to reversible circuits generated using some synthesis approaches.

1.3 Motivation and threat model

1.3.1 Motivation

Emerging technologies based on reversible computing require embedding the target Boolean function into a reversible circuit, which typically adds ancillary inputs and garbage outputs to the circuit, despite the underlying technology/application. A reversible circuit can be reconfigured to support different functions other than the target function by varying the value of the ancillary inputs. The number of embedded functions can be even larger if we consider any input as an ancillary input and any output as a garbage output. The upper bound of the number of embedded functions, referred to as *the number of embeddings*, is defined as following:

$$(2^n - 1) \times \prod_{i=1}^{n} \left(\sum_{j=0}^{k_i} C(k_i, j) \times 2^j \right) \tag{1.1}$$

where $\sum_{j=0}^{k_i} C(k_i,j) \times 2^j$ is the number of embedded functions in an output y_i, where k_i is the number of inputs that drive y_i but not previously traversed output y_p $(1 \leq p < i)$. The binomial coefficient $C(k_i,j)$ is the number of all possible ancillary input combinations. The number of embedded functions is computed for all possible output combinations. There are $2^n - 1$ possible output combinations.

Example 1.8. *The number of embeddings of the reversible circuit in Figure 1.1 is 189, where $n = 3$, $k_1 = 2$, $k_2 = 1$, and $k_3 = 0$. One possible embedding is $f = y_3 = x_1\overline{x_2}$, when x_3 is the only ancillary input with value 0, and y_1 and y_2 are the garbage outputs.*

Yet, if the number of embeddings is very large, why are reversible circuits vulnerable to IP/IC piracy attacks that reveal the target function? One glaring reason is that synthesis approaches that generate reversible circuits leave telltale signs on these circuits, which limit the search space of the target function. An attacker does not have to consider all possible ancillary inputs and garbage outputs combinations. The number of embeddings can be reduced to 2^n for known location of the ancillary inputs and garbage outputs, where n is the number of ancillary inputs. The number of embeddings can be further reduced to 1 if the value of all the ancillary inputs can be recovered.

1.3.2 Threat model

We consider an attacker who has access to the gate level implementation of the reversible circuit. The goal of the attack is to reveal the target function of the reversible circuit by launching IP piracy attacks. The number of embeddings can be used as a security metric to assess the difficulty of the attack. While the analysis provided in this chapter focuses on the gate level representation of the reversible circuit, which provides the core concept for several emerging applications of reversible computing, we discuss the threat model for some possible applications.

Reversible circuits that are realized using physical gates and follow the same design and fabrication approaches as conventional CMOS circuit are susceptible to IP/IC piracy attack. The attacker is an un-trusted **foundry** and possesses the following additional assumptions:

- The attacker has no access to a functional chip considering national agencies such as the government and DoD, which apply a strictly controlled distributed process.
- A reversible circuit inputs and outputs are connected to the input and output pins of the chip, respectively.

Reversible logic is also used for building quantum oracles, which is a pivotal building block of quantum circuits and requires specialized expertise and means to design. Quantum oracles are utilized in oracle-based quantum algorithms such as Grover's search algorithm [42]. The quantum circuit is realized using a set of instructions (logical gates) that are executed on quantum hardware as gate-pulses instead of using physical gates. The current business model suggests outsourcing the quantum circuit to a third-party quantum server (e.g., IBM quantum computers [43]). A quantum oracle is susceptible to eavesdropping. The attacker is the **man in-the-middle** who tries to intercept the communication to identify the IP of the oracle. The attacker can also be someone **with access to an online quantum compiler**, which optimizes and decomposes the quantum circuit to elementary quantum gates supported by the quantum hardware. In both attack scenarios on the quantum circuits, the following assumption holds:

- The initial value of the ancillary inputs is unknown to the attacker. In the case of eavesdropping attacks, we assume that the initial value is sent separately to the server[†] through secure channels.

1.4 IP/IC piracy attacks

Reversible circuits are subject to attacks that reveal their functionality. In order to thwart these attacks, a detailed understanding of how an attacker would determine the

[†]Our threat model does not consider a malicious server.

target function of a reversible circuit is crucial. To this end, we show how the target function of reversible circuits generated using state-of-the-art synthesis approaches can be recovered [17–21]. To launch an IP/IC piracy attacks, the synthesis approach should be identified first, followed by an attack procedure that uses the telltale signs of the synthesis to reveal the ancillary inputs and garbage outputs, and thus, identify the target function. These steps are illustrated in Figure 1.6.

1.4.1 Machine learning-based classification

An attacker can identify the reversible circuit synthesis approach using a machine learning-based scheme. Telltale signs of the synthesis approaches are formalized as continuous and discrete features, which are extracted from the reversible circuits and used in the machine learning-based scheme. Different supervised machine learning models can be applied, such as the decision tree [44], the random forest [45], the support vector machine [46], and the logistic regression [47] models, which are trained using reversible circuits with known synthesis approaches. A machine learning model predicts the synthesis approach of an unknown reversible circuit given its features.

Table 1.2 shows the mapping of most of the telltale signs of BDD, QMDD, ESOP, and TBS synthesis, in Section 1.2.2, to features. Not only the four synthesis approaches are considered but also their optimized versions. BDD sub-circuits-only evaluates to 1 if the reversible circuit consists of only predefined BDD sub-circuits and 0 otherwise. BDD optimization [48] introduces new predefined sub-circuits, which can also be incorporated and tested under BDD sub-circuits-only feature. Interference ratio provides the ratio of common reachable gates between each pair of circuit lines, which is computed as $Normalize\{\sum_{i,j} |gate(i) \cap gate(j)|\}$, where $|gate(i) \cap gate(j)|$ indicates the number of common gates between each pair. BDD-based reversible

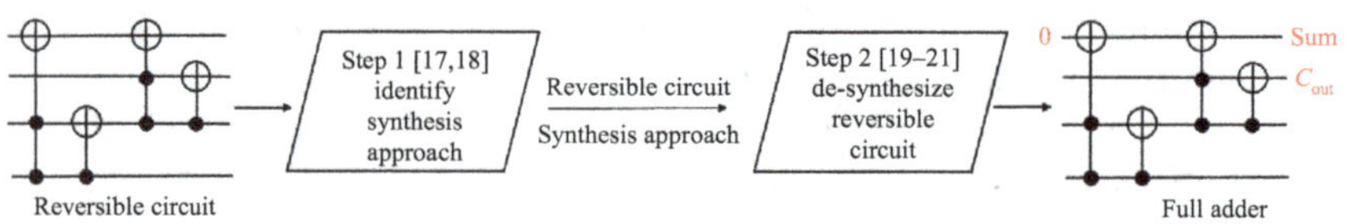

Figure 1.6 Steps of IP/IC piracy attacks on full-adder reversible circuit

Table 1.2 Features of reversible synthesis approaches

Synthesis	Sign	Features for reversible synthesis
BDD	Sign I	BDD sub-circuits-only
	Sign II	Interference ratio between different circuit lines
QMDD	Sign I	QMDD identity transformation
ESOP	Sign I	(Ratio) control-only and target-only lines
TBS	Sign I	Ratio of the number of control lines reduction

circuits exhibit a low interference ratio. QMDD identity transformation is set to 1 if the reversible circuit is divided into regions, where each region has a unique variable used as either a control or a target line for all the gates in the region, and 0 otherwise. QMDD-based optimization tends to minimize redundancy while preserving its telltale sign [30]. Control-only and target-only feature is set to 1 if the ESOP-based telltale sign is satisfied. However, as target lines can also be used as control lines under optimization [49], the corresponding feature is refined to report the ratio of control-only and target-only lines instead, which is expected to be high for an ESOP-based reversible circuit. According to the telltale sign of TBS, the reduction ratio in the number of control lines of adjacent gates from the input to the output side of the circuit is expected to range from high, when the original synthesis is applied, to medium, when the bidirectional TBS‡ is applied [28].

Example 1.9. *To show how the synthesis features are computed, we extract the features provided in Table 1.2 from all the reversible circuits in Section 1.2.2 that realize a function* $f = x_1x_2\overline{x}_3x_4 + x_1x_2x_3$ *using different synthesis approaches. Table 1.3 summarizes their values. As expected, sub-circuit-only feature is evaluated to 1 in BDD-based reversible circuit, only. Furthermore, the interference ratio of BDD-based reversible circuit is very low compared to other circuits. While QMDD-identity feature seems to be ineffective in identifying the QMDD synthesis, experimental results in [17,18] prove the ability of identifying the QMDD-based synthesis using this feature. Two reasons explain the odd results of QMDD-identity in Table 1.3, which are the small target function and the presence of a single primary output. QMDD-identify feature is more powerful in the presence of larger circuits with multiple primary outputs. ESOP-based reversible circuit provides the highest ratio of the control-only target-only lines, while the reversible circuit generated using TBS has the highest ratio of the number of control lines reduction.*

Table 1.3 Features of different reversible circuits that realize function $f = x_1x_2\overline{x}_3x_4 + x_1x_2x_3$

Reversible circuits	Sub-circuit-only	Interference	QMDD identity	Control-only target-only	Control lines reduction
BDD-circuit (Figure 1.2(b))	**1**	**0.3**	1	0.7	0.25
QMDD-circuit (Figure 1.3)	0	1	1	0.6	0.3
ESOP-circuit (Figure 1.4)	0	1	1	**1**	0
TBS-circuit (Figure 1.5)	0	1	1	0.6	**0.5**

Note: The numbers in bold indicate the effective features that determine the corresponding synthesis approach.

‡Reversible gates can be added to the input or the output side of the circuit to minimize the gate cost.

While we shed light on a subset of the reversible synthesis approaches, the machine learning-based scheme can be applied to reversible circuits generated using other optimized synthesis approaches.

1.4.2 De-synthesis of reversible circuits

De-synthesis refers to the process of reversing the synthesis approach of the reversible circuit to reveal the target function. The telltale signs of a synthesis approach are used not only to identify the synthesis itself but also to recover the target function. The attacker exploits the telltale signs of the synthesis approach to de-synthesize the reversible circuit. To demonstrate the attack, de-synthesis is applied to reversible circuits generated using BDD, QMDD, and ESOP-based synthesis. Similar to QMDD-based synthesis, TBS is preceded by an embedding procedure that determines the ancillary inputs and the garbage outputs. However, the embedding is conducted based on truth tables that suffer from poor scalability due to the exponential growth of the data structure. Thus, we omit the discussion on the TBS, despite its resistance to IP/IC piracy attack. The de-synthesis of reversible circuits generated using different synthesis approaches is illustrated as given below:

- **BDD-based synthesis:** Using the telltale signs of BDD-based synthesis, an attacker can recover not only the ancillary inputs value but also the location of the ancillary inputs and garbage outputs. The second telltale sign of the BDD-based synthesis determines the location of the primary inputs (control-only lines) that are directly connected to garbage outputs. Intermediate results of reversible circuits are also connected to garbage outputs. In other words, a target line of a Toffoli gate, which is used as a control line for a subsequent gate, is also connected to a garbage output. Once the locations of the primary (ancillary) inputs and primary (garbage) outputs are known, the attacker uses the first telltale sign of the BDD synthesis to recover the value of the ancillary inputs. All the applied predefined sub-circuits are extracted using pattern matching to determine the associated ancillary input value.

  Example 1.10. *To illustrate the de-synthesis of BDD-based reversible circuits, we consider the circuit in Figure 1.2(b). The control-only lines determine the primary inputs (x_1, x_2, x_3, x_4) and some of the garbage outputs (y_1, y_2, y_3, y_4). The remaining garbage outputs (y_5, y_6) are connected to the intermediate results of the circuit. The reversible circuit is partitioned into three regions. Each one maps to a predefined sub-circuit, which is associated to a certain value of the ancillary inputs ($x_5 = 0, x_6 = 0, x_7 = 0$).*

- **QMDD-based synthesis**: In QMDD-based synthesis, the embedding procedure determines the initial value of the ancillary inputs. Therefore, the attacker targets the embedding to recover the ancillary input value. The second telltale sign of the QMDD-based synthesis, inspired by the embedding procedure, indicates that the functional assignment to the ancillary inputs activates the largest number of Toffoli gates. By examining the input combinations that activate the largest number of Toffoli gates, referred to as *golden patterns*, an attacker can identify the location of the ancillary inputs (stable inputs) and some primary inputs (unstable inputs) in addition to the value of the ancillary inputs. Golden patterns are generated using automatic test pattern generation (ATPG) tools for missing target line fault model in reversible circuits [50]. An ATPG tool generates test patterns that activate each Toffoli gate, and thus, invert the value of the target lines. These patterns are sorted in descending order based on the number of activated missing target line faults. The top test patterns with the maximum number of activated faults are selected as the golden patterns, which share the functional assignment to the ancillary inputs.

Example 1.11. *To illustrate the de-synthesis process of QMDD-based reversible circuits, we consider the circuit in Figure 1.3. The golden patterns are* $x_1x_2x_3x_4x_5 = 11000$ *and* $x_1x_2x_3x_4x_5 = 10xx0$*, where x indicates a don't care value. The golden patterns activate two Toffoli gates. We observe bit-flip at* x_2*,* x_3*, and* x_4 *of the golden patterns, which indicates the location of most of the primary inputs. On the other hand, there are two potential stable ancillary inputs, which are* $x_1 = 1$ *and* $x_5 = 0$*. While some of the results are false positive (*x_1 *is a primary input), the attacker can still determine the ancillary input value (*$x_5 = 0$*). The embedding procedure does not reveal the exact location of the primary outputs. If the location of the ancillary inputs and garbage outputs is known to the attacker, the attack success rate will be 100%.*

- **ESOP-based synthesis:** According to the telltale sign of ESOP-based synthesis, ancillary inputs are initialized to zero and considered as target lines for all the reversible gates. Thus, not only the value of the ancillary inputs is known but also the location of both ancillary inputs and garbage outputs.

Example 1.12. *By revisiting ESOP-based reversible circuit in Figure 1.4, we can verify that the control-only line is connected to the ancillary input (*$x_5 = 0$*) and the primary output (*y_5*).*

1.5 Countermeasures

As a response to IP/IC piracy attacks on reversible circuits, we discuss and revise two potential countermeasures that hide the target function of the reversible circuit.

1.5.1 Insertion of redundant inputs/outputs

Telltale signs of the synthesis approach enable identifying the ancillary inputs and garbage outputs added by the synthesis/embedding. However, randomly inserted redundant inputs/outputs prior to the synthesis/embedding will no longer exhibit their telltale signs. From the attacker's perspective, these inputs/outputs behave as primarily inputs/outputs of the circuit. Even if their location is known to the attacker, their value is still hidden. The downsides of this approach are the high cost of the reversible circuit and the failure to utilize the redundancy of the reversible circuit toward security. The former issue is due to the arbitrary embedded functions in the redundant inputs/outputs to support reversibility. The later issue is due to the attacker ability to extract the ancillary inputs and garbage outputs added by the synthesis/embedding. While the problem of the high cost of the reversible circuit can be resolved by inserting the redundant inputs/outputs post-synthesis, their location as well as the location and the value of the ancillary inputs and garbage outputs can still be recovered by an attacker. The insertion of redundant inputs post-synthesis to hide the target function is analogue to logic obfuscation techniques of conventional circuits [51–53].

Example 1.13. *To illustrate the cost and the benefits of inserting redundant inputs/outputs, we consider QMDD-based reversible circuit in Figure 1.3. Two redundant inputs are added. Figure 1.7(a) shows the QMDD-based reversible circuits when* x_5 *and* x_6 *are inserted prior to the embeddings, while Figure 1.7(b) shows the corresponding reversible circuit when* x_6 *and* x_7 *are added post-synthesis. We launch the de-synthesis attack on both circuits according to the telltale sign of the QMDD-based synthesis. The golden patterns that activate the largest number of Toffoli gates in Figure 1.7(a) are* $x_1x_2x_3x_4x_5x_6x_7 = 1110100$, 1111010, *and* 1101000, *which reveal the value of the ancillary input* x_7, *despite the hidden value of* x_5 *and* x_6. *On the other hand, the golden pattern in Figure 1.7(b) is* $x_1x_2x_3x_4x_5x_6x_7 = 10xx011$, *which still reveals the value of the ancillary input* (x_5). *However, it fails to assign the functional value to some of the redundant inputs as the target function in Figure 1.7(b) is activated when* $x_6x_7 = 01$. *In other words, inputs added post-synthesis do not follow the telltale*

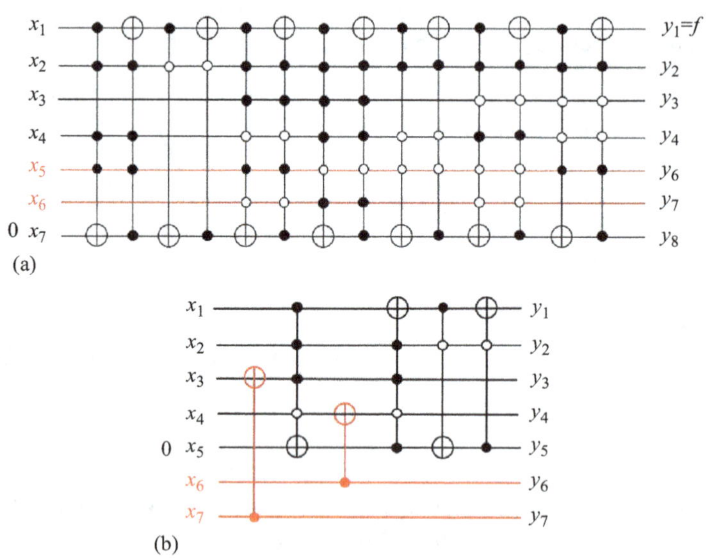

Figure 1.7 QMDD-based reversible circuit in which two redundant inputs are inserted (a) prior to embeddings and (b) post-synthesis

signs of the synthesis. The reversible circuit cost when applying pre-embedding insertion of redundant inputs (cost = 998) is significantly higher than the corresponding cost when redundant inputs are added post-synthesis (cost = 70).

1.5.2 Insertion of redundant reversible gates

Solving the IP/IC piracy problem at minimum cost requires utilizing the redundancy of the reversible circuit for security. Ancillary inputs and garbage outputs should be exploited to hide the target function. Thus, telltale signs of the synthesis used to recover the value or the location of ancillary inputs/garbage outputs should be destroyed, making it harder to extract the target function.

A telltale sign of the synthesis approach can be classified as structural or functional telltale sign. A structural telltale sign describes the structure of the reversible circuit generated using a given synthesis approach such as the BDD predefined reversible sub-circuits and the ESOP control-only lines. A functional telltale sign relays on the embedding procedure of the target function such as the number of activated gates in the functional mode of the reversible circuit. In both categories, telltale signs (1) identify the synthesis of the reversible circuit, (2) distinguish between ancillary and primary inputs and garbage and primary outputs, and (3) reveal the ancillary inputs value. Reversible gates should be added to destroy the telltale signs such that an attacker no longer differentiates between ancillary and primary inputs and garbage and primary outputs and fails to recover ancillary inputs value. The insertion of the reversible gates should be controlled by not only the telltale signs of the synthesis approach but also post-synthesis optimization to minimize the cost of the reversible circuits. In best-case scenario, additional reversible gates create optimization opportunities and destroy the telltale signs of the synthesis, which reduce the cost of the original reversible circuit. The following steps are applied to erase the telltale signs of the synthesis:

1. Reorder reversible gates to construct sub-circuits that partially match a design rule for optimization. A Toffoli gate $TOF(C_i, t_i)$ can be swapped with the adjacent Toffoli gate $TOF(C_{i+1}, t_{i+1})$ if $t_{i+1} \notin C_i$ and $t_i \notin C_{i+1}$.
2. For each sub-circuit, check if nonfunctional reversible gates[§] can be added to the circuit to reduce the number of telltale signs and at the same time create a design rule for optimization. If the number of telltale signs remains the same, keep the original sub-circuit.
3. For the remaining telltale signs, add nonfunctional reversible gates to erase them.

The procedure of adding reversible gates as described earlier can be customized to different synthesis approaches, where certain optimization techniques are more effective than others. In BDD-based synthesis, for example, reversible gates are reordered to create sub-circuits with nonunique ancillary input value and sub-templates for optimization [28,36]. For the remaining known ancillary inputs, nonfunctional reversible gates are added to a randomly selected subset of known ancillary inputs. The goal is to

[§]Nonfunctional gate indicates that the gate is activated by nonfunctional value of the ancillary input.

violate the structure of randomly selected sub-circuits while creating new optimization opportunities if possible. In QMDD-based synthesis, reversible gates are reordered, if possible, to increase the number of common control lines between adjacent gates while preserving the embedded target function [37–39]. New nonfunctional gates are added next to balance the number of activated gates for functional and nonfunctional modes and reduce the circuit cost due to optimization (e.g., sharing common control lines or generating templates for optimization).

Example 1.14. *Let us consider again QMDD-based reversible circuit in Figure 1.3. To remove telltale sign II of QMDD-based synthesis, we follow the three steps mentioned earlier. As no Toffoli gate can be reordered without violating the functionality of the reversible circuit, the order of the reversible gates is preserved. We recall that the golden patterns obtained by de-synthesis attack are* $x_1x_2x_3x_4x_5 = 11000$ *and* $10{-}00$, *which activate two Toffoli gates. A new nonfunctional reversible gate is added to the circuit such that a nonfunctional input pattern activates two Toffoli gates. The nonfunctional input pattern that yields a reduction in the gate cost is selected. In this example, the Toffoli gate TOF* $(\{x_2, \overline{x_3}, \overline{x}_4, x_5\}, x_1)$ *is added to the circuit as shown in Figure 1.8(a), which is activated by nonfunctional input pattern* $x_1x_2x_3x_4x_5 = 11001$. *Thus, the set of golden patterns of the updated reversible circuit includes a nonfunctional input pattern* $x_1x_2x_3x_4x_5 = 11001$, *which activates two Toffoli gates despite the nonfunctional value of the ancillary input. The new reversible gate also creates a template that can be further optimized to reduce the cost of the reversible circuit as shown in Figure 1.8(b). The cost of the original reversible circuit in Figure 1.3 is 68, while the cost of the corresponding updated and optimized reversible circuit in Figure 1.8(b) is dropped to 41.*

An attacker may easily recover the location of the ancillary inputs and garbage outputs of the reversible circuits in some applications. Furthermore, advanced synthesis approaches can utilize a minimum number of ancillary inputs. The end result is a small number of embeddings, rendering a smaller search space of the target function. A combination of the two countermeasures listed earlier (redundant inputs/outputs and reversible gates) can be applied to further harden the IP/IC piracy attacks at minimum cost.

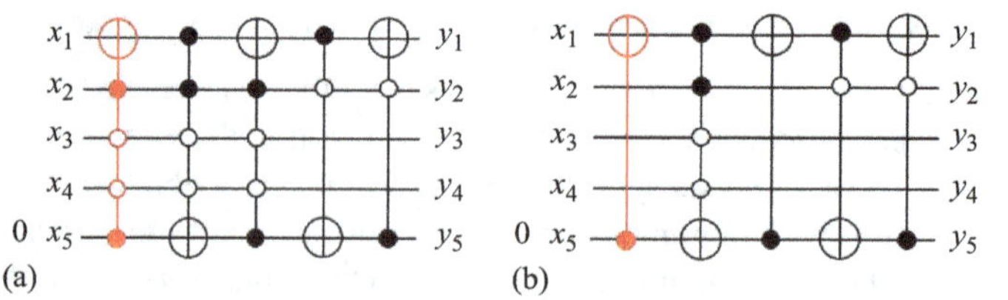

Figure 1.8 Adding redundant gates to the QMDD-based reversible circuit in Figure 1.3: (a) balancing functional and nonfunctional input patterns; (b) applying post-synthesis optimization

1.6 Summary

While the technologies for designing reversible circuits for different applications remain somehow unclear, the security risks of reversible circuits in the logic level can still be addressed. There exist opportunities to bolster the security of reversible computing outside of traditional approaches. To understand and develop techniques that leverage redundancy in reversible circuits to enhance security, it is important to demonstrate their security implications. Toward this end, this chapter sheds light on IP/IC piracy threats of reversible circuits.

As several functions are embedded into the reversible circuit, activating the target function requires the knowledge of the ancillary inputs and the garbage outputs determined by the synthesis approach. Thus, the goal of the IP/IC piracy attack is to reveal the ancillary inputs and garbage outputs. An attacker takes advantage of the telltale signs of the synthesis approaches to recover the target function. We describe the IP/IC piracy attacks on reversible circuits generated using state-of-the-art synthesis. To thwart IP/IC piracy attacks, cost-effective defense mechanisms guided by telltale signs of the synthesis approaches are illustrated to leverage redundant inputs/outputs/gates to hide the target function of the reversible circuit. These countermeasures can be integrated with the synthesis approaches to design piracy-aware synthesis approaches.

References

[1] Bennett CH. Logical Reversibility of Computation. IBM Journal of Research & Development. 1973;17(6):525–532.

[2] Berut A, Arakelyan A, Petrosyan A, *et al.* Experimental Verification of Landauer's Principle Linking Information and Thermodynamics. Nature. 2012;483:187–189.

[3] Zulehner A, Frank MP, and Wille R. Design Automation for Adiabatic Circuits. In: Asia and South Pacific Design Automation Conference. New York, NY: ACM; 2019. p. 669–674.

[4] Athas WC and Svensson LJ. Reversible Logic Issues in Adiabatic CMOS. In: Proceedings of Workshop on Physics and Computation. New York, NY: IEEE; 1994. p. 111–118.

[5] Frank MP. The Future of Computing Depends on Making It Reversible. IEEE Spectrum. 2017.

[6] Takeuchi N, Yamanashi Y, and Yoshikawa N. Reversible Logic Gate Using Adiabatic Superconducting Devices. In: Scientific Reports; 2014.

[7] Nielsen MA and Chuang IL. Quantum Computation and Quantum Information. Cambridge: Cambridge University Press; 2000.

[8] Shor PW. Polynomial-Time Algorithms for Prime Factorization and Discrete Logarithms on a Quantum Computer. SIAM Journal on Computing. 1997;26(5):1484–1509.

[9] Tehranipoor M and Koushanfar F. A Survey of Hardware Trojan Taxonomy and Detection. IEEE Design & Test of Computers. 2010;27(1):10–25.

[10] Rajendran J, Gavas E, Jimenez J, *et al.* Towards a Comprehensive and Systematic Classification of Hardware Trojans. In: Proceedings of IEEE International Symposium on Circuits and Systems. New York, NY: IEEE; 2010. p. 1871–1874.

[11] Torrance R and James D. The State-of-the-Art in Semiconductor Reverse Engineering. In: Proceedings of ACM/EDAC/IEEE Design Automation Conference. New York, NY: IEEE; 2011. p. 333–338.

[12] Bhunia S, Hsiao MS, Banga M, *et al.* Hardware Trojan Attacks: Threat Analysis and Countermeasures. Proceedings of the IEEE. 2014;102(8):1229–1247.

[13] Kocher PC, Jaffe J, and Jun B. Differential Power Analysis. In: Proceedings of the International Cryptology Conference on Advances in Cryptology. Berlin, Heidelberg: Springer-Verlag; 1999. p. 388–397.

[14] Kocher PC. Timing Attacks on Implementations of Diffie-Hellman, RSA, DSS, and Other Systems. In: Proceedings of the International Cryptology Conference on Advances in Cryptology. Berlin, Heidelberg: Springer; 1996. p. 104–113.

[15] Cui X, Saeed SM, Zulehner A, *et al.* On the Difficulty of Inserting Trojans in Reversible Computing Architectures. IEEE Transactions on Emerging Topics in Computing. 2018:1.

[16] Cui X, Saeed SM, Zulehner A, *et al.* On the Difficulty of Inserting Trojans in Reversible Computing Architectures. CoRR. 2017;abs/1705.00767. Available from: http://arxiv.org/abs/1705.00767.

[17] Saeed S, Mahendran N, Zulehner A, *et al.* Identifying Reversible Circuit Synthesis Approaches to Enable IP Piracy Attacks. In: Proceedings of IEEE International Conference on Computer Design. New York, NY: IEEE; 2017. p. 537–540.

[18] Saeed SM, Zulehner A, Wille R, Drechsler R, and Karri R, Reversible Circuits: IC/IP Piracy Attacks and Countermeasures. In: IEEE Transactions on Very Large Scale Integration (VLSI) Systems. 2019;27(11):2523–2535. Available from: http://doi: 10.1109/TVLSI.2019.2934465.

[19] Saeed SM, Cui X, Wille R, *et al.* Towards Reverse Engineering Reversible Logic. CoRR. 2017;abs/1704.08397.

[20] Saeed SM, Zulehner A, Wille R, Drechsler R, and Karri R, Reversible Circuits: IC/IP Piracy Attacks and Countermeasures. In: IEEE Transactions on Very Large Scale Integration (VLSI) Systems. 2019;27(11):2523–2535. Available from: http://doi: 10.1109/TVLSI.2019.2934465.

[21] Saeed SM, Cui X, Zulehner A, *et al.* IC/IP Piracy Assessment of Reversible Logic. In: Proceedings of the International Conference on Computer-Aided Design. ICCAD '18. New York, NY, USA: ACM; 2018. p. 5:1–5:8.

[22] Toffoli T. Reversible Computing. In: de Bakker W, van Leeuwen J, editors. Automata, Languages and Programming. Heidelberg: Springer; 1980. p. 632.

[23] Maslov D, Dueck GW, Miller DM, *et al.* Quantum Circuit Simplification and Level Compaction. IEEE Transactions on CAD. 2008;27(3): 436–444.

[24] Miller DM, Wille R, and Sasanian Z. Elementary Quantum Gate Realizations for Multiple-Control Toffoli Gates. In: 2011 41st IEEE International Symposium on Multiple-Valued Logic. New York, NY: IEEE; 2011. p. 288–293.

[25] Amy M, Maslov D, Mosca M, *et al.* A Meet-in-the-Middle Algorithm for Fast Synthesis of Depth-Optimal Quantum Circuits. IEEE Transactions on Computer-Aided Design of Integrated Circuits and Systems. 2013;32(6): 818–830.

[26] Fazel K, Thornton MA, and Rice JE. ESOP-Based Toffoli Gate Cascade Generation. In: Proceedings of IEEE PacRim. New York, NY: IEEE; 2007. p. 206 –209.

[27] Wille R and Drechsler R. BDD-Based Synthesis of Reversible Logic for Large Functions. In: Proceedings of ACM/IEEE Design Automation Conference. New York, NY: IEEE; 2009. p. 270–275.

[28] Miller DM, Maslov D, and Dueck GW. A Transformation Based Algorithm for Reversible Logic Synthesis. In: DAC. New York, NY: IEEE; 2003. p. 318–323.

[29] Soeken M, Wille R, Hilken C, *et al.* Synthesis of Reversible Circuits With Minimal Lines for Large Functions. In: Proceedings of Asia and South Pacific Design Automation Conference. New York, NY: IEEE; 2012. p. 85–92.

[30] Zulehner A, and Wille R. Make It Reversible: Efficient Embedding of Non-reversible Functions. In: Proceedings of Design Automation and Test in Europe. New York, NY: IEEE; 2017. p. 458–463.

[31] Bryant RE. Graph-Based Algorithms for Boolean Function Manipulation. IEEE Transactions on Computers. 1986;35(8):677–691.

[32] Niemann P, Wille R, Miller DM, *et al.* QMDDs: Efficient Quantum Function Representation and Manipulation. IEEE Transactions on Computer-Aided Design of Integrated Circuits and Systems. 2016;35(1):86–99.

[33] Prasad AK, Shende VV, Markov IL, *et al.* Data Structures and Algorithms for Simplifying Reversible Circuits. Journal on Emerging Technologies in Computing Systems. 2006;2(4):277–293. Available from: http://doi.acm.org/10.1145/1216396.1216399.

[34] Maslov D, Dueck GW, and Miller DM. Techniques for the Synthesis of Reversible Toffoli Networks. ACM Transactions on Design Automation of Electronic Systems. 2007;12(4).

[35] Maslov D and Saeedi M. Reversible Circuit Optimization Via Leaving the Boolean Domain. IEEE Transactions on Computer-Aided Design of Integrated Circuits and Systems. 2011;30(6):806–816.

[36] Datta K, Rathi G, Wille R, *et al.* Exploiting Negative Control Lines in the Optimization of Reversible Circuits. In: Proceedings of the 5th International Conference on Reversible Computation. RC'13. Berlin, Heidelberg: Springer-Verlag; 2013. p. 209–220. Available from: http://dx.doi.org/10.1007/978-3-642-38986-3_17.

[37] Deb A, Wille R, Drechsler R, *et al.* An Efficient Reduction of Common Control Lines for Reversible Circuit Optimization. In: 2015 IEEE International Symposium on Multiple-Valued Logic. New York, NY: IEEE; 2015. p. 14–19.

[38] Miller DM, Wille R, and Drechsler R. Reducing Reversible Circuit Cost by Adding Lines. In: 2010 40th IEEE International Symposium on Multiple-Valued Logic; 2010. p. 217–222.

[39] Wille R, Soeken M, Otterstedt C, *et al.* Improving the Mapping of Reversible Circuits to Quantum Circuits Using Multiple Target Lines. In: 2013 18th Asia and South Pacific Design Automation Conference (ASP-DAC). New York, NY: IEEE; 2013. p. 145–150.

[40] Maslov D, Dueck GW, and Miller DM. Toffoli Network Synthesis With Templates. IEEE Transactions on Computer-Aided Design of Integrated Circuits and Systems. 2005;24(6):807–817.

[41] Saeedi M, Wille R, and Drechsler R. Synthesis of Quantum Circuits for Linear Nearest Neighbor Architectures. Quantum Information Processing. 2011;10(3):355–377.

[42] Grover LK. A Fast Quantum Mechanical Algorithm for Database Search. In: Theory of Computing. New York, NY: ACM; 1996. p. 212–219.

[43] IBM. IBM QX Backend Information; 2017. Available from: https://github.com/Qiskit/qiskit-backend-information.

[44] Quinlan JR. Induction of Decision Trees. Machine Learning. 1986;1(1): 81–106.

[45] Ho TK. Random Decision Forests. In: ICDAR. vol. 1. New York, NY: IEEE; 1995. p. 278–282.

[46] Cristianini N and Shawe-Taylor J. An Introduction to Support Vector Machines: And Other Kernel-Based Learning Methods. Cambridge: Cambridge University Press; 2000.

[47] Freedman DA. Statistical Models: Theory and Practice. Cambridge: Cambridge University Press; 2009.

[48] Wille R and Drechsler R. Effect of BDD Optimization on Synthesis of Reversible and Quantum Logic. Electronic Notes in Theoretical Computer Science. 2010;253(1):57–70.

[49] Sanaee Y and Dueck GW. Generating Toffoli Networks From ESOP Expressions. In: Proceedings of IEEE Pacific Rim Conference on Communications, Computers and Signal Processing. New York, NY: IEEE; 2009. p. 715–719.

[50] Wille R, Zhang H, and Drechsler R. ATPG for Reversible Circuits Using Simulation, Boolean Satisfiability, and Pseudo Boolean Optimization. In: VLSI (ISVLSI), 2011 IEEE Computer Society Annual Symposium on. IEEE; 2011. p. 120–125.

[51] Yasin M, Sengupta A, Nabeel MT, *et al.* Provably-Secure Logic Locking: From Theory to Practice. In: Proceedings of the 2017 ACM SIGSAC Conference on Computer and Communications Security. CCS'17. New York, NY, USA: ACM; 2017. p. 1601–1618. Available from: http://doi.acm.org/10.1145/3133956.3133985.

[52] Roy JA, Koushanfar F, and Markov IL. Ending Piracy of Integrated Circuits. Computer. 2010;43(10):30–38.

[53] Yasin M, Rajendran JJ, Sinanoglu O, *et al.* On Improving the Security of Logic Locking. IEEE Transactions on Computer-Aided Design of Integrated Circuits and Systems. 2016;35(9):1411–1424.

Chapter 2

Improvements and recent updates of persistent fault analysis on block ciphers

Fan Zhang[1,2,3], Bolin Yang[4], Guorui Xu[1], Xiaoxuan Lou[4], Shivam Bhasin[5], Xinjie Zhao[6], Shize Guo[6], and Kui Ren[3]

Persistence is an intrinsic nature of many errors yet has not been caught enough attractions for years. In this chapter, the feature of persistence is applied to fault attacks (FAs), and the *persistent FA* is proposed. Different from traditional FAs, adversaries can prepare the fault injection stage before the encryption stage, which relaxes the constraint of the tight-coupled time synchronization. The persistent fault analysis (PFA) is elaborated on different implementations of AES-128, specially fault-hardened implementations based on dual modular redundancy (DMR). Our experimental results show that PFA is quite simple and efficient in breaking these typical implementations. To show the feasibility and practicability of our attack, a case study is illustrated on a few countermeasures of masking. This work puts forward a new direction of FAs and can be extended to attack other implementations under more interesting scenarios.

2.1 Introduction

FA is a class of implementation level attacks on embedded systems [1], which is usually used to attack different ciphers such as RSA, AES, PRESENT [2], LED [3], Piccolo [4]. FA is an active attack that disturbs the operation of target device. The disturbance is realized by forcing the device in a non-nominal operating condition. Common methods include changing the power-supply voltage, changing the frequency of the external clock, varying the temperature or exposing the circuits to lasers during the key-dependent computations such as encryptions [5–9]. The idea of FA was first reported on RSA-CRT by Boneh *et al.* in 1996 [10]. Later, Biham

[1]College of Computer Science and Technology, Zhejiang University, Hangzhou, China
[2]State Key Laboratory of Cryptology, Beijing, China
[3]Alibaba-Zhejiang University Joint Institute of Frontier Technologies, Hangzhou, China
[4]College of Information Science and Electronic Technology, Zhejiang University, Hangzhou, China
[5]Temasek Labs, Nanyang Technological University, Singapore, Singapore
[6]The Institute of North Electronic Equipment, Beijing, China

and Shamir proposed the differential fault analysis (DFA) attack on the block cipher DES, which combines an FA with differential cryptanalysis [11]. Since then, DFA has been used to break various block ciphers [12–14]. Apart from breaking cryptographic systems, FA is also used for other attacks, such as bypassing security checks in smart cards [5].

DFA operates in a differential setting, i.e., exploiting the difference of correct and faulty ciphertexts for a fixed input. Later, other fault analysis techniques were introduced. Some analysis techniques exploited the algebraic structure of the algebraic fault analysis (AFA [15]) in a differential setting. Other analysis methods exploited statistical biases introduced due to fault injection [16,17]. These biases could be exploited either in a differential setting or with faulty ciphertexts only.

Most, if not all, of the proposed fault analyses are developed with a transient fault assumption. A fault is injected during a target computation, while all other computations remain unaffected. Fault models, such as bit flips, random byte, are often used in a transient fault setting. Alternately, some fault analysis techniques assume permanent faults. Permanent faults are equivalent to device defects that stay during the lifetime. As the fault is fixed, the bias introduced is exploited by statistical means. Stuck-at fault model is a common example of this kind.

In this chapter, we develop fault analysis for a third kind of fault model, called as *persistent fault*. This fault falls between transient and permanent. While the fault persists from one encryption to another, it disappears when the target device reboots. We propose a statistical technique to exploit such faults, called as PFA.

For aforementioned analysis techniques to work, it is desirable that the fault is injected in the last few rounds of the cipher. If the fault is injected much deeper into the cipher (middle rounds), the analysis becomes too complex and does not give much advantage over a simple brute force search. Moreover, majority of the known attacks can only handle a single-fault injection. This requirement puts several restrictions on the attacker's capability, as he is expected to inject single faults in later rounds only.

On the contrary, PFA assumes that a persistent fault might be always present. A common example is when the fault is injected into an algorithm constant stored in memory (ROM), e.g., one element in an S-box. Unless the ROM is refreshed, the fault will persist for all the subsequent encryptions; however, only the rounds accessing that particular faulty S-box element will be affected. A DFA (or other aforementioned techniques) cannot be applied in this setting for two reasons. First, fault can be multiple and present in earlier rounds. Second, it is not possible to acquire correct and faulty ciphertext in the presence of persistent fault for a given plaintext, which prevents any differential analysis. The proposed PFA is developed to exploit such cases, where a fault is persistent and can affect multiple rounds. From a practical aspect, PFA also relaxes some constraints enforced on the adversary. The adversary does not need to synchronize the fault injection with later rounds in run time and also does not require re-encryption for correct/faulty ciphertext pairs. The attack target can be perturbed beforehand, by injecting the fault that persists and exploited later. When the victim encrypts on the plaintext, the adversary observes the resultant ciphertext and performs PFA to retrieve the secret key.

PFA is also capable of compromising some of the widely used fault countermeasures under its basic fault model. We target DMR, which performs redundant operations followed by comparison to detect faults. This countermeasure is also used for reliability verification and thus widely adopted in commercial products. It is believed to be provably secure against single-fault injection. As we show later, such countermeasures can be easily broken by PFA. While some versions of DMR are broken by design, others can be broken with a higher number of available ciphertexts. For example, in AES, the attack roughly needs 10× more ciphertexts than an unprotected design.

Masking [18] is the most studied countermeasure against side-channel attacks. The key idea behind masking is to mask the side-channel activity of a sensitive intermediate value in a cryptographic algorithm by mixing it with a random value. Each encryption call requires fresh randomness to totally remove dependency between sensitive value and side-channel activity. Randomness is sometimes updated several times within an encryption to avoid sophisticated attacks such as higher order attacks. Theoretically, masking does not prevent against FAs; however, due to randomness involved, the fault analysis can be complicated.

The main contributions of this work are summarized as follows:

- We target a new category of injected faults, called persistent fault. Based on persistent fault, we develop a fault analysis technique called PFA and explain its working mechanism. Unlike common fault analysis technique, PFA is not differential and it uses statistical means for key recovery. Thus, it is a faulty ciphertext-only attack.
- We extend PFA to work in a multiple-fault setting.
- We first validate PFA on S-box- and T-box-based AES-128 on Virtex-5 FPGA. Xilinx data2mem software is used for emulation of persistent faults.
- Then we show that PFA could break fault countermeasures based on DMR. Different variants of the countermeasure were broken with 2–10× extra ciphertexts compared to unprotected designs.
- We also validate PFA on a few public implementations of masking. The key advantage is that PFA only requires one fault injection and multiple encryptions.

The rest of the chapter is organized as follows. Section 2.2 introduces the related work. Section 2.3 highlights the core idea, the process of PFA, the complexity and the comparison with other FAs. Section 2.4 extends PFA with multiple faults. Section 2.5 gives the background of AES implementations. Section 2.6 evaluates the PFA on different scenarios, especially on those with FA countermeasures. Section 2.7 gives a case study of PFA on different masking countermeasures. Section 2.8 concludes the chapter.

2.2 Related works

Fault duration: Faults in electronic circuits can be either *permanent* or *transient*. A *permanent fault* is caused by intentional or unintentional defects in the chip [19], which permanently modifies its functionality. In contrast, a *transient fault* [5] only influences the device for a very short time. A common application of transient faults is corrupting a single execution of an encryption. In this work, we are more interested in the third category called as *persistent fault*. The term "persistent" refers to the characteristic of a new type of faults whose duration may not be permanent and typically can last for several encryptions, for example, a few minutes or up to a few hours. Sometimes it might be persistent till the device is reset. An example of such fault is a modification of a stored constant, like an S-box entry, using rowhammer injection techniques [20]. Rowhammer injection techniques are used in some previous works to attack ciphers [21–23]. With such faults, all the rounds of all encryptions will be affected. A reboot or refresh of the affected memory will restore the original functionality.

Fault analysis: Most fault analysis techniques are differential in nature. They require a correct and faulty computation with same inputs, to exploit the difference of outputs for key recovery. Typical techniques include DFA [11], algebraic fault analysis (AFA) [15], fault rate analysis [24] and more. Other techniques are statistical in nature and sometimes exploit faulty ciphertexts only. Common examples are statistical fault analysis (SFA) [16,17] and fault sensitivity analysis (FSA) [25]. In practical FAs, adversaries usually face a ciphertext-only scenario, with no or limited control on inputs. In addition, some countermeasures restrict multiple encryptions with the same inputs by using techniques such as random value padding [26]. In this chapter, we are interested in exploiting FAs that can be conducted under ciphertext-only attack scenario. A thorough comparison of PFA against other common analysis techniques is drawn in Section 2.3.5.

Persistent faults: The notion of *persistent FA* is not new. In [27], ultraviolet light was used to erase the contents of a microcontroller, particularly lookup tables. However, the precision of the attack was limited. Also the offline analysis in [27] was differential and not developed particularly to exploit persistent faults. In [28], the AES lookup table implemented on the Xilinx FPGA using embedded block random access memories (BRAMs) was tampered, where the persistent attack is first mounted on the hardware implementation. However, the attack model was too strong and corresponding offline persistent analysis in [28] was very simple due to their assumption that the entire AES table was set as all zeros and the last round key can be directly output as the ciphertext.

Masking: Masking has come under the scanner of FAs in few previous works. Boscher and Handschuh [29] showed that masking does not protect against classical differential FAs. While the analysis was a bit more restrictive in terms of the fault model and the number of faults that are required, the key recovery was possible with increased attack effort. A new kind of fault analysis called FSA was shown to break masking by Li *et al.* [30]. FSA used some side-channel information with FA to achieve the goal, again with increased effort as compared to unprotected implementation. FSA was further combined with collision attack to enhance its power leading to stronger attack on several countermeasures, including masking and threshold implementation [31]. Use of randomness was recommended as a fault countermeasure prerequisite by Lomné *et al.* [32]. Recently in CHES 2018, a special class of FAs called statistical ineffective FA (SIFA [33]) were used to target and break masking countermeasure at any masking order. SIFA requires several ineffective fault injections to statistically determine the key. In this work, we assess the security of several public implementations of masking countermeasure under PFA. As shown later, PFA on masking requires only one fault injection and breaks masking at any order d.

2.3 Persistent fault attack

This section provides details about the proposed *PFA* method.

2.3.1 Fault model

The assumed fault model is as follows:

- The adversary can inject faults before the encryption of a block cipher. Typically, these faults alter a stored algorithm constant.
- The injected faults are persistent, i.e., the affected constant stays faulty unless refreshed. Thus, all iterations are computed with the faulty constant.
- The adversary is capable of collecting multiple ciphertext outputs. Thus, a watchdog counter on detected faults is considered out of scope.

In this section, we first show the analysis with single-fault injection. Exploitation of multiple-fault injections is discussed in the next section.

2.3.2 Core idea

As stated in the fault model, the fault persists over several computations. In the case of block ciphers, the prime target of this attack, these computations refer to the round function. Each encryption is composed of several calls of a round function. The injected fault persists over several encryptions (thus round function calls), but the faulty value may not be accessed. For example, if the fault exists in an S-box element, the round computation is only faulty if the faulty S-box element is accessed. Otherwise, the injected fault does not impact this round computation. If the faulty value is not accessed during an encryption, the resultant ciphertext will be correct, otherwise incorrect. We further exploit the statistical distribution of correct and incorrect ciphertexts to reveal key-dependent information. We call this newly developed fault analysis technique as PFA. The corresponding attack is called as persistent FA.

The complete persistent FA is composed of three stages. In the fault injection stage, the persistent fault is injected before the first encryption. Unlike traditional DFA, identification of exact round timing or precise location is not required. In the encryption stage, the adversary (denoted as $\mathscr{A}$) then waits for the victim (denoted as $\mathscr{V}$) to start the encryptions. $\mathscr{A}$ can then observe the produced ciphertexts, few of which are correct while others are incorrect due to the persistent fault. In the fault analysis stage, $\mathscr{A}$ analyzes the mixture of correct and faulty ciphertexts with PFA to recover the secret key. As shown later, PFA can be applied on unprotected implementation as well as some state-of-the-art fault-hardened implementations.

2.3.3 Persistent fault analysis

In this part, we further detail the analysis technique of a persistent FA. While the analysis technique remains generic, we start to explain with an example of SPN block cipher targeted for last round key recovery.

Let us assume a typical SPN construction with L words of b bits. A b-bit input is processed by a substitution component (typically, S-box), followed by linear permutation layer (LP) and a round key addition. Next, we take PFA on the last round of SPN block cipher as an example to describe the technique. As LP is linear, we remove it for a simpler analysis. Let x_j and y_j denote the jth word of last round, at input and output of S-box, respectively. y_j, when mixed with jth word of last round key, produces jth word of ciphertext c_j. Then, it satisfies $y_j \oplus k_j = c_j$, which is equivalent to the following:

$$k_j = y_j \oplus c_j \tag{2.1}$$

Let $Pr(y_j)$ and $Pr(c_j)$ denote the distribution probability of y_j and c_j, respectively. As for the correct encryption, due to the avalanche effect, for each candidate of y_j, $Pr(y_j)$ is close to 2^{-b}. Let us assume that the fault is injected in the first S-box element, where correct value $S[0] = v$ is altered to $S'[0] = v^*$ and $v \neq v^*$. The same is illustrated in Figure 2.1. The fault injection makes $Pr(y_j = v)$ as zero. As the number of observed encryption increases, $Pr(y_j = v^*)$ approaches to 2^{1-b}. For all other values of y_j, $Pr(y_j)$ converges to 2^{-b}. This difference in probabilities can be statistically distinguished, thus requiring multiple ciphertexts for conducting PFA. As k_j is fixed, the probability distribution of y_j also relates to distribution of c_j. With the collected ciphertext, the adversary can build the distribution of c_j, to retrieve information on y_j, eventually allowing the recovery of the key k_j. This analysis can be formally written as

$$Pr(c_j) = Pr(y_j \oplus k_j) \tag{2.2}$$

For the jth word of ciphertext, each possible value of c_j is denoted as $t, 0 \leq t < 2^b$. The adversary can collect N ciphertexts and calculate the appearance of t denoted as

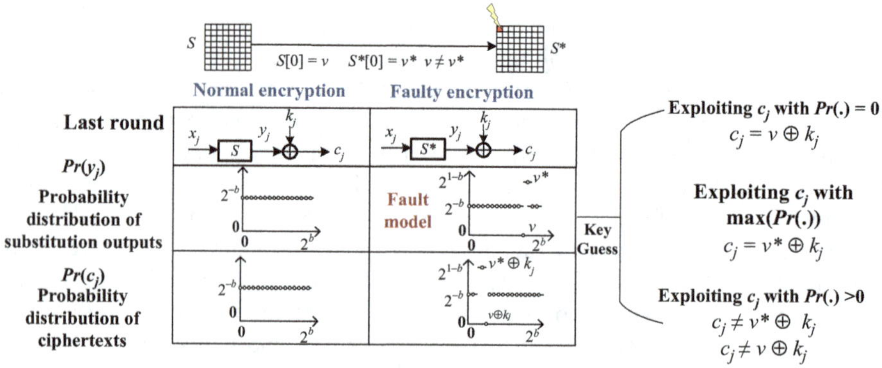

Figure 2.1 Overview of persistent fault analysis

Counts(t). *Counts*(t) is a function to count the number of ciphertexts, where $c_j = t$. Suppose *arg_mincounts*(t) and *arg_maxcounts*(t) are two respective functions to find the value of t with the minimal and maximal number of counts. The corresponding value of t is denoted as $t_{\min}$ and $t_{\max}$, respectively, which can be computed as following:

$$t_{\min} = arg\_mincounts(t) \triangleq \{t|\forall s : Counts(t) \leq Counts(s)\} \quad (2.3)$$

$$t_{\max} = arg\_maxcounts(t) \triangleq \{t|\forall s : Counts(t) \geq Counts(s)\} \quad (2.4)$$

Then, three cryptanalysis strategies can be applied to recover the secret key.

Strategy 1: Exploiting $t_{\min}$.
Since the adversary can calculate the statistic distribution of each element in ciphertexts, he is aware of the value of $t_{\min}$. He also knows v that is publicly known as the original value of the element in the S-box. If N is large enough, k_j can be directly deduced as

$$k_j = v \oplus t_{\min} \quad (2.5)$$

Strategy 2: Exploiting impossible values for $t \neq t_{\min}$.
For the other values t of c_j, where $t \neq t_{\min}$, the adversary can use these values to eliminate impossible candidates of k_j, which can be denoted as

$$k_j \neq v \oplus t \quad (2.6)$$

Strategy 3: Exploiting $t_{\max}$.
The adversary can also try to find the value of $t_{\max}$ whose frequency is approaching 2^{1-b}. If the adversary knows v^*, i.e., the faulty value of the element with the persistent fault, k_j, can be easily computed as

$$k_j = v^* \oplus t_{\max} \quad (2.7)$$

All three strategies can be used in exploiting the fault for key recovery. However, based on the application scenario, one of the strategies might be better suited. Strategies 1 and 2 are accurate analyses. As long as the probability of t is nonzero, the value of $v \oplus t$ is the impossible candidate for k_j and can be eliminated. Both strategies require the value of v to be known.

Strategy 3 is a statistical analysis. Only when the total number of ciphertexts N is increased to a representative value, the probability of $t_{\max}$ (approaching to 2^{1-b}) can be obviously distinguished from other cases. As a result, $k_j = v^* \oplus t_{\max}$ can be recovered. Strategy 3 requires the additional knowledge that the adversary should also know v^*, the value of the faulty element in the lookup table.

Algorithm 2.1 describes the pseudo code of PFA on the last round key. The attack eliminates the impossible candidates for each key element until it identifies all L words of the last round key. In the attack, a two-dimensional array, *Counts*[u][t], $0 \leq u \leq L$, $0 \leq t \leq 2^b$, is initialized to all zeros. Then, the value of specific element of *Counts*[u][t] is updated by counting the appearance of each ciphertext element. If the total number of counts for *Counts*[u][t] is nonzero, then $(t \oplus v)$ can be discarded as an impossible candidate for k_u. Otherwise $(t \oplus v)$ can be kept as a possible candidate for k_u. In the end, only one nonzero value remains for one of the *Counts*[u] at indice t',

Algorithm 2.1: Pseudo code of PFA on the last round of a general block cipher

```
for u = 0; u < L; u++ do
    for t = 0; t < 2^b; t++ do
        Counts [u][t]=0;
    end
end
for u = 0; u < L; u++ do
    for n = 0; n < N; n++ do
        Counts [u][c_{u,n}]++ ;                // c_{u,n} is c_u in the nth ciphertext
    end
end
for u = 0; u < L; u++ do
    for t = 0; t < 2^b; t++ do
        if Counts [u][t] > 0 then
            Discard candidate k_u = t ⊕ v;
        end
    end
end
```

which reveals the value of $k_u = t' \oplus v$. Note that this analysis reveals all key elements in one enumeration of all N ciphertexts, which is quite simple and efficient.

2.3.4 Complexity analysis

Recall N is the total number of ciphertexts that are available and n is the number of ciphertexts already used for analysis. For any c_j, let θ_n denote the "average" number of different outputs of table lookups using n ciphertexts. When n is small, the lookup outputs might be pairwise different, so $\theta_n = n$. When n is large, θ_n will converge to the value $\eta = 2^b - 1$, assuming single fault is injected. The adjective "average" means that the analysis of θ_n is conducted in a probabilistic way, which will give the estimated value of N for finding out the only impossible value v for c_j.

Let $\Delta\theta_n = \theta_n - \theta_{n-1}$. When $n = 0$, $\theta_0 = 0$. When $n = 1$, $\theta_1 = 1$ and $\Delta\theta_1 = 1$. For $n = 2$, we have $\Delta\theta_2 = (\eta - \theta_1)/\eta$, assuming that the remaining possible values of the lookup output θ_1 will satisfy the uniform distribution. Then, we can easily deduce the following:

$$n = 2, \quad \Delta\theta_2 = \frac{(\eta - \theta_1)}{\eta}, \quad \theta_2 = \theta_1 + \Delta\theta_2 = 1 + \frac{(\eta - \theta_1)}{\eta} \tag{2.8}$$

$$n = 3, \quad \Delta\theta_3 = \frac{(\eta - \theta_2)}{\eta}, \quad \theta_3 = \theta_2 + \Delta\theta_3 = 1 + \frac{(\eta - \theta_1)}{\eta} + \left(\frac{\eta - \theta_1}{\eta}\right)^2 \tag{2.9}$$

$$\ldots$$

According to the observations in (2.8), we can infer the formula of θ_n by using the mathematical induction of geometric progression. Thus to compute θ_n, we have

$$\theta_n = \frac{1 - q^n}{1 - q}, \quad \text{where} \quad q = \frac{\eta - 1}{\eta} \tag{2.10}$$

The value of v can never be output in the encryption. Let ε_n denote the average number of the remaining possible key candidates associated with c_j using n ciphertexts. We have $\varepsilon_n = 2^b - \theta_n = 2^b - ((1-q^n)/(1-q))$. When n increases and θ_n is approaching $\eta = 2^b - 1$, η impossible candidates of k_j, that is, $v \oplus t$, can be gradually eliminated. Finally, only the correct candidate of k_j, i.e., $v \oplus t_{\min}$, remains.

Figure 2.2 describes the relationship between $\log_2 \varepsilon_n$ and n, where $b = 8$ (for AES). We can see that $\log_2 \varepsilon_n$ decreases with the increase of n. When n is greater than 1,400, $\log_2 \varepsilon_n$ can be reduced to 1, which implies that two possible key candidates exist. In fact, when $n \geq 2{,}000$, $\log_2 \varepsilon_n$ can be reduced to 0, which gives a unique key candidate. Note that Figure 2.2 only gives a theoretical estimation of $\log_2 \varepsilon_n$ for better understanding on the limit of N. In practice, the adversary needs more ciphertexts to statistically get the correct key.

The estimation of n (when $\log_2 \varepsilon_n$ is reduced to 1) is popularly known as the coupon collector's problem [34] where it needs $255 \times (1/1 + 1/2 + \cdots + 1/255) \simeq 1{,}561$ tries to collect 255 coupons. Similarly, it roughly requires 1,561 trials to see all 255 occurring ciphertexts at least once. If the cipher is composed of smaller S-boxes (like $b = 4$), the analysis needs fewer ciphertexts ($n \simeq 49$). PFA can be applied to DES like ciphers where S-boxes are not bijective. However, if the S-boxes are not identical, only a part of key bits will be recovered with one fault. PFA is currently applied on the last round, which, in the case of AES, extracts 16 key bytes for S-box implementation with one injection. PFA could be on the last but one round at the cost of complex analysis as fault might affect both last two rounds.

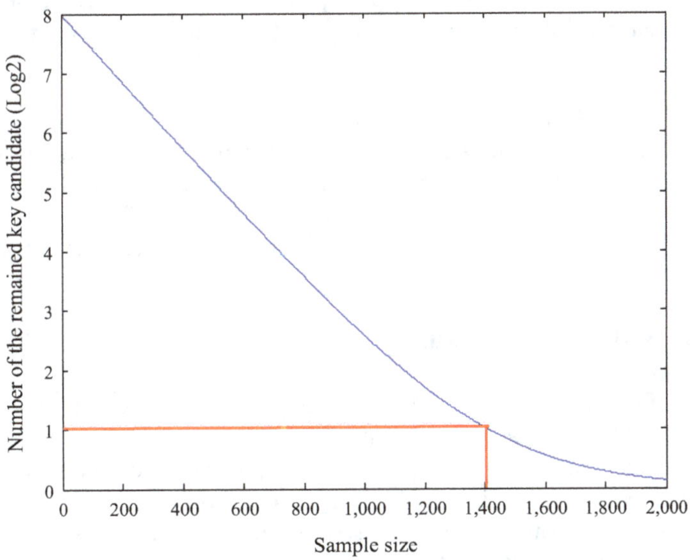

Figure 2.2 Relationship between $\log_2 \varepsilon_n$ *and n for one element of master key where* $b = 8$

2.3.5 Comparison with other fault analysis

Most fault analyses, such as DFA, AFA are differential in nature. They exploit difference between correct and faulty ciphertexts, produced from a fixed plaintext.

Other fault analyses, such as SFA [16,17] or FSA [25], require extra requirements like biased fault or side-channel information. SFA needs multiple biased or constant faults, while PFA needs a single random fault (in ROM). Depending on attackers' capability, one of PFA/SFA might be preferred. The advantages and disadvantages of PFA against other fault analysis techniques can be summarized as follows.

2.3.5.1 Advantages

- The attack is not differential in nature and thus the control over the plaintext is not required. It exploits the statistical properties, which can be built directly upon faulty ciphertexts.
- The adversary does not necessarily need live synchronization to inject a fault at a sensitive moment. The adversary can inject a persistent fault beforehand and wait for the victim to start encryption. Such setting implies that remote, powerful but slow, injection techniques such as rowhammer can be well suited.
- The fault model remains relaxed compared to statistical attacks such as SFA and FSA. While SFA assumes biased fault injection, FSA needs side-channel information for the analysis. PFA is built upon ciphertexts with a random fault model only without any biases. No side-channel information is required.
- As shown later, PFA can also be applied in multiple-fault setting.
- By its nature, PFA can bypass some redundancy-based countermeasures.
- Some circuits deploy fault injection sensors to detect injection attempts. For energy saving, these sensors are only operated in the so-called sensitive mode. An adversary can always inject the persistent fault before the victim is switched to the sensitive mode, rendering the protection ineffective.

2.3.5.2 Disadvantages

- As the analysis technique is statistical, it needs a much higher number of ciphertexts as compared to DFA, which in some cases can be as low as 1 or 2 ciphertext pair.
- Persistent faults can be detected by some built-in health test mechanism.

2.4 PFA with multiple faults

Unlike other fault analysis techniques, PFA can also be applied in a multiple-fault injection setting. By multiple faults, we mean the adversary modifies several elements in a persistent manner. If we take the previous example, the adversary modifies several elements of the S-box. Analysis of multiple injections is becoming more realistic when the technology node is shrinking much faster than the fault injection capability. Thus, the adversary is likely to affect a larger area, like multiple memory locations [27].

Let us assume that the adversary injects faults into λ elements of the S-box. The fault model remains similar to the previous case. There are at least $(2^b - \lambda)$ possible output values from the S-box. For simplicity, we do not consider the linear layer in our equations. Thus, we have $(2^b - \lambda)$ candidate values for each ciphertext word c_j. Suppose S_v denotes the set of the corrupted S-box output elements $(v_0, v_1, \ldots, v_{\lambda-1})$. For each v_i in S_v,

$$Counts(t) = 0, \quad k_j = v_i \oplus t, \quad (0 \leq i < \lambda, \quad 0 \leq j < L) \tag{2.11}$$

In the attack, for each possible value t in each ciphertext element, we calculate $Counts(t)$ for 2^b candidates of t. If $Counts(t) = 0$, we can keep $v_i \oplus t$ as candidates of k_j. If $Counts(t) \neq 0$, we can discard $v_i \oplus t$. With an enough number of ciphertexts, at most λ candidates of k_j can be kept after PFA on each element of ciphertext. Then, the maximal residual entropy of the last round key can be calculated as $L \times \log_2 \lambda$, where L is the number of ciphertext words, i.e., for AES $L = 16$, 16 words of one byte ($b = 8$) each.

Figure 2.3 shows the relationship between the average residual key entropy corresponding to the sample size N and the number of multiple faults λ. We set $1 \leq \lambda \leq 16$ (no. of persistent faults) and $1 \leq N \leq 5{,}000$ (no. of ciphertexts). For each value of λ, we repeat PFA 1,000 times on different data sets, and the final result is averaged over all experiments. We increase the number of samples one by one and calculate the residual entropy of the last round key as a function on N.

From Figure 2.3, we can see that, for each value of λ, the residual key entropy of the block cipher with $b = 8$ can be at most reduced to $16 \times \log_2 \lambda$ after PFA on the last round. Also, at $N = 2{,}000$, the attack saturates and the key entropy cannot be further reduced without extra information. Normally, an adversary can try brute force of the remaining candidates. If the key entropy is beyond brute force search, the adversary can extend PFA to the last but one round and further reduce the key entropy.

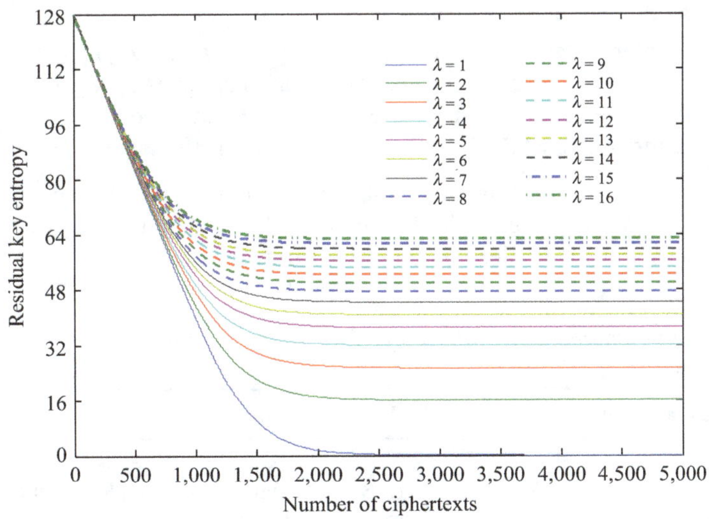

Figure 2.3 Relationship among the residual key entropy, N and λ

2.5 Validation of PFA on AES-128

2.5.1 AES implementation

In this section, we check validation of our attack on AES-128. AES-128 [35] is a symmetric block cipher algorithm with a block size of 128 bits. Suppose the ith round is denoted as R_i. The initial nine rounds R_i, $1 \leq i \leq 9$ comprise of four major operations: SubByte(SB), ShiftRow(SR), MixColumn(MC) and AddroundKey(AK). There is an additional key whitening before R_1 and no MixColumn operation in R_{10}. The SB function is the only nonlinear operation of the block cipher consisting of an substitution box (S-box) lookup applied to the state. AES-128 uses an 8×8 S-box, i.e., 256 bytes. Moreover, the operations in AES-128 are byte oriented, which makes $b = 8$ or $\eta = 2^b - 1 = 255$.

Three exemplary AES-128 implementations are considered. All implementation are based on lookup tables that are stored in the memory. An adversary can inject faults in the memory to corrupt a particular entry of the lookup table, through various available injection means. The details of the tested AES-128 implementations are as follows:

I1: S-box implementation has an S-box lookup table in the encryption and an inverse table S^{-1} in the decryption. There are 256 elements in both S and S^{-1} tables and are used in all ten rounds.

I2: T-table implementation optimizes performance by merging SB, SR and MC operations in one T-table lookup. There are four T tables in the encryption, denoted as T_0, T_1, T_2, T_3. Each T_i is an 8×32 lookup, i.e., 256 elements of 32 bits. In each round $R_i (1 \leq i \leq 10)$, each table of T_0, T_1, T_2, T_3 is accessed for four times. In the decryption, the four inverse lookup tables are denoted as $T_0^{-1}, T_1^{-1}, T_2^{-1}, T_3^{-1}$. Since R_{10} has no MC operation, the required output is extracted from the same T-tables to optimize memory requirements.

I3: T-table implementation is similar to I2. Each of T_0, T_1, T_2, T_3 is accessed for four times in the first nine rounds. Four additional tables T_0', T_1', T_2', T_3' are used in R_{10}. Each element of T_i' also has 32 bits. However, three bytes of the element in T_i' are just zeros, while the only nonzero byte has the same value as that of S.

Table 2.1 summarizes the types of implementations that are offered to apply the AES using S-box and T-box with the lookups in each round and table size. Note that I3 is also a realistic implementation whose assembly code could be found in the

Table 2.1 Different implementations of AES-128 encryptions

Type	Lookups in each round	Table size	Notes
I1	R_{1-10}: S	S:256B	Typical S-box implementation
I2	R_{1-10}: T_0, T_1, T_2, T_3	T_i:1KB	Typical T-box implementation
I3	R_{1-9}: T_0, T_1, T_2, T_3 R_{10}: T_0', T_1', T_2', T_3'	T_i:1KB T_i':1KB	Code can be found in rijndael-amd64.S in the library Libgcrypt 1.6.3

file `rijndael-amd64.S` of the shared library `Libgcrypt 1.6.3`. Similar to Openssl, Libgcrypt is also a well-known cryptographic library that provides numerous cryptographic building blocks.

I3 is analyzed in the next section using rowhammer attacks on modern processors. We test I1 and I2 on Virtex-5 FPGA (xc5vlx50). The area results of I1 and I2 are summarized in Table 2.2. Since there are 16 S-box and 16 S^{-1}-box consumption for each complete AES, each BRAM (RAMB36) can accommodate 4 S-box or S^{-1}-box. The BRAM cost for S-box-based AES is 8. By contrast, the BRAM cost for T-box-based AES is 4.

For the validation of the attack, we use a *data2mem* tool [36] provided by Xilinx. It can update the BRAM content without the need of re-flashing a new bitstream. Thus, it allows us to emulate the persistent fault injection.

2.5.2 PFA on vulnerable S-box implementation (I1)

It is assumed that an adversary affects AES S-box with a fault injected on one element to collide with another existing element, resulting in only 255 distinct elements. The correct value victim v of the eth element does not appear instead the effected faulty value v^* appears twice. All other elements in the S-box remain correct and appear with a probability of $1/256$. When only one fault is injected to the table, the discrete probability distribution function of y can be described as follows:

$$Pr(y = v) = 0; \quad Pr(y = v^*) = \frac{2}{256}; \quad Pr(y \neq v \wedge y \neq v^*) = \frac{1}{256} \tag{2.12}$$

The goal for adversary $\mathscr{A}$ is to extract the last round key K^{10}. Here Algorithm 2.1 can be directly applied to handle this situation where $b = 8$. However, the number of N samples needs to be carefully chosen in order to successfully launch the attack based on the theoretical estimation.

2.5.2.1 Attack result

Figure 2.4 shows the result of an exemplary PFA on AES S-box implementation. In the attack, the first element of S-box is injected with the persistent fault, $v = 0x63$. After the fault injection, only the 7th bit of v is flipped from one to zero, $v^* = 0x61$. In Figure 2.4, 2×10^4 ciphertexts are collected. For each ciphertext byte, for example, c_1 in Figure 2.4(a) and c_2 in Figure 2.4(b), the probability for each value is plotted as one curve, which is calculated as the counts of appearances for that specific value divided by the number of ciphertexts already used in the analysis. In both subfigures, two curves are obviously distinct from the rest. The red curve at the bottom is for $t_{\min}$, which never appears in c_j. The blue curve on the top is for $t_{\max}$, whose probability is converged

Table 2.2 Cost of S-box and T-box AES implementations

AES style	RAMB36	Slice LUTs	Slice registers	Occupied slices
S-box	8	2,630	2,469	2,131
T-box	4	4,526	2,492	1,370

to 2/256 and significantly larger than that of other values. Since $c_1 = 0x2e$, k_1 can be computed as $k_1 = 0x61 \oplus 0x2e = 0x4f$. Similarly, $k_2 = 0x61 \oplus 0x0a = 0x6b$. The attack values of k_1 and k_2 are the same as the one in K^{10}.

2.5.2.2 Residual key entropy for different sample size

Let $\phi_t(n)$ denote the theoretical estimation of the residual key entropy for K^{10} when n ciphertexts are used for analysis. Note that the similar analysis can be applied to all 16 key bytes simultaneously. $\phi_t(n)$ can be approximately calculated as $16 \times \log_2 \varepsilon_n$, where $\varepsilon_n = 256 - ((1 - q^n)/(1 - q))$ and $q = 254/255$.

Let $\phi(n)$ denote the actual residual key entropy when n ciphertexts are used for analysis. The value of $\phi(n)$ can be calculated according to Algorithm 2.1, assuming the attacks on each byte k_j are equivalent. Figure 2.5(a) shows how $\phi(n)$ and $\phi_t(n)$

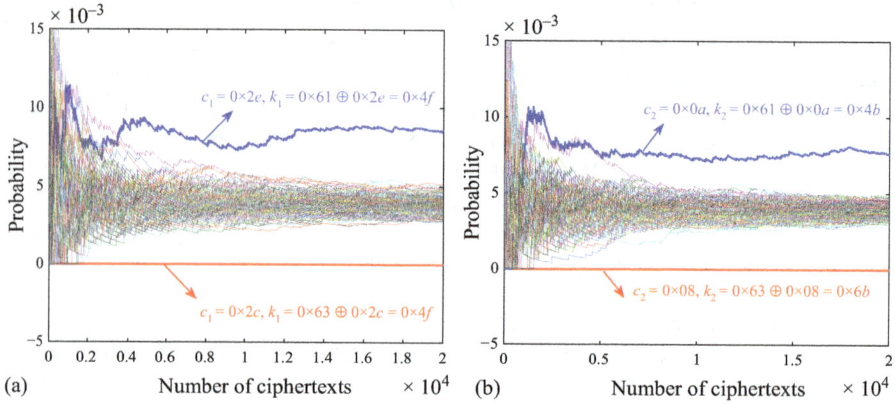

Figure 2.4 Exemplary PFA on AES-128 using distributions of ciphertext values: (a) extract k_1 using the distribution of c_1, (b) extract k_2 using the distribution of c_2

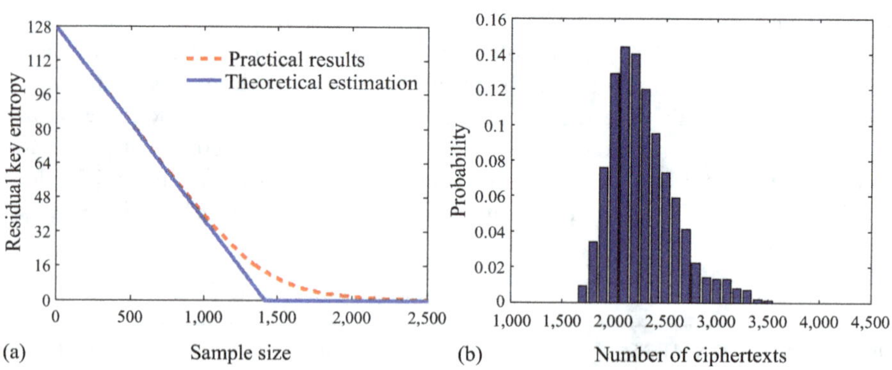

Figure 2.5 Attack result of PFA on unprotected AES: (a) $\phi(n)$ vs. $\phi_t(n)$, (b) distributions of N_f

decrease when n is increased. In Figure 2.5(a), $\phi(n)$ is very close to the theoretical value of $\phi_t(n)$. When $n \geq 1{,}240$, $\phi_t(n) \leq 16$. When $n \geq 1{,}360$, $\phi(n) \leq 16$, which implies that there are at most 2^{16} candidates of K^{10}. When $n \geq 1{,}405$, $\phi_t(n) \leq 1$. And when $n \geq 2{,}148$, $\phi(n) \leq 1$, which implies that there are at most two candidates for the full K^{10}.

2.5.2.3 Sample size distributions for full key recovery

In order to guarantee the success rate of our PFA method, the FAs are conducted again ξ times with the random plaintext. For each attack, we increase the number of ciphertexts n in the analysis, until all the 16 key bytes are recovered. Suppose N_f denotes the number of ciphertexts that is required when the adversary $\mathscr{A}$ can successfully extract all the 16 bytes of AES for the first time in one specific attack.

Figure 2.5(b) describes the distributions of N_f for $\xi = 1{,}000$ attacks. It is clear that $1{,}678 \leq N_f \leq 3{,}504$, it depicts that at least 1,678 and at most 3,504 samples are required to recover the full master key. The average value of N_f is 2,281.

2.6 Defeating fault attack countermeasures with PFA

In this section, we enhance the PFA to AES-128 protected against FAs.

2.6.1 Countermeasures against fault attacks

DMR is a mechanism of using two redundant modules to prevent FAs [1]. DMR has characteristics of reliability and security and also offers robustness to the systems in order to detect the error. This countermeasure is readily adopted in commercial solutions due to its properties, thus enhances the reliability.

As shown in Figure 2.6, there are two modules in the DMR scheme: Module 1 and Module 2. If both modules are performing encryption, the countermeasure is named as redundant encryption-based DMR (REDMR), where Module 2 is functionally equivalent to Module 1. Provided that the resultant ciphertexts of two modules are the same ($C' = C''$), the security check for REDMR is passed. Thus, the ciphertext

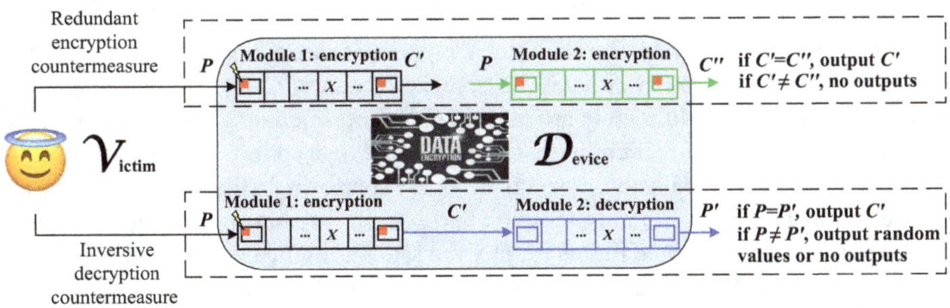

Figure 2.6 Countermeasures against fault attacks: REDMR and IDDMR

is considered true and can be sent to display the output. REDMR passes the security check and is considered secure against single-fault injection. In order to defeat this countermeasure, one has to inject either same fault in both the modules (two fault injections are required) or a fault in one module and bypass the security check.

An alternative strategy is adopted to make the same fault in two modules harder. If Module 1 is an encryption type and Module 2 is decrypting the ciphertext from Module 1, the corresponding countermeasure is named as inversive decryption-based DMR (IDDMR). If the decrypted plaintext from the module is the same as the original plaintext ($P' = P$), the ciphertext of Module 1 is considered as true and can be sent to display the output. As both the modules are performing different operations and have different architectures, injecting complementary fault is harder. Next, we focus on IDDMR, the stronger of the two countermeasures. The same analysis also applies to REDMR.

Based on the reaction to failed security check, three countermeasures can be classified.

C1: No ciphertext output (NCO). If $P' \neq P$ is detected, the victim $\mathscr{V}$ will not display the incorrect ciphertext C'.

C2: Zero value output (ZVO). If $P' \neq P$ is detected, the victim $\mathscr{V}$ will display the output that is a ciphertext $C' = 0$.

C3: Random ciphertext output (RCO). A ciphertext with total random values will appear at output when $P' \neq P$ is detected. The major benefit of this scheme is to embed the incorrect ciphertexts into a large pool of randomized ciphertexts, resulting in the adversary's difficulty of directly differentiating the fault leakages. Different from NCO and ZVO, $\mathscr{A}$ cannot distinguish the correct encryption from faulty encryption.

Note that for REDMR, if both the modules use shared memory, i.e., common lookup tables, all three countermeasures will fail to detect a fault in lookup tables, that is, target of PFA. In the following, we do not consider this case but a stronger implementation where each module has independent memory, for instance, IDDMR.

2.6.2 *PFA on S-box (I1) with NCO and ZVO*

With NCO and ZVO, only a fraction of the N ciphertexts are available while the rest of ciphertexts are suppressed by either NCO or all zero values (ZVO). This is because the injected faults in the lookup table are used in some intermediate computations in the encryption, and the incorrect ciphertext is used in the decryption, which further leads to the faulty output P'. Thus, the adversary $\mathscr{A}$ has to perform further more encryptions in order to have a significant number of ciphertexts that are not suppressed. However, the analyzing methodology remains exactly the same as in the unprotected case, once an enough number of ciphertexts are available. Each AES encryption involves 160 S-box calls (16 in each of 10 rounds). Note that the 40 lookups in the key schedule are not considered. The probability p that one plaintext can bypass IDDMR, that is, the probability that all 160 S-box lookups do not access the faulty element in the S-box table, can be calculated as

$$p = \left(1 - \frac{1}{256}\right)^{160} \approx 0.5346 \tag{2.13}$$

Thus, only $p \times N$ ciphertexts can be used for the attack. In this case, the attacker would need around $N/p \approx 1.8706 \times N$ encryptions, instead of N encryptions to perform full key recovery.

To investigate the case of PFA on NCO/ZVO-based IDDMR, we repeat the attack for $\xi = 1{,}000$ times. Figure 2.7 describes the distribution of N_f. The statistic shows that $3{,}042 \leq N_f \leq 7{,}141$, which means that at least 3,042, at most 7,141 ciphertexts are required to recover the full master key. The average value of N_f is 4,234. In the attack, if we set $n > 7{,}200$, the success rate is 100%.

2.6.3 PFA on S-box (I1) with RCO

Contrary to Section 6.2, in the presence of RCO, the adversary cannot distinguish between correct ciphertext and random ciphertext. Each ciphertext byte can take all possible 256 values. Strategy 2, which depends on the impossible value of ciphertexts, cannot defeat this countermeasure scheme. Whether Strategies 1 and 3 (exploiting $t_{\min}$ or $t_{\max}$) can be applied or not depends upon the probability distribution for different values of the ciphertext byte. As all the ciphertexts are generated either by a correct output when $P' = P$ or by a random output when $P' \neq P$, the discrete probability distribution function for the ciphertext byte y can be computed in two parts accordingly. One is the output of the S-box whose distribution is already described in

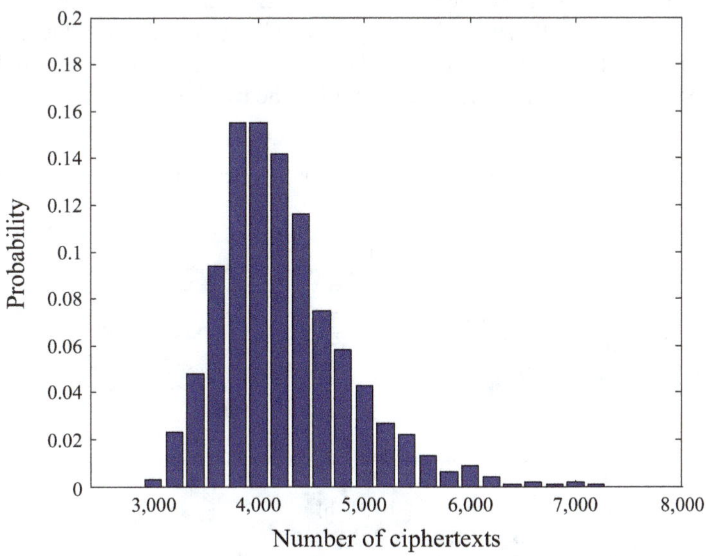

Figure 2.7 Distribution of N_f for PFA on AES with NCO/ZVO

(2.2). However, the constraint that the ciphertext is correct should be added, whose probability is given in (2.13). The other is the random output that satisfies the uniform distribution. The specific distribution is listed in (2.14), where p is the probability that one encryption does not access to any faulty element in S-box.

$$Pr(y = v) = 0 \times p + \frac{1}{256} \times (1 - p) = \frac{0.4654}{256} \quad (2.14)$$

$$Pr(y = v^*) = \frac{2}{256} \times p + \frac{1}{256} \times (1 - p) = \frac{1.5346}{256} \quad (2.15)$$

$$Pr(y \neq v \wedge y \neq v^*) = \frac{1}{256} \times p + \frac{1}{256} \times (1 - p) = \frac{1}{256} \quad (2.16)$$

From (2.14), $y = v$ is still with the minimal probability and $y = v^*$ is with the maximal probability. In this case, both Strategies 1 and 3 can be adopted to extract the secret key, with updated specific probability value $Pr(y = v)$ and $Pr(y = v^*)$. Figure 2.8 clearly demonstrates the probability distribution of two illustrative ciphertext bytes, c_1 in Figure 2.8(a) and c_2 in Figure 2.8(b), which are the first and the second byte of AES ciphertext for S-box implementation protected with RCO. For each ciphertext byte, the probability for each value corresponding to the increase of N is plotted as one curve, which is calculated as the counts of appearances for that specific value divided by the number of ciphertexts already used in the analysis.

In Figure 2.8, 4×10^4 ciphertexts under the IDDMR are collected. The red curve at the bottom is for $t_{\min}$, whose probability is converged to 0.4654/256 and significantly smaller than that of other values. The blue curve on the top is for $t_{\max}$, whose probability is converged to 1.5346/256 and obviously larger than that of other values. In both the subfigures, two curves are obviously distinctive from the rest.

Algorithm 2.2 describes the analysis on AES-128 (Implementation I1) that extracts K^{10} when IDDMR with RCO is applied. The attack sets two threshold values τ_1 and τ_2 to filter the correct key candidates. The choice of τ_1 and τ_2 is based on the empirical experience, in order to distinguish the two biased bytes from other 254

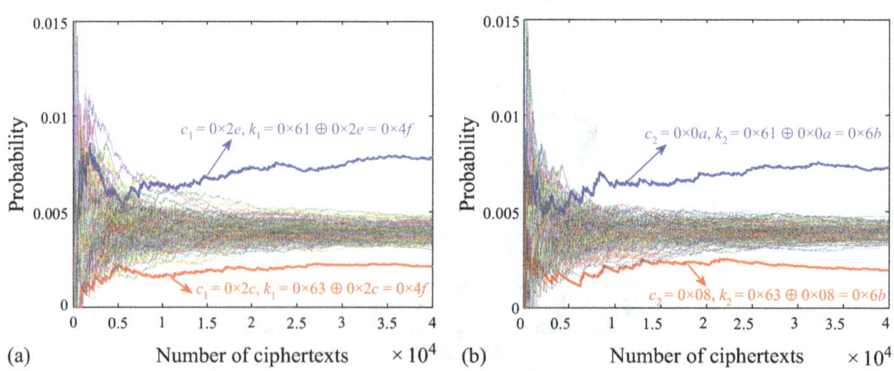

Figure 2.8 Attack result on AES-128 S-box implementation with RCO: (a) extract k_1 using the distribution of c_1, (b) extract k_2 using the distribution of c_2

unbiased bytes. For the sake of simplicity, τ_1 is set as $0.9 \times (1.5346/256)$ and τ_2 is set as $1.1 \times (0.4654/256)$, in order to identify the key candidate with the maximal and minimal probability, respectively.

Similar to Algorithm 2.1, the two-dimensional $Counts[u][t]$ is updated by counting the appearance of each ciphertext byte. For Strategy 1, if the total number of counts for $Counts[u][t]$ divided by N is smaller than τ_1, $(t + v)$ can be kept as the possible candidate for k_u and inserted into a candidate set $\Phi_{u,1}$. For Strategy 2, if $Counts[u][t]$ divided by N is larger than τ_2, $(t + v^*)$ can be kept and inserted into a candidate set $\Phi_{u,2}$.

According to Algorithm 2.2, we conduct two experiments to estimate $\phi(n)$, the residual key entropy, by using τ_1 and τ_2. Let $\phi_{u,1}$, $\phi_{u,2}$ denote the residual key entropy ($\log_2$-based) of $\Phi_{u,1}$, $\Phi_{u,2}$, respectively. $\Phi_{u,1}$ and $\Phi_{u,2}$ are the residual key search space when analyzing the uth byte of the last subkey in R_{10} merely with only τ_1 or τ_2, respectively.

In the first experiment, the residual key entropy $\phi_1(n)$ is calculated as the sum of all $\phi_{u,1}$. In the second experiment, $\phi_2(n)$ is the sum of all $\phi_{u,2}$. In each experiment, 1,000 attacks are performed. $\phi_1(n)$ and $\phi_2(n)$ are depicted in Figure 2.9(a), respectively, showing how the residual key entropies change with the number of ciphertexts used in analysis. We can see that $\phi_1(n)$ in PFA using τ_1 is much less than that using τ_2. On average, the required sample size is about 9,280 and 12,660 for using different thresholds (τ_1, τ_2), respectively, when the residual key entropy is less than 16.

2.6.4 PFA on T-tables (I2) with RCO

In this section, we enhance our FA to T-tables-based implementation (I2) protected with additional RCO. Even though the use ofT-tables-based implementation is limited

Algorithm 2.2: Attack on AES-128 to extract the round key in R_{10} under IDDMR

```
for u = 0; u < 16; u++ do
    for t = 0; t < 256; t++ do
        Counts [u][t]=0;
    end
end
for u = 0; u < 16; u++ do
    for n = 0; n < N; n++ do
        Counts [u][c_{u,n}]++; ;                 // c_{u,n} is c_u in the nth ciphertext
    end
end
for u = 0; u < 16; u++ do
    for t = 0; t < 256; t++ do
        if Counts[u][t]/N < τ1 then
            Insert t + v into Φ_{u,1}; ;        // Strategy 1
        end
        if Counts[u][t]/N > τ2 then
            Insert t + v* into Φ_{u,2}; ;       // Strategy 2
        end
    end
end
```

and discouraged due to huge memory requirement and vulnerability to cache attacks, it is still used in some libraries. Compared to the S-box implementation, in this type of implementation, the attack requires more effort in fault injection. For T-tables (I2), each table T_i is accessed four times in each AES round and 40 times in one encryption. Let k_j denote the jth byte of the last round key K^{10}, $0 \leq j \leq 15$.

In the case of a straightforward attack, the adversary is required to modify four entries, one in each T-table to make sure the correspondence of single-fault injection in S-box implementation. For T-tables stored in memory, four byte faults are injected into T_0, T_1, T_2, T_3 simultaneously. For instance, the adversary can modify the following four elements: the first element of T_0 (the third byte, $0xc66363a5 \rightarrow 0xc66361a5$) and the first element of T_1 (the fourth byte, $0xa5c66363 \rightarrow 0xa5c66361$) the first element of T_2 (the first byte, $0x63a5c663 \rightarrow 0x61a5c663$), and the first element of T_3 (the second byte, $0x6363a5c6 \rightarrow 0x6361a5c6$). Then, p'', the probability that one plaintext can bypass IDDMR, is still $(1-(1/256))^{160} \approx 0.5346$.

In our attack, we set $\tau_1 = 0.9 \times (1.5346/256)$ and $\tau_2 = 1.1 \times (0.5346/256)$. Figure 2.9(b) shows the relationship between $\phi(n)$ and the sample size n. It is observed that in this scenario, $\phi(n)$ in PFA using τ_1 is much less than using τ_2. To reduce the residual key entropy to be less than 16, the average sample size is 8,840 and 12,870 for τ_1 and τ_2, respectively.

2.6.5 Discussion

It is important to mention that the 40 S-box lookups in the key schedule of AES do not affect our analysis that much. In the scenario of REDMR or IDDMR, the probability of outputting a correct ciphertext will be calculated as $(1-(1/256))^{200}$. The analysis will remain as the same and quite straightforward. In addition, considering the multiple faults on IDDMR, it is also possible that two faults are injected to S and S^{-1} separately. Intuitively, the fault in S^{-1} might cancel some faulty propagation during the reversible

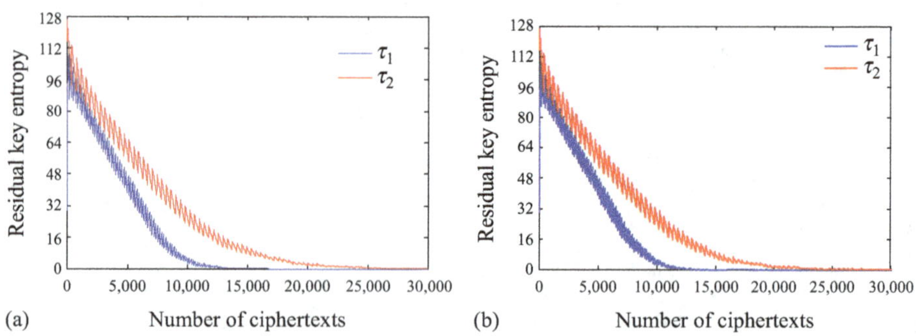

Figure 2.9 $\phi_1(n)$, $\phi_2(n)$ *vs. the number of samples, when different thresholds* (τ_1, τ_2) *are used in PFA on IDDMR countermeasures;* $\phi_1(n) = \sum_{u=0}^{15} \phi_{u,1}$, $\phi_2(n) = \sum_{u=0}^{15} \phi_{u,2}$*: (a) S-box implementation with RCO, (b) T-box implementation with RCO*

decryption process, resulting a more number of ineffective ciphertexts for analysis. However, our preliminary analysis shows the second fault to S^{-1} does not improve our PFA. It will require the same number of ciphertexts as in PFA on the normal IDDMR. It should be noted that, to further reduce the data complexity and the key search space, our PFA can be extended to more rounds besides the last round, e.g., the ninth round of AES-128.

2.7 Case studies: breaking public implementation of masking schemes with single fault

2.7.1 General idea

Block ciphers are composed of repetitions of a round function. In PFA, we are mainly concerned about the final round as it is directly related to the ciphertext. The last round of cipher with basic Boolean masking can be written as follows:

$$c = (L(S'(x \oplus m) \oplus m') \oplus k) \oplus L(m') \tag{2.17}$$

where c denotes the ciphertext, L denote some linear functions (typically permutations), x denotes the last round input, and m and m' denote penultimate and last round masks, respectively. k denotes the round key and $S'(x)$ denotes the masked S-box that can be calculated as $S'(x) = S(x \oplus m)$. Note that the higher order masking can also be included in this analysis, where m can be calculated as $m = m_1 \oplus m_2 \oplus \cdots \oplus m_d$ with d as the masking order.

In our attack model against masking block ciphers, we assume the original (unmasked) S-box is stored for lookup and a persistent fault is injected. The analysis scheme remains generic as illustrated in the previous section. For each S-box call in the encryption, ideally a fresh set of masks are drawn and a new masked S-box S' is computed. This is popularly known as the recomputation method.

If faulty value x' is injected to the ith element of S where the original value $S(i) = x \neq x'$, it leads to the faulty element in the correspondingly calculated masked S-box, where $S'(i \oplus m) = x' \oplus m'$. Consequently, the $x \oplus m$ element is missing in the S', and the $x' \oplus m$ element is doubled. With this knowledge, the adversary can deduce that $c_* = L(x \oplus m') \oplus L(m') \oplus k = L(x) \oplus k$ will not appear in the output ciphertexts. Similarly, $c'_* = L(x') \oplus k$ will be doubled. Since the computation of c_*, c'_* does not depend on either m or m', the attack is equivalent to attacking an unmasked implementation. Even for d-order masking, m and m' can be written as the combination of d mask, which eventually gets canceled out to compute the ciphertext, making the complexity constant even when increasing order d. We target a few public implementations of masking in this section. The key advantage of PFA is that it requires only one fault injection and multiple encryptions, thus limiting the practical effort of injecting the faults. The required fault model is described before, and several works have been practically validated in a range of devices. In the following, we focus on developing the analysis technique with simulations under compatible fault models.

2.7.2 Bytewise masking AES

We apply PFA to the public implementation of bytewise masking available at [37]. It is a typical implementation that follows the general idea illustrated in the previous section. In this case, six randomly generated masks denoted by $m, m', m_1, m_2, m_3, m_4$ are involved in each encryption, where m_i, $1 \leq i \leq 4$ correspond to four rows of AES, respectively. For the MixColumns operation $MC(col_1, col_2, col_3, col_4)$, four output-masks have to be calculated in advance accordingly, denoted by m'_1, m'_2, m'_3, m'_4, such that $(m'_1, m'_2, m'_3, m'_4) = MC(m_1, m_2, m_3, m_4)$.

When all ten masks are generated, a masked AES S-box denoted by S' is pre-calculated prior to the encryption as

$$S'_{m,m'}(x) = S(x \oplus m) \oplus m' \tag{2.18}$$

where S denotes the original AES S-box in which a persistent fault will be injected, and m and m' are the generated masks. With a persistent fault in S, every S' will contain a fault, irrespective of mask values. One single fault would be enough to reveal the key with the statistical method.

The algorithm of this bytewise masking AES is shown in Algorithm 2.3. The operations directly affected by the persistent fault are shown in red. Here we apply the $t_{\min}$ strategy. We use the available code for our analysis. We injected one persistent fault in S, by randomly changing one S-box element. The attack was repeated 100 times, and the average of all results is computed. By coupon collector's problem, the minimum number of ciphertext required is ≈1,560. In the experiments, we found that with 1,500 ciphertexts the attacker has on average less than two key byte candidates to test and a unique key with little over 2,000 ciphertexts. The analysis remains exactly the same to recover all the bytes independently from same set of ciphertext, thus revealing the last round key and eventually the master key.

Algorithm 2.3: Bytewise masking AES

```
Input: plaintext p = (p1, p2, p3, p4), where pi, 1 ≤ i ≤ 4 represent the ith column vector of p, key k
Output: ciphertext c
rk ← KeySchedule(k)
(m, m', m1, m2, m3, m4) ←$ (F_{2^8}, ... )
(m'1, m'2, m'3, m'4) ← MixColumn(m1, m2, m3, m4)
S' ← GetMaskedSbox(S, m, m')                                   // (2.18)
x ← p ⊕ (m, ... ) ⊕ rk[0]
for i = 1; i < 10; i + + do
    x ← S'(x)
    x ← ShiftRows(x)
    x ← x ⊕ (m1, m2, m3, m4) ⊕ (m', ... )
    x ← MixColumn(x)
    x ← x ⊕ rk[i] ⊕ (m'1, m'2, m'3, m'4) ⊕ (m, ... )
end
x ← S'(x)
x ← ShiftRows(x)
c ← x ⊕ rk[10] ⊕ (m', ... )
```

2.7.3 Coron's higher order masking of lookup tables [38]

In Eurocrypt 2014, Coron presented a method to securely compute lookup tables in a block cipher, secure at any order d [38]. This scheme is an ideal target for PFA as it uses lookup tables by design, which is vulnerable to persistent faults. We target the publicly available implementation of AES protected with this scheme, provided by the author [39].

The key feature of Coron's countermeasure [38] is table recomputation. It uses independent masks with additional refresh of the masks between every successive shift of the input. One can view every line u of the randomized table as a n-dimensional vector of elements in $\{0, 1\}^k$, and for all inputs $u \in \{0, 1\}^k$:

$$T(u) = (s_{u,1}, s_{u,2}, \dots, s_{u,n})$$

where initially each vector $T(u)$ is an n-Boolean sharing of the value $S(u \oplus x_1)$. The vectors $T(u)$ of the randomized table are then progressively shifted for all $u \in \{0, 1\}^k$, first by x_2 and so on until x_{n-1}. Eventually the evaluation of $T(x_n)$ gives a vector of n output shares that corresponds to $S(x)$.

To refresh the masks between successive shifts, one can generate a random n-sharing of 0, that is, $a_1, \dots, a_n \in \{0, 1\}^k$ such that $\bigoplus_{i=1}^{n} a_i = 0$, and XOR the vector $T(u)$ with $(a_1, \dots, a_n)$, independently for every u. More concretely, we can use the **RefreshMasks** procedure in Algorithm 2.4 from [18], which gives a masking of y as $y = y_1 \oplus \cdots \oplus y_n$ by XORing both y_1 and y_i with $r_i \leftarrow^{\$} \mathbb{F}_{2^k}$, in an iterative manner from $i = 2$ to n, where the original value of y_1 is y. The full description of the procedure of Coron's higher order masking of lookup tables is provided in Algorithm 2.5.[1]

Algorithm 2.5 uses two temporary tables T and T' in RAM. Both are generated on the basis of the lookup table $S : \{0, 1\}^k \to \{0, 1\}^k$. We show that, however, with as few as one single-faulty element in table S, the following masking provides no protection against PFA. The operation marked in red in Algorithm 2.5 denotes the one directly involving injected persistent fault. It results in a faulty table S', which is the same as table S but one element.

Algorithm 2.4: RefreshMasks

Input: shares $(x_i)_i$ satisfying $\bigoplus_i x_i = x$
Output: shares $(x'_i)_i$ satisfying $\bigoplus_i x'_i = x$
$(z'_0, z'_1, \dots, z'_d) \leftarrow (z_0, z_1, \dots, z_d)$
for $i = 1; i < d + 1; i++$ **do**
  $r_i \leftarrow^{\$} \mathbb{F}_{2^k}$
  $z'_0 \leftarrow z'_0 \oplus r_i$
  $z'_i \leftarrow z'_i \oplus r_i$
end

---

[1] For simplicity, we assume both the input and output of $S(x)$ are words of k bits.

The attack is performed on AES implementation available at [39], which follows Algorithm 2.5. For each attack, a single fault is injected into S, and PFA is applied for $d = 1$. The masking offers no resistance against PFA as it reduces to the generic case presented in Section 2.7.1, where the key recovery remains independent of the mask. This results in the attack similar to unprotected AES with key recovery with around 2,000 ciphertexts. The increase in masking order d has no impact on the attack because the combination of d different masks can be reduced to a single equivalent mask as $m = m_1 \oplus m_2 \oplus \cdots \oplus m_d$.

Next, we target other masking schemes that do not directly use the S-box and thus making the analysis more complicated, yet possible.

2.7.4 Rivain and Prouff's masking [18]

In CHES 2010, Rivain and Prouff [18] proposed an efficient method to mask the AES S-box processing at any order. Specifically, the authors use the algebraic structure of the AES S-box, which is the composition of an affine function over $\mathbb{F}_2^8$ with the power function $x \mapsto x^{254}$ over $\mathbb{F}_{256}$, and they showed that it can be expressed as a sequence of operations involving a few linear functions over $\mathbb{F}_2^8$, which is easy to mask, and four multiplications over $\mathbb{F}_{256}$. If this computation is performed completely on the fly without any lookup tables, PFA does not apply in principle.

Now, we look at the public implementation of this scheme available at [39]. Let us focus on the S-box masking part, where component affine transformation is realized through the lookup table [39]. The additive part of the affine transformation is 0*x*63; thus it can be checked that

$$Af(x_0) \oplus \cdots \oplus Af(x_d) = \begin{cases} Af(x) & \text{if } d \text{ is even,} \\ Af(x) \oplus 0x63 & \text{if } d \text{ is odd,} \end{cases} \tag{2.19}$$

Algorithm 2.5: Coron's masked computation of $y = S(x)$

Input: shares $x_1, \ldots, x_n$ such that $\bigoplus_i x_i = x$
Output: shares $y_1, \ldots, y_n$ such that $\bigoplus_i y_i = y = S(x)$

```
for all u ∈ F_{2^k} do
    T(u) ← (S(u), 0, ..., 0) ∈ (F_{2^k})^n                 // ⊕(T(u)) = S(u)
end
for i = 1 to n − 1 do
    for all u ∈ F_{2^k} do
        for j = 1 to n do
            T'(u)[j] ← T(u ⊕ x_i)[j]                        // T'(u) ← T(u ⊕ x_i)
        end
    end
    for all u ∈ F_{2^k} do
        T(u) ← RefreshMasks(T'(u))                         // ⊕(T(u)) = S(u ⊕ x_1 ⊕ ··· ⊕ x_i)
    end
end
// ⊕(T(u)) = S(u ⊕ x_1 ⊕ ··· ⊕ x_{n−1}) for all u ∈ F_{2^k}
(y_1, ..., y_n) ← RefreshMasks(T(x_n))                    // ⊕(T(x_n)) = S(x)
```

where $x = x_0 \oplus x_1 \oplus \cdots \oplus x_d$, for a d-order masking. The vulnerable lookup table is highlighted in red in Algorithm 2.6, which we target by PFA.

However, we need to update the strategy of PFA to target this implementation. Recall that the main idea of PFA is to make a distinct disturbance, which is predictable or observable for the adversary, on the distribution of the output. The previous cases are ideally vulnerable to PFA since the output of the target function (S-box) is linearly dependent on one single lookup of a permutation table; thus it is rather easy to produce distinguishable and predictable faulty outputs with one single fault. When multiple lookup operations are involved in the target function, as in the case of Rivain–Prouff's S-box, we show that the output is still distinguishable and predictable with one random fault injection for any masking order, to allow PFA.

Consider a random variable $r(v, v^*, \delta) \in \{0, 1\}^b, b \in \mathbb{N}^+$ whose probability is

$$Pr(r = k) = \begin{cases} \frac{1}{2^b} + \delta & k = v^*, \\ \frac{1}{2^b} - \delta & k = v, \\ \frac{1}{2^b} & \text{else}, \end{cases} \tag{2.20}$$

where $v, v^* \in \{0, 1\}^b$ and $0 < \delta \leq 1/2^b$. Therefore, for independent $r_0(v, v^*, \delta)$ and $r_1(v, v^*, \Delta)$, we have

$$Pr(r_0 \oplus r_1 = k) = \begin{cases} \frac{1}{2^b} + 2\delta\Delta & k = 0, \\ \frac{1}{2^b} - 2\delta\Delta & k = v \oplus v^*, \\ \frac{1}{2^b} & \text{else}. \end{cases} \tag{2.21}$$

So $r_0(v, v^*, \delta) \oplus r_1(v, v^*, \Delta)$ is equivalent to $r(v \oplus v^*, 0, 2\delta\Delta)$. Similarly, we can show that $r_0(v, v^*, \delta) \oplus r_2(v \oplus v^*, 0, \Delta)$ is equivalent to $r(v, v^*, 2\delta\Delta)$.

With one persistent random fault injection into the Af table, when the random input x is under uniform distribution, the output of the faulted table $Af'(x)$ is equivalent to the random variable r as shown earlier as $r(v, v^*, 1/2^8)$, where v denotes the original value of the element where the fault is injected, and v^* denotes the faulty value.

Algorithm 2.6: Rivain and Prouff's secure AES S-box

Input: shares x_i satisfying $\bigoplus_i x_i = x$
Output: shares y_i satisfying $\bigoplus_i y_i = y = S(x)$
$(y_0, \ldots, y_d,) \leftarrow Exp254(x_0, \ldots, x_d)$
for $i = 0; i \leq d; i++$ **do**
  $y_i \leftarrow Af(y_i)$
end
if $d \bmod 2 = 1$ **then**
  $y_0 \leftarrow y_0 \oplus 0x63$
end

For masking order $d = 1$, by (2.21), we have

$$Pr(Af'(x_0) \oplus Af'(x_1) = k) = \begin{cases} \frac{1}{2^8} + 2 \times (\frac{1}{256})^2 & k = 0, \\ \frac{1}{2^8} - 2 \times (\frac{1}{256})^2 & k = v \oplus v^*, \\ \frac{1}{2^8} & \text{else}, \end{cases} \tag{2.22}$$

which is equivalent to $r(v \oplus v^*, 0, 2 \times (1/256)^2)$. The bias is much lower as compared to previous cases, requiring more samples for the attack.

For any odd masking order d, we can decompose $\bigoplus_{i=0}^{d} Af'(x_i) = \bigoplus_{i=0}^{\frac{d-1}{2}} (Af'(2i) \oplus Af'(2i+1))$ to $(d+1)/2$ pairs of independent outputs of Af'. Each pair is equivalent to $r(v \oplus v^*, 0, 2 \times (1/256)^2)$. By applying (2.21) $(d+1)/2$ times, we have $\bigoplus_{i=0}^{d} Af'(x_i)$ is equivalent to $r(v \oplus v^*, 0, 2^d \times (1/256)^{d+1}) = r(v \oplus v^*, 0, 2^{-7d-8})$. For any even masking order d, we consider it as a combination of the $(d-1)$-order masking and $Af'(x_d)$, whose probability should be the same with $r_{(d-1)}(v \oplus v^*, 0, 2^{d-1} \times (1/256)^d) \oplus r_d(v, v^*, 1/2^8)$, which is equivalent to $r(v, v^*, 2^d \times (1/256)^{d+1}) = r(v, v^*, 2^{-7d-8})$. In Figure 2.10, we apply this strategy to the public implementation of [39], where key k can be extracted with both $t_{\max}$ and $t_{\min}$ strategy, when d is odd.

However, since $\delta = 2^{-7d-8}$, it decreases exponentially as masking order d increases, and thus more ciphertexts are required to perform PFA. In order to make an estimation of the number of ciphertexts required with higher masking order d, we study the case of AES. For each ciphertext byte, it has the probability of $1/256$ of appearing, so with n ciphertexts, the total number c of its appearance obeys binomial distribution as $c \sim \mathscr{B}(n, p)$, where $p = 1/256$. Therefore, the

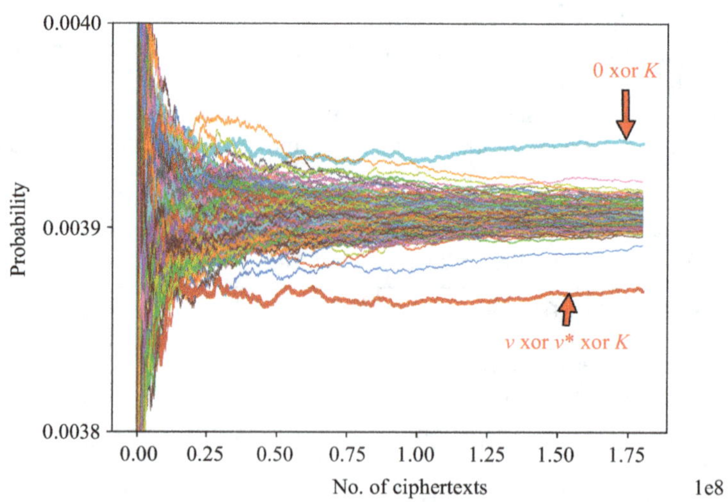

Figure 2.10 Key extraction for Rivain and Prouff scheme [39] with d = 1

variance of c/n is $p(1-p)/n$, and by central limit theorem, c/n approximately follows normal distribution $\mathcal{N}(p, p(1-p)/n)$. To perform PFA successfully, we need $(\sqrt{p(1-p)/n})/2^{-7d-8} \propto$ constant. Therefore, we have $n \propto 2^{14d}$, which means n grows exponentially as d increases.

2.7.5 *Software threshold [40]*

Sasdrich *et al.* [40] extended the widely used threshold implementation (TI [41]) for software targets. They use PRESENT cipher as a case study, showing a first-order secure implementation. Interested readers can refer to [40] for details on software TI implementation of PRESENT. As public source code is not available, we implemented it in *C* language.

We implemented Algorithm 2.7. It uses a lookup table:

$$T(x^i, x^j) = A''(f_{\mathcal{Q}_{12}}(A(x^i), A(x^j))$$

which is composed of 256 elements of four bits. Targeting at T is not optimal, as each element stands a much less chance of being accessed in the process of encryption. Instead, we target at the smaller lookup table A''' : 8FDACB9E43160752, which is an affine permutation of four-bit elements and already marked in red in Algorithm 2.7.

Intuitively, one single fault seems insufficient for PFA since each access of the faulted table is relevant to only one share of all three. However, we can use the same

---

Algorithm 2.7: First-order threshold implementation of PRESENT

Input: $\bar{x} = (x^1, x^2, x^3)$: shared plaintext
k: cipher key
Output: $\bar{y} = (y^1, y^2, y^3)$: shared ciphertext
$rk \leftarrow KeySchedule(k)$
for $i = 1; i \leq 31; i{+}{+}$ **do**
  $x^1 \leftarrow x^1 \oplus rk[i]$
  $t^3 \leftarrow T(x^1, x^2)$
  $t^2 \leftarrow T(x^3, x^1)$
  $t^1 \leftarrow T(x^2, x^3)$
  $t^3 \leftarrow A'''(t^3)$
  $t^2 \leftarrow A'''(t^2)$
  $t^1 \leftarrow A'''(t^1)$
  $x^3 \leftarrow T(t^1, t^2)$
  $x^2 \leftarrow T(t^3, t^1)$
  $x^1 \leftarrow T(t^2, t^3)$
  $x^1 \leftarrow P(x^1)$
  $x^2 \leftarrow P(x^2)$
  $x^3 \leftarrow P(x^3)$
end
$y^1 \leftarrow x^1 \oplus rk[32]$
$y^2 \leftarrow x^2$
$y^3 \leftarrow x^3$

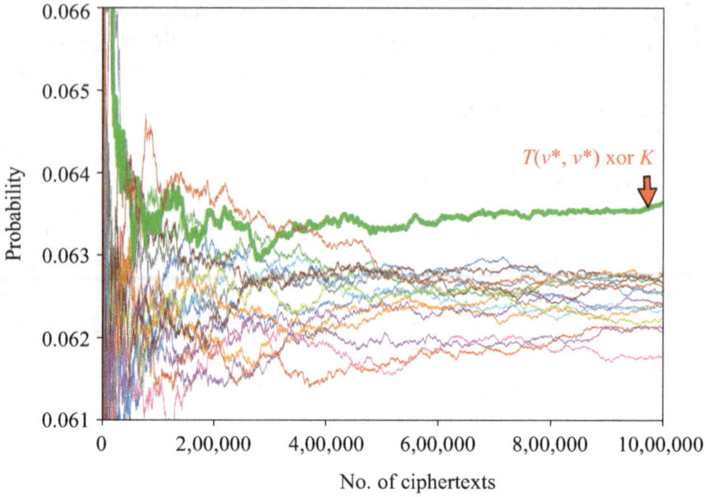

Figure 2.11 Key extraction with t_{max} strategy on Software TI [40]

model with the Rivain–Prouff's AES S-box to estimate the probability distribution of the final output of threshold implementation.

For example, a faulty value 0 is injected into the first element of table A''', whose original value is 8. This injects a bias in the input of T (see Algorithm 2.7). While an input of 8 will never arrive, input 0 is doubled. In this condition, the probability distribution of the outputs of function T is biased as well. Let T' denote the biased T. The truth table of T' shows probability of 6 being the output $9/256$, and the probability of 12 is $23/256$, while all the others have probabilities that are much closer or equal to $16/256$.

We can use the same analysis model in the Rivain–Prouff's case and calculate the probability distribution of $T'(x_0, x^1) \oplus T'(x_2, x^0) \oplus T'(x_1, x^2)$. Note that for any fault injection with a random fault f, $T(v^*, v^*)$ will have the maximal probability to appear at output of T'. Correspondingly, with v denoting the original value where the fault is injected, $T(v, v)$ will always be the one with minimal probability. Therefore, either $t_{\max}$ or $t_{\min}$ strategy can be applied to extract the key k. In Figure 2.11, we show how $t_{\max}$ strategy can be applied to recover the key with less than 400,000 ciphertexts.

2.8 Conclusion

In this chapter, we propose the PFA, a novel FA based on persistent fault model. The power of the proposed PFA is that it can even attack general block ciphers hardened with certain countermeasures against FAs. The attack is first validated in an FPGA environment on AES-128 hardened with DMR countermeasures, to recover the last round key. The proposed attack opens an alternate analysis technique that motivates to break various fault countermeasures under this threat model. As a countermeasure, built-in health test with fault counters can be integrated to verify the functionality of the algorithm before performing block encryptions and limit the number of faulty ciphertexts. We also show that only one persistent fault is enough to break masking at any masking order *d*. Moreover, it motivates to research new configurations of DMR countermeasures and other novel countermeasures to resist PFA.

References

[1] Joye M. Fault Analysis in Cryptography. Berlin, Heidelberg: Springer; 2012.

[2] Bogdanov A, Knudsen LR, Leander G, *et al.* PRESENT: An Ultra-Lightweight Block Cipher. In: International Workshop on Cryptographic Hardware and Embedded Systems. Berlin, Heidelberg: Springer; 2007: 450–466

[3] Guo J, Peyrin T, Poschmann A, *et al.* The LED Block Cipher. In: International Workshop on Cryptographic Hardware and Embedded Systems. Berlin, Heidelberg: Springer; 2011: 326–341.

[4] Shibutani K, Isobe T, Hiwatari H, *et al.* Piccolo: An Ultra-Lightweight Blockcipher. In: International Workshop on Cryptographic Hardware and Embedded Systems. Berlin, Heidelberg: Springer; 2011: 342–357

[5] Bar-El H, Choukri H, Naccache D, *et al.* The Sorcerer's Apprentice Guide to Fault Attacks. Proceedings of the IEEE. 2006;94(2):370–382.

[6] Alderighi M, Casini F, D'Angelo S, *et al.* Evaluation of Single Event Upset Mitigation Schemes for SRAM Based FPGAs Using the FLIPPER Fault Injection Platform. In: IEEE International Symposium on Defect and Fault-Tolerance in Vlsi Systems; 2007. p. 105–113.

[7] Torrance R and James D. The State-of-the-Art in IC Reverse Engineering. Springer Berlin Heidelberg, Berlin Heidelberg; 2009.

[8] InspectorFI. https://www.riscure.com/security-tools/inspector-fi/.

[9] InspectorSCA. https://www.riscure.com/security-tools/inspector-sca/.

[10] Boneh D, DeMillo RA, and Lipton RJ. On the Importance of Checking Cryptographic Protocols for Faults. In: International Conference on Theory and Application of Cryptographic Techniques. Berlin, Heidelberg: Springer; 1997: 37–51

[11] Biham E and Shamir A. Differential Cryptanalysis of the Data Encryption Standard. Crystal Research & Technology. 2006;17(1):79–89.

[12] Biham E and Shamir A. Differential Fault Analysis of Secret Key Cryptosystems. In: Annual International Cryptology Conference. Berlin, Heidelberg: Springer; 1997: 513–525.

[13] Dusart P, Letourneux G, and Vivolo O. Differential Fault Analysis on A.E.S. In: International Conference on Applied Cryptography and Network Security. Berlin, Heidelberg: Springer; 2003: 293–306.

[14] Tunstall M, Mukhopadhyay D, and Ali S. Differential Fault Analysis of the Advanced Encryption Standard Using a Single Fault. Community Mental Health Journal. 2011;49(6):658–667.

[15] Courtois NT, Jackson K, and Ware D. Fault-Algebraic Attacks on Inner Rounds of DES. In: E-Smart'10 Proceedings: The Future of Digital Security Technologies. Strategies Telecom and Multimedia; 2010.

[16] Rivain M. Differential Fault Analysis on DES Middle Rounds. In: Cryptographic Hardware and Embedded Systems – CHES 2009, International Workshop, Lausanne, Switzerland, September 6–9, 2009, Proceedings; 2009. p. 457–469.

[17] Fuhr T, Jaulmes E, Lomne V, *et al.* Fault Attacks on AES With Faulty Ciphertexts Only. In: 2013 Workshop on Fault Diagnosis and Tolerance in Cryptography. IEEE; 2013: 108–118.

[18] Rivain M and Prouff E. Provably Secure Higher-Order Masking of AES. In: International Workshop on Cryptographic Hardware and Embedded Systems. Berlin, Heidelberg: Springer; 2010: 413–427.

[19] Skorobogatov S. Optical Fault Masking Attacks. In: Fault Diagnosis and Tolerance in Cryptography. New York, NY: IEEE; 2010. p. 23–29.

[20] Kim Y, Daly R, Kim J, *et al.* Flipping Bits in Memory Without Accessing Them: An Experimental Study of DRAM Disturbance Errors. ACM SIGARCH Computer Architecture News. 2014;42(3): 361–372.

[21] Bhattacharya S and Mukhopadhyay D. Curious Case of Rowhammer: Flipping Secret Exponent Bits Using Timing Analysis. Springer Berlin Heidelberg, Berlin Heidelberg; 2016.

[22] Razavi K, Gras B, Bosman E, *et al.* Flip Feng Shui: Hammering a Needle in the Software Stack. In: 25th USENIX Security Symposium, USENIX Security 16, Austin, TX, USA, August 10–12, 2016; 2016. p. 1–18. Available from: https://www.usenix.org/conference/usenixsecurity16/technical-sessions/presentation/razavi.

[23] Xiao Y, Zhang X, Zhang Y, *et al.* One Bit Flips, One Cloud Flops: Cross-VM Row Hammer Attacks and Privilege Escalation. In: 25th USENIX Security Symposium, USENIX Security 16, Austin, TX, USA, August 10–12, 2016; 2016. p. 19–35. Available from: https://www.usenix.org/conference/usenixsecurity16/technical-sessions/presentation/xiao.

[24] Wang A, Chen M, Wang Z, *et al.* Fault Rate Analysis: Breaking Masked AES Hardware Implementations Efficiently. IEEE Transactions on Circuits & Systems II Analog & Digital Signal Processing. 2013;60(8):517–521.

[25] Li Y, Sakiyama K, Gomisawa S, *et al.* Fault Sensitivity Analysis. In: Cryptographic Hardware and Embedded Systems, CHES 2010, International Workshop, Santa Barbara, CA, USA, August 17–20, 2010. Proceedings; 2010. p. 320–334.

[26] Baksi A, Bhasin S, Breier J, *et al.* Protecting Block Ciphers Against Differential Fault Attacks Without Re-keying. In: 2018 IEEE International Symposium on Hardware Oriented Security and Trust (HOST). New York, NY: IEEE; 2018. p. 191–194.

[27] Schmidt JM, Hutter M, and Plos T. Optical Fault Attacks on AES: A Threat in Violet. In: Fault Diagnosis and Tolerance in Cryptography. New York, NY: IEEE; 2010. p. 13–22.

[28] Aldaya AC, Sarmiento AJC, and Sánchez-Solano S. AES T-Box Tampering Attack. Journal of Cryptographic Engineering. 2016;6(1):31–48.

[29] Boscher A and Handschuh H. Masking Does Not Protect Against Differential Fault Attacks. In: Fault Diagnosis and Tolerance in Cryptography, 2008. FDTC'08. 5th Workshop on. New York, NY: IEEE; 2008. p. 35–40.

[30] Li Y, Sakiyama K, Gomisawa S, *et al.* Fault Sensitivity Analysis. In: International Workshop on Cryptographic Hardware and Embedded Systems. Berlin, Heidelberg: Springer; 2010: 320–334.

[31] Moradi A, Mischke O, Paar C, *et al.* On the Power of Fault Sensitivity Analysis and Collision Side-Channel Attacks in a Combined Setting. In: International Workshop on Cryptographic Hardware and Embedded Systems. Berlin, Heidelberg: Springer; 2011: 292–311.

[32] Lomné V, Roche T, and Thillard A. On the Need of Randomness in Fault Attack Countermeasures-Application to AES. In: Fault Diagnosis and Tolerance in Cryptography (FDTC), 2012 Workshop on. New York, NY: IEEE; 2012. p. 85–94.

[33] Dobraunig C, Eichlseder M, Gross H, *et al.* Statistical Ineffective Fault Attacks on Masked AES With Fault Countermeasures. In: Cryptology ePrint Archive, Report 2018/357, 2018. https://eprint. iacr. org/2018/357;.

[34] Blom G, Holst L, and Sandell D. Problems and Snapshots from the World of Probability. Springer, Berlin, Heidelberg: Springer Verlag; 1994.

[35] AES. http://nvlpubs.nist.gov/nistpubs/FIPS/NIST.FIPS.197.pdf.

[36] Data2MEM User Guide (UG658); 2009. https://www.xilinx.com/support/documentation/sw_manuals/xilinx11/ise_n_data2mem_user_guide.htm

[37] Masked-AES-Implementation. Available from: https://github.com/Secure-Embedded-Systems/Masked-AES-Implementation.

[38] Coron JS. Higher Order Masking of Look-Up Tables. In: Annual International Conference on the Theory and Applications of Cryptographic Techniques. Berlin, Heidelberg: Springer; 2014. p. 441–458.

[39] Higher Order Countermeasures for AES and DES. Available from: https://github.com/coron/htable.

[40] Sasdrich P, Bock R, and Moradi A. Threshold Implementation in Software. In: International Workshop on Constructive Side-Channel Analysis and Secure Design. Berlin, Heidelberg: Springer; 2018. p. 227–244.

[41] Nikova S, Rijmen V, and Schläffer M. Secure Hardware Implementation of Nonlinear Functions. Journal of Cryptology. 2011:292–321.

Chapter 3

Deployment of EMC techniques in design of IC chips for hardware security

Makoto Nagata[1]

3.1 Overview

Semiconductor integrated circuit (IC) chips are used in diversified electronic applications. Safety and security of IC chips are of critical importance in the fields such as automotive, aerospace/aviation, healthcare and medical technologies (Figure 3.1). Cryptographic functions protect control and private data from falsification and sustain their authenticity, throughout the data transmission through in-vehicle or in-vessel networks as well as the data transfer externally to roadside units and base stations. The data integrity is to be guaranteed by cryptographic processors for data encryption and/or digital signatures.

Cryptographic algorithms are mathematically proven to be resilient against the attempts by an adversity to decrypt data without keys. However, there are known vulnerabilities to side-channel attacks (SCAs) once they are physically implemented in cryptographic engines on semiconductor IC chips [1–3].

Electromagnetic (EM) waves are radiated from an IC chip when it is in operation, as shown in Figure 3.2. Power current flows from an external power source to circuits on a die through power paths in series of power lines on a printed circuit board (PCB),

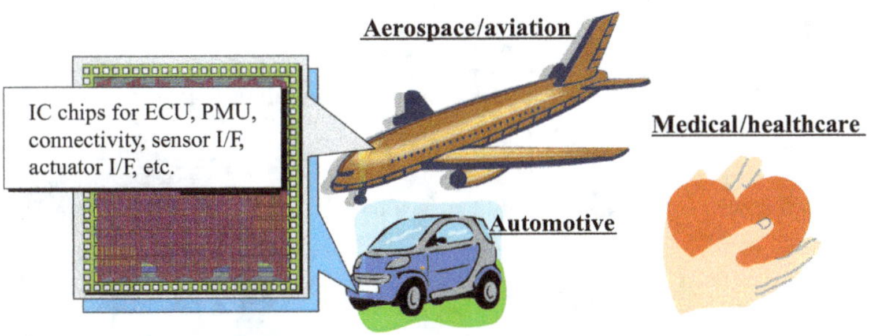

Figure 3.1 Semiconductor IC chips in mission critical applications [4]

[1]Graduate School of Science, Technology and Innovation, Kobe University, Kobe, Japan

power pins of an IC chip package, power pads and power wirings in an IC chip. Power current is consumed by circuits in operation and dynamically changes according to the contents of circuit operation. Power current interacts with parasitic impedance over the power paths, where EM waves are emitted from inductive elements as parasitic antennas. The size of an IC chip is typically as small as 5 mm^2, where the whole of active circuits are located. However, the power current flows over the entire power paths on a PCB of the 100 mm or larger, and the EM waves are propagated in a free space and reaches to an observer even at 1,000 mm or more in distance from an IC chip. The whole process is generally dictated in an electromagnetic interference (EMI) [4].

An adversary remotely captures and analyzes EM waves that reflect the change of power current according to the operation of ICs, as shown in Figure 3.3. A tiny antenna is scanned over the surface of PCB, including IC chips. The EM wave is sensed by the antenna and recorded in the time-domain by an oscilloscope [5,6]. There are several techniques to relate the measured EM waves with information processing internally in an IC chip—called side-channel (SC) analysis or SCA. Or, there is a fact where

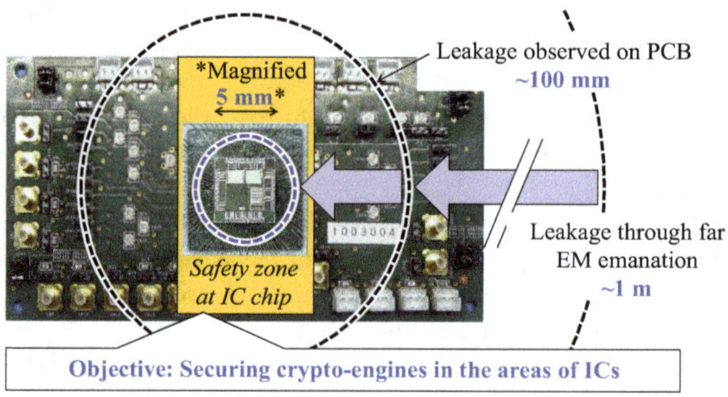

Figure 3.2 Electromagnetic wave radiation from IC chip [4]

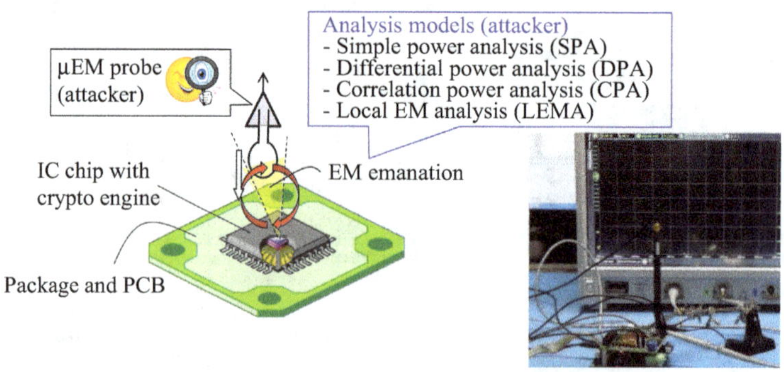

Figure 3.3 Side-channel passive attacks [4]

EM wave delivers secret information of an IC chip to an observer located away from the chip—called SC information leakage.

On the other hand, an adversary may intentionally induce faults, or, namely, erroneous bits within cryptographic processors by irradiating large-power EM waves from an external signal source. There is another class of SC leakage analysis techniques that derive internal information from the difference between the correct cypher (without faults) and the altered output (due to injected faults) both processed by a cryptographic engine [7,8]. The physical mechanisms to fault injection are dictated in an electromagnetic susceptibility (EMS) problem. It is noted that the disciplines of electromagnetic compatibility (EMC) involve bidirectional aspects of EMI and EMS problems and relevant countermeasure solutions [9].

The goal of this chapter is to summarize the deployment of EMC techniques in the design of an IC chip with security functionality. The EMI of an IC chip will be discussed from the viewpoint of hardware security (HWS) and specially emphasized with simulation techniques on SC information leakage [10].

3.2 EMC simulation technique

EM radio wave is emitted from an active IC chip and potentially interfered with other electronic devices in the case of EMI. An IC chip, as modeled in Figure 3.4, has multiple power domains (PDs). A circuit (Ckt) draws dynamic power current from an external power source when it operates. The power current flows through a power delivery network (PDN) on a PCB and interacts with parasitic antennas associated with metallic paths on the PDN, which leads to the EM radio wave emission. The PDs are prepared for a core digital circuit such as a cryptographic engine, digital data input/output (I/O) interface circuitry, analog functionality, including power converters, analog-to-digital and digital-to-analog converters, a clock signal generator and so on. Each EM wave is characterized with the functionality of circuits supplied

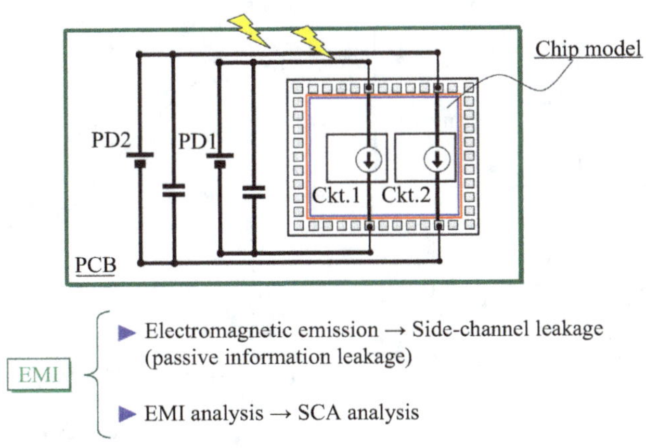

Figure 3.4 Electromagnetic interference simulation model [4]

in the corresponding PDs. The EM wave associated with the PD of a cryptographic engine reflects its internal operation and thus potentially leaks information through SC leakage. Once the model is set up for EMI analysis [11], it will be usable to the analysis of SCA.

The parasitic impedances on PDN of an IC chip are sketched in Figure 3.5. The power (V_{DD}) and return or ground (V_{SS}) lines form a closed circuit for power current to flow between a power source and active circuits. A board (PCB) is typically captured in an S-parameter model by using filed solver software for its frequency-domain response of impedance. A package model is often simplified with series inductances associated with bonding wires or soldering bumps. An IC chip is represented by a chip model, including resistive and capacitive networks of V_{DD}, V_{SS} and silicon substrate (V_{SUB}) [12,13].

An example PDN is fabricated and captured in a model, as shown in Figure 3.6. The impedance seen from the power source side (Z_{DD}) is measured by an impedance

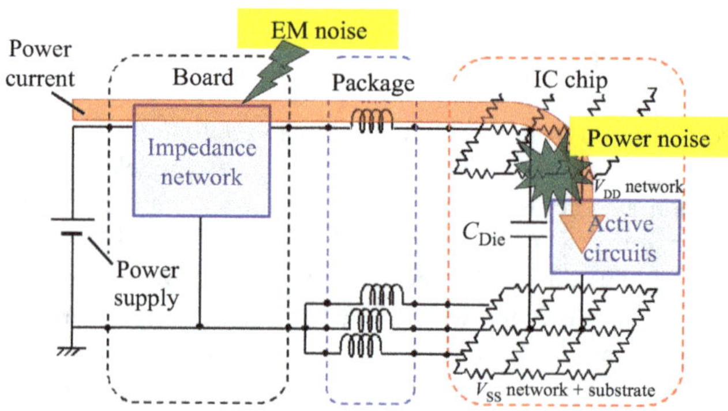

Figure 3.5 Power delivery network simulation model [17]

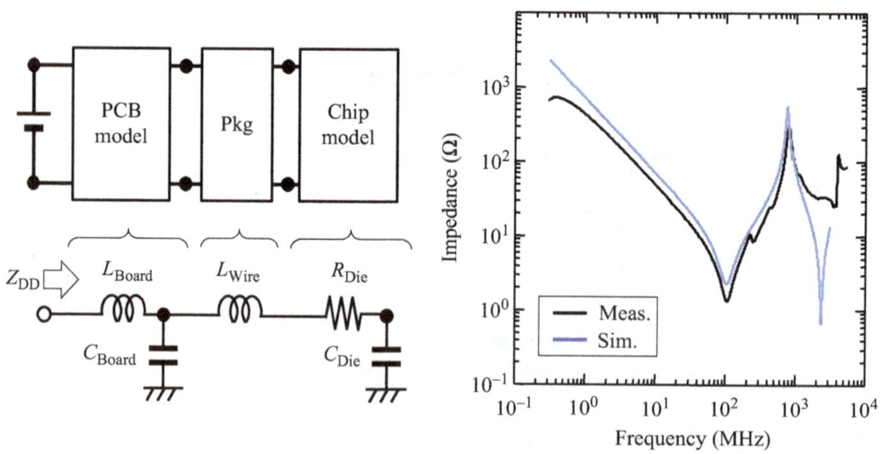

Figure 3.6 Power delivery network impedance simulation [4,11,13]

analyzer while simulated with the equivalent circuit model. The model consists of chip, package and PCB sub-models connected in series to form PDN—the whole network is called a chip–package–system board (C–P–S) model.

As the frequency response of Z_{DD} is given in Figure 3.6 (right), the C–P–S model finely captures the resonating frequencies of the PDN. The V_{DD} impedance exhibits a series LCR resonance, as the end of PDN path is openly terminated with the total capacitance of ICs, C_{Die}. The first resonance frequency approximately at 120 MHz comes mainly from C_{Die} =175 pF and L_{Wire} of 10 nH from the typical length of bonding wires. The board capacitance, C_{Board}, is smaller than 5 pF and negligible. Once we include a decoupling capacitor, C_{Decap} of 10 uF in parallel to C_{Board}, the first resonance is suppressed while shifted in its frequency.

When an active power current model is included in a C–P–S model, the whole network can be analyzed by a circuit simulator for evaluating time-domain power noise waveforms. If we assume a digital circuit as an array of shift registers operating in parallel and synchronously to a clock signal, the V_{DD} voltage exhibits resonating fast decaying wavelets regularly induced at every edge of the clock signal, as shown in Figure 3.7 (left). Once the waveform is analyzed in its frequency components as in Figure 3.7 (right), the most significant component is found at around the first resonating frequency of 120 MHz. This is naturally concluded from the fact that power noise waveforms are characterized by the parallel resonance of PDN.

For C–P–S modeling and power noise simulation, the active and passive part of an IC chip model is described in Figures 3.8 and 3.9, respectively.

The active part represents power current of a digital circuit. It is assumed that a digital circuit is coded in register-transfer level description and synthesized with the collection of logic cells prepared in a standard cell library of a given semiconductor technology. Every standard cell is pre-characterized for its power current, $I(t)$, by transistor-level circuit simulation (SPICE). For instance, the operation of an inverter logic (NOT) is simulated in the time domain with its transistor-level circuit netlist, when the logical value at an input node is changed either from “0” to “1” or from “1” to “0.” The power current flowing from V_{DD} to V_{SS} is then derived as the node

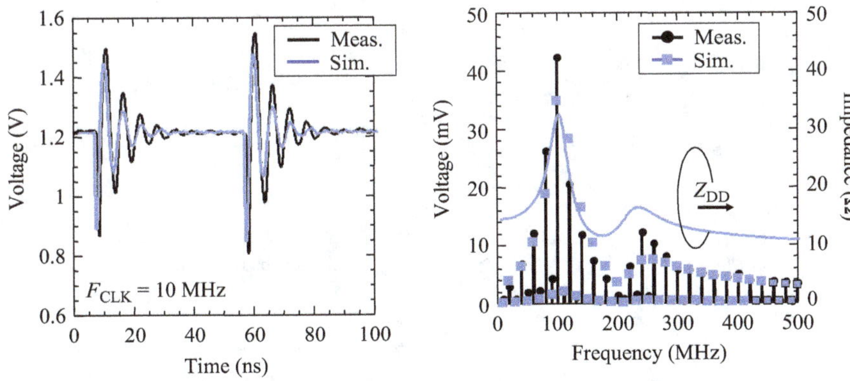

Figure 3.7 Power noise simulation [4,11,13]

current computed by SPICE simulator. The power current is iteratively simulated and collected for a complete set of input and output logic values according to a logical lookup truth table of each logic cell. The active part is finally represented as Norton equivalent current source, where $I(t)$ is accompanied with the path in parallel to it, that is formed by equivalent resistance (R_{ESR}) and capacitance (C_{ESC}) connected in series. These elements are also pre-characterized from the physical layout of each logic cell.

The V_{DD} and V_{SS} wirings that are locally connected to power nodes of a logic cell are included in respective resistive meshes in the passive part of a model. In addition, the capacitor, C_{Well} represents the capacitive coupling between the V_{DD} and V_{SS} wirings through diode junctions that are typically formed in a logic cell at the interface of n-type wells and p-type Si substrate.

The passive part represents the whole PDN network in a large mesh of resistive, capacitive and inductive elements (Figure 3.9) that are parasitic to metallic wirings of power grids (V_{DD} and V_{SS}), diode junctions among integrated semiconductor devices and electronic materials of packaging structures. Power grids have resistive meshes corresponding to the high-side (V_{DD}) and low-side (V_{SS}) domains and provide the routes of power current flowing from V_{DD} to V_{SS} pads that are typically distributed

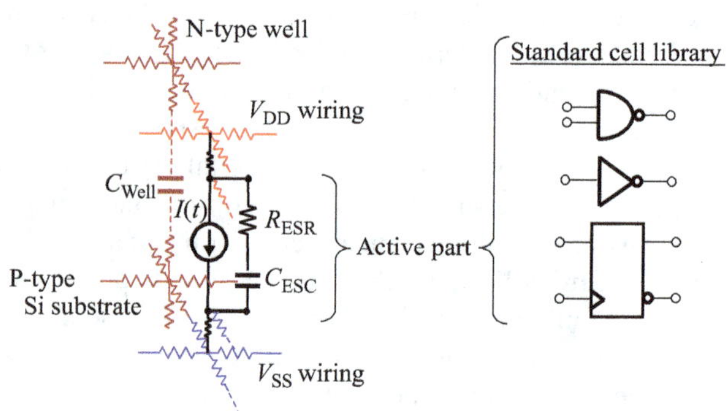

Figure 3.8 Power current simulation model

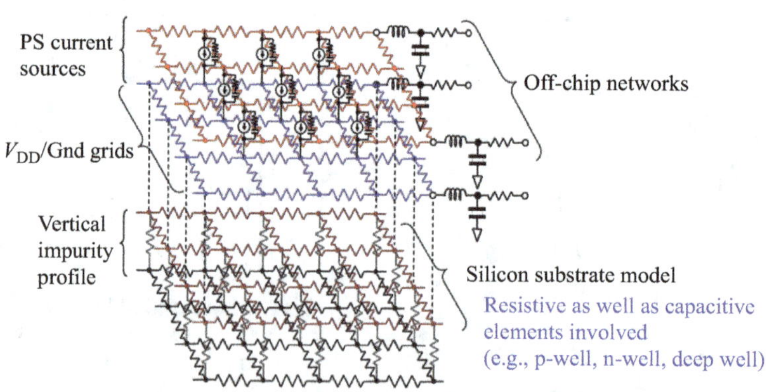

Figure 3.9 Parasitic impedance network model [4,11]

in the periphery of an IC chip. Logic cells are represented by the active models (Figure 3.8) and inserted between V_{DD} and V_{SS} nodes at the respective location in an IC chip.

The passive part also involves Si substrate sub-model that captures resistive and capacitive networks parasitic to diode junctions (e.g., n-type well, p-type deep well and p-type Si substrate) reflecting the vertical impurity profile of a given Si technology. In addition, the V_{DD} and V_{SS} pads are connected to an off-chip network, mostly consisting of package models of inductors associated with bonding wires or soldering bumps and capacitors coupled to the system ground.

The whole model of an IC chip with a very large volume of passive and active models is analyzed by a highly efficient solver for the voltage and current among the limited number of nodes of interest. The interaction of power current with parasitic impedance creates voltage variations on the V_{DD} and V_{SS} domains that are observed as power noise as well as substrate noise and considered as the equivalent entity of EM noise to be used for SC information leakage analysis.

A silicon test vehicle of Figure 3.10 is exemplified for the power noise analysis using the IC-chip-level model [14]. The chip was designed and fabricated in a 65 nm CMOS technology with nine layers of metal for power and signal wirings. Advanced encryption standard (AES) cores are involved in the chip as cryptographic engines for SC leakage analysis. In addition, on-chip monitor (OCM) circuits are included, which enable in-place capturing of time-domain power noise waveforms on V_{DD} and V_{SS} wirings within AES cores and also some points on Si substrate (V_{SUB}) [15,16]. The AES core with the size of 150 μm × 200 μm is located within the whole chip area of 3 mm × 4 mm. The chip model is created for the whole of test vehicle IC chip, according to the passive and active modeling.

The power noise waveforms on V_{DD} and V_{SS} nodes in the AES core are compared in Figure 3.11, for the C–P–S simulation and OCM measurements. The C–P–S simulation model integrates the chip model with the package and PCB models derived from our experimental setup. The overall shape of waveforms and the size of peak drops are almost consistent [17].

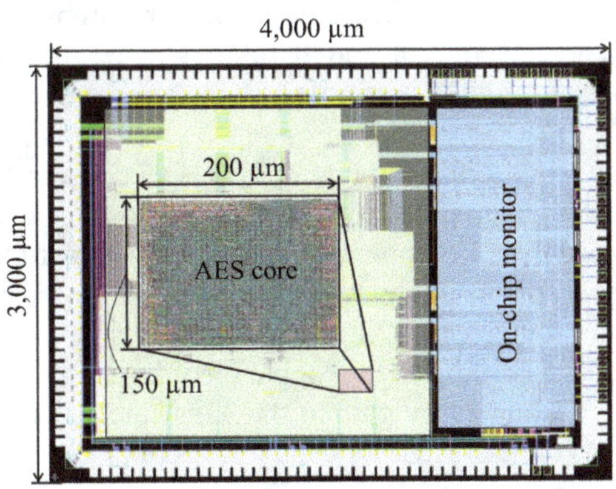

Chip summary*	
Process	65-nm CMOS
Metal	9-layer Cu metal
Security core	AES cores with different S-box implementation
On-chip monitor	Measuring on-chip V_{DD}, V_{SS} and V_{SUB}

***SPACESexplorer chip,** for security evaluation of physically attacked cryptoprocessors in embedded systems

Figure 3.10 SC leakage test vehicle [4]

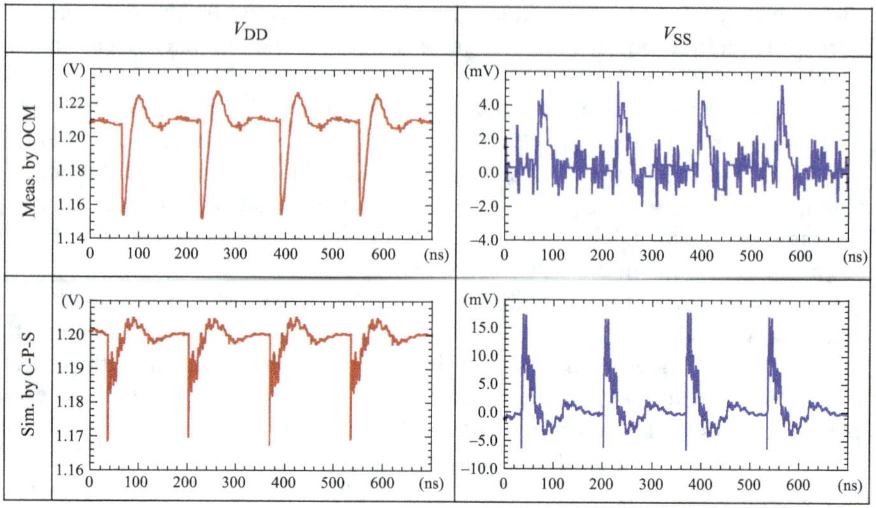

Figure 3.11 Power noise simulation and measurements [4,17]

3.3 SC leakage analysis

The operation flow of AES core is outlined in Figure 3.12, applying byte-wise crypto computation. The whole flow is operating in parallel among 16 bytes when the AES core is designed for a 128-bit secret key length. The flow adopts the round-based architecture, where 11 clock cycles are needed in the whole AES encryption processing. Here, all the rounds except the last include the mixture of key shuffling.

One of the most known SC leakage models assumes the correlation among the secret key byte and Hamming distance in the data register in the last round of AES processing, which is free from the mixture of shuffled key. Hamming distance represents the number of bits flipped in the output data register before and after the last round of AES processing. From the power noise viewpoint, the larger Hamming distance leads to the larger power current consumption and therefore induces the higher voltage drop. This fact directly relates the C–P–S power noise simulation with the SC leakage analysis.

The SC leakage simulation flow is described in Figure 3.13. Once a secret key (target key) is given, a set of plain texts is prepared to set up test vectors for the SC leakage simulation of AES core. Since the AES core is designed with standard logic cells, a gate-level logic simulator analyzes its operation and records all the toggling (switching) actions during the whole sequence of AES processing. The active part of a chip model can also be produced according to the toggle records using the precharacterized power current models among every gate in the AES core. It is noted that the active part of the model needs to be created for each test vector. On the other hand, once the passive part is derived from the whole of an IC chip, it can be reusable for the set of active models. The power noise waveforms are then simulated with the

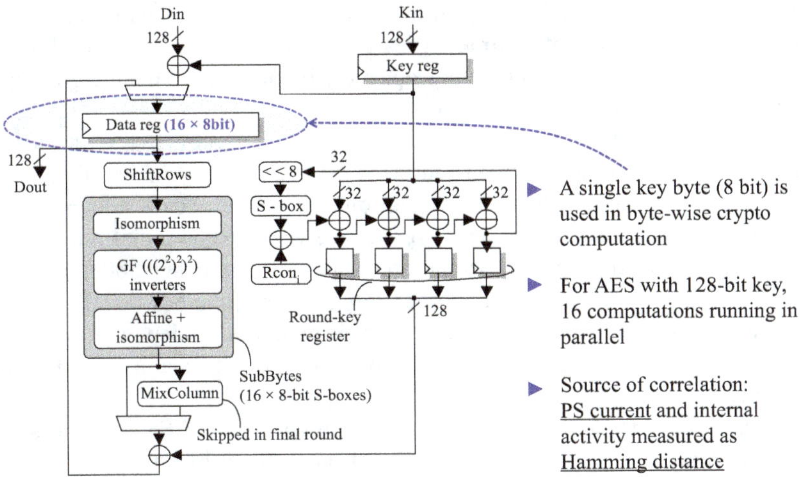

Figure 3.12 SC leakage model of AES [4]

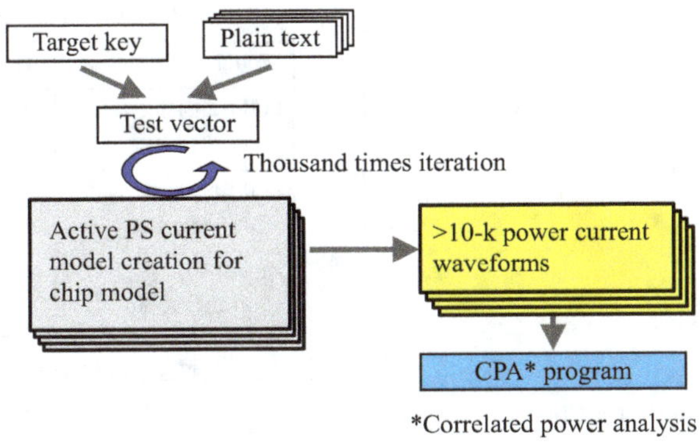

Figure 3.13 SC leakage simulation flow of AES [4]

C–P–S model of AES core for the set of test vectors and correlated with the secret key bytes [18–21].

One of the important parameters in the analysis is the number of test vectors (with different plain texts as payloads) to finely correlate the voltage drops in the last round of AES processing with the secret key bytes. This number can be a security indicator of a given cryptographic engine. The larger number requires the larger simulation costs and also the higher cost burdened to the adversary.

The test vehicle of Figure 3.10 is tested with the SC leakage simulation flow. The simulated power noise waveforms in the clock cycle of interest of AES processing for 10,000 payloads are pictured in Figure 3.14. The simulation becomes so exhaustive if we use a full-transistor-level netlist of the AES core that the inclusion of parasitic passive components will not be feasible. It is estimated to need more than 3 months for

10,000 test cases with the netlist only with transistors (without parasitic components). On the other hand, the simulation time is greatly reduced once we apply the chip model technique (Figures 3.8 and 3.9) that appropriately involves the parasitic components and even in the C–P–S model integration.

The simulated waveforms clearly reveal the secret key bytes as shown in Figure 3.15. The horizontal and vertical axes represent the number of traces with different test vectors and the level of correlation for each candidate value among 256 potential numbers in a secret key byte, respectively. It is shown that one particular number (red) exhibits the highest correlation in case with more than 500 traces (different test vectors), in comparison to the other 255 numbers (in black lines). The results are compared with measurements and exhibit very similar tendency. It is important to notice that the smaller number of test vectors is required to evaluate SC leakage in simulation.

The SC leakage is analyzed among the different design of AES cores, as summarized in Figure 3.16. It is obviously shown that the number of traces (with different

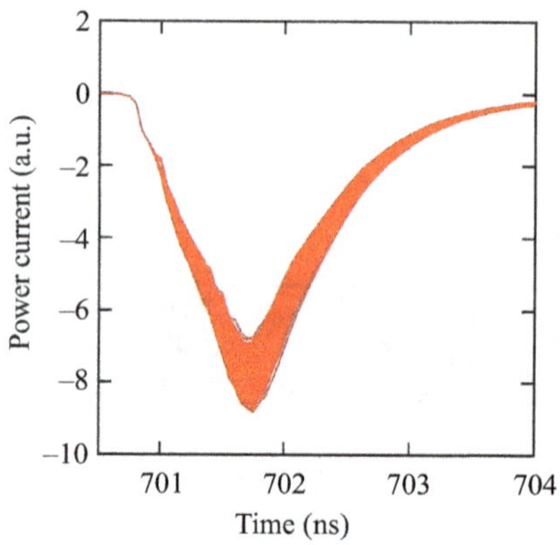

Cost of simulation for 10,000 plain texts

Model	Cost
Full transistor (pre-layout)	115 days
Full transistor (post-layout)	Unlikely
Active PS current model	10 h

280 times acceleration is achieved

Figure 3.14 Power current simulation for SC leakage analysis [4]

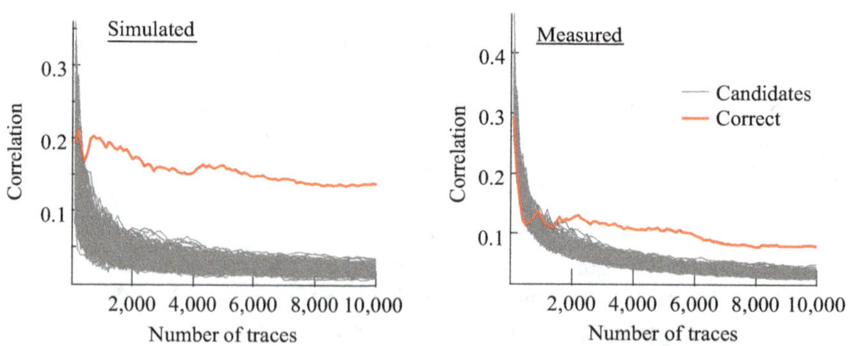

Figure 3.15 SC leakage simulation and measurements on AES core [4,19]

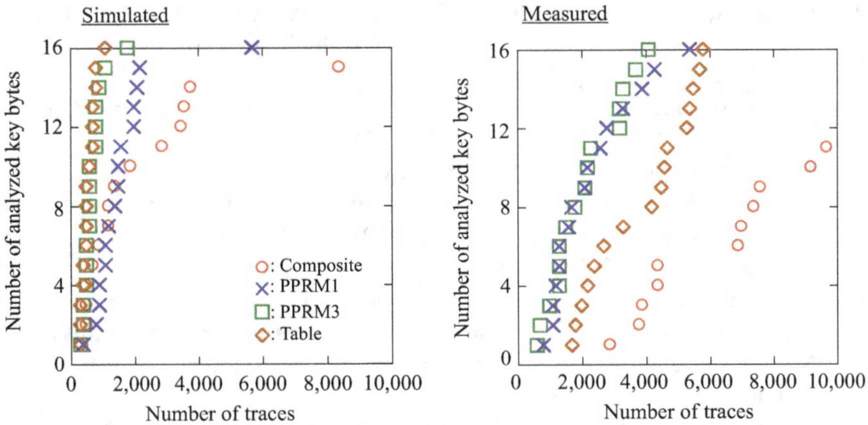

Figure 3.16 SC leakage analysis among different AES designs [4,19]

test vectors) to correlate all the 16 secret key bytes depends on the design styles and physical architectures. Again, the number of test vectors is much smaller than measurements. This implies that the C–P–S simulation purely captures SC leakage mechanisms inherent to a given design and technology.

3.4 Conclusion

The efficacy and accuracy of C–P–S simulation technique for full-chip analysis of a secure IC chip is demonstrated. The modeling and simulation flow of SC leakage from HWS viewpoints are based on the knowledge of EMC simulation. The technique can be applicable to the exploration of cryptographic engines with the stronger SCA resilience, in the design stage of IC chips. The technique will also be evolved for more sophisticated passive SC analysis as well as for active SC leakage with intentional fault injection.

Acknowledgments

This work was in part based on results obtained from a project, JPNP16007, commissioned by the New Energy and Industrial Technology Development Organization (NEDO). The authors would like to deeply thank K. Matsuda, A. Tsukioka, Dr. D. Fujimoto, Dr. T. Miki and Prof. N. Miura for their valuable help and scientific discussions.

References

[1] P. Kocher, “Timing Attacks on Implementations of Diffie-Hellman, RSA, DSS, and Other Systems,” CRYPTO 1996, Lecture Notes in Computer Science. Berlin, Heidelberg: Springer-Verlag, vol. 1109, pp. 104–113, 1996.

[2] P. Kocher, J. Jaffe, and B. Jun, "Differential Power Analysis," CRYPTO 1999, Lecture Notes in Computer Science. Berlin, Heidelberg: Springer-Verlag, vol. 1666, pp. 388–397, 1999.

[3] I. Verbauwhede, Ed. "Secure Integrated Circuits and Systems," New York, NY: Springer, 2010 (ISBN 978-0-387-71829-3).

[4] M. Nagata, "Power Noise Simulation of IC Chips for Hardware Security," in 2019 IEEE International Symposium on Electromagnetic Compatibility, Signal & Power Integrity (EMC+SIPI 2019), New Orleans, LA: IEEE, Tutorial, FR-AM-3-2, 2019.7.26.

[5] IEC 61967-6: Integrated circuits – Measurement of electromagnetic emissions, 150 kHz to 1 GHz – Part 6: Measurement of conducted emissions – Magnetic probe method. Geneva: IEC. Available from: https://webstore.iec.ch/publication/6193.

[6] M. Nagata, D. Fujimoto, N. Miura, N. Homma, Y. Hayashi, and K. Sakiyama, "Protecting Cryptographic Integrated Circuits With Side-Channel Information," IEICE Electronics Express, vol. 14, no. 2, pp. 1–13, 2017.

[7] E. Biham and A. Shamir, "Differential Fault Analysis of Secret Key Cryptosystems," CRYPTO, Lecture Notes in Computer Science, vol. 1294, pp. 513–525, 1997.

[8] D. Karaklajic, J.-M. Schmidt, and I. Verbauwhede, "Hardware Designer's Guide to Fault Attacks," IEEE Transactions on Very Large Scale Integration (VLSI) Systems, vol. 21, no. 12, pp. 2295–2306, 2013.

[9] M. Nagata, T. Miki, and N. Miura, "On-Chip Physical Attack Protection Circuits for Hardware Security (Invited)," in Proc. IEEE Custom Integrated Circuits Conference (CICC 2019), Austin, TX: IEEE, #15-5, pp. 1–6, Apr. 2019.

[10] M. Nagata, "Deployment of EMC-Compliant IC Chip Techniques in Design for Hardware Security (Invited)," in The 5th France-Japan Cybersecurity Workshop, Kyoto, 2019.5.24. http://project.inria.fr/FranceJapanICST/files/2019/04/190424_Nagata_to_Japan_French_Workshop.pdf.

[11] M. Nagata, "T8: Noise Simulation in Mixed-Signal SoCs," in IEEE Intl. Solid-State Circuits Conference, Tutorial, Feb. 2016. https://resourcecenter.sscs.ieee.org/short-courses-and-tutorials/2016-isscc-short-courses-and-tutorials/SSCSTUT20160120.html.

[12] H. H. Park, S.-H. Song, S.-T. Han, T.-S. Jang, J.-H. Jung, and H.-B. Park, "Estimation of Power Switching Current by Chip-Package-PCB Cosimulatoin," IEEE Transactions on Electromagnetic Compatibility, vol. 52, no. 2, pp. 311–319, 2010.

[13] K. Yoshikawa, Y. Sasaki, K. Ichikawa, Y. Saito, and M. Nagata, "Measurements and Co-Simulation of On-Chip and On-Board AC Power Noise in Digital Integrated Circuits," in Proc. 8th International Workshop on the Electromagnetic Compatibility of Integrated Circuits (EMC Compo 2011). Dubrovnik: IEEE, pp. 76–81, Nov. 2011.

[14] D. Fujimoto, D. Tanaka, N. Miura, *et al.*, "Side-Channel Leakage on Silicon Substrate of CMOS Cryptographic Chip," in Proc. 2014 IEEE International

Symposium on Hardware-Oriented Security and Trust (HOST 2014), #3-3, pp. 32–37, May 2014.

[15] K. Noguchi and M. Nagata, "An On-Chip Multi-Channel Waveform Monitor for Diagnosis of Systems-on-Chip Integration," IEEE Transactions on VLSI Systems, vol. 15, no. 10, pp. 1101–1110, 2007.

[16] T. Hashida and M. Nagata, "On-Chip Waveform Capture and Application to Diagnosis of Power Delivery in SoC Integration," IEEE Journal of Solid-State Circuits, vol. 46, no. 4, pp. 789–796, 2011.

[17] A. Tsukioka, N. Yamamoto, R. Korenaga, *et al.*, "Active Power Noise Modeling Toward Design for EMI Compliance of IC Chips," in 2017 DesignCon. London: Informa, Jan. 2017.

[18] D. Fujimoto, M. Nagata, T. Katashita, A. Sasaki, Y. Hori, and A. Satoh, "A Fast Power Current Analysis Methodology Using Capacitor Charging Model for Side Channel Attack Evaluation," in 2011 IEEE Intl. Symp. Hardware-Oriented Security and Trust (HOST 2011), #P35, pp. 87–92, Jun. 2011.

[19] D. Fujimoto, T. Katashita, A. Sasaki, A. Satoh, and M. Nagata, "A Fast Power Current Simulation of Cryptographic VLSI Circuits for Side Channel Attack Evaluation," IEICE Transactions on Fundamentals, vol. E96-A, no. 12, pp. 2533–2541, 2013.

[20] A. Tsukioka, M. Nagata, K. Srinivasan, *et al.*, "A Full System Simulation Technique of Power-noise Side Channel Leakage in Cryptographic Integrated Circuits," ACM/IEEE Design Automation Conference (DAC 2019), Designer Track Reviewed #18.6, Jun. 2019.

[21] M. Nagata, "Physical Attack Protection of IC Chips for Hardware Security," IEEE Solid-State Circuits Society, 2019. https://www.youtube.com/watch?v=o86YGdLxFHQ.

Part II

Design for security

Chapter 4
Hardware obfuscation for IP protection

Abdulrahman Alaql[1], Md Moshiur Rahman[1], Tamzidul Hoque[1], and Swarup Bhunia[1]

The horizontal business model adopted by the semiconductor industry has resulted in a manufacturing process that has integrated various foreign entities focusing on individual production steps. However, serious trust issues have emerged with the internationalization of the semiconductor supply chain, and the trust concerns grow further with the proliferation of theft, reverse engineering (RE), and piracy of hardware intellectual property (IP). Unlike the software industry, the semiconductor supply chain cannot benefit from traditional cryptographic solutions due to the requirement of white-box accessibility to the hardware IP. Therefore, hardware obfuscation has originated as a potential countermeasure against dreaded supply chain attacks. The obfuscation process aims to hide the design intent and lock functionality by introducing structural and functional transformations to the IP, which increases the resistance against unauthorized use and potential exploitation of the design by an adversary. In this chapter, we investigate hardware obfuscation techniques that have been developed in the past decade and the limitations and challenges of these techniques. In particular, we explore the evolution of different techniques that are applicable at various supply chain stages of the semiconductor life cycle, the development of obfuscation benchmarks, and the research trends in the area of obfuscation. We also expand on the innovative attacks that have been developed to compromise the obfuscation techniques, as well as the countermeasures that are introduced to combat them. This chapter also sheds light on the lessons learned from the decade-long research on obfuscation by discussing the attack trends and unsolved challenges. Finally, we conclude by presenting a road map that aims to improve the existing solutions and provides improved security metrics for systematic evaluation.

4.1 Introduction

4.1.1 IP protection in globalized supply chain

The ability of the semiconductor industry to design and manufacture innovative market products in an ever-decreasing production cycle time and reduced cost has resulted

[1]Department of Electrical & Computer Engineering (ECE), University of Florida, Gainesville, FL, USA

in the rapid rise of human dependency on electronic systems. System on chip (SoC) is being commonly adopted in various electronic systems, especially in smart and connected devices that are required to facilitate a multitude of functionalities in a small form factor. Meanwhile, the SoC industry has moved toward a design paradigm that commonly incorporates hardware IP cores from third-party vendors to enable a faster design cycle and reduced development risk that lead to significant financial gain. Furthermore, most of the design houses manufacture the integrated design in external foundries. While the distributed design and manufacturing process of semiconductor IPs has deducted cost and time-to-market, it came at the expense of security and trust in the supply chain. For the purpose of integration, verification, and fabrication, IP consumers and foundries must receive white-box accessibility of the design. Hence, the security of the IP has become a serious concern as access to the valuable IP core facilitates theft, overuse, RE, and malicious alteration. Piracy relates to the illegal use or exchange of the design with little or no modification. IP overuse indicates the act of exceeding the approved number of uses without any knowledge of the IP vendor. RE is performed to extract a more reusable or comprehensible representation of the design. From a given design abstraction and available RE resource, an IP could be reversed to various representations (e.g., layout, gate-level, register transfer-level (RTL), and specification). RE is considered a serious threat even in the field where resourceful end users can try to extract design-oriented information from a fabricated device through invasive techniques. The associated threats at different stages of the globalized supply chain are illustrated in Figure 4.1. From an IP consumer's perspective, integration of hidden malicious functionality in designs obtained from a third-party IP vendor is a serious concern. However, from the IP vendor's standpoint, the IP is vulnerable to theft, overuse, and RE in the facilities of IP consumer, design-for-test vendor, and foundry.

4.1.2 IP infringement cases

IP piracy and RE efforts have raised a serious impact on the semiconductor industry as estimates suggest a loss of $250 billion due to counterfeit components each year. IP piracy is a well-known threat that has been commonly known to impact the market of many consumer products such as clothing, medicine, fashion items, and cosmetics. With the rapid increase in the use of computing systems in our everyday lives, the

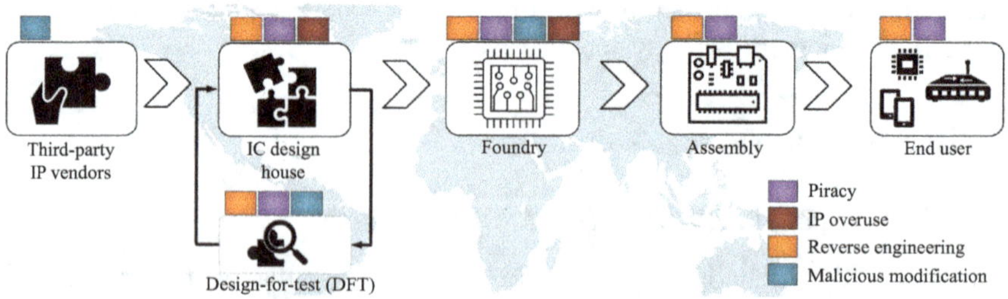

Figure 4.1 The horizontal supply chain model of semiconductor devices and associated security and trust issues at various stages

software and hardware industry is also facing the challenge of IP theft. Two former employees of a financial technology company Flow Trader were indicted by the NY State Supreme Court in 2013 for allegedly stealing source code from their company [1]. One of the largest vendors of reconfigurable hardware, Xilinx, filed a complaint the same year against one of their distributors, Flextronics, stating that Flextronics marketed their products to customers at unauthorized prices [2]. According to the lawsuit, Flextronics purchased components from Xilinx at a lower price by lying about the intended consumer, as Xilinx offers discounted rates only for specific end users. There exist countless examples of such cases of IP violation in both software and semiconductor industry, which has led to a massive loss of revenue and reputation.

4.1.3 Encryption and watermarking for IP protection

Encryption of digital IP is a useful solution for preventing white-box accessibility. However, the application of encryption relies on a software framework that allows the decryption of the IP before the intended use. For protecting digital media content and software, encryption can be easily integrated, since human intervention could be avoided during their use. For example, a music file can be downloaded in encrypted form and can be played through a music player that decrypts the file prior to its use. For hardware IP, it is not possible to keep the design encrypted throughout the supply chain. For example, RTL designs can be protected through encryption, as the synthesis and simulation tools that use the RTL code can be augmented with decryption functionality. However, most of the supply chain steps after synthesis until fabrication requires human intervention that eliminates the feasibility of hiding the unencrypted design throughout the process. Again, EDA tools that conceal the IP during synthesis and simulation can be compromised to leak the encryption key or the decrypted IP [3]. Researchers have also explored watermarking as a viable solution for protecting IP rights [4]. Watermarking primarily provides the ability to prove the ownership of an IP or IC. However, it does not introduce functional locking or structural obfuscation. Hence, watermarking does not prevent someone from using the IP, causing modifications, or performing RE effort.

4.1.4 Hardware obfuscation

Hardware obfuscation has emerged as a potential solution against piracy, RE, and malicious alteration. Obfuscation methods aim to introduce various functional and structural transformations to the design to lock its functionality and obscure the design intent, respectively. As shown in Figure 4.2, these transformations are often applied based on a key that becomes a required input to the obfuscated design to enable the original functionality. Hardware obfuscation can be applied in different stages of the hardware IP's life cycle. As shown in Figure 4.3, pre-synthesis or RTL-obfuscation techniques are applicable for obfuscating behavioral description of a design. After synthesis, the RTL design is converted to a structured gate-level netlist. Most of the existing obfuscation techniques are applicable at the gate-level. The obfuscation techniques applicable at the RTL and gate-level are primarily focused on performing functional obfuscation that locks the normal operation of the IP based on a

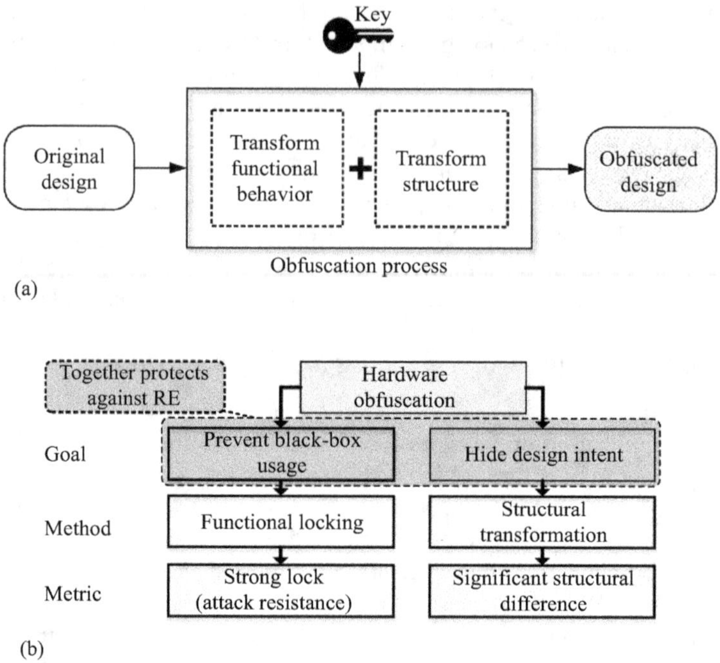

Figure 4.2 Overview of hardware obfuscation: (a) the original IP goes through functional and/or structural transformations that create an obfuscated version. This process is often guided by a secret key; (b) transformations are applied with one or more specific security goals in mind. The quality of obfuscation is evaluated using metrics that often focus on the structural and functional discrepancies between the original and obfuscated design or quantifies the security against different attacks on the obfuscation method.

secret key. Such functional locking techniques mainly prevents piracy and malicious modification. Obfuscation techniques are also applicable at the layout level. Since the design goes through various transformations during the obfuscation process, an additional verification step must be introduced to ensure that the obfuscated design provides equivalent functionality once unlocked and the impact on the performance is acceptable.

4.1.5 Difference from software obfuscation

Software obfuscation focuses on hiding the implementation of algorithms used in the system, which are represented as code, rather than locking its functionality. Different techniques can be used to obfuscate these codes, ranging from simple white space addition to the alteration of the program's control flow. Many of these transformations may not even be applicable to hardware obfuscation. While software obfuscation is typically performed on the software code, hardware obfuscation could be required

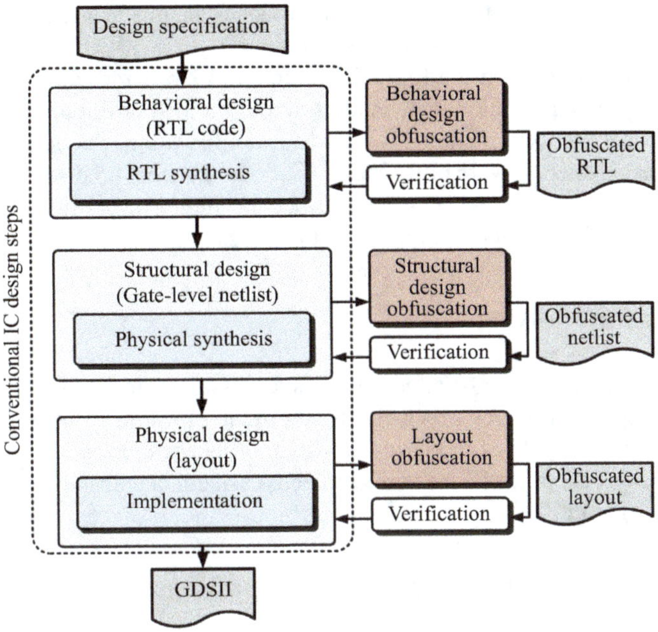

Figure 4.3 Obfuscation in design flow

at a number of design abstractions, including RTL description, gate-level netlist, or physical layout. Besides, while most of the techniques in hardware obfuscation are key-based, software obfuscation techniques usually do not incorporate additional key inputs as new variables to the obfuscated program.

4.1.6 Outline of the chapter

In this chapter, we discuss various aspects of the hardware obfuscation domain that has evolved during the past decade. While several promising and innovative obfuscation approaches have been proposed in the literature, vulnerabilities inherent to these techniques have also been discovered to compromise them. The chapter covers the following key aspects:

- The chapter provides a study on the evolution of hardware obfuscation domain, techniques, benchmarks, and vulnerabilities.
- It presents a taxonomy of attacks on obfuscation with associated countermeasures to prevent them.
- It summarizes the present status of this domain using a quantitative analysis of the publication trends and identifies possible future directions to solve the critical challenges.

We believe that setting the right path forward for hardware obfuscation requires a clear knowledge of the achievements and challenges of current obfuscation methods.

The directive evaluation of obfuscation approaches in this work can help one to improve current methods and inspire new ones.

This chapter consists of the following: Section 4.2 provides an overview of the attack space and the possible threat models across the IP's development cycle. Section 4.3 explores the used methods for hardware obfuscation. Section 4.4 describes the known attacks on hardware obfuscation methods. Section 4.5 discusses the trends of hardware obfuscation. Section 4.6 describes the future of hardware obfuscation. Finally, Section 4.7 presents a summary of this chapter.

4.2 Threat models

The development and usage of attack models are essential to identify the use cases of different countermeasures precisely. Hence, it is essential to observe all attack scenarios in the life cycle of the hardware IP. In order for these attack models to be defined, the life cycle of the IP, the varied objectives of attackers, and their capabilities should be considered.

4.2.1 Threat at different stages of the supply chain

The IP is vulnerable to different sets of threats at different stages of the supply chain. Next we discuss the potential attacks in major supply chain stages.

4.2.1.1 Untrusted SoC developer/IC design house

SoC developers have access to IPs that are purchased from third-party vendors. These IPs are sometimes procured to be explicitly used in a specific number of devices (for reconfigurable hardware) or to manufacture a limited number of instances. However, rightful owners can always overuse the IP without the knowledge of the vendor. They can also advertise the IP as their own and sell it to other design houses. An untrusted consumer can even reverse engineer the IP to understand the design intent. Even though a design house could alter the design, the motivation would not be to insert hardware Trojans. Instead, a malicious consumer could be interested in bypassing security restrictions or watermarks embedded in the IP.

4.2.1.2 Untrusted foundry

The foundry typically obtains an IP from the design house for fabrication in the form of GDSII. An untrusted foundry may easily use the GDSII file to overproduce the IC and supply the market with counterfeits. They can also sell the GDSII to other vendors as their own IP. Moreover, the foundry can also attempt to reverse engineer the design to obtain a more reusable version of the IP (e.g., gate-level/RTL) or to modify the circuits to insert hardware Trojans.

4.2.1.3 Untrusted end user

Resourceful attackers in the field with RE capability could extract the physical layout by applying decapsulation, delayering, and imaging process on the fabricated IC.

Table 4.1 An overview of all possible attack models of the SoC development cycle

Attack model	SoC developer	Foundry	End user
Model A	Trusted	**Untrusted**	**Untrusted**
Model B	**Untrusted**	Trusted	**Untrusted**
Model C	**Untrusted**	**Untrusted**	Trusted
Model D	Trusted	Trusted	**Untrusted**
Model E	Trusted	**Untrusted**	Trusted
Model F	**Untrusted**	Trusted	Trusted
Model G	**Untrusted**	**Untrusted**	**Untrusted**

Extraction of the layout may lead to IP piracy or RE of more high-level abstraction to understand the design intent. If the IC is fabricated from an obfuscated design that is locked using a key, the attacker could perform a dynamic analysis on the IC to leak the obfuscation key.

4.2.2 Comprehensive attack models

As untrusted parties can be involved in the development process at any stage, one can assume that the same threat can exist in multiple phases in the IP's life cycle. Hence, each party should assume that other parties may not be trusted. For example, an IP vendor should always assume that the SoC developer, the foundry, and the end user are untrusted, and secure their IPs on that basis. Table 4.1 shows an overview of the attack models. Model A assumes that the SoC developer is the only trusted entity, Model B assumes that foundry is only trusted, and Model C assumes that all but the end users are untrusted. Models D–F assume that the end user, the foundry, and the SoC developer, respectively, are not trusted, while Model G assumes that all parties in the development cycle are untrusted.

4.3 Hardware obfuscation techniques

The goal of hardware obfuscation is to lock the original design that is available in the form of RTL, netlist, or layout. In the literature, different obfuscation techniques that are applicable to a different abstraction of the design have been proposed. In this section, we will discuss some of the obfuscation techniques that are applicable in post-synthesis design (i.e., netlist). These techniques are broadly categorized as logic obfuscation or logic locking. Logic obfuscation techniques aim to transform the design such that it gets locked by a key. The obfuscated design will only generate correct output when the correct key is inserted. Therefore, access to the obfuscated design no longer allows an adversary to use it or sell it. Logic obfuscation techniques insert key-gates in the design to incorporate the key-dependent unlocking of functionality. The type of key-gates used and their placement dictates the corruption at the output of the design when the key is not correct. Figure 4.4(a) shows an original design with three gates. Figure 4.4(b) and (c) shows that the output of this design is locked by inserting two different types of key-gates using XOR and MUX, respectively. For XOR key-gate, if the key input is provided with an incorrect value, the output of the design will get inverted. The key input for the MUX key-gate is applied at its select input. Hence, depending on the key value applied, the output of the MUX passes either the state of the actual output of the design in (a) or the state of a different signal, generating an incorrect functionality. Next, we briefly introduce some of the logic obfuscation techniques.

4.3.1 Random insertion

A simple approach to logic obfuscation is to randomly insert different key-gates across the design without following a specific strategy [5]. The level of protection in this method for each instance is not consistent as the selected location for key-gate insertion changes every time the obfuscation process is executed.

4.3.2 Secure logic locking (SLL)

During random insertion of key-gates, the impact of one key-gate in corrupting the output may mute the impact of other [6]. This vulnerability is exploited by key sensitization attacks (KSAs) that leak the correct key value from an unlocked IC by applying judiciously selected test patterns. Secure logic locking (SLL) aims to prevent the insertion of such mutable key-gates and prevent KSA. However, the algorithm for identifying a suitable location for key-gate insertion in SLL is of high complexity that could suffer from scalability issues for a large design.

4.3.3 Logic cone size (CS) obfuscation

The logic cone size (CS) method aims at scanning every net in the design to find the largest cones in order to place the key-gates [7]. The CS is calculated as the total number of gates in the fan-in and fan-out of the net. The goal of such an insertion process is to provide high output corruptibility with the least number of key-gates. However, it does not guarantee any security against attacks such as KSA or satisfiability (SAT).

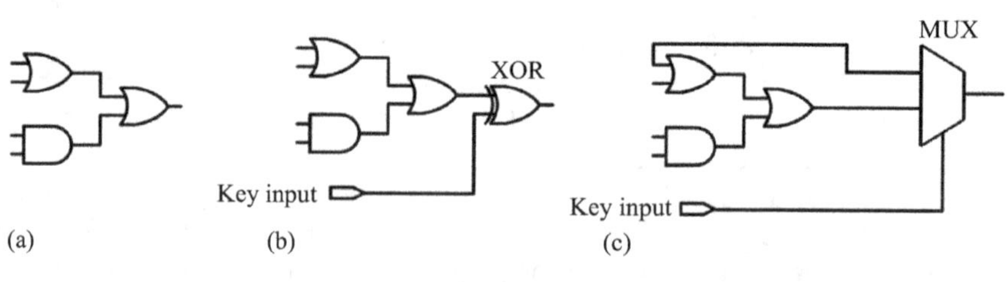

Figure 4.4 (a) An original circuit, (b) an obfuscated version of the original circuit using XOR/XNOR obfuscation, and using MUX-based obfuscation (c)

4.3.4 *Binary decision diagram (BDD) obfuscation*

This method focuses on applying the obfuscation to the binary decision diagram (BDD) representation of the design [8]. The obfuscation process adds key inputs to the early branches of the diagram of the design and a set of dummy decisions that will lead to an invalid output. The limitation of this method is scalability and overhead (Figure 4.5).

4.3.5 *Logic obfuscation for reconfigurable hardware*

Logic obfuscation can also be applied in a technology-mapped netlist in reconfigurable hardware. This method leverages unused portions in look-up-table (LUT) to insert key inputs that either select the original function in the LUT or a dummy function [9]. Apart from enabling functional locking, this technique has been leveraged to hide critical functions in the netlist [10]. Other forms of bitstream obfuscation require architectural modification to the reconfigurable components to incorporate key inputs [11].

4.3.6 *Finite state machine (FSM) obfuscation*

A finite state machine (FSM) obfuscation scheme is a technique to lock the functional behavior of design in the FSM level of abstraction of a design. An FSM is a control logic that stores the state of operation of the design and provides a control signal to the datapath of the design. In [12], Chakraborty and Bhunia have introduced HARPOON, where the transition path of the FSM states is modified appending some additional states (locked states) during power-up of a design that can only be traversed knowing a specific (key) sequence. Without the knowledge of the key, original states cannot be reached. The additional states control several nodes in the design where a locked state causes these nodes to flip, and only the correct input sequence to the FSM allows correct functionality. In HARPOON, the authors have improved the security

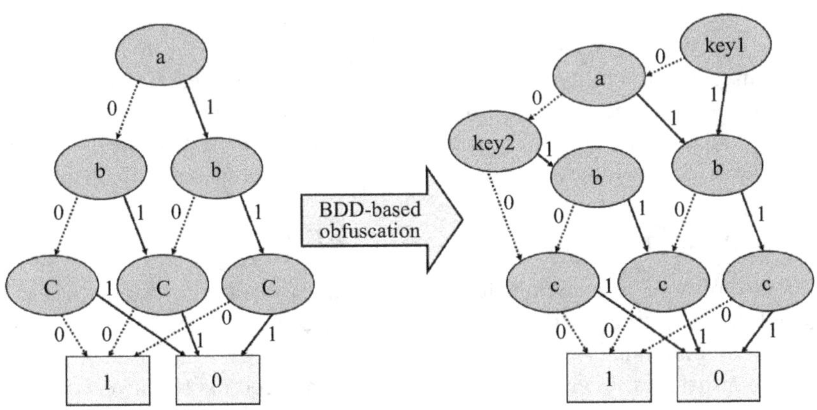

Figure 4.5 BDD obfuscation

and applicability of FSM obfuscation by focusing on gate-level locking. The method offers authentication and obfuscation to the target system, where an initialization FSM is inserted to the circuit, and a logic structure modification is applied to some nodes in the design.

4.4 Attacks on hardware obfuscation

As obfuscation methods have been proposed in the past decade, attacks on obfuscation have also emerged. In this section, we will discuss the attacks that have been introduced in the literature and their impact on current obfuscation methods.

4.4.1 Boolean satisfiability (SAT) attack

Boolean SAT attack is a powerful tool to significantly reduce the key search space in a short time [13]. Unlike brute-force attacks, SAT employs advanced SAT algorithms to help one to provide a set of decisions that can quickly resolve the key and break the obfuscation. The attack requires the locked design and the unlocked IC (which could be bought from the market) with correct functionality. The attacker needs access to the scan-chain to perform the attack if the design is sequential.

4.4.2 Key sensitization attack (KSA)

KSA is designed to leak the correct key value to the output without being corrupted or affected by other key-gates [6]. The attack requires a reverse-engineered obfuscated netlists as well as a functional system with valid outputs. The attacker tries to determine an input pattern that can avoid any other key-gate interference while providing a valid output. This pattern can be used to identify a subset of the key, and the process is repeated with different patterns until the key is resolved.

4.4.3 Structural analysis using machine learning attack (SAIL)

SAIL is a structural-based machine learning attack that aims to restore the gate-level netlist into its pre-synthesis representation [14]. The attack's goal is to reveal how key-gates are inserted, which, in turn, can reveal the key value of each input. The attack captures a subgraph around the key input port and predicts whether that subgraph has changed after the synthesis process. Additionally, the attack predicts the pre-synthesis subgraph form if the synthesis process is believed to have caused it to change.

4.4.4 Constant propagation attack (SWEEP)

SWEEP is a constant propagation attack that leverages the optimization aspect of synthesis tools to determine whether a key prediction is correct [15]. The attack performs the analysis on individual key inputs by assigning it to logic 0 in an instance, and logic 1 in the other instance. In each instance, the resultant circuit is synthesized and optimized, and all functional and structural features (such as area, power consumption, and critical path delay) are extracted and compared. The attack performs a thorough analysis to establish a correlation between the extracted features and the correctness of the key through the use of training data sets. The established correlation is then applied to the circuit under attack, and the correct key is extracted.

4.5 The trends of hardware obfuscation

In parallel to the evolution of threats to the supply chain of hardware, researchers proposed a number of obfuscation techniques over the last decade, which have been discussed in earlier sections. In this section, we will provide some quantitative facts regarding hardware obfuscation to have a glimpse of the trend and present status of the research in this subfield of hardware security. We relied on *IEEE Xplore* to investigate the facts by using keyword "hardware obfuscation" for all the publications on hardware obfuscation. We believe that this will cover most of the scopes of research in the field of hardware obfuscation. Based on the search, we will discuss the progression of research and its outcomes in the field of obfuscation in the last decade.

4.5.1 Evolution of obfuscation research

In the last decade, a good number of technical articles have been published on the topic of hardware obfuscation. These publications explored different types of attack models, obfuscation–deobfuscation schemes, their flaws and improvizations, surveys, and so on. Figure 4.6 depicts the quantitative publication trend until the beginning of the year 2018. Around 160 papers and articles have been published, although the accelerating phase of publications seems to take place between 2007 and 2010. After 2010, there has been a less number of published articles. We believe it happened for two explicit reasons: (1) new obfuscation schemes evolution was slow as the existing attacks at that time were so robust to break the obfuscation schemes; (2) new attacks on those new schemes were not that much promising in that period. However, after 2013, there has been a dramatic rise in terms of number of publications, and 2015 and 2017 have been the most productive years of research in the field of hardware

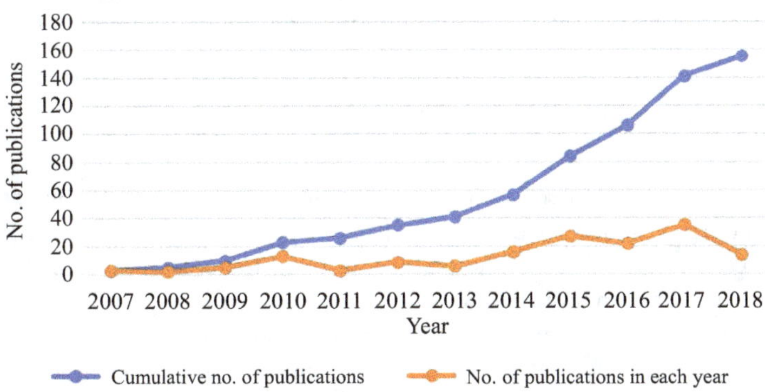

Figure 4.6 Quantitative analysis of trend in publications

obfuscation. It also indicates that the engagement and contribution to the hardware obfuscation research field are of huge impact in recent years, and it is creating future research opportunities as well. Therefore, research on hardware obfuscation is still on track, and there is a lot to improve.

4.5.1.1 Evolution of obfuscation techniques

Research on hardware obfuscation has grown up in a matured level over the last decade. However, it took a lot of effort to come to this stage and still a lot more to do. If we classify the obfuscation approaches in the following three types: (1) sequential obfuscation, (2) combinational obfuscation, and (3) miscellaneous, then the first proposed outcome of research on sequential obfuscation was in 2007, which was active hardware metering [16]. Afterward, to introduce obfuscation in IP level to prevent IP piracy, a new approach of netlist obfuscation for protecting hardware IP was proposed [17], which was implemented in SoC security later and it was named HARPOON [12]. Some security metrics to measure the strength of obfuscation were introduced as well, which worked as a pioneer to open up the field to work on hardware obfuscation. In 2009, obfuscation was first recognized as security against hardware Trojan [18]. New obfuscation scheme using a modification of control and data flow came into light in 2010 [4] and obfuscating FPGA bitstream added a new dimension in the scope of hardware obfuscation in 2011 [19].

In 2012, researchers came up with a new scheme of combinational obfuscation by the insertion of key-gates, which they named logic encryption [20]. This technique was improvised later to SAT attack resistant locking and by the insertion of anti-SAT block to make the design resistant to SAT attack [21]. Furthermore, gate camouflaging [22], logic permutation [23], etc. came out as a defense mechanism to prevent RE at chip level and board level. After every instance of new attack revealing, these techniques were either enhanced or became less efficient in terms of performance, security, overhead, or cost.

Other than combinational and sequential obfuscations, some emerging integration approaches such as split manufacturing came into light in 2014 [24]. This technique acquired the latest technology of 2.5D/3D IC integration. However, due to the susceptibility to proximity attack and larger overheads, split manufacturing could not emerge, and there is a lot to improve by reducing the cost without compromising the security.

4.5.1.2 Evolution of attacks on obfuscation

With the progression of obfuscation techniques, vulnerabilities of those obfuscation techniques have also been exploited. Researchers proposed various attacks on obfuscation during the period of the last decade. In 2012, KSA on key-based logic encryption was disclosed [25]. The major attack on logic obfuscation, SAT attack, became so much prominent in 2015 [13], and till now, it is a great threat to most of the key-based logic obfuscation schemes. Although anti-SAT approaches have been taken, researchers proved that the insertion of anti-SAT block does not potentially make the design resistant to SAT attack as removal of anti-SAT attack was proposed in 2017 [26]. In 2017, a new form of attack referred to as CycSAT [27] was proposed, which exploited the obfuscation scheme of cyclic logic encryption. The most recent attacks on combinational obfuscation are Structural Analysis using Machine Learning (SAIL) attack [14] and structural functional (SURF) attack [28], which are proven to recover most of the key values from a locked design.

Figure 4.7 gives the bird's eye view of the evolution of research on hardware obfuscation techniques and attacks. The figure illustrates that, initially, most of the efforts were given by the researchers to come up with new techniques to withstand possible threats in IC life cycle. In subsequent years, new threat models were analyzed, and the quest for more robust schemes got much more importance. At the same time, new attacks on those obfuscation approaches were revealed, which further increased the demand for attack resilient features of the existing or new approaches. Figure 4.7 also depicts that there has been a variety in the trend of research on obfuscation.

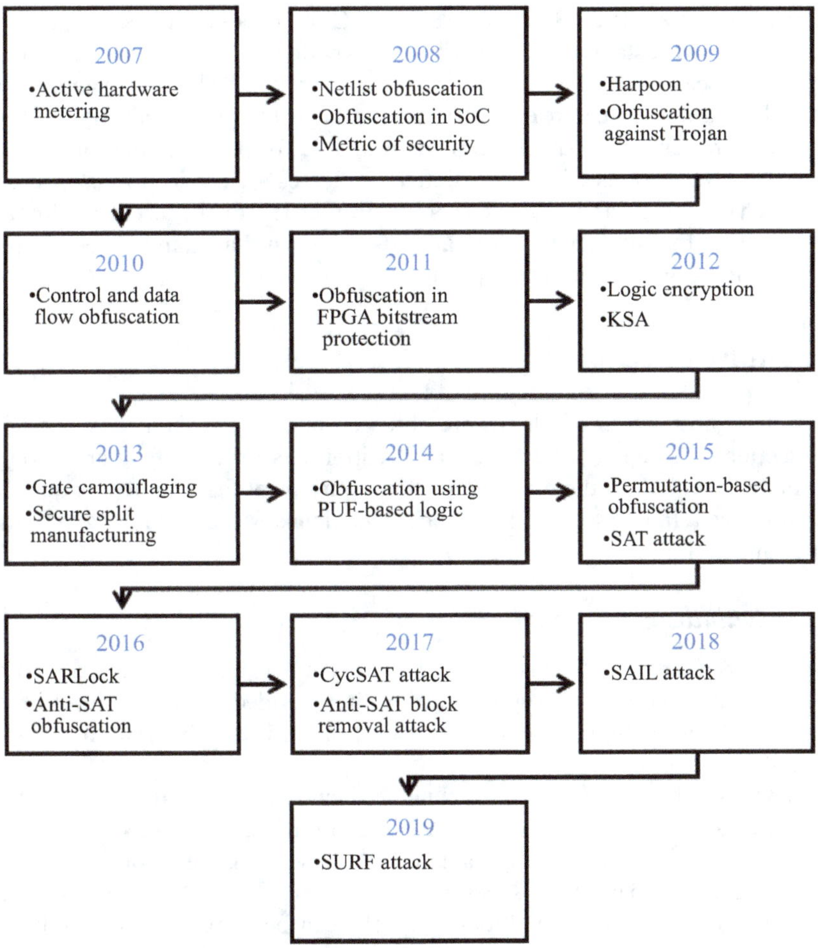

Figure 4.7 Timeline of research on hardware obfuscation: techniques and attacks

However, recent research is mostly based on SAT resistivity or analysis of a new obfuscation scheme that can thwart SAT attack or make it difficult for SAT solver to find the correct key(s). It is evident from the trend that the more obfuscation techniques append the current stack, the more attacks will be imposed on those new schemes. That is why it is important to propose a comprehensive solution that can thwart all known attacks and can avoid leaving traces or vulnerabilities in the implementation algorithm.

4.5.2 Evolution of obfuscation benchmarks

The chronological advancement in the research field of obfuscation incorporates the implementation of different obfuscation approaches on different sequential and combinational circuits. In order to show the adequacy of claims in terms of attack and defense techniques, researchers often use benchmark circuits. Some popular benchmark suites are ISCAS'85, ISCAS'89, ITC'99, open-source benchmarks, etc. Figure 4.8 shows the status of benchmark circuits collected from different sources and has been obfuscated over the last decade. It is clear from the plot that the obfuscation research field is not yet mature enough to deploy the techniques in industry level since a few no. of benchmarks have been made available for public critics and analysis. More obfuscated designs with robust defense schemes are needed to make sure that the research is on track, and the design community can rely on the fact that obfuscation techniques have the potential to defend all sorts of known attacks and can be integrated into the design cycle of different applications.

4.6 Future direction

The levels of abstraction that hardware obfuscation covers make it unique to other IP protection mechanisms. Hardware obfuscation has many pitfalls and promises that can be observed. In the previous sections, various obfuscation techniques have been introduced. In this section, all of these techniques will be evaluated in terms of security and performance.

4.6.1 Evaluation of security

As each obfuscation technique is designed to prevent a specific set of attacks, various properties can be compared when analyzing the provided protection level of each technique. The evaluation of the security aspect of hardware obfuscation is outlined in Table 4.2.

All post-synthesis logic locking techniques increase the difficulty for RE attacks, even if the gate-level netlist is obtained, obfuscation gates and key values are not easily identifiable. In the reconfigurable logic obfuscation, the reconfigurable slice used to redirect all datapaths is designed to prevent unauthorized access, which makes RE difficult to achieve. Probing attacks can be applied if the key is internally stored/generated, given that attackers have prior knowledge about the location of key tracks. Probing the reconfigurable slice is extremely difficult due to its volatile

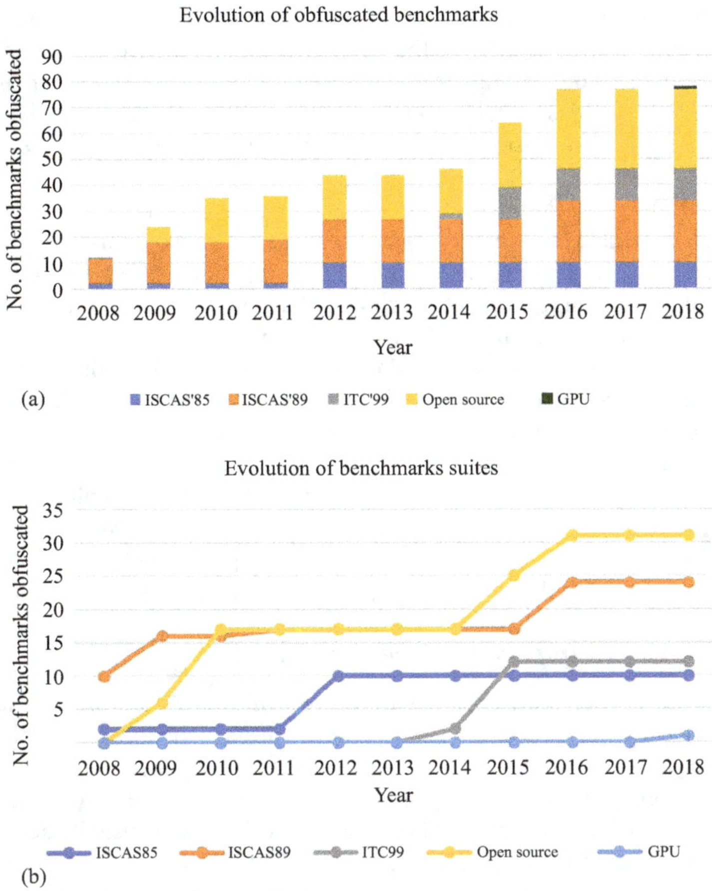

Figure 4.8 Evolution of benchmarks: (a) no. of benchmarks from different suites used for implementing defense and attack schemes (b) trend in using different benchmarks from different IP repositories

Table 4.2 Attack resiliency provided by some hardware obfuscation techniques

Obfuscation technique	**Attack resiliency**			
	SAT	**KSA**	**SAIL**	**SWEEP**
Random insertion	Weak	Weak	Weak	Weak
SLL	Weak	Strong	Weak	Weak
Logic CS	Weak	Weak	Weak	Weak
Reconfigurable logic obfuscation	Weak	Weak	Weak	Weak
BDD locking	Weak	Weak	Strong	Weak
FSM obfuscation (HARPOON)	Strong	Strong	Strong	Strong

state. Reconfigurable logic obfuscation is not based on key insertion, which keeps it out of the scope of the SAT attack. Otherwise, all logic locking techniques are vulnerable to SAT attacks though mitigation techniques that prevent the SAT from resolving the key can be utilized. Random insertion and logic CS obfuscation techniques are vulnerable to KSAs. Opposing KSA, SLL is designed by rearranging key-gates to prevent sensitization.

For the BDD obfuscation technique, altering the BDD makes the structure of the design unrecognizable, and the key-controlled part of the diagram is fused with the original functions, which prevents identifying the function using structural analysis. Similar to logic locking techniques, a probing attack is achievable if the attacker can identify the location of key tracks. BDD approach is also vulnerable to SAT attacks unless SAT mitigation blocks are added to the design. Depending on the correlation of the key-gates, KSAs may be able to deduce the key of the logic-locked circuit.

FSM obfuscation is designed to mix existing sequential parts with locking mechanisms, which makes identifying obfuscation gates and RE extremely difficult. Probing can be applied to capture the sequential patterns that are used to unlock the circuit, though it does not help to obtain the unlocked gate-level netlist. SAT attacks are not able to break sequential circuits. As long as direct dependency between the outputs and the keys are observed, KSAs can be applied even to sequential circuits.

4.6.2 Evaluation of performance and overheads

Every obfuscation technique comes with the cost of degradation in performance or an increase in resource allocation. Physical obfuscation techniques are applied at the end of the development cycle, where the modification is applied to the layout of the circuit. These techniques can be applied to any type of design, and they are scalable to any size. The overhead of physical obfuscation techniques is usually acceptable.

Post-synthesis techniques are applied to the synthesized gate-level netlist, where they target combinational circuits. Random insertion obfuscation is the only scalable technique, whereas SLL and logic-CS-based approaches are not applicable to industrial-sized IPs. The overhead for logic locking techniques is low and easily controllable since most obfuscation tools give the user the option to choose the amount of allowed overhead.

Pre-synthesis techniques are applied to the RTL abstraction of the design, and the presence of sequential elements is required for the FSM obfuscation to be applied. BDD obfuscation is not easily scalable as the size of the decision tree grows exponentially as the number of gates increases. Additionally, the overhead for the BDD obfuscation is very large compared to other techniques. A comparative evaluation of all obfuscation techniques in terms of performance and overhead is summarized in Table 4.3.

4.6.3 The future of hardware obfuscation

The future of hardware obfuscation depends on the emergence of a wide range of new techniques, as well as improving current techniques in terms of security and performance. In the past decade, promising obfuscation approaches have been presented.

Table 4.3 Performance evaluation of some hardware obfuscation techniques

Obfuscation technique	**Performance evaluation**			
	Implementation	**Applicability**	**Scalability**	**Overhead**
Random insertion	Gate-level netlist	Combinational circuits	Scalable	Low
SLL	Gate-level netlist	Combinational circuits	Not scalable	Low
Logic CS	Gate-level netlist	Combinational circuits	Scalable	Low
Reconfigurable logic obfuscation	Datapath rerouting	Any design	Scalable	Moderate
BDD locking	RTL modification	Any design	Not scalable	High
FSM obfuscation	RTL modification	Sequential elements	Scalable	Moderate

However, many sophisticated attacks were developed to bypass the level of security provided by these techniques. Additionally, many limitations in current techniques have made it difficult for IP vendors to apply these techniques. These limitations include the presented weaknesses of current obfuscation techniques, the lack of diverse and industrial-scaled benchmarks, and the lack of a good metric that evaluates the added level of security by each technique. In this section, we highlight the main areas of research that need to be further explored.

4.6.3.1 Stronger obfuscation techniques

Stronger obfuscation approaches are always needed as more powerful attacks have emerged. The security that is provided by current techniques is designed to prevent specific attacks at specific developmental stages. However, a wide range of protection coverage has not been achieved yet. Therefore, a stronger approach that can enhance the system's confidentiality level is essential.

4.6.3.2 Robust security assessment framework

In order for the obfuscation method to provide robustness against known attacks, a thorough security assessment process is needed. There is a need for a comprehensive approach to evaluate and quantify how secure the design is after it is being obfuscated.

4.6.3.3 High-level obfuscation

Most obfuscation techniques are designed to target gate-level netlist and layouts, which are the lowest levels of abstraction. However, there is a need for obfuscation techniques that target a higher level of abstraction (such as RTL).

4.6.3.4 Scalability

Most existing obfuscation techniques are not scalable to industrial size IPs. The issue of scalability creates a major limitation that prevents IP vendors form considering hardware obfuscation. Hence, the research area of scalable obfuscation techniques should be considered.

4.6.3.5 Better metrics

Lack of a good evaluation metric for hardware obfuscation limits the ability to provide a compelling argument to IP vendors in applying hardware obfuscation. Without comprehensive and detailed metrics, obfuscation techniques may not be accurately evaluated. Existing metrics such as keyspace size and the failure rate of formal testing can help one to measure the level of security in functional obfuscation techniques. However, there is still a lack of generalized metric that can accurately evaluate all techniques.

4.7 Summary

In this chapter, we have attempted to provide an overview of hardware obfuscation that aims at protecting hardware IP blocks against various attacks, primarily piracy and RE. We have presented the evolution of hardware obfuscation techniques in the past decade. We have categorized and described each technique and provided illustrative examples, and attacks and countermeasures have also been discussed and thoroughly evaluated. We have presented a comprehensive list of attack models as well as a description of their goals and capabilities. The research trends of obfuscation techniques, attacks, and benchmarks have been studied and analyzed in this chapter. Finally, we have addressed the limitations and challenges in existing techniques and suggested a road map for future research.

References

[1] Bloomberg, "Ex-Flow Traders Employee Pleads Guilty in Code-Theft Case," 2014. [Online]. Available: https://www.bloomberg.com/news/articles/2014-12-15/ex-flow-traders-employee-pleads-guilty-in-code-theft-case.

[2] EPSNews, "Xilinx Sues Flextronics for Alleged Fraudulent Chip Resale," 2013. [Online]. Available: https://epsnews.com/2013/12/13/xilinx-sues-flextronics-alleged-fraudulent-chip-resale/

[3] A. Chhotaray, A. Nahiyan, T. Shrimpton, D. Forte, and M. Tehranipoor, "Standardizing bad cryptographic practice: A teardown of the IEEE standard for protecting electronic-design intellectual property," in *Proceedings of the 2017 ACM SIGSAC Conference on Computer and Communications Security*, Dallas, TX: Association for Computing Machinery, 2017, pp. 1533–1546.

[4] R. S. Chakraborty and S. Bhunia, "RTL hardware IP protection using key-based control and data flow obfuscation," in *VLSI Design, 2010. VLSID'10. 23rd International Conference on*. Bangalore: IEEE, 2010, pp. 405–410.

[5] J. A. Roy, F. Koushanfar, and I. L. Markov, "EPIC: Ending piracy of integrated circuits," in *Proceedings of the Conference on Design, Automation and Test in Europe*, Munich: Association for Computing Machinery, 2008, pp. 1069–1074.

[6] M. Yasin, J. J. Rajendran, O. Sinanoglu, and R. Karri, "On improving the security of logic locking," *IEEE Transactions on Computer-Aided Design of Integrated Circuits and Systems*, vol. 35, no. 9, pp. 1411–1424, 2015.

[7] Y.-W. Lee and N. A. Touba, "Improving logic obfuscation via logic cone analysis," in *2015 16th Latin-American Test Symposium (LATS)*. Puerto Vallarta: IEEE, 2015, pp. 1–6.

[8] X. Xu, B. Shakya, M. M. Tehranipoor, and D. Forte, "Novel bypass attack and BDD-based tradeoff analysis against all known logic locking attacks," in *International Conference on Cryptographic Hardware and Embedded Systems*. Taipei: Springer, 2017, pp. 189–210.

[9] R. Karam, T. Hoque, S. Ray, M. Tehranipoor, and S. Bhunia, "Robust bitstream protection in FPGA-based systems through low-overhead obfuscation," in *2016 International Conference on ReConFigurable Computing and FPGAs (ReConFig)*. Cancun: IEEE, 2016, pp. 1–8.

[10] T. Hoque, K. Yang, R. Karam, *et al.*, "Hidden in plaintext: An obfuscation-based countermeasure against FPGA bitstream tampering attacks," *ACM Transactions on Design Automation of Electronic Systems (TODAES)*, vol. 25, no. 1, pp. 1–32, 2019.

[11] R. Karam, T. Hoque, S. Ray, M. Tehranipoor, and S. Bhunia, "Mutarch: Architectural diversity for FPGA device and IP security," in *2017 22nd Asia and South Pacific Design Automation Conference (ASP-DAC)*. Chiba: IEEE, 2017, pp. 611–616.

[12] R. S. Chakraborty and S. Bhunia, "HARPOON: An obfuscation-based SoC design methodology for hardware protection," *IEEE Transactions on Computer-Aided Design of Integrated Circuits and Systems*, vol. 28, no. 10, pp. 1493–1502, 2009.

[13] P. Subramanyan, S. Ray, and S. Malik, "Evaluating the security of logic encryption algorithms," in *Hardware Oriented Security and Trust (HOST), 2015 IEEE International Symposium on*. Washington, DC: IEEE, 2015, pp. 137–143.

[14] P. Chakraborty, J. Cruz, and S. Bhunia, "SAIL: Machine learning guided structural analysis attack on hardware obfuscation," in *2018 Asian Hardware Oriented Security and Trust Symposium (AsianHOST)*. Hong Kong: IEEE, 2018, pp. 56–61.

[15] A. Alaql, D. Forte, and S. Bhunia, "Sweep to the secret: A constant propagation attack on logic locking," in *2019 Asian Hardware Oriented Security and Trust Symposium (AsianHOST)*. Xi'an: IEEE, 2019.

[16] Y. Alkabani and F. Koushanfar, "Active hardware metering for intellectual property protection and security." in *USENIX Security Symposium*, Boston, MA: USENIX Association 2007, pp. 291–306.

[17] R. S. Chakraborty and S. Bhunia, "Hardware protection and authentication through netlist level obfuscation," in *Proceedings of the 2008 IEEE/ACM International Conference on Computer-Aided Design*. San Jose, CA: IEEE Press, 2008, pp. 674–677.

[18] R. S. Chakraborty and S. Bhunia, "Security against hardware Trojan through a novel application of design obfuscation," in *Computer-Aided Design-Digest of Technical Papers, 2009. ICCAD 2009. IEEE/ACM International Conference on*. San Diego, CA: IEEE, 2009, pp. 113–116.

[19] S. Gören, O. Ozkurt, A. Yildiz, and H. F. Ugurdag, "FPGA bitstream protection with PUFs, obfuscation, and multi-boot," in *Reconfigurable*

Communication-Centric Systems-on-Chip (ReCoSoC), 2011 6th International Workshop on. Montpellier: IEEE, 2011, pp. 1–2.

[20] J. Rajendran, Y. Pino, O. Sinanoglu, and R. Karri, "Logic encryption: A fault analysis perspective," in *Proceedings of the Conference on Design, Automation and Test in Europe*. Dresden: EDA Consortium, 2012, pp. 953–958.

[21] Y. Xie and A. Srivastava, "Mitigating sat attack on logic locking," in *International Conference on Cryptographic Hardware and Embedded Systems*. Taipei: Springer, 2016, pp. 127–146.

[22] J. Rajendran, M. Sam, O. Sinanoglu, and R. Karri, "Security analysis of integrated circuit camouflaging," in *Proceedings of the 2013 ACM SIGSAC Conference on Computer & Communications security*. Berlin: ACM, 2013, pp. 709–720.

[23] Z. Guo, M. Tehranipoor, D. Forte, and J. Di, "Investigation of obfuscation-based anti-reverse engineering for printed circuit boards," in *Proceedings of the 52nd Annual Design Automation Conference*. San Francisco, CA: ACM, 2015, p. 114.

[24] K. Vaidyanathan, B. P. Das, E. Sumbul, R. Liu, and L. Pileggi, "Building trusted ICs using split fabrication," in *Hardware-Oriented Security and Trust (HOST), 2014 IEEE International Symposium on*. Arlington, VA: IEEE, 2014, pp. 1–6.

[25] J. Rajendran, Y. Pino, O. Sinanoglu, and R. Karri, "Security analysis of logic obfuscation," in *Proceedings of the 49th Annual Design Automation Conference*. San Francisco, CA: ACM, 2012, pp. 83–89.

[26] M. Yasin, B. Mazumdar, O. Sinanoglu, and J. Rajendran, "Security analysis of anti-sat," in *Design Automation Conference (ASP-DAC), 2017 22nd Asia and South Pacific*. Chiba: IEEE, 2017, pp. 342–347.

[27] H. Zhou, R. Jiang, and S. Kong, "CycSAT: Sat-based attack on cyclic logic encryptions," in *Computer-Aided Design (ICCAD), 2017 IEEE/ACM International Conference on*. Irvine, CA: IEEE, 2017, pp. 49–56.

[28] P. Chakraborty, J. Cruz, and S. Bhunia, "SURF: Joint structural functional attack on logic locking," in *2019 IEEE International Symposium on Hardware Oriented Security and Trust (HOST)*, McLean, VA: IEEE, 2019, pp. 181–190.

Chapter 5
Formal verification for SoC security

Jiaji He[1], Xialong Guo[2], Yiqiang Zhao[3] and Yier Jin[4]

The wide adoption of third-party hardware intellectual property (IP) cores, including those from untrusted vendors, has raised security concerns for system designers and end users. Existing approaches to ensure the trustworthiness of individual IPs rarely consider the entire system-on-chip (SoC) design, especially the IP interactions through SoC bus. These methods can hardly identify malicious logic (or design flaws) distributed in multiple IPs, whereas individual IPs fulfill security properties and can pass the security testing/verification. One possible solution is to treat the SoC as one IP core and try to verify security properties of the entire design. This method, however, suffers from scalability issues due to the large size of SoC designs with multiple IP cores integrated. In this chapter, we present a scalable SoC bus verification framework trying to verify the security properties of SoC bus implementation where the bus protocol plays the role of the golden reference. More specifically, finite state machine (FSM) models will be constructed from the bus implementation, and the trustworthiness will be verified based on the property set derived from the bus protocol and potential security threats. Along with IP level formal verification solutions, the proposed framework can help one to ensure the security of large-scale SoCs. Experimental results on ARM advanced micro-controller bus architecture (AMBA) bus demonstrate that our approach is applicable and scalable to prevent information leakage and denial-of-service (DoS) attack by verifying security properties.

5.1 Introduction

With the rapid development of modern integrated circuit (IC)-fabrication techniques, the complexity of ICs is increasing rapidly. In order to reduce the time-to-market, circuit designers try to integrate different IP cores into a single chip to form an SoC. To coordinate interactions among different IPs, bus protocols for data and control transfers are used. For example, the AMBA [1] from ARM serves as a leading example among these bus protocols. However, the possibility of inserting hardware Trojans

[1]Institute of Microelectronics, Tsinghua University, Beijing, China
[2]Department of Electrical and Computer Engineering, Kansas State University, Manhattan, NY, USA
[3]School of Microelectronics, Tianjin University, Tianjin, China
[4]Department of Electrical and Computer Engineering, University of Florida, Gainesville, FL, USA

at the design stage in an IP by an untrusted third-party vendor has raised security concerns in the SoC design industry [2,3]. To ensure the trustworthiness of an SoC design, two types of methods, runtime detection and static verification solutions, have been developed targeting the SoC bus and IPs. In the runtime method, potential security vulnerabilities are identified during the SoC operation. It enables an SoC integrator to enhance security of existing bus protocols by adding additional runtime-monitoring circuits or enhancing the existing protocols. The static solution uses testing and verification techniques to either validate the security of SoC design and bus protocols or detect any design flaws that may impact the security of the SoC operations before implementation.

As shown in Figure 5.1, an SoC designer builds a system using IPs from third-party vendors, and some of the IPs in the SoC are untrusted. To ensure the security of the design, in this chapter, we propose a comprehensive verification framework that is scalable for large SoCs. The interactions between IPs and the system bus will be considered and checked. Anomalies in normal communications between untrusted IPs and the rest of the SoC design can be detected. Our method only focuses on verification of digital circuits. If there are analog parts in the SoC system, the analog circuits will not be modeled. FSMs are first extracted from the SoC utilizing gate-level netlist, and the formal model is established of the design. Then security specifications are proposed based on the knowledge of possible attacks through untrusted IPs connected to the bus. Finally, the formalized model of the SoC is verified against security properties using an automated model-checking tool. If an SoC design satisfies the security specifications, then the security of the SoC is proved.

The main contributions of this chapter are listed as follows:

- Formal security specifications at the SoC level are developed with consideration of real bus operations and possible Trojans attacks.
- The FSM of the whole SoC is extracted utilizing gate-level netlist. All behaviors of the circuit are modeled without performing any abstraction, and we ensure that any malicious behavior is not lost during modeling process. Our modeling strategy is scalable for large SoCs.
- We focus on securing the entire SoC from hardware Trojans considering possible attacks utilizing normal bus communications.

Figure 5.1 Interactive of components in proposed framework

The rest of this chapter is organized as follows. In Section 5.2, we discuss previous work on SoC design protection and verification. Section 5.3 describes the threat model and provides background of the tools used in our framework. In Section 5.4, we introduce a methodology in detail. In Section 5.5, we carry out a series of experiments to demonstrate our method. The experiment results are shown in Section 5.6, and a discussion of applying information-flow tracking (IFT) in detecting bus Trojans is in Section 5.7. Conclusions are drawn in Section 5.8. Finally, Section 5.9 presents discussions regarding the research work presented in this chapter and potential future research directions.

5.2 Related work

In this section, we will introduce both runtime and static solutions for hardware system protection.

5.2.1 Runtime methods

Many runtime hardware approaches were developed for providing secure assurances of IPs and SoC. In [4], IP sandboxing technique was utilized to defeat drone-jamming-based attacks. A resilient SoC security architecture was proposed in [5], and this architecture implements certain security policies and also ensures trusted information flow within the blocks. This solution provides a comprehensive solution in the secure-architecture-based SoC protection. However, it still needs a central security policy checker and security wrappers. Also there are other techniques such as gate-level IFT [6] to ensure the information security of the system bus. Authors in [7] gave a solution for hardware runtime formal verification of security properties on IPs depending on proof-carrying hardware (PCH) approach. A hardware-based SAT solver was developed to perform validating of fabricated hardware. Still, high overhead in runtime PCH causes scalability issue in practical use.

Further, runtime methods enhance security of bus protocols with a wrapper or monitor and require specific design modifications to achieve particular security purposes. In [8], correctness of modules was checked using an IP interface monitor, which was developed using high-level specifications. The interface monitor cannot detect system-level malicious behaviors if the modules follow the normal bus protocol. Lin *et al.* [9] designed a monitor for bus compliance testing and for observing bus signals in an SoC system. Although they verified the data part of the bus transfer against necessary properties systematically, their method cannot detect malicious behaviors. A security-enhanced communication architecture was designed in [10] to enhance bus communications, but this architecture is not applicable to other security vulnerabilities. Kim *et al.* [11] developed a secure bus architecture by constructing new structures around core modules of a bus, and the architecture was able to provide runtime protection against hardware Trojans. Although the secure architecture accomplished protection without incurring high costs in terms of bus resources and performance, it can handle security issues related to one single module only.

Even if runtime methods can provide a very satisfactory security protection, these methods all require hardware modifications of original circuits or additional modules in the systems. Moreover, the modifications usually require the designers of the security solution have a well understanding of the circuits and systems. Further, implementing these methods will inevitably introduce area and power overhead of the original design and will have impact on the performance.

5.2.2 Static methods

Static methods, which are used prior to design deployment in the pre-layout stage, neither cause any hardware overhead nor require modifications to original designs. Among the static methods, there are many approaches targeted at pre-layout hardware Trojan detection, including feature analysis and formal verification. The feature analysis Trojan detection methods focus on extracting particular features from pre-layout design files directly to find anomalies. In [12], an HTChecker was proposed to detect hardware Trojans in VLSI design based on static characteristics of gate-level netlist. Trojan features are extracted from gate-level netlist in [13], and the random forest classifier is utilized to select the best set of Trojan features. The author in [14] proposed to apply the intercluster distance to distinguish Trojan gates and genuine gates in a design based on the controllability and observability characteristics. In [15], a framework is proposed to combine flip-flop level and combinational logic level features for analyzing to achieve both high efficiency and accuracy. Feature-analysis-based methods directly detect hardware Trojans in individual IPs; however, the trustworthiness of the whole system composed of multiple IPs may not be guaranteed.

Formal methods do not directly detect HTs but try to evaluate the trustworthiness of IP cores integrated in the SoC systems. Meanwhile, formal verification can develop security properties with the consideration of potential system-level threats. Formal verification with trusted models was used in the Algorithm for Resisting Trojans of [16] as part of a series of steps to provide complete protection of design from Trojan payloads, but their methods only focused on individual IPs. In [17], security critical objectives were selected to divide the system into smaller security-sensitive assets, and security properties were developed and checked accordingly. Although they improved the scalability of formal verification by breaking down the SoC into subsets of security-sensitive assets, some critical information may be lost during the decomposition process. A SAT-based formal verification method was proposed in [18], and the authors proposed a series of properties to detect the information-leaking Trojan. However, their methods cannot detect system-level Trojans. Additionally, formal verification has been widely used for checking functional and security properties on SoC bus. An abstract model of a commercial field bus was developed in [19], and a number of imperfections in the protocol logic itself were found. An arbiter core of IBM Core Connect bus was verified by Goel and Lee [20] and they found defects of the protocol itself during functional checking. In [21], an AMBA AHB protocol model was developed according to the specification of the protocol, and the non-starvation property was checked for the master. Guo *et al.* [22] used a hybrid approach in PCH framework to alleviate the scalability issue in formal verification of large designs. However, the previous verification still only focused on the function checking.

IFT [23] is a powerful approach for preventing sensitive information leakage. In this approach, labels representing secrecy/trust levels are assigned to data and operations on data are extended to include operations on their labels, based on predefined information flow policies. Access or propagation of data with sensitive labels is thereby restricted to trusted segments of the code or system and is forced to abide by the desired information flow policies. SecVerilog [24,25] and QIF-Verilog [26] are some of the static solutions based on IFT.

From the previous discussion, current formal verification methods usually focus on single IP verification and do not consider the interconnections among various IPs through certain bus protocols. It is important to take the bus protocols into consideration when verifying the security of the whole system. Although some methods can detect system-level Trojans or security threats, they only build an abstract model of the SoC design prior to verification. As a result, some details of the design, including malicious logic, may be lost during the abstraction process. So a comprehensive model containing all logic operations of the system is required.

5.3 Background and preliminary

5.3.1 Threat model

Hardware Trojans can be inserted at many different stages of the IC lifecycle. In this chapter, we consider the rogue agent at third-party design house who can access the hardware design and make modifications to the RTL code and netlist files. Therefore, the availability of gate-level netlist is required. In practice, soft IPs are available in the form of RTL code, and firm IPs are in the form of gate-level netlist. While hard IPs are available in layout format, the gate-level netlist can be extracted using reverse engineering methods [27]. The proposed Trojan detection approach targets at detecting digital hardware Trojans that utilize the bus interface to monitor the triggering signals and perform malicious functions. As there are many SoC assets that could be targeted by an attacker, it is highly possible that security threats really exist. With more third-party IPs integrated in an SoC system, system designers are having less control over the security of the SoC design. As all the IPs mounted in an SoC have access to the bus, a highly representative case is that a Trojan hidden in untrusted IPs can be triggered either by specific data on the bus or by a certain operation. Moreover, a malicious IP can typically affect system-level behaviors through bus communications, control flow, and data transfer. Upon activation, a Trojan may leak sensitive information from the chip, modify functionality to tamper the integrity of the data, or cause a DoS to the system. The security properties are developed by considering possible Trojan attacks and the operations of bus protocol. These properties must resort to validation of the whole system behavior. Accordingly, the correctness of properties indicates the absence of malicious logic. Please note that the proposed method in this chapter does not provide protection against attacks that are not described by the set of security properties.

5.3.2 Model checking

Model checking [28] is an automated method for verifying models in software and hardware applications. The model checkers are developed based on either symbolic approaches or satisfiability solving. Symbolic model checking is one of the initial approaches used for hardware system verification.

UPPAAL [29] is developed based on constraint-solving and on-the-fly mechanism. It provides a tool box for modeling, simulation, and verification of real-time systems. In this model checker, the real-time system is first modeled as a network of timed automata and then the property is checked using forward reachability analysis. Widely used for verification of hardware and protocol designs, UPPAAL helps developers find errors and gain confidence in a design by achieving the goal of a bug-free system. In this chapter, model checker UPPAAL is utilized to verify the SoC against security properties.

5.3.3 Reverse engineering finite state machine

Reverse engineering FSM (REFSM) [30] leverages gate-level netlist to extract a high-level description of the control logic. REFSM performs extraction efficiently through logic registers' simulation. If a more accurate model is required, external parameters can be adjusted to take into account extra temporal logic. The state exploration is performed using a breadth first search over the state space. Once a state is selected for transition exploration, REFSM uses a modified 3-SAT algorithm to find all possible transitions. These transitions are partially compressed using a branch and bound method on the decision tree. An optional FSM decomposition process can be used to break a large FSM into several graph factors. The decomposition process is implemented using a polynomial time heuristic. The heuristic uses the information of the registers to make estimations on the final graph factors. In this chapter, the decomposition function is turned off to allow the generation of FSMs that fully emulate the entire FSM interactions within the SoC design. The abstraction level of our method is down to behavior level of the design, so any malicious behavior should not be missed out. In this chapter, REFSM is utilized to generate a non-simplified FSM of the SoC design from gate-level netlist.

5.3.4 Noninterference and information-flow tracking

Confidentiality of a system is defined as follows. Given that the inner structure of a system is treated as a black box, an adversary learns no extra information of the system executions besides that from the direct observations on system outputs. We define a state of system as S, and then the state S is divided into low and high parts, such as S_H and S_L. The low part states that S_L is in public domain, which can be observed by an adversary, while the high part states that S_H is the one we do not want an adversary to learn. Here, inner structure of the system is treated as a black box.

When the system is executing, initial states $S_{H1} \in S_H$ and $S_{L1} \in S_L$ are its inputs. Consequently, the output states of the system are $S'_{H1} \in S_H$ and $S'_{L1} \in S_L$. When there is a second execution of the system, the low-part states input–output, S_{L1} and S'_{L1} remain the same, whereas the high-part states change value, i.e., $S_{H2} \in S_H$ and $S'_{H2} \in S_H$ are new inputs and outputs of the system, respectively. A description of this procedure is given in Figure 5.2.

In the previous system, since adversaries cannot learn the difference between the $\{S_{H1}, S'_{H1}\}$ and $\{S_{H2}, S'_{H2}\}$, we define the indistinguishable relation under observation of S_L states as $\sim_L$. Further, let $[[s]]$ be behaviors starting from states S, like execution starting from S. Then, the noninterference given by [31] can be represented as follows:

$$S_1 \sim_L S_2 \Rightarrow [[S_1]] \approx_L [[S_2]] \tag{5.1}$$

where $\approx_L$ defines similar indistinguishable relations in behaviors. Equation (5.1) means that starting from two states S_1 and S_2, if the initial states are indistinguishable, then their behaviors will also be indistinguishable. Actually, different confidentiality properties can be captured by properly choosing between $\sim_L$, $[[\bullet]]$, and $\approx_L$. For language-based IFT solutions, $[[\bullet]]$ stands for semantics of design or program, while the two relations $\sim_L$ and $\approx_L$ imply what an attacker can observe.

All the previous language-based IFT methods set up sensitive levels from a partially ordered set and determine the security policy. A commonly followed practice is to design two levels H (high sensitive, private) and L (low sensitive, public), then restrict H-labeled signals flowing to L-labeled signals, while the reverse is allowed. We will propose an IFT-based approach enforcing the noninterference in the SoC bus system.

5.4 Methodology

In this section, we will demonstrate how an SoC is formalized from gate-level netlist and how security properties are specified and mapped to a design. For the reason that the proposed approach follows the process of static formal verification methods and focuses on security verification of a design before the application stage, there is no overhead in area and time in runtime. Workload of the method is only caused

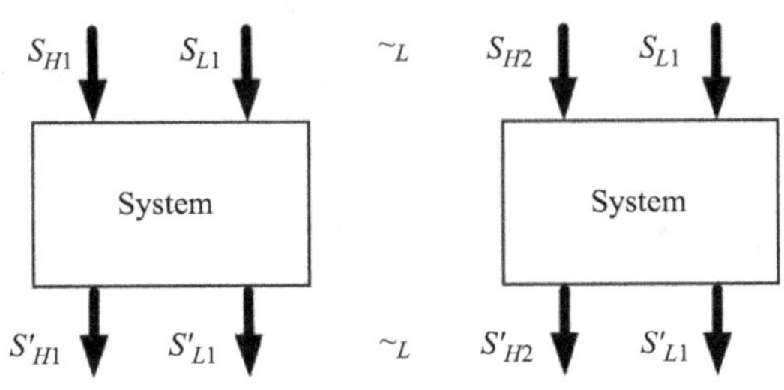

Figure 5.2 Noninterference model

by model formalization and property checking in design stage. The advantages of the proposed framework are as follows: (1) The SoC formalization method is able to extract behavioral models from gate-level netlist in the SoC formalization process, and all behaviors are modeled without any abstractions; (2) Security policies applied in the verification process are proposed based on the analysis of potential system security vulnerabilities and interpretation of bus protocols.

As shown in Figure 5.3, the framework includes three steps: (1) REFSM is used to extract the model of an SoC from gate-level netlist; (2) security properties are developed by specifying the proper states of the FSM to state formulae and path formulae; and (3) the extracted model against security properties is verified in UPPAAL.

5.4.1 SoC formalization

Suppose an SoC design is in the format of gate-level netlist N, and the netlist of each individual IPs of SoC, including I/O interfaces, wires, and gates (G), is denoted as $n \in N$. The bus interconnect module [1] belongs to N and is denoted as n_{inner}. The states of I/O interfaces and wires are defined as *Sig*. To extract FSMs from the netlist, a transformation f is carried out of N with consideration of keeping design details. The extracted FSM is denoted as M, and the FSM of each individual IP is denoted as m. The individual FSMs have connections among each other, then m forms M. In addition, a configuration file is derived from the bus interconnect module defined as *Syn*, which contains the synchronization information among FSMs within M. The total states in the FSMs are denoted as S and individual state is denoted as s. The transformation of netlist to FSMs of each IPs can be given by (5.2) and transformation of the entire SoC is given by (5.3).

$$n \xrightarrow{f} m \tag{5.2}$$

$$N \xrightarrow{f} M \wedge Syn \tag{5.3}$$

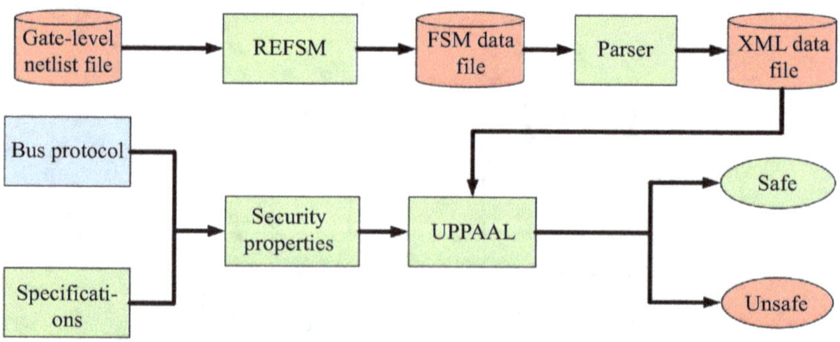

Figure 5.3 The working procedure of the verification process

Equation (5.4) shows that N is made up of n_{inner} and netlist of multiple individual IPs, and (5.5) shows that the FSM of the SoC is composed of many individual FSMs, where $k \in \mathbb{Z}$.

$$N := n_{inner} \wedge n_1 \wedge n_2 \wedge \cdots \wedge n_k \tag{5.4}$$

$$M := m_1 \cup m_2 \cup \cdots \cup m_k \tag{5.5}$$

Further, m is constructed by using states and transfers between states as shown in (5.6)–(5.8), where $l \in \mathbb{Z}$, and *cons* represents the conditions for the transfers to happen. A transition ω is defined by (5.7), where $s_{pre} \in S$ represents the source prestate, and $s_{post} \in S$ represents the target post-state. The corresponding transition condition of ω is shown in (5.8), where $sig \in Sig$ and $j \in \mathbb{Z}$. A transfer is driven by changes of signals in circuit.

$$m := \omega_1 \wedge \omega_2 \wedge \cdots \wedge \omega_l \tag{5.6}$$

$$\omega := s_{pre} \xrightarrow{cons} s_{post} \tag{5.7}$$

$$cons := sig_1 \wedge sig_2 \wedge \cdots \wedge sig_j \tag{5.8}$$

Considering specific gates in the design, behavior B of a specific gate $g \in G$ is described as a collaboration of input value *gin*, type of gate *gtype*, and corresponding output value *gout* and is denoted as (5.9). Compositions of B for G produces corresponding states, which is denoted as (5.10), where $i \in \mathbb{Z}$, *gin* and *gout* are both binary vector, and *gtype* belongs to an enumerable gate library.

$$B := gin \xrightarrow{gtype} gout \tag{5.9}$$

$$B_1 \wedge B_2 \wedge \cdots \wedge B_i \longrightarrow s \tag{5.10}$$

5.4.2 Security specification

The main purpose of a model checker is to verify whether a model satisfies required specifications. The informal specifications, which are written in natural language, should be translated to a formal language prior to verification. For UPPAAL, the properties should be specified in Timed Computation Tree Logic that consists of state and path formulae. State formulae describe individual states, and path formulae quantify over paths or traces of the model [32]. As shown in Figures 5.1 and 5.3, after receiving the required information, the SoC designer and verifier can agree upon a set of security properties that the SoC system and IPs should satisfy.

All masters and slave IPs integrated on bus must obey the bus protocol and follow certain sequence of operations. The composition of states is given by (5.11), where $s_1 \in m_1, s_2 \in m_2, \ldots, s_k \in m_k$ and $k \in \mathbb{Z}$.

$$s := s_1 \wedge s_2 \wedge \cdots \wedge s_k \tag{5.11}$$

Hardware Trojans in untrusted IPs may pass the functional testing and verification and can get activated through interactions among IPs when mounted on the bus. Thus, a malicious state set $S_{bad} \in S$ exists in the formal model accordingly. The components in the state set S_{bad} are states related with the Trojans' trigger and malicious payload.

Then the security specification is designed as whether there exits at least one *Path* that is denoted as (5.12) that makes S_{bad} reachable as shown in (5.13), where $p \in \mathbb{Z}$. The states in the state set S_{bad} should be reached in a particular order following the *Path* to activate the Trojans and then perform certain malicious function.

$$Path := cons_1 \wedge cons_2 \wedge \cdots \wedge cons_p \tag{5.12}$$

$$s \xrightarrow{path} S_{bad} \tag{5.13}$$

5.5 Implementations

In this section, the effectiveness of our framework for SoC system protection is validated using practical cases.

5.5.1 Attack vectors

Typically, an SoC design can suffer from different types of attacks, such as information leakage and DoS. While some of the threats come from one single malicious Trojan-infected IP, other threats arise as a collaboration of several system-level malicious IPs [5]. These attacks are very hard to detect because the Trojans hidden in untrusted IPs all act like non-malicious circuits and follow the normal bus operation. Following are two typical cases for a representative SoC implementation. The first is the information leakage attack performed by one single malicious Trojan-infected IP, and the second is the DoS attack performed by several system-level malicious IPs. In addition, a more general possible threat that may come from either individual or multiple IPs is introduced in this chapter.

5.5.1.1 Information leakage attack

Information leakage attack is one of the most common attacks for SoC designs. A hardware Trojan can be inserted in one of the slave modules. The Trojan is triggered by monitoring normal bus operation, and the trigger condition is a specific set of data transmitted on bus. Once triggered, the Trojan can leak data that belong to other modules. This kind of attack is very hard to detect using normal method because the Trojan's behavior follows the normal bus protocol and leaks information anonymously.

5.5.1.2 Denial-of-service attack

DoS attack can cause system failure at key points. A typical case is that a hardware Trojan can be inserted in one of the master modules. The Trojan's trigger can be hidden in one of the slave modules. Once the Trojan is activated, it will continuously occupy the system bus and block other master modules from using the system bus, thus causing the system's malfunction or DoS. Although there are some strategies that can help preventing this type of DoS attack, such as round-robin arbitration [33] in AMBA bus systems, the Trojan can still cause DoS in masters with lower priority settings [34].

5.5.1.3 Integrity tampering attack

The integrity of the data assets on an SoC is of great importance for the system to function normally. The integrity tampering attacks aim to maliciously change the functionalities or alter the core data processed on an SoC. Typical cases are that Trojans are inserted into either slave or master modules to monitor the control and signal data on the bus and then perform the integrity tampering attacks. The Trojans can be triggered by presetting data sequences or combinations, and the trigger conditions can also be precisely controlled to keep the stealthy nature of the Trojans. With the consideration that this kind of attacks are of various kinds and the functionalities are distinct between different SoCs, the security properties should be carefully developed according to specific circumstances.

5.5.1.4 Malicious modifications in IP wrapper

The popularity of SoC design increases the reuse of individual IPs. However, many of the existing IPs are designed in a stand-alone way or for a specific bus protocol. Therefore, in order to reuse the existing hardware IPs, instead of redesigning the whole IP library, modern SoC integrators adopt a wrapper-based solution. That is, a wrapper will be designed as the interface between the hardware IP and the targeting bus protocol. The development of IP wrapper not only helps increase the reusability of IP cores but also increases the attack surface. Supported by the IP wrapper, attackers do not need to understand the internal logic of individual IPs but can maliciously modify the IP wrapper for hardware Trojan insertions. For example, a large set of the inserted hardware Trojans in the Hack@DAC competition [35] is at the IP wrapper layer.

5.5.2 Modeling process

An SoC design testbed is built upon six IPs, which are two masters, advanced eXtensible interface (AXI) interconnect, and three slaves. As shown in Figure 5.4, one of the slaves, whose name is *Slave Trojan*, is infected with a hardware Trojan and this Trojan will leak secret information after activated. Also one of the masters, whose name is *Master Trojan*, is infected with a hardware Trojan and this Trojan will block other masters from using the bus once triggered. The bus protocol in the SoC design is the AXI that is one of the most widespread AMBA interface [1], and AXI4 bus protocol is the latest version of AXI bus. A bus system with third-party IPs based on

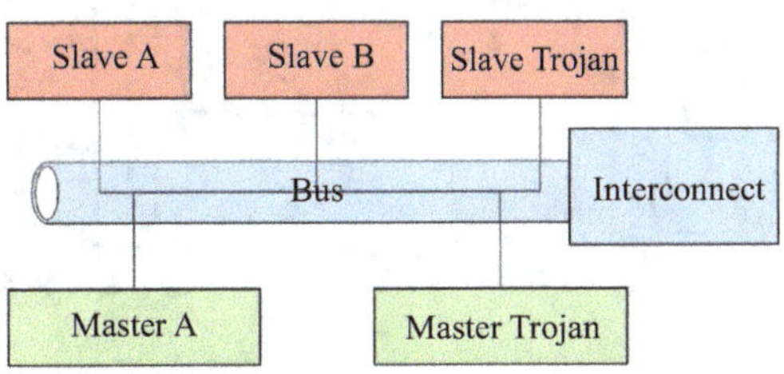

Figure 5.4 Untrusted IPs mounted on bus

the AXI4-Lite interface is established, and the AXI interconnect is configured in the shared access mode [34], as shown in Figure 5.5. In the shared access mode, the write and the read channels are shared and connected to all slaves so that slaves can access all data on the bus channel. Also, addresses provided by the masters are broadcast to all slaves.

The details of modeling process are shown in Figure 5.6. The first step is to get gate-level netlist for the entire SoC design. The system is developed utilizing RTL code according to AXI4-Lite interface specifications. The SoC design is synthesized to gate-level netlist using Design Compiler and the SAED_EDK3228_CORE Digital Standard Cell Library from Synopsys. During the synthesis process, the original hierarchy of the design is retained. The hardware Trojan showed in Figure 5.4 is inserted at RTL and also synthesized into gate-level netlist. The synthesis results and gate counts for these modules are shown in Table 5.1. The Master A interface has 1,926 gates, while the Master Trojan interface has 176 more gates than Master A. The difference in the gate count is caused by the Trojan in the master module, because the function of the two masters is identical except for the address range and Trojan insertion. Slave A interface has 1,542 gates and Slave B interface has 1,513 gates, while Slave Trojan interface has 1,512 gates. The difference in the gate count between Slave B and Slave Trojan is also caused by the Trojan in the slave module.

After the synthesis process, FSMs are extracted from the netlist using REFSM tool [30], the details of this process are shown in Figure 5.6. The data structure of a state transition is composed of three parts: *prestate*, *post-state*, and *conditions*. *The prestate* stands for the source state of a transition, whereas *post-state* stands for the target state. Each state in FSMs is mapped to corresponding states of SoC circuit and *conditions* imply the conditions that trigger transitions between different states. In the extracted FSMs, a state is actually a reflection of one specific circuit status, and the running hardware will lead to state changes in FSMs. Accordingly, circuit's behaviors, such as receiving data or triggering the Trojan, are mapped to one or a sequence of state transitions. For each slave module, there are over 700 states

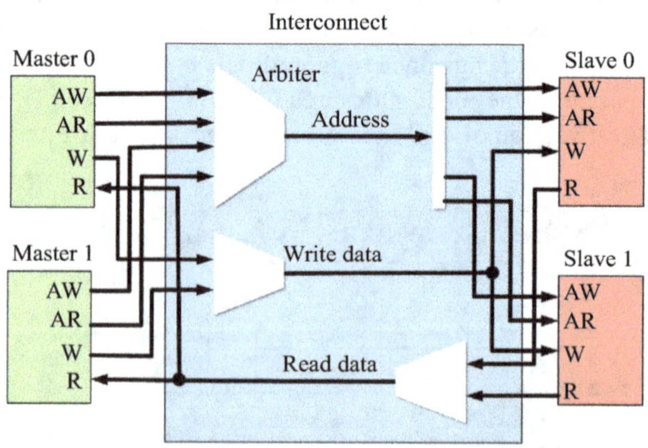

Figure 5.5 Shared access mode of AXI interconnect

and 60,000 transitions. Also there are over 1,000 states and 10,000 transitions for the master modules. The interconnect module is designed to be a combination of connections between masters and slaves. The interconnect module has the arbiter integrated inside in the experiment.

The next stage is to model FSMs as the input of UPPAAL. In general, manually drawing schematic through graphical user interface is used to facilitate modeling in UPPAAL. However, this manual modeling approach suffers from scalability problems when there are too many states and transitions, so a parser is developed to map all FSMs into XML files which UPPAAL can read. Note that a stand-alone command line verifier of UPPAAL *verifyta* [32] is leveraged to improve the data processing capability. UPPAAL can take XML file as input with specific data format as shown in Figure 5.6. In the format, *template* indicates the module, which is one formal model of an IP, and a *transition* in the *template* means an FSM state transition. Furthermore, the *prestate*s are mapped as *source*, the *post-state*s are mapped as *target*, and the *conditions* are mapped as transitions in UPPAAL and tagged with different *label*s. These *label*s help one to distinguish different kinds of transitions, such as guard, synchronization, update, or broadcast [29]. Finally, all the modules are assembled together as a group using the broadcast feature of UPPAAL.

5.5.3 Property development

According to the formal models, security properties are developed in natural language first, then formal specifications are constructed with the knowledge of FSMs of the SoC and AXI4-Lite bus protocol.

Table 5.1 Model parameter counts for the bus system

Modules	Gates	States	Transitions
Master A	1,926	1,244	14,141
Master Trojan	2,102	1,244	14,141
AXI Interconnect	1,359	81	290
Slave A	1,542	768	66,880
Slave B	1,513	768	66,880
Slave Trojan	1,512	768	66,880

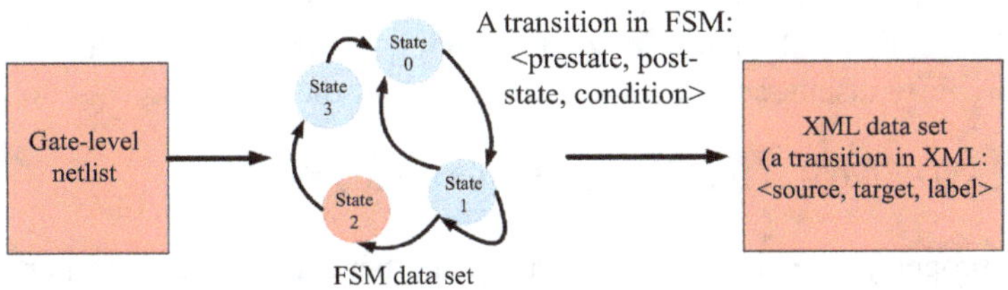

Figure 5.6 Data structure and parsing in formalization process

5.5.3.1 Information leakage Trojan detection

For the information leakage attack, Master A first reads data from Slave A and then writes the data to Slave B, and the data contain sensitive information that should not be leaked to other slaves. The Trojan is triggered by a specific write address, and Slave Trojan steals the data that belong to Slave B and will not give out any response signal. Properties are developed to prevent information leakage attack. In natural language, the property is stated as *"Information written to Slave B should never be leaked to other IPs."* During the property development and verification process, the first verification should be carried out to check whether there is any risk in the design, and the second step is supposed to be locating where the risk comes from. If there are more than one Trojan-infected slave modules in the system, the source of the risk can still be located by printing out the search paths, denoted as (5.12), by UPPAAL, because the Trojans are not identical on either of triggering conditions or of payload functions.

Slave B first recognizes that the address in the address channel belongs to its own address range, then Slave B begins to receive data from the data channel. During this operation, Slave B reaches a data-receiving-related state in its FSM. At the same time, Slave Trojan also recognizes the address in the address channel and the Trojan in Slave Trojan gets activated, then the Trojan steals the data from the bus that belongs to Slave B. Because Slave Trojan can pass the normal functional checking, then it should follow the normal bus protocol, thus Slave Trojan also reaches a similar data-receiving state in its FSM. However, in order to prevent information leakage, the previous two states cannot be reached at the same time. More specifically, only one slave is permitted to receive data from the data channel when one write address is sent to the address channel. Therefore, the previous security property can be constructed as (5.14). In this expression, state formula *slaveX_selected* represents the state of slave X that is ready to receive data from the data channel. The path formula $E <> \phi$ represents that there at least exits a path that satisfies ϕ.

$$E <> (slaveT\_selected \text{ } and \text{ } slaveB\_selected) \tag{5.14}$$

In the SoC model of UPPAAL, states are labeled with the name of modules and binary values. More specifically, before the binary values, *Par*, *Pm*, *Pmt*, *Pa*, *Pb*, and *Pt* represent Arbiter, Master A, Master Trojan, Slave A, Slave B, and Slave Trojan modules in UPPAAL, respectively. Further, in the established model, the state relevant to data receiving is numbered as 00111100000. As all the modules in the system follow the same AXI4-Lite interface, the same binary value *id*00111100000 is utilized to represent the ready state of receiving data for all three slaves. Then the security property discussed in (5.14) is elaborated as the following equation:

$$E <> (Pt.id00111100000 \text{ and } Pb.id00111100000) \tag{5.15}$$

If the property given in (5.14) is violated, another set of security properties is utilized to identify the source of attacks. From the description of attacks and the stealthy nature of Trojans, Slave B provides response after receiving data from Master A while Slave Trojan does not. This response given by a slave X is mapped to a state, *slaveX_response*, of the FSM. When the state *slaveX_selected* is reached, the behavior of response is equivalent to the judgment *"the state slaveX_response is reachable from slaveX_selected."* Hence, the above security property is formally stated as the following equation:

$$\begin{aligned} &slaveT\_selected \rightarrow slaveT\_response \\ &slaveB\_selected \rightarrow slaveB\_response \end{aligned} \tag{5.16}$$

In this expression, the path formula $\phi \rightarrow \psi$ represents that when the state formula ϕ is true, then the state formula ψ will also be true. Also, the states *slaveX_selected* and *slaveX_response* are numbered as *id*00111100000 and *id*00011100001, respectively. The corresponding security properties discussed in (5.16) are elaborated as (5.17).

$$\begin{aligned} &Pt.id00111100000 \rightarrow Pt.id00011100001 \\ &Pb.id00111100000 \rightarrow Pb.id00011100001 \end{aligned} \tag{5.17}$$

5.5.3.2 Denial-of-service attack detection

In the normal operations of an SoC system, the master modules are granted permission by arbiter only when the masters need to access the bus. During a DoS attack, Master Trojan keeps requesting the permission from Arbiter, and blocks other masters from using the bus for an indefinite period of time. However, there are possibilities that the normal operated master uses the bus for a certain period of time, which could cause a high false positive rate if only one master's behavior is considered. To address this issue, all master modules in the system are taken into account in developing the system-level security properties. In natural language, the property is that *"The bus permission should never be given to only one specific master continuously while other masters are requesting permission."* During the property development and SoC verification, the first step is to check whether there exists a master that keeps requesting permission from the arbiter, and the next step is to check whether other masters require the permission simultaneously.

As a normal operation, a master first requests permission from the arbiter and then the arbiter decides whether to give the master access to the bus. During functional testing, the Trojan is not activated, and Master Trojan only requests permission when needed according to the normal function. However, when the Trojan in Master Trojan module is activated, Master Trojan module keeps requesting permission from Arbiter. If Master Trojan has a high priority settings, Arbiter will not permit the requests from other master modules, like Master A, thus causing a DoS or malfunction of the entire SoC. In order to prevent such attack, security properties needs to be developed to check whether there are infinite loops that involve between the request-permission pair of Master Trojan module and Arbiter. The above security property is constructed as (5.18). In this expression, the state formula *masterX_request* represents the state that master X is requesting permission from Arbiter, and *masterX_permission* represents that Arbiter gives the permission to master X. The path formula $E[]\phi$ represents that there is at least an execution path in which ϕ holds for all states in the path. The *deadlock* indicates that whether there is any infinite loop in the model.

$$\begin{aligned}&E[]\ deadlock\ imply\\&(masterX\_request\ and\ masterX\_grant)\end{aligned} \tag{5.18}$$

Restricted by the expression ability of the query language [29] of UPPAAL, the property demonstrated in (5.18) is decomposed into several specific formal properties. In the established model, the state when Arbiter gives out permission signal is denoted as $idar11010010$. The permission signal-receiving states in Master A and Master Trojan are denoted as $Pm.idm10010000000$ and $Pmt.idmt10010000000$, respectively. The decomposed security properties are given in (5.19). The reachability of each permission signal-receiving state is checked first, then the loop between permission request and response is checked to determine the trustworthiness of the modules.

$$\begin{aligned}&Par.idar11010010 \rightarrow Pm.idm10010000000\\&Pm.idm10010000000 \rightarrow Pm.idm10010000000\\&Par.idar11010010 \rightarrow Pmt.idmt10010000000\\&Pmt.idmt10010000000 \rightarrow Pmt.idmt10010000000\end{aligned} \tag{5.19}$$

If the property given in (5.18) is violated, it means that there exists a situation where one master keeps occupying Arbiter. The next step is to check whether there are any other masters trying to request the permission at the same time. The supplementary property is demonstrated in (5.20). In this expression, the state formula *master_request* represents the state that a master is requesting the permission from Arbiter, and $A <> \phi$ represents that in the whole system, the state ϕ is reachable. After verifying this property in UPPAAL, traces for state transitions are printed out for checking according to the bus protocol, and the relevant states can also be verified.

$$A <> master\_request \tag{5.20}$$

5.5.3.3 Integrity tampering Trojan detection

In this chapter, the verification focus is on the Trojans that leverages the normal bus communications to perform malicious functionalities. Although the integrity tampering Trojans are not embedded into the SoC testbed, the potential security properties utilized for Trojan detection are discussed. The basic verification principles for detecting information leakage and DoS Trojans are validating the states based on the bus protocols. However, for integrity tampering Trojans, the verification principles will be checking additional states that correspond to malicious functionalities. With the consideration that the Trojans that can change the functionalities are digital Trojans, in natural language, the property is that *"When vital control signals are broadcast to the bus, there should not exist any additional operations other than the slaves or masters that should operate correspondingly."* The UPPAAL can print all the states, so if there are any additional states, manual inspection is required. For detecting the Trojans targeting the altering of data transmitted on the bus, the security properties should focus on additional states except for the data-processing-related slaves or masters, which can be located through analyzing the control signals on the bus. In natural language, the property is that *"For any additional logic operation that happens while the data is transmitted other than the normal operations, the additional logic operations should not correlate with the data on bus."*

5.6 Experimental results

After a design is formalized and properties are proposed, the model checker verifier of UPPAAL *verifyta* [32] is utilized for verification on a server. The server is equipped with 4 CPU cores and 8 GB memory, using Linux Red Hat distribution as the operating system. Totally, five properties are designed for information leakage Trojan detection, four properties are designed for DoS attack detection. The results are shown in Table 5.2. Note that in Table 5.2, Slave T represents Slave Trojan and Master T represents Master Trojan.

5.6.1 Information leakage Trojan detection results

Three properties are designed for identifying the Trojan's existence and two properties are designed for Trojan localization. Property details and experimental results are given in Table 5.2. For Trojan detection properties, satisfaction of property in the first line of Table 5.2 means that the data-receiving state can be reached at the same time for Slave Trojan and Slave B. In the second and third lines, properties are not satisfied, which implies that the data-receiving state cannot be reached at the same time for Slave A and Slave B, and for Slave A and Slave Trojan. These results show that (1) an attack occurred, and (2) either Slave Trojan or Slave B is the malicious IP. To validate the satisfaction of these properties, the system states given in Table 5.3 are checked. During the information leakage attack, Slave Trojan's state is the same with Slave B, meaning that both Slave Trojan and Slave B are in the data-receiving state, while Slave A's state proves that Slave A is indeed not in the data-receiving state.

Next, two Trojan localization properties are checked after Slave B and Slave Trojan reach the data-receiving state at the same time. The satisfaction result implies that the response state is reached in Slave B, while the dissatisfaction result demonstrates that the response state is unreachable in Slave Trojan. So the conclusion is that the attack is caused by Slave Trojan because of Slave Trojan's malicious behavior. Our method can also print out the states of signals under certain conditions, so to validate the results of Trojan localization, the relevant signals when the attack occurred are checked. *s00_axi_awvalid* indicates that Master A is signaling a valid write address and control information. *slave_axi_wvalid* indicates that valid write data are available. *s00_axi_awaddr* represents the write address. Additionally, the write address allocated to Slave B in bus system is 0111. By analyzing the value of those signals

Table 5.2 Security property checking results

Attacks	Properties	Formal property	Results	Time (s)
Information leakage	Concurrent data read by Slave T and B	E<> (Pt.id00111100000 and Pb.id00111100000)	Satisfied	97.82
	Concurrent data read by Slave A and B	E<> (Pa.id00111100000 and Pb.id00111100000)	Not satisfied	100.04
	Concurrent data read by Slave T and A	E<> (Pa.id00111100000 and Pt.id00111100000)	Not satisfied	98.42
	Slave T's response exists after data receiving	Pt.id00111100000 –> Pt.id00011100001	Not satisfied	99.43
	Slave B's response exists after data receiving	Pb.id00111100000 –> Pb.id00011100001	Satisfied	101.69
Denial-of-service	Loop between Master A request and Arbiter	Par.idar11010010 –> Pm.idm10010000000	Satisfied	88.70
	permission	Pm.idm10010000000–> Pm.idm10010000000	Non-loop	N/A*
	Loop between Master T request and Arbiter	Par.idar11010010 –> Pmt.idmt10010000000	Satisfied	87.07
	permission	Pmt.idmt10010000000–> Pmt.idmt10010000000	Loop	N/A*
	Master A permission state reachability	A<> Pm.idm10010000000[†]	Not satisfied	N/A*

*Checking time not applicable.
[†] Under the condition where the DoS attack happens.

Table 5.3 System states

Attacks	Modules	States ID
Information leakage	Slave A	Pa.id00000000000
	Slave B	Pb.id00111100000
	Slave Trojan	Pt.id00111100000
Denial-of-service (DoS)	Arbiter	Par.idar11010010
	Master A	Pm.idm10010000000
	Master Trojan	Pmt.idmt10010000000

in (5.21), it is confirmed that the data were supposed to be written to Slave B legally when the attack occurred. In other words, information of Slave B is leaked under attack.

$$\begin{aligned} &s00\_axi\_awvalid = 1 \\ &slave\_axi\_wvalid = 1 \\ &s00\_axi\_awaddr = 0111 \end{aligned} \tag{5.21}$$

5.6.2 Denial-of-service attack detection

Four properties are proposed for DoS attack detection, and one manual check is performed. Property details and experimental results are given in Table 5.2. Two sets of properties are designed to check whether there exists an infinite loop that Master A or Master Trojan keeps occupying the bus. Within each set of the properties, first we check the normal operation that the arbiter gives permission to either masters, then we check the loop between the master request and arbiter permission. In the experiments, all paths that involve the master request and arbiter permission states are checked. From the results, it is possible that Master Trojan constantly requests permission from Arbiter, and it is confirmed that the Trojan's trigger is involved in this situation after printing out the searching path. These results show that (1) a DoS attack happened, and (2) the normal operation of Master A module is not influenced when Master Trojan is not triggered. The system states when the DoS attack happens are shown in Table 5.3. The Arbiter is in the state granting the permissions to master modules, while both Master A and Master Trojan are in the state waiting for permission.

Then we apply the same conditions in our model and check whether Master A can get the permission from Arbiter when the DoS attack happens. To validate the results of DoS attack, the relevant signals when the attack happens are checked. More specifically, *txn_done*1 and *txn_done*2 indicate that Master A and Master Trojan are permitted, respectively. *m_axi_awvalid*1 and *m_axi_awvalid*2 represent the permission request in Master A and Master Trojan, respectively. The states are demonstrated in (5.22). By analyzing the value of those signals, there exists conflict between Master A and Master Trojan, and the permission is only issued to Master Trojan. More specifically, during normal bus operations, if the DoS attack happens, Master A will never enter a bus permission state during the attack. A DoS Trojan is detected in the SoC system.

$$\begin{aligned} &txn\_done1 = 0 \\ &txn\_done2 = 1 \\ &m_a xi_a wvalid1 = 1 \\ &m_a xi_a wvalid2 = 1 \end{aligned} \tag{5.22}$$

5.7 Information-flow tracking-based detection

In this section, we discuss the security analysis of the SoC bus system using IFT. In the IFT, sensitive data flow is explored through tracking information flow from sensitive/taint sources to sensitive/taint targets. Information flow policies are defined to configure the taint source, taint target, and tracking rules for taint propagation. In the discussion, we also use the attack vectors and properties developed in Section 5.5.1—information leakage, DoS, and integrity tampering—and then analyze them through setting up the information flow policies. Through detecting those attack vectors, confidentiality, integrity, and availability are enforced in the bus system.

5.7.1 Information leakage analysis

For detecting the information leakage attack, the entire SoC bus is formalized as a data flow graph (DFG). Each variable or signal in the design is defined as the node in the DFG. The sensitive information from the source IP Slave A is designed as the taint source. The input signals of all the other IPs in the bus system are defined as the taint targets. The label is propagated with the sensitive signal. Except Slave B, if any taint target receives the label, we treat it as a violation of the security property.

5.7.2 Denial-of-service attack analysis

In detecting the DoS attack, the control-flow graph is generated as the formal model for the design. We define the module Slave A who requests access to the Arbiter as both the taint source and taint target. Then a timer setup and reduced depending on each propagation/clock cycle. If the timer overflows and the label is still not propagated back to the taint target, the DoS warning would be raised.

5.7.3 Integrity tampering attack analysis

In this attack, we assume that the Trojan Slave writes data to the victim Slave B by simulating signals of the Master Slave A. Again, the bus communication is formalized as the DFG for detecting the integrity tampering attack. The slave who starts the writing request is defined as the taint source, while other slaves are taint targets. Labels are propagated with the signal. For this case, we apply the multi-label to detect the integrity tampering. In the bus system, we give a specific type of label to each slave and its accessible slaves. In other words, labels are given depending on the slaves' privilege. Once the label conflicts are detected, it stands for the threats of integrity violation.

5.8 Conclusions

In this chapter, a formal verification approach is proposed to protect the entire SoC system from threats of untrusted third-party IPs. FSMs are first extracted from gate-level netlist of an SoC, then security properties are constructed by analyzing possible attacks from Trojans in IPs and bus protocols. Finally, a model checker is used to verify whether these properties are satisfied. For demonstrating the effectiveness of our method, we consider an SoC with an AMBA bus and Trojans in some of the slave and master modules. Results show that our framework can detect and localize information leakage attack and DoS attack launched by hardware Trojans hidden in IPs of an SoC.

5.9 Discussions and future research directions

The work presented in this chapter focuses on the verification of the SoC system before the implementation stage and no hardware modification of the original circuit is needed. Compared with the state-of-the-art formal verification-based techniques, our method performs better in several aspects: scalability, flexibility, and physical overhead. Even though the proposed approach is scalable for large SoC designs and the scalability is guaranteed by the model methods and verification strategies, there are still requirements to improve the verification speed. In future work, we may explore proper algorithms that are able to distinguish and remove normal states before the verification process, thus alleviating potential state explosion issue and speeding up the verification process.

As the interactions between different IPs at system level are considered and all vulnerabilities along with hardware designs are mapped to states in our model, this chapter provides a flexible solution to address a wide range of threats in design stage. When facing other types of potential hardware Trojan threats, the behaviors of Trojans' trigger conditions and payload functions will be analyzed first according to the bus protocols. Security properties will then be developed to specify the Trojans' malicious logic utilizing the states in the formalized model. Please note that in future formalization process, except for sequential logic status, potential combinatorial logic states within the circuit under evaluation should also be considered. Finally, the security properties are verified for Trojan detection.

At the meantime, there is still room for further improvements in our framework. In detail, the success in detecting the security vulnerabilities relies on several aspects. First is the integrity of the security properties. For a security threat, there exist several aspects that can be utilized for property development. Theoretically, any security properties that have been verified should be comprehensive enough for judging whether there are any threats. However, for threats that are not included in the properties, the method may not find the vulnerabilities. Second is the checking engine and verification strategy, the UPPAAL verification tool resorts to a width first strategy, so every time when the property satisfies, the tool returns a path for the results. The success of Trojan detection is based on the verification of well-defined security properties. In this chapter, we have covered almost every aspects of the mentioned vulnerabilities. We also set the tool to search and verify every possible paths until all paths that satisfy the property are returned. Third is that some of the properties cannot be fully expressed using query language, we should exploit other formal language in the future for better expression ability to cover all possible aspects of potential attacks.

Although the research focus in this chapter is the detection of malicious Trojans, there are other system-level threats that are not induced by Trojans, including design flaws and system bugs. Thus, a library of potential threats needs to be developed and then a security property library will also be developed to cover different SoC peripherals. Currently, the verification process requires a lot of manual effort, and for future work, we can develop an automatic way of modeling and verification.

References

[1] AMBA Specifications. (accessed 21 July 2020). https://www.arm.com/products/system-ip/amba-specifications.

[2] Mosensoson G. Practical approaches to SoC verification. In: Proceedings of DATE User Forum. CiteSeer; 2002. p. 05–08.

[3] Salmani H and Tehranipoor M. Analyzing circuit vulnerability to hardware Trojan insertion at the behavioral level. In: Defect and Fault Tolerance in VLSI and Nanotechnology Systems (DFT), 2013 IEEE International Symposium on. New York, NY: IEEE; 2013. p. 190–195.

[4] Mead J, Bobda C, and Whitaker TJ. Defeating drone jamming with hardware sandboxing. In: Hardware-Oriented Security and Trust (AsianHOST), IEEE Asian. Yilan: IEEE; 2016. p. 1–6.

[5] Basak A, Bhunia S, Tkacik T, *et al.* Security assurance for system-on-chip designs with untrusted IPs. IEEE Transactions on Information Forensics and Security. 2017;12(7):1515–1528.

[6] Oberg J, Hu W, Irturk A, *et al.* Information flow isolation in I2C and USB. In: Proceedings of the 48th Design Automation Conference. New York, NY: ACM; 2011. p. 254–259.

[7] Guo X, Dutta RG, He J, *et al.* PCH framework for IP runtime security verification. In: Asian Hardware Oriented Security and Trust (AsianHOST). Beijing: IEEE; 2017. p. 79–84.

[8] Oliveira MT and Hu AJ. High-level specification and automatic generation of IP interface monitors. In: Proceedings of the 39th Annual Design Automation Conference. New York, NY: ACM; 2002. p. 129–134.

[9] Lin H, Yen C, Shih C, *et al.* On compliance test of on-chip bus for SOC. In: Proceedings of the 2004 Asia and South Pacific Design Automation Conference. Yohohama: IEEE Press; 2004. p. 328–333.

[10] Coburn J, Ravi S, Raghunathan A, *et al.* SECA: security-enhanced communication architecture. In: Proceedings of the 2005 International Conference on Compilers, Architectures and Synthesis for Embedded Systems. New York, NY: ACM; 2005. p. 78–89.

[11] Kim LW, Villasenor JD, and Koc CK. A Trojan-resistant system-on-chip bus architecture. In: Military Communications Conference, 2009. Boston, MA: IEEE; 2009. p. 1–6.

[12] Shen H and Zhao Y. HTChecker: Detecting hardware Trojans based on static characteristics. In: 2017 IEEE International Symposium on Circuits and Systems (ISCAS). Baltimore, MD: IEEE; 2017. p. 1–4.

[13] Hasegawa K, Yanagisawa M, and Togawa N. Trojan-feature extraction at gate-level netlists and its application to hardware-Trojan detection using random forest classifier. In: 2017 IEEE International Symposium on Circuits and Systems (ISCAS). Baltimore, MD: IEEE; 2017. p. 1–4.

[14] Salmani H. COTD: reference-free hardware Trojan detection and recovery based on controllability and observability in gate-level netlist. IEEE Transactions on Information Forensics and Security. 2017;12(2):338–350.

[15] Chen X, Liu Q, Yao S, *et al.* Hardware Trojan detection in third-party digital intellectual property cores by multilevel feature analysis. IEEE Transactions on Computer-Aided Design of Integrated Circuits and Systems. 2018;37(7): 1370–1383.

[16] Waksman A, Sethumadhavan S, and Eum J. Practical, lightweight secure inclusion of third-party intellectual property. IEEE Design & Test. 2013;30(2):8–16.

[17] Portillo J, John E, and Narasimhan S. Building trust in 3PIP using asset-based security property verification. In: VLSI Test Symposium (VTS), 2016 IEEE 34th. Las Vegas, NV: IEEE; 2016. p. 1–6.

[18] Rajendran J, Dhandayuthapany AM, Vedula V, *et al.* Formal security verification of third party intellectual property cores for information leakage. In: VLSI Design and 2016 15th International Conference on Embedded Systems (VLSID), 2016 29th International Conference on. Kolkata: IEEE; 2016. p. 547–552.

[19] David A and Yi W. Modelling and analysis of a commercial field bus protocol. In: Real-Time Systems, 2000. Euromicro RTS 2000. 12th Euromicro Conference on. Stockholm: IEEE; 2000. p. 165–172.

[20] Goel A and Lee WR. Formal verification of an IBM CoreConnect processor local bus arbiter core. In: Proceedings of the 37th Annual Design Automation Conference. New York, NY: ACM; 2000. p. 196–200.

[21] Roychoudhury A, Mitra T, and Karri S. Using formal techniques to debug the AMBA system-on-chip bus protocol. In: Proceedings of the Conference on Design, Automation and Test in Europe (DATE). Munich: IEEE; 2003. p. 828–833.

[22] Guo X, Dutta RG, Mishra P, *et al.* Scalable SoC trust verification using integrated theorem proving and model checking. In: Hardware Oriented Security and Trust (HOST), 2016 IEEE International Symposium on. McLean, VA: IEEE; 2016. p. 124–129.

[23] Myers AC and Liskov B. A decentralized model for information flow control. In: Proceedings of ACM Symposium on Operating Systems Principles (SOSP). Saint-Malo: ACM; 1997. p. 129–142.

[24] Zhang D, Askarov A, and Myers AC. Language-based control and mitigation of timing channels. ACM SIGPLAN Notices. 2012;47(6):99–110.

[25] Zhang D, Wang Y, Suh GE, *et al.* A hardware design language for timing-sensitive information-flow security. ACM SIGPLAN Notices. 2015;50(4): 503–516.

[26] Guo X, Dutta RG, He J, *et al.* QIF-Verilog: quantitative information-flow based hardware description languages for pre-silicon security assessment. In: 2019 IEEE International Symposium on Hardware Oriented Security and Trust (HOST). McLean, VA: IEEE; 2019. p. 91–100.

[27] Torrance R and James D. The State-of-the-Art in IC Reverse Engineering. In: CHES. vol. 5747. Berlin, Heidelberg: Springer; 2009. p. 363–381.

[28] Clarke EM, Grumberg O, and Peled D. Model Checking. Cambridge, MA: MIT Press; 1999.

[29] UPPAAL. (accessed 21 July 2020). http://www.uppaal.org/.

[30] Meade T, Zhang S, and Jin Y. Netlist reverse engineering for high-level functionality reconstruction. In: Design Automation Conference (ASP-DAC), 2016 21st Asia and South Pacific. Macau: IEEE; 2016. p. 655–660.

[31] Smith G. On the foundations of quantitative information flow. In: International Conference on Foundations of Software Science and Computational Structures. Berlin, Heidelberg: Springer; 2009. p. 288–302.

[32] A Tutorial on UPPAAL 4.0; 2006. (accessed 21 July 2020). http://www.it.uu.se/research/group/darts/papers/texts/new-tutorial.pdf.

[33] Shin ES, Mooney III VJ, and Riley GF. Round-robin arbiter design and generation. In: Proceedings of the 15th International Symposium on System Synthesis. Kyoto: ACM; 2002. p. 243–248.

[34] AXI Interconnect. (accessed 21 July 2020). https://www.xilinx.com/support/documentation/ip_documentation/axi_interconnect/v2_1/pg059-axi-interconnect.pdf.

[35] Hack@DAC; 2020. Https://hackat.events/dac20/.

Chapter 6
Silicon-based true random number generators

Yuan Cao[1], Egbochukwu Chukwuemeka Chidiebere[1], Chenkai Fang[1], Mingrui Zhou[1], Wanyi Liu[1], Xiaojin Zhao[2], and Chip-Hong Chang[3]

True random number generator (TRNG) is an essential hardware security primitive that has been used in a variety of security, hazard games, gambling, computer simulation, statistical testing and industrial labeling applications. In cryptology, they can be used to produce a digital key stream for encrypting or signing messages, generate nonces for authentication in secure communication or for chip manufacturing and as seed for generation of device-specific identities. Monte Carlo simulations of physical phenomena and neural networks also employ TRNGs to generate many initial parameters. Due to the economy of scale, standard CMOS-compatible or solid-state TRNGs with full entropy, high throughput and small footprint are attractive for integration into security solutions of commercial products. This chapter presents a few modern solid-state TRNG circuits. The design and their performances are detailed. Besides, on-chip post-processing techniques to improve the randomness and security of TRNGs are reviewed and discussed. Furthermore, we also illustrate the randomness evaluation methods in the literature.

6.1 Introduction

Randomness sits at the heart of various applications such as cryptography, Monte Carlo simulations, lottery, statistical research and mathematical modeling. Although it is not easy to generate a genuine random number, the search for the true randomness has never been stopped in the human history. Since ancient times, people toss a coin or throw a dice for a bet. There is a 50-50 chance of winning or losing, and it is impossible to predict the result accurately. However, these slow methods cannot meet the requirements of modern electronic systems. New methods based on optical chaos [1], thermal noise [2], telegraph noise [3] and computational algorithms [4] have been developed to

[1]College of Internet of Things Engineering, Hohai University, Changzhou, China
[2]College of Electronics and Information Engineering, Shenzhen University, Shenzhen, China
[3]School of Electrical and Electronic Engineering, Nanyang Technological University, Singapore, Singapore

generate random and uniformly distributed numbers at extremely fast speed. Besides, various magical random number generation technologies are continuously reported in the news. Last year, American web-infrastructure and website-security titan company Cloudflare started a project "League of Entropy" to provide a verifiable, decentralized randomness beacon [5]. Cloudflare took a video of a wall of lava lamps. These unpredictable visual data of floating blogs were converted to true random numbers [6]. More interestingly, it is reported that truly random numbers can even be created from growing crystals [7].

Generally, the device in an electronic system that is responsible for harvesting the randomness is named random number generator (RNG). It can generate random and patternless numbers with no predictability. There are two different kinds of RNG: pseudo RNG (PRNG) and TRNG. A PRNG is essentially a mathematical algorithm or formula whose output is completely determined by its initial state, more often known as a "seed," and has a finite periodicity bounded by its number of states. Albeit statistically sound and easily implemented in logics, PRNG is vulnerable and can be easily predicted once the seed is known [8,9]. The famous example for the attack on the PRNG is the Hot Lotto fraud scandal [10]. An Iowa man orchestrated one of the most complex heists of a public lottery in US history by tampering with a randomness beacon's algorithm that generated random numbers, allowing him to determine the winning numbers at precise time in the future. As opposed to PRNG, a TRNG is a hardware primitive that can generate independent and identically distributed (i.i.d.) random numbers from a fundamentally nondeterministic physical process. TRNG is nonperiodic and secure for the most demanding white box cryptography because its output is unpredictable even when all the design information such as algorithms, parameters, schematics and operations are known to the adversaries [11]. TRNGs are now widely used in not only cryptography, but also deep neural network modeling, Markov Chain Monte Carlo analysis, testing, and even gambling and gaming, where a number of deterministic PRNGs are found to exhibit artifacts that make them less reliable than expected to produce adequate results. The design of the TRNG remains a hot topic in the integrated circuit (IC) design community. Driven by the IoT and 5G industry, the implementation of safe, low-cost and energy-efficient TRNG on edge devices is fundamentally important to secure the communication channels for data transfer.

This chapter provides an overview of the design and research problems of TRNGs. The remaining sections of this chapter are organized as follows. The pros and cons of PRNGs are discussed as they are still used by most computer systems to generate random numbers. Next, a comprehensive review of the state-of-the-art silicon-based TRNGs is provided. These TRNGs are classified into four categories based on the different entropy sources, elaborated with examples in each category. In order to improve the randomness of the final output bitstreams, post-processing is often employed. In Section 6.4, five post-processing techniques are presented with design details. They are used to remove the bias of random number distribution, decorrelate the successive bits and increase the throughput. The randomness test is an important topic and hard question in the TRNG community. Although there are some standard test suites to evaluate the randomness of TRNG, there is still uncertainty and controversy on the randomness even if a TRNG passes all these tests. The mainstream methods to evaluate the randomness will be presented in Section 6.5. Finally, conclusion is drawn in Section 6.6.

6.2 Pseudo random number generators

Essentially, PRNG is an algorithm or a mathematical formula that can be used to produce a sequence of random numbers from an initial state called a seed. A "good" PRNG should:

- be unpredictable (without knowing the seed);
- have a very long period;
- have good statistical properties; and
- be efficient in generating the random numbers.

There are two types of PRNGs: linear and nonlinear. Linear PRNGs are low cost in hardware implementation with a high throughput. However, the major drawback of these generators is that they are easy to be predicted due to their linear structure. Examples of linear PRNGs are linear feedback shift registers (LFSRs) [12,13], linear congruential generators (LCGs) [14,15] and multiple recursive generators [16,17]. While the nonlinear PRNGs in [4,18–20] can overcome the predictability problem, they are very expensive in hardware implementation and have lower throughput. Some of them have limited states and period [21].

6.2.1 Linear congruential generator PRNG

LCG is perhaps the most commonly used algorithm for PRNG [22]. It generates a stream of pseudo randomized numbers with a discontinuous piecewise linear equation. It is fast and easy to be implemented in computer hardware that supports modulo arithmetic by storage-bit truncation. The generator is defined by the following recurrence relation:

$$X_{i+1} = (aX_i + b) \bmod m \tag{6.1}$$

where X_i, a, b and m are the ith output pseudo random value, multiplier, increment and modulus, respectively.

Usually, the seed for (6.1) is denoted as X_0, which is an integer constant smaller than m. The algorithm begins with X_0. The next random number X_{i+1} is then generated from the current X_i with the combination of constant parameters, a, b and m. If $b = 0$, then m must be a prime or power-of-prime number to be used as a PRNG. LCG with $a = 0$ is known as multiplicative congruential generator. The pseudo randomness property of LCG is due to the piecewise linearity of modulo arithmetic. As the modulus and the multiplier determine the spacing between the hyperplanes where the points chosen by the LCG are lying, the quality of the pseudo randomness is highly sensitive to the parameters m and a. With a large m, the choice of a becomes less critical. The problem with the poor choices for a is the correlation of successive numbers in the generated sequence and the lack of uniformity of the distribution for large quantities. An example of the poor choice of LCG parameters is the ill-conceived PRNG called RANDU [23]. It is developed by IBM and used since the 1960s. The correlation in the random bitstream generated by RANDU caused many research results that used RANDU as the PRNG at that time to become very suspicious.

Nevertheless, LCGs are fast and require minimal amount of memory to implement, making them attractive for memory constrained non-cryptographic applications. With appropriate choice of parameters, the period of LCG is sufficiently long. For example, a well-designed 64-bit LCG can produce over two millions of good random samples. With larger modulus, the number of states produced is sufficiently large to pass even the stringent statistical tests. By keeping only the high-order bits of the states, LCGs can offer high-quality randomness. Even with a very long period, LCGs and many other known PRNGs are still not secure enough for cryptographic applications.

6.2.2 Cryptographically secure PRNG

PRNGs that are suitable for the cryptographic applications are called cryptographically secure PRNGs (CSPRNGs). The tests (evaluations) for CSPRNGs are stricter than other PRNGs. First, CSPRNGs should pass the next-bit test. Passing the next-bit test means the attacker who knows the first k bits of a random sequence cannot precisely predict the $(i + 1)$th bit within a polynomial time. It has been proven that the RNG passing the next-bit test can pass all other polynomial-time statistical tests for randomness [24]. Second, CSPRNG is more resilient to the "state compromise extensions" attack. This means that the attacker cannot reconstruct all the random bitstreams if some states have been revealed. This is because CSPRNG is usually implemented with many-to-one function to make the guess of the reverse mapping very hard. CSPNGs can be designed with cryptographic primitives or some hard mathematical problems. Some CSPRNGs even introduce additional entropy to prevent attacks when the initial state is compromised. A few official standards have been published to recommend the design of CSPRNGs, such as Federal Information Processing Standard (FIPS) 186-4, National Institute of Standard and Technology (NIST) SP 800-90A and American National Standard Institute X9.17-1985 Appendix C. However, it is reported that American National Security Agency has embedded a kleptographic backdoor in fourth generator Dual_EC_DRBG included in NIST SP 800-90A in 2007. A revised version of NIST SP 800-90A that removes Dual_EC_DRBG was published in June 2015 [25]. Most recently, it was reported in [26] that image compression algorithm such as WEBP [27] can produce highly entropic output to fool FIPS 140-2 test. One example is the image of Albert Einstein on a Long Island beach in 1939, as shown in Figure 6.1. It passes all FIPS 140-2 randomness tests with zero failures. If such glaringly obvious biases are not detected by the statistical tests of randomness, the risk is a PRNG that passes the test can be backdoored by making use of WEBP images to leak key information recoverable with the knowledge of the offset at which they occur in the output of the PRNG.

Figure 6.1 Albert Einstein on a Long Island beach in 1939

CSPRNG cannot be used to generate a true random bitstream because its output is completely determinable by the seed. It was well said of John von Neumann that "*Anyone who considers arithmetical methods of producing random digits is, of course, in a state of sin.*" [28]. If the random numbers can be calculated exactly by a deterministic Turing machine, the cryptographic strength of the protocol against eavesdropper is questionable. For this reason, PRNGs are more widely used in simulation than in cryptography due to their fast speed and reproducibility.

6.3 True random number generators

A TRNG is a hardware security primitive that yields unpredictable true random numbers even if the internal design details (e.g., schematic, algorithm, timing and operations) are all known. A TRNG essentially produces the randomness from a physical phenomenon or process rather than a mathematic algorithm. The physical process can either be microscopic such as photoelectric effect, chaos and quantum effect or be macroscopic such as dice tossing and mouse clicks. In contrast to PRNG, there is no period for TRNG. This implies that the output binary sequence of a TRNG can never be repeated. According to the well-known Kerckhoff principle named after the famous Netherlands cryptographer Auguste Kerckhoff who postulated that "*A cryptosystem should be secure even if everything about the system, except the key, is public knowledge*", TRNGs provide higher security property for modern cryptography applications than PRNGs. In this chapter, we focus on TRNG design that originated from solid-state devices or can be easily implemented and integrated on-chip with other circuit modules for mass production. Most of the silicon TRNGs are designed based on the four classic methods to harvest the physical or environmental randomness. They are noise-based, chaos-based, jitter-based and metastability-based TRNGs. The design principles and examples are presented in the following subsections.

6.3.1 Noise-based TRNG

Electronic noise is the unpredictable and uncontrollable disturbance of signals that carry no useful information during acquisition, processing and transmission. It is often undesired in an electrical system as it can corrupt the useful signal but is otherwise useful for the design of TRNG. For instance, thermal noise in devices is a good source of randomness because it is frequency-independent and technology-invariant [2,3,29]. However, as thermal noise is often very weak, a wide-bandwidth amplifier is required to boost it up, which can consume significant power and area. Alternatively, Matsumoto *et al.* [30] proposed to add a silicon nitride (SiN) layer in a standard CMOS process to raise the thermal noise to a measurable level. The amplifier area is reduced with no quality checker at the cost of an extra expensive SiN mask.

Recently, a high-speed TRNG based on thermal noise amplification is proposed by Bae *et al.* [31]. The idea is to harvest the thermal noise introduced from the biasing circuit for a common-mode operating comparator and the sampling uncertainty of a D flip-flop (DFF), as depicted in Figure 6.2. It consists of a biasing circuit, a common-mode comparator, a differential-to-single (D2S)-ended amplifier, a slicer, a DFF and an output driver. The biasing circuit is essentially a beta-multiplier voltage reference. Both inputs of the comparator are connected together to the output of the biasing circuit, which makes the comparator work in the common mode. The thermal noise is added up in the comparator and amplified by the following D2S amplifier. The amplified noise is fed into the slicer to generate a full swing output asynchronously. The DFF samples this unpredictable output with an external 3 GHz clock. Depending

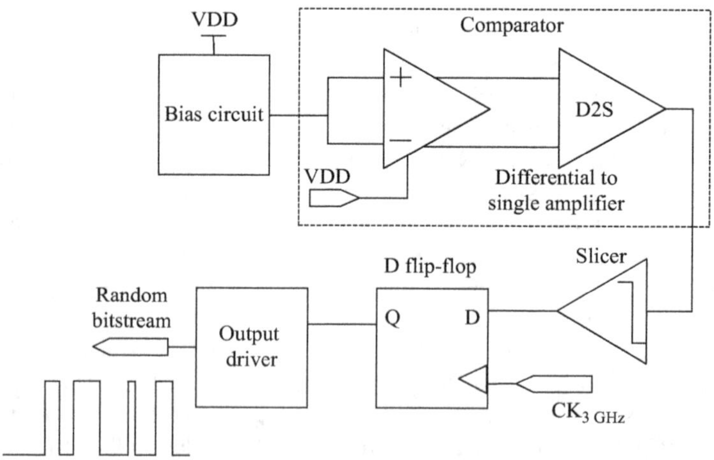

Figure 6.2 Block diagram of the noise-based TRNG in [31]

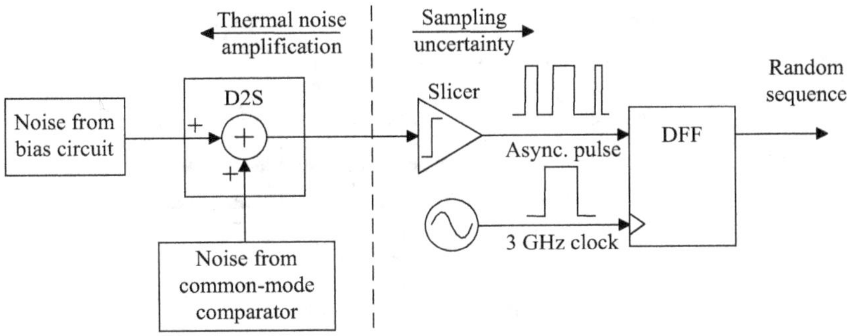

Figure 6.3 The design concept of the noise-based TRNG in [31]

on the arrival time of the input signal at port D, there is an uncertainty during sampling. The principle of this design can be illustrated in Figure 6.3. The randomness of the TRNG is enhanced by combining the two entropy sources: thermal noise and the sampling uncertainty. It can work at a very high speed, i.e., 3 Gb/s, but the power consumption is as high as 5 mW. Besides, the external high-speed 3 Gb/s sampling clock generator is itself very power hungry.

6.3.2 Chaos-based TRNG

The second category of TRNGs is based on chaotic system described by deterministic equations. Chaos is essentially a deterministic system. Its behavior is well described as "*when the present determines the future, but the approximate present does not approximately determine the future*" in [32]. As the apparently random states of disorder are originated from deterministic rules, it sounds ironic to extract true randomness from a deterministic system. One main reason behind this is that chaotic system is very hard to be modeled mathematically. Defined using a chaotic map and a bit generation function, these TRNGs can provide high-quality randomness for cryptographic applications because they are extremely sensitive to the initial conditions that can be triggered by environmental noise or noise generated by a nondeterministic physical system, resulting in a long-term unpredictability [33–35].

There are many methods to implement the chaotic maps. The most widely used map is the 1-D linear Markov map [36]. It satisfies:

$$x_{n+1} = f(x_n) = ax_n + b, (x_n \in R) \tag{6.2}$$

where n is the time sequence, x_n is the n-th system state, and a and b are the system constants. The mapping function $f(x)$ has the same domain as x. The initial state of the system is x_0, which is sensitive to the environmental noise.

Unfortunately, the chaotic map characteristics are susceptible to PVT variations, resulting in the degradation of randomness in real-time operation. The bit rate of the deterministic post-processing functions that can generate nearly i.i.d. bits from this process is also limited by its entropy rate. It turns out that the optimal bit generation

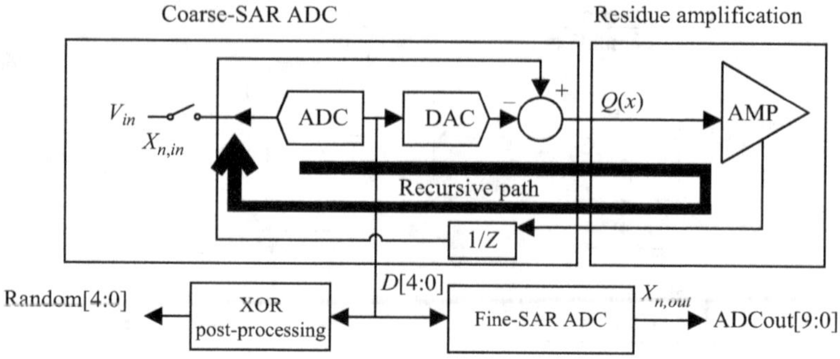

Figure 6.4 Block diagram of the chaotic map-based TRNG of [35]

function for achieving the highest possible entropy rate from a map function is not trivial for hardware implementation, and such TRNGs usually occupy large silicon area and consume great power.

Figure 6.4 shows the block diagram of the chaos-based TRNG proposed in [35]. It consists of a 10-bit fine-SAR ADC, a 5-bit coarse-SAR ADC, a dynamic residue amplifier, and an XOR post-processing block. The ADC structure implicitly realizes a discrete time chaotic map that can recursively amplify the initial state of the system with environmental noise. The design is very sensitive to the initial state of the design, which is dominated by the quantization error in the coarse-SAR ADC. Moreover, the coarse-SAR ADC is also responsible for reducing the switching power of the fine-SAR ADC by detecting and skip switching. In this manner, the power and area are greatly reduced. Additionally, it employs a novel dynamic residue amplifier and an adaptive reset comparator to further save the power by 45% and 58%, respectively, over conventional designs. The entire design consumes only 82 nW of power. The energy efficiency can reach 0.3 pJ/bit, which is the best reported result for this category of TRNGs. Its produced bitstreams can pass all the NIST tests.

6.3.3 Jitter-based TRNG

Another entropy source is the clock jitter of free running ring oscillators (ROs) [11,37]. Jitter-based TRNGs offer high flexibility and simplicity in their extractor implementation. Conventional jitter-based TRNG uses the slow jittery frequency clock to sample the fast clock. It is simple and easy to implement, but additional power-hungry clock generators are required to provide adequate jitter variations. Yang *et al.* [38] proposed a jitter-based TRNG that uses the oscillation collapse in a double-edge-injected RO, which exhibits good randomness and good process variation tolerance. The problem is that it needs several (8 for the reported results) selectable inverters per stage to provide enough tuning space to achieve robustness against process variations, and a number of stages (32 for their reported results) to increase the possibility of finding configurations with the most reliable output random bit.

The following analysis theoretically explains the principle of TRNG design based on jitter noise. The total jitter of the RO in the strong inversion can be written as [39]

$$\sigma_\tau^2 = \frac{2kT}{If_o}\left(\frac{\gamma_N + \gamma_P}{V_{DD} - V_{th}} + \frac{1}{V_{DD}}\right) \tag{6.3}$$

where k, T, I, f_o, γ_N, γ_P, V_{th} and V_{DD} are the Boltzmann constant, absolute temperature, saturation current, oscillation frequency, noise coefficients of NMOS and PMOS transistors, threshold voltage and power supply voltage, respectively.

On the other hand, the total jitter generated by the inverters conducted in the weak inversion (subthreshold region) is given by [40]

$$\sigma_\tau^2 = \frac{q}{If_o}(1 + e^{-V_{DD}/2V_t}) \tag{6.4}$$

The jitter variance depends on the operating region of CMOS transistors. Transistors in subthreshold region contribute significantly higher noise current as the diffusion currents and transconductances are very small [41,42]. To maximize the randomness for TRNG design, the RO should be biased in the subthreshold region. To conserve energy, the transistors of the RO can be biased to operate at minimum current to conserve energy.

The delay t_d of a single inverter in an RO is given by

$$t_d = \frac{CV_{DD}}{2I} \tag{6.5}$$

where C is the total load capacitance.

The oscillation frequency f_o can be expressed as

$$f_o = \frac{1}{M \cdot 2t_d} = \frac{I}{CMV_{DD}} \tag{6.6}$$

where M is the number of inverter stages.

Substituting (6.6) into (6.4), we have:

$$\sigma_\tau = \frac{1}{I}\sqrt{qCMV_{DD}(1 + e^{-V_{DD}/2V_t})} \tag{6.7}$$

This expression shows that the total jitter of the RO in the weak inversion depends on the power supply voltage, the load capacitance, the number of stages and the drain current. M should be sufficiently large to ensure that the sampling edge of the faster regular inverter RO will always fall in the differential jitter duration.

To this end, the randomness defined in [40] can be adopted to assess the quality Q of the jitter-based TRNG. It is defined as the jitter to period ratio, i.e.,

$$Q = \frac{\sigma_\tau}{\tau} = \sigma_\tau f_o = \sqrt{\sigma_\tau^2 f_o^2} = \sqrt{\frac{q}{CMV_{DD}}(1 + e^{-V_{DD}/2V_t})} \tag{6.8}$$

Q measures the amount of uncertainty in a period of RO. A larger Q implies more random bits can be extracted in a single period τ. A larger number of inverter stages M lower f_o, which increases the jitter amplitude but reduces the randomness quality Q

defined in (6.8) at the same time. Besides, RO with larger M consumes more silicon area but less power also due to its lower oscillation frequency.

Based on the previous analysis, Cao *et al.* [43] proposed a TRNG to extract the randomness from the jitter noise between two independent current-starved ROs (CSROs). The CSRO provides the flexibility to adjust the working current and the running frequency, trading off the clock rate for jitter noise. The architecture and the timing diagrams at nodes A, B, C, D and N are depicted in Figure 6.5. The proposed TRNG is lightweight and consists of only two short CSROs with an identical layout, a three-stage regular RO and an L-bit counter. In order to maximize its randomness, the inverters in the two CSROs are biased in the weak inversion region and the inverters in the regular RO are operating in the strong inversion region. When the EN signal becomes high, the two CSROs start to ring simultaneously. The outputs, A and B, of the two matched CSROs are XOR-ed to generate a random pulse at C. Its pulse width is determined by the jitters from the two CSROs. The random pulse is used to gate the regular inverter-based RO. The number of oscillation cycles at node D is proportional to the pulse width of C. Random bits can be extracted by clocking a counter with the output D of the regular RO. The counter output N increases with each cycle of regular RO and automatically resets upon reaching its maximum count value determined by the length of the counter. The digitized output of the counter is read out at each cycle of A (or B) when the value of D has stabilized. If the length L of the counter is too long, there will be temporal correlation in the bit patterns as several higher order bits of the counter output remain unchanged for multiple read cycles, as shown in the penultimate waveform N in Figure 6.5. To break the monotonic correlation, at least one reset must occur within one cycle of waveform A (or B), as shown in the last waveform N of Figure 6.5.

This jitter-based TRNG was fabricated with a compact layout in a standard 65 nm 1.2 V CMOS process. Measured results from the fabricated chips showed that the TRNG consumes only 260 μW at a bit rate of 52 Mbps. It outperforms many on-chip jitter-based TRNGs with the best figure-of-merit of 5 pJ/bit and the smallest footprint of 366 μm^2. Its generated bit sequence passes the statistical randomness tests, including NIST test, ACF test and bias test with native entropy/bit of 0.999998940380628 measured up to 15 decimal places of accuracy.

6.3.4 Metastability-based TRNG

Metastability refers to a state of apparent equilibrium that can be attained by an electronic circuit or system for an undefined or unbounded duration before it is changed into a more stable state. It is an unstable equilibrium state of a dynamical system other than the system's state of least energy. Metastability can be explained by a ball staying in a concave on a slope as shown in Figure 6.6. The ball is somehow stable in the concave in the left (Region A) if only it is slightly pushed. However, it will fall into the lower concave (Region C) with a stronger push. Metastability in cross-coupled inverters, latches, DFFs and SRAMs that cause the circuit to behave in an unpredictable way can be utilized to produce a random bitstream at a high bit rate [44–46].

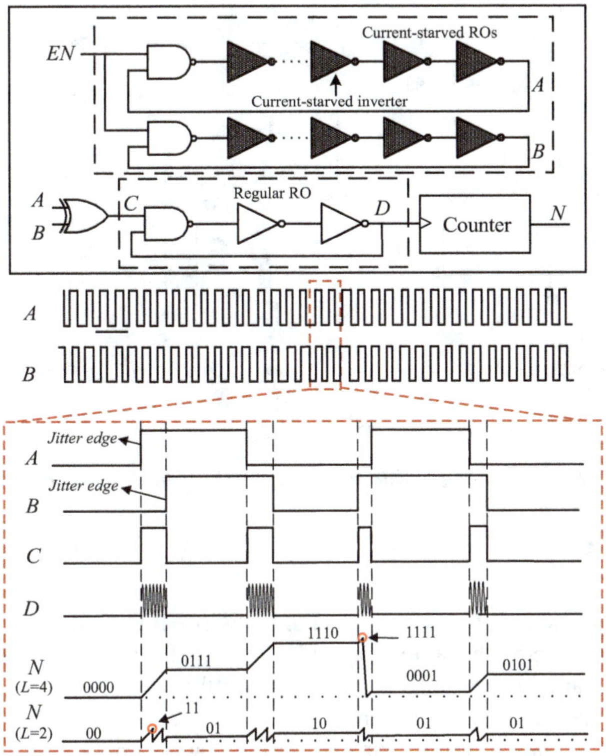

Figure 6.5 Building blocks and timing diagrams of the TRNG proposed in [43]

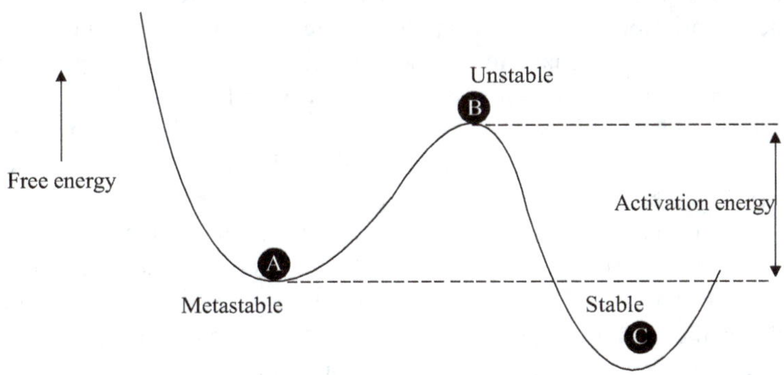

Figure 6.6 Example of the metastability

The major problem with this random source is that a complex post-processing unit is required to eliminate the systematic bias due to the process variations.

One typical metastability-based TRNG is proposed in [47]. It can produce raw random output with high entropy and without any post-processing. The core structure

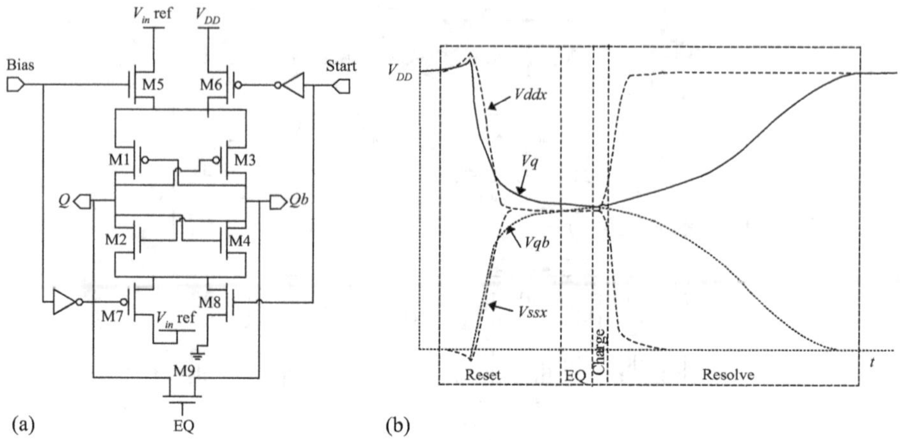

Figure 6.7 (a) Metastable latch circuit [47] and (b) its operation

is the metastable latch circuit as shown in Figure 6.7(a). It is designed from two cross-coupled inverter pairs with a symmetrical layout. The PMOS and NMOS transistors in the inverter pairs are carefully sized to ensure that the rising and falling delays are equal. The idea and the operation of the design are shown in Figure 6.7(b). The output voltages V_q and V_{qb} are initialized to the power supply voltage and ground, respectively. By asserting the *EQ* signal, the outputs are equalized to half V_{DD}. After equalization, *EQ* is released. Additional charge at node q is injected by the dynamic control module. This charge is used to compensate the static voltage difference caused by the process variations. Finally, the latches are activated to resolve the output to full swing by asserting *Start* signal and releasing the bias voltage. The output is a random value depending on the thermal noise inside the two latches.

The high randomness is ensured by checking the settling time t_d of the metastable events. In an ideal system without any noise, t_d should be infinity and both outputs are half V_{DD}. If the voltage difference introduced by the process variations increases, t_d becomes smaller. Therefore, it is possible to tune the latch into the metastable region to resist the process variations by observing the values of t_d. The implementation of the metastability-based TRNG in [47] is shown in Figure 6.8. It mainly consists of three modules: the random bit generation module, the control module and the grading module. The random bit generation module involves the abovementioned metastability latch, a time-to-digital converter (TDC) and a completion detector. In each cycle, a random bit is produced by the metastability latch and its corresponding settling time t_d is digitized by the TDC. The statistical information is extracted by the control module for a number of measured t_d, for example, 128 in [47]. Based on the information, the control module calculates the feedback value to tune the metastable latch. Finally, the grading module compares the measured t_d with a predefined threshold t'_d. Based on the conjecture that the output bit with the larger t_d has higher entropy,

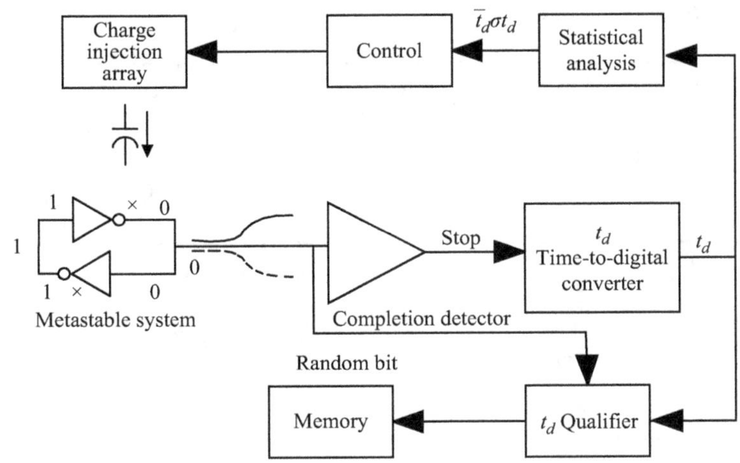

Figure 6.8 Block diagram of metastability-based TRNG proposed in [47]

the comparison result will be stored provided that t_d is larger than the threshold. The design constraints lie in the metastability latch. The switching speed cannot be too fast to prevent t_d from falling below the time resolution of the TDC. Besides, the latch size and its load have to be properly designed so that thermal noise dominates flicker noise. This is because despite its random nature, the colored distribution observed in the power spectral density of flicker noise insinuates potential entropy leakage due to the presence of correlation between adjacent samples.

6.4 Post-processing

Post-processing techniques aim to reconcile the statistically imperfect distribution of the raw output stream generated by the entropy source, such as low min-entropy or systematic bias, to make them cryptographically useful. It not only masks serious defects of the TRNG at run-time but also matches the bit rate, as some applications may require a bit rate higher than can be achieved by the bitstream extracted directly from the entropy source. Many state-of-the-art silicon-based TRNGs that have impressively low energy consumption, silicon area and high throughput may have less than perfect or barely satisfactorily min-entropy. In fact, NIST special publication 800-90B [48] actually recommends employing one of the proven post-processing methods based on cryptographically secure primitives, i.e., block ciphers or cryptographic hash functions. One main drawback of post-processing is that they can be costly in hardware implementation. The extra area and power overheads may not be justifiable for lightweight and resource constrained devices [49,50]. In this section, five post-processing techniques to improve the imperfect solutions of TRNG are introduced.

Table 6.1 VNC operation

Input1	Input2	Operation
0	0	Discard bit
0	1	1
1	0	0
1	1	Discard bit

6.4.1 Simple correctors

Simple correctors can be easily implemented in ASIC or FPGA to effectively filter the bias in the output bitstream of the entropy source, such as intermittent long runs of "1"s and "0"s induced by fluctuating operating conditions. Examples for the simple correctors are block-wise XORing [51,52], LFSRs [53], von Neumann corrector (VNC) [54] and so on. The most popular simple corrector for the post-processing of TRNG is the VNC. VNC equalizes the percentage of "1" and "0" in the bitstream by either discarding a dibit or converting a dibit into a single bit, as shown in Table 6.1. Hence, this post-processing method, despite its simplicity, reduces the throughput of the TRNG by four times. The limitation for the VNC is that it requires the probability of "1"s in the output to be always constant and there is no correlations in the bitstream. Let us look at an extreme example: 0101010101.... The output after VNC is 00000.... The bias is even increased. Another simple corrector is XORing two consecutive random bits to remove the correlation between neighboring bits [55]. Mathematical proof is also provided in this work. The drawback is that the throughput is reduced to half.

6.4.2 Cryptographic hash functions

Cryptographic hash function is a one-way function that maps an arbitrary number of input bits to a fixed number of output bits. A good hash function should meet the four criteria:

- It must reproduce the same hash value for the same input message.
- It must be computationally infeasible to reverse the output hash value to the input message.
- It must have a high collision resistance, i.e., the probability of finding two messages with the same hash value should be extremely low.
- A very small change in a message must result in a significant change in hash value which is totally different and uncorrelated with the original hash value. This property is known as avalanche effect.

It has been proved that if the input has high entropy, the output of the cryptographic hash function will have an almost ideal uniform distribution. However, since most of these hash functions are nonlinear, the hardware implementation may not be very efficient [56]. More importantly, there is still no convincing theoretical proof of randomness improvement by using the cryptographic hash functions.

Cryptographic secure hash functions for post-processing of TRNG bitstream can be divided into unkeyed and keyed hash functions. The main difference between these two types of hash functions is that the output of the unkeyed hash function depends only on the input message, whereas the output of a keyed hash function depends on both the input message and a secret key. For both types of hashed functions, there is no easier means than a brute-force search to find an input that hashes to a specific output value or two colliding input messages. The unkeyed hash function can provide good pre-image resistance and collision resistance properties [57]. On the one hand, a good keyed hash function has, in addition, the diffusion and confusion properties. Diffusion hides the statistical properties between the output value and the input message such that a single bit change of an input affects approximately half of the output bits and vice versa. Confusion, on the other hand, hides the relation between the output and the key by making each bit of the hash value dependent on several bits of the key [58–60]. Consequently, without the knowledge of the secret key, the chance of finding the correct output for any input to the keyed hash function is no better than a random guess. It should be noted, however, that using unkeyed hash function to extract a secure one-time-pad from a "single input" of partially random data is a misguided construction. This is because it has been proven by McInnes and Pinkas [61] that it is impossible to compress a weakly random source (i.e., one with less than perfect min-entropy) into a perfectly random output with full min-entropy.

6.4.3 Extractor functions

Another widely used post-processing method for the TRNG output stream is the entropy extractor. It is basically a mathematical function that extracts the randomness from the raw source bitstream and ideally produces a uniform distribution irrespective of the original bitstream distribution. The primary entropy extractor solutions are seeded or multiple-input extractors, which can be implemented with block ciphers and cryptographic keyed hash functions, such as Advanced Encryption Standard, Secure Hash Algorithms and Barak–Impagliazzo–Wigderson [62]. However, they suffer from the long latency typically around tens of cycles, and large area overhead around 1K–40K logic gates. Another drawback is that these extractors can only work at a fixed bit rate regardless of the quality of the input stream. Besides, they cannot filter out long runs of "0"s or "1"s. It should be noted that secure seeded entropy extractor requires a uniform random seed as a second input, which has a major drawback of inadequate seed sources. It is proven in [63] that HMAC and CBC-MAC, with the fixed length key used as the seed input, are secure entropy extractors. They are also one of the three extractors (another one is CMAC) approved by SP800-90B standard. The NIST specification further requires that the output-to-input entropy ratio must be at least two, which means that the input bitstream must have a minimum entropy rate of 0.5 to guarantee the target entropy in the output bitstream.

A lightweight multiple-input extractor leveraging on the small footprint and inherent capability of VNC to eliminate biased input bitstream is proposed in [62]. To overcome the high throughput penalty of VNC due to the discarded dibits, the discarded dibits are distinguished from the other dibits by introducing a secondary output called the residue bit. The residue bit is "1" if the primary output bit is not discarded and "0" if the primary output bit is "-", i.e., it is marked for deletion. Figure 6.9 shows a trilevel hierarchical construction of seven VNCs operating in parallel. Each VNC block can process a native bitstream to produce two bitstreams, one from its primary output and one from its residual output. The primary bit sequence will be directly appended to the final output stream due to its full-entropy. The residual stream detects transition in consecutive dibits. The residual outputs from two adjacent VNCs in the first level are input to a VNC in the next level. The latter will again output a primary bitstream of full-entropy but at half the bit rate of the input stream. Similarly, the partially random bitstreams from the residue outputs of two adjacent VNCs are further combined to produce a useful full-entropy bitstream at half the bit rate of its input residue bitstreams. This way, the parallel VNC multiple-input entropy extractor can achieve high efficiency with 100% utilization of all VNCs. The third stream shown in Figure 6.9(a), called the helper stream, is not used for random bitstream generation but for regenerating each original input stream from the corresponding primary and residual streams if necessary. The helper output of a VNC is "-" if the input dibit is "01" or "10" (i.e., the residue output is "1"), and "0" or "1" if the input dibit is "00" or "11", respectively. The helper stream has low entropy content and low bit rate. If reconstruction of the original bitstream is not required, it can be omitted completely to reduce the extractor cost. Figure 6.10 shows the transistor level implementation of a VNC circuit. It only consists of two flip-flops and a few logic gates. The internal nodes of one of the flip-flops is XORed to produce the residual stream to save the circuit area by 38%.

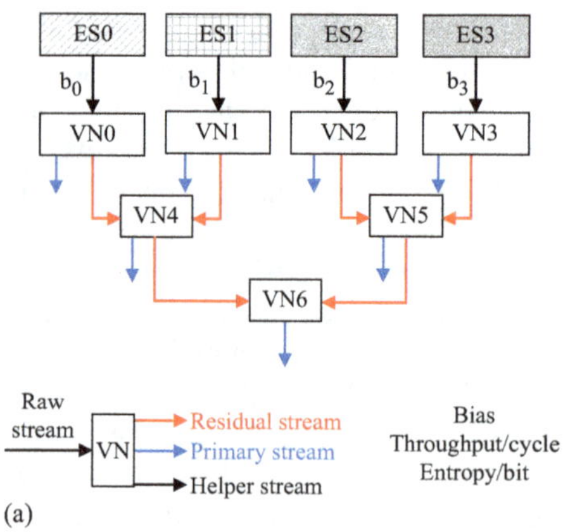

	Raw stream	Primary stream	Residual stream	Helper stream
	0	-	-	-
	1	1	1	1
	1	-	-	-
	1	-	0	1
	1	-	-	-
	1	-	0	1
	0	-	-	-
	1	1	1	-
	1	-	-	-
	0	0	1	-
	0	-	-	-
	0	-	0	0
Bias	10%	0%	2%	19%
Throughput/cycle	1	0.24	0.5	0.26
Entropy/bit	0.971	1	0.9988	0.8905

(b)

Figure 6.9 (a) A VNC with three outputs, (b) the multiple-input entropy extractor constructed with a hierarchy of VNCs and (c) primary, residual and helper stream generated from an example raw bitstream [62]

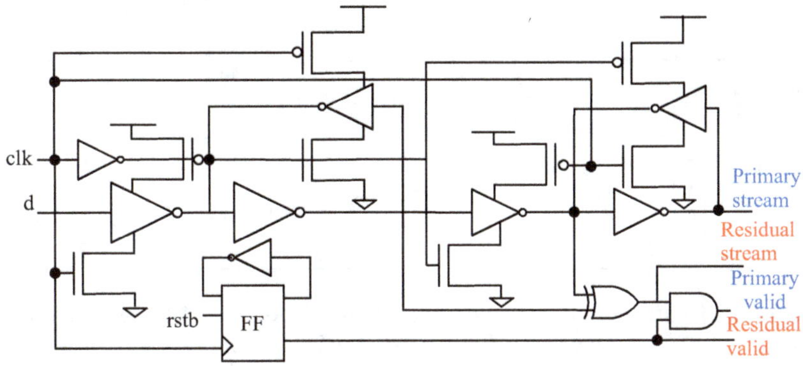

Figure 6.10 VN extractor circuit [62]

6.4.4 Resilient functions

An (n, m, t)-resilient function is an n-input m-output function: $F^n \rightarrow F_m$. It outputs m-tuple with equal probability when the arbitrary inputs are fixed and the remaining $n - k$ bits are randomly chosen. The resilient function can enhance the randomness by filtering the deterministic bits and removing the bias in a random bitstream. For example, a resilient function is proposed to help the jitter-based TRNG to generate provably random bits at the throughput of a few megabits per second in [64]. The function is also used to resist some adversarial manipulation. The key idea is the "filling rate" that measures the fraction of time in which the analog output signal is arguably random. By introducing the resilient function, random samples can only be extracted from a signal with moderate fill rate, resulting in less ROs in the design. A framework was developed to accurately quantify the resilience against the fault inject attack. Unfortunately, the resilient function sometimes reduces the bit rate and is more vulnerable to some active attacks in the output stream than the entropy extractor [65].

6.4.5 PUF-based entropy pump

In [66], physical unclonable function (PUF) technology is employed to build a low-cost entropy pump, which boosts the entropy of keys generated from low-entropy sources. PUF is an innovative circuit that can extract digital fingerprint from the uncontrollable and unpredictable manufacturing process variations for ICs [67]. Due to its low power consumption and the tiny area it occupies, PUF is often used to authenticate a single chip or generate a secret key for encryption. A typical example is the RO-based PUF [68]. The RO PUF consists of a number of ROs with identical layout. Due to the process variations, the output frequency of each RO is slightly different. A binary stream (challenge) is used to select two of the ROs for comparison. The frequency comparison result is a random value. Wang and Qu [66] implement the entropy pump based on a highly flexible RO. The structure of the RO is shown in Figure 6.11. For simplicity, only one ring has been drawn in the figure. The low

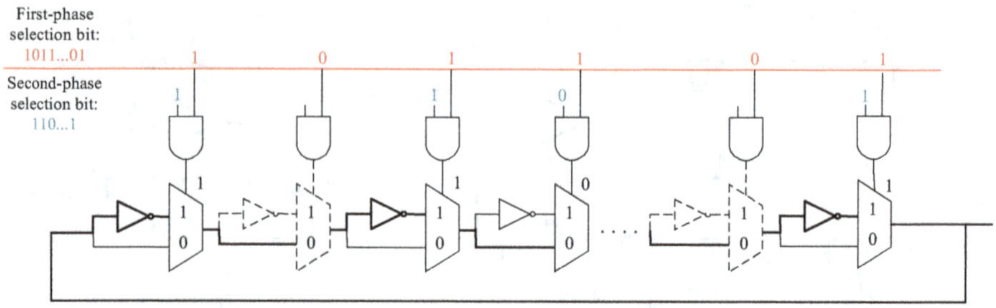

Figure 6.11 Architecture of two-phase configurable RO in [66]. The bold line shows the selection path in this configuration bit.

entropy source is input as the phase selection bitstream, which configures the delay length in the RO. If the selection bit is "1," the corresponding inverter in the stage would be used in the ring; otherwise, the inverter will be bypassed through the wire. This design has some significant differences from the basic RO PUF. It can increase the entropy per bit by 50%, while the hardware cost is much smaller than the conventional hash functions or cryptographic protocols.

6.5 TRNG randomness tests

The test (evaluation) of randomness for a TRNG is not easy. It is even harder than the design of TRNG itself in a sense. In this section, three mainstream methods to test the randomness, standard statistical tests, entropy estimate and attack analysis are discussed.

6.5.1 Standard tests

Batteries of standard statistical tests are used to evaluate the randomness of the bit sequences produced from a TRNG. These tests basically analyze the distribution of the input bit sequences. A good random stream should have an equal percentage of "1"s and "0"s. However, this is just a necessary but not sufficient condition. For example, a sequence whose first half bits are "0"s and the rest are all "1"s is obviously not truly random. Thus, these tests will also check if an input bitstream is patternless. Besides, no correlation should also be present in a random sequence. Standard randomness tests based on statistical analysis such as Crypt-X [69], ENT utility [70], TestU01 [71], Diehard [72] and NIST statistical test suite [73] are used in the evaluation of the randomness of TRNGs. Among them, Crypt-X is only for commercial purpose and ENT utility from Fourmilab provides only very basic randomness tests. They are not as popular as the remaining three tests. The NIST statistical test suite is most well known for the evaluation and certification of RNGs for cryptographic applications. The Diehard test intends to test RNGs but not files of possibly random numbers as the latter is a fallacious view of what it means to be random. However, Diehard does not implement all the NIST tests and it is very stringent for the setup of the test parameters (e.g., sample size). TestU01 is a software library that provides a set of tools for the empirical randomness testing of TRNGs. Some state-of-the-art randomness tests are implemented in TestU01 despite not very efficient. Besides, TestU01 only processes sample size of 32 bits and interpret them in the interval (0,1), making it more sensitive to the most significant bits than the least significant bits.

We take the NIST test as an example to explain the principle of the statistical randomness tests. The NIST test suite consists of 15 tests (as shown in Table 6.2) that are developed to evaluate the randomness of an arbitrarily long binary sequence. It is important to emphasize that some statistical tests in NIST test suite can successfully indicate randomness for an input sequence, while other tests can fail for the same sequence. Each test has different targets and vulnerabilities. Generally, NIST test is performed by means of a hypothesis test. It assumes that the stream in tests is random. Under this assumption, the distribution of the test statistic is determined. Judgment in the hypothesis test is taken through the p-value, which is computed using the theoretical distribution of the test statistics. Small p-value that is less than the significant level is interpreted as that the sequence is unlikely to be random. A commonly used significant level is 0.01. If the p-value in a test exceeds 0.01, the bitstream is considered as passing that NIST test with a confidence of 99%.

These standard tests check one or more statistical properties of the long random bitstreams. However, the most important problem for these tests is the lack of certainty. It means that even if the generator passes all the known statistical tests, one can get a high confidence but cannot absolutely guarantee that it will pass the new tests in the future [11]. Moreover, the tests are usually much slower than the generation of the random numbers. If one can expect certain flaws in the RNG, he can only employ special tests that are sensitive to these flaws to increase the test efficiency.

Table 6.2 The 15 tests of NIST 800-22 statistical test suite

1	The frequency (monobit) test
2	Frequency test within a block
3	The runs test
4	Tests for the longest run-of-ones in a block
5	The binary matrix rank test
6	The discrete Fourier transform (spectral) test
7	The nonoverlapping template matching test
8	The overlapping template matching test
9	Maurer's "Universal Statistical" test
10	The linear complexity test
11	The serial test
12	The approximate entropy test
13	The cumulative sums (Cusums) test
14	The random excursions test
15	The random excursions variant test

6.5.2 Entropy estimate

Another method to evaluate the randomness is to measure the entropy of the TRNG's output stream. In information theory, entropy refers to disorder or uncertainty. It measures the amount of information conveyed by each event. The international standard ISO/IEC 18031 [74] and Germany standard AIS 31 [75] recommend to evaluate the randomness of a TRNG by building up a mathematical model based on its stochastic attributions. With some rationale assumptions, the mathematical model can accurately describe the process of extracting and digitizing the random noise. The entropy can be calculated by substituting the TRNG design parameters, such as the environmental temperature, jitter amplitude and sampling frequencies [76]. Detailed examples of modeling the TRNGs can be found in [77–79].

Generally, the estimate of information entropy (Shannon entropy) can be calculated by [80]:

$$H = -\sum_{i=1}^{N} p(x_i)\log_2 p(x_i) \tag{6.9}$$

where $p(x_i)$ and N are, respectively, the probability of and total number of possible data values of x_i.

In a binary random bitstream, the maximum entropy is 1 when $p(1) = p(0) = 1/2$. Usually, min-entropy, which is the worst case entropy, is considered. It is formulated as the negative logarithm of the largest probability of the possible data value. Min-entropy is the most conservative way of measuring the unpredictability of a TRNG output sequence.

6.5.3 Attack analysis

Attack analysis can also be used for the evaluation of the TRNG randomness. The security of cryptographic systems is strongly related to randomness. The output of these systems must be seen by any adversary as a sequence of random values carrying the secret but revealing absolutely no clues about the precious information. If the randomness can be easily compromised by the attacker, this TRNG is considered insecure. Besides, TRNG is a particularly attractive target to attackers, since it is typically a single isolated hardware block that can be easily located. There have been many reported attacks on the RNGs, such as cryptanalytic attack, machine learning attacks, side-channel attacks, fault injection attacks and other physical attacks. Additionally, with the rapidly advancing machine learning algorithms, it is very hard to predict the new threats they pose.

Side-channel attack is based on the information gained from the physical implementation of a TRNG, rather than by brute-force exploration or by exploiting theoretical weaknesses in the design. The exploitable side-channels of a device can be the path delay, power consumption or electromagnetic (EM) radiation. Bayon *et al.* [81] have shown the vulnerability of RO-based TRNG to EM active noninvasive attack. They proved that it is possible to locate the ROs in the chip and to measure the corresponding frequencies through the side-channel signals as shown in Figure 6.12. The obtained information will help one to predict the output of the TRNG through the active attack.

Fault attack introduces deliberate faults into the computation of the cryptographic function and exploits the faulty results to extract information about the secret key. For example, most oscillator-based TRNGs are susceptible to frequency injection attack [82,83]. The setup of the attack is shown in Figure 6.13. By injecting frequencies

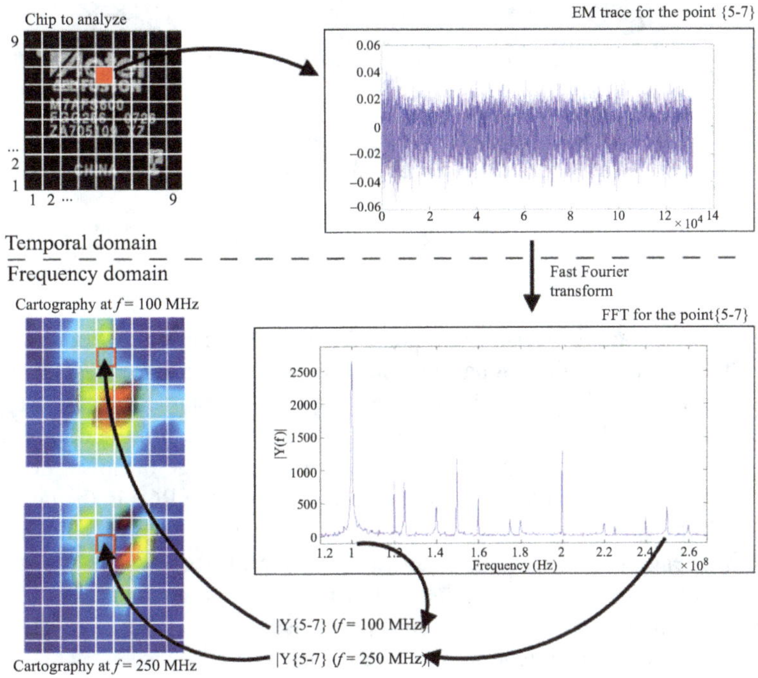

Figure 6.12 EM side-channel attack using frequency analysis [81]

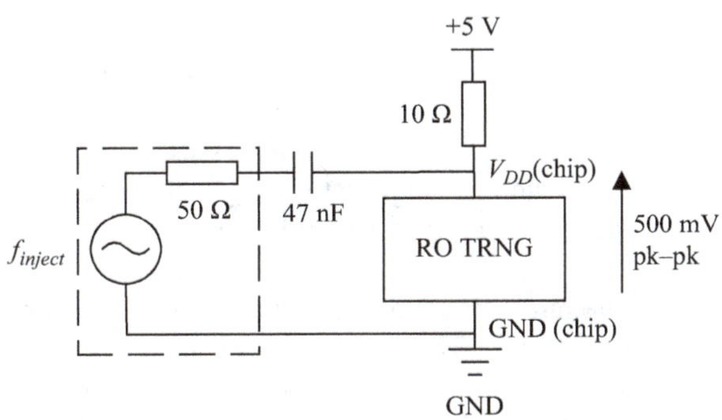

Figure 6.13 The setup of the frequency injection attack for the jitter-based TRNG

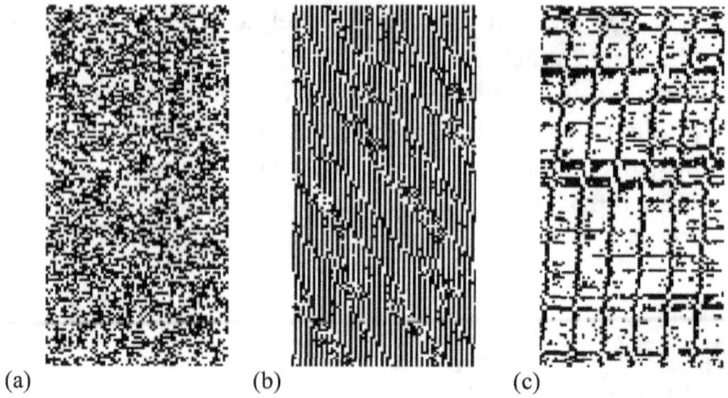

Figure 6.14 The frequency injection attack result on the jitter-based TRNG [83]: (a) no injection, (b) 1.822880 MHz injection and (c) 1.929629 MHz injection

through the power rail, the ROs can be excited to lock at the injected frequency. Once locked, jitter in the injection signal will propagate equally to all rings. As a result, the entropy source, i.e., the jitter, is destroyed. Figure 6.14 shows the attack result on a TRNG in automated teller machines. Typical patterns appear in the output random bits. To protect the design from the attack over a larger range of noise, decoupling capacitors can be placed close to the power rail or around the ROs in the layout [37]. The decoupling capacitor acts as a low-pass filter to filter out the possible injection of resonant frequency in the circuitry. In terms of countermeasures, Bahadur *et al.* [84] proposed a side-channel attack resilient jitter-based TRNG using Galois RO. Yang *et al.* [85] also presented an on-chip entropy measurement circuitry to detect such attacks and failure to generate random bit sequence.

6.6 Conclusion

In this chapter, we present four categories of silicon-based TRNGs. Examples with design details are provided and discussed. Depending on the applications, each of these methods have their own strength and weaknesses in entropy harvesting. To enhance the randomness of the raw output streams, post-processing techniques are often applied. VNC emerges as the most popular and lightweight post-processing method to remove the bias and improve the randomness at the cost of lowering bit rate. The evaluation of the randomness of a TRNG is also very important, or at least as significant as the design of the TRNG itself. Some standard tests with computation tools can be used to evaluate the randomness, statistically. However, passing these tests does not guarantee that the TRNG is "truly" random nor is foolproof in terms of security. In the new era of the IoT, there is an urge to gear up the research into lightweight and low-cost TRNGs with high entropy and security against attacks.

Acknowledgments

This research is supported by the Fundamental Research Funds for the Central Universities (2019B22714), Open Research of the State Key Laboratory of Computer Architecture (CARCH201901), Natural Science Foundation of Jiangsu Province (BK20191160) and Zhejiang Lab's International Talent Fund for Young Professionals, the Fundamental Research Foundation of Shenzhen (JCYJ20190808151819049) and the Singapore Ministry of Education Tier 1 grant MOE2018-T1-001-131 (RG87/18).

References

[1] Wang A, Li P, Zhang J, *et al.* 4.5 Gbps high-speed real-time physical random bit generator. Optics Express. 2013;21:20452–20462.

[2] Mathew S, Johnston D, Newman P, Satpathy S, and Krishnamurthy, R. μRNG: A 300-950 mV, 323 Gbps/W all-digital full-entropy true random number generator in 14 nm FinFET CMOS. IEEE Journal of Solid-State Circuits. 2016;51(7):1695–1704.

[3] Brederlow R, Prakash R, Paulus C, and Thewes R. A low-power true random number generator using random telegraph noise of single oxide-traps. In: 2006 IEEE International Solid State Circuits Conference – Digest of Technical Papers, San Francisco, CA; 2006: 1666–1675.

[4] Blum L, Blum M, and Shub M. A simple unpredictable pseudo random number generator. SIAM Journal on Computing. 1986;15(2):364–383. Available from: https://doi.org/10.1137/0215025.

[5] Available from: https://www.cloudflare.com/leagueofentropy/.

[6] Limer E. Why a Wall Full of Lava Lamps Is a Terrific Random Number Generator; 2017. Available from: https://www.popularmechanics.com/technology/security/news/a28921/lava-lamp-security-cloudflare/.

[7] Lee EC, Parrilla-Gutierrez JM, Henson A, Brechin EK, and Cronin L. A crystallization robot for generating true random numbers based on stochastic chemical processes. Matter. 2020.

[8] Kelsey J, Schneier B, Wagner D, *et al.* Cryptanalytic attacks on pseudorandom number generators. In: Proc. International Workshop on Fast Software Encryption. Paris: Springer; 1998. p. 168–188.

[9] Dichtl M. How to predict the output of a hardware random number generator. In: Proc. Workshop Cryptograph. Hardw. Embedded Syst. Cologne, Germany. vol. 2779; 2003. p. 181–188.

[10] Rogers G. Tipton Brothers Plead Guilty in Iowa Lottery Rigging Scandal; 2017. Available from: https://www.desmoinesregister.com/story/news/crime-and-courts/2017/06/29/tipton-pleads-guilty-iowa-lottery-rigging-scandal/438039001/.

[11] Stipcevic M and Ko CK. True random number generators. In: Open Problems in Mathematics and Computational Science. Berlin, Heidelberg: Springer; 2014. p. 275–315.

[12] Knuth D. The Art of Computer Programming, 2nd ed. Reading, MA: Addison-Wesley; 1981.
[13] Klapper A and Goresky M. Feedback shift registers, 2-adic span, and combiners with memory. Journal of Cryptology. 1997;10(2):111–147.
[14] Lehmer DH. Mathematical methods in large-scale computing units. In: Proceedings of a Second Symposium on Large-Scale Digital Calculating Machinery. Cambridge, MA: Harvard University Press; 1951: 141–146.
[15] Wu P-C. Multiplicative, congruential random-number generators with multiplier $\pm 2^{k1} \pm 2^{k2}$ and modulus $2^p - 1$. ACM Transactions on Mathematical Software. 1997;23(2):255–265.
[16] Deng LY and Xu H. A system of high-dimensional, efficient, long-cycle and portable uniform random number generators. ACM Transactions on Modeling Computer Simulation. 2003;13(4):299–309.
[17] Deng LY, Li H, Shiau JJH, *et al.* Design and Implementation of Efficient and Portable Multiple Recursive Generators with Few Zero Coefficients. Berlin, Heidelberg: Springer Berlin Heidelberg; 2008.
[18] Gammel BM, Gottfert R, and Kniffler O. An NLFSR-based stream cipher. In: Proc IEEE Int Symp Circuits Syst. Munich: Infineon Technologies AG; 2006. p. 2917–2920.
[19] Mukhopadhyay D, Chowdhury DR, and Rebeiro C. Theory of composing non-linear machines with predictable cyclic structures. In: Cellular Automata, 8th International Conference on Cellular Automata for Research and Industry, ACRI 2008, Yokohama, Japan, September 23–26, 2008. Proceedings; 2008.
[20] Mukhopadhyay D. Group properties of non-linear cellular automata. Journal of Cellular Automata. 2010;5(1):139–155.
[21] Shujun LI, Chen G, and Mou X. On the dynamical degradation of digital piecewise linear chaotic maps. International Journal of Bifurcation Chaos. 2011.
[22] Fischer V and Drutarovsky M. True random number generator embedded in reconfigurable hardware. In: Cryptographic Hardware and Embedded Systems (CHES), San Francisco Bay, CA, USA; 2002. vol. 2523; 2002. p. 415–430.
[23] Compaq Computer Corporation. Compaq Fortran Language Reference Manual, Houston, Texas. 1999. Available from: http://h30266.www3.hpe.com/odl/unix/progtool/cf95au56/dflrm.htm.
[24] Yao AC. Theory and application of trapdoor functions. In: 23rd IEEE Annual Symposium on Foundations of Computer Science Conference. Chicago, IL, USA; 2008.
[25] Barker EB and Kelsey JM. SP 800-90A. Recommendation for Random Number Generation Using Deterministic Random Bit Generators. Gaithersburg, MD: National Institute of Standards and Technology; 2007.
[26] Hurley-Smith D, Patsakis C, and Hernandez-Castro J. On the unbearable lightness of FIPS 140-2 randomness tests. IEEE Transactions on Information Forensic and Security. 2020.

[27] Ginesu G, Pintus M, and Giusto DD. Objective assessment of the WebP image coding algorithm. Signal Processing: Image Communication. 2012; 27(8):867–874. Special issue on: pervasive mobilemultimedia. Available from: http://www.sciencedirect.com/science/article/pii/S0923596512000203.

[28] von Neumann J. Various Techniques used in Connection with Random Digits. vol. 5. Oxford: Pergamon Press; 1963. Von Neumanns Collected Works.

[29] Petrie CS and Connelly JA. A noise-based IC random number generator for applications in cryptography. IEEE Transactions on Circuits and Systems I: Fundamental Theory and Applications. 2000;47(5):615–621.

[30] Matsumoto M, Yasuda S, Ohba R, *et al.* 1200 μm^2 physical random-number generators based on SiN MOSFET for secure smart-card application. In: Proc. IEEE ISSCC Dig. Tech. Papers, San Francisco, CA, USA; 2008. p. 414–624.

[31] Bae S, Kim Y, Park Y, *et al.* 3-Gb/s high-speed true random number generator using common-mode operating comparator and sampling uncertainty of D flip-flop. IEEE Journal of Solid-State Circuits. 2017;52(2):605–610.

[32] Danforth CM. Chaos in an Atmosphere Hanging on a Wall; 2013. Available from: http://mpe.dimacs.rutgers.edu/2013/03/17/chaos-in-an-atmosphere-hanging-on-a-wall/.

[33] Stojanovski T, Pihl J, and Kocarev L. Chaos-based random number generators-Part II: Practical realization. IEEE Transactions on Circuits and Systems I. 2001;38(3):281–288.

[34] Pareschi F, Setti G, and Rovatti R. Implementation and testing of high-speed CMOS true random number generators based on chaotic systems. IEEE Transactions on Circuits and Systems I. 2010;57(12):3124–3137.

[35] Kim M, Ha U, Lee KJ, *et al.* A 82-nW chaotic map true random number generator based on a sub-ranging SAR ADC. IEEE Journal of Solid-State Circuits. 2017;52(7):1953–1965.

[36] Stojanovski T and Kocarev L. Chaos-based random number generators—Part I: Analysis [cryptography]. IEEE Transactions on Circuits and Systems I: Fundamental Theory and Applications. 2001;48(3):281–288.

[37] Tang Q, Kim B, Lao Y, *et al.* True random number generator circuits based on single- and multi-phase beat frequency detection. In: Proc. IEEE 2014 Custom Integrated Circuits Conference. San Jose, CA; 2014. p. 1–4.

[38] Yang K, Blaauw D, and Sylvester D. An all-digital edge racing true random number generator robust against PVT variations. IEEE Journal of Solid-State Circuits. 2016;51(4):1022–1031.

[39] Abidi AA. Phase noise and jitter in CMOS ring oscillators. IEEE Journal of Solid-State Circuits. 2006;41(8):1803–2006.

[40] Guler U and Dundar G. Modeling CMOS ring oscillator performance as a randomness source. IEEE Transactions on Circuits and Systems I. 2014; 61(3):712–724.

[41] Sarpeshkar R, Delbruck T, and Mead CA. White noise in MOS transistors and resistors. IEEE Circuits and Devices Magazine. 1993;9(6):23–29.

[42] Sansen WM. Analog Design Essentials. Berlin, Heidelberg: Springer; 2007.

[43] Cao Y, Chang C, Zheng Y, *et al.* An energy-efficient true random number generator based on current starved ring oscillators. In: Proc. 2017 IEEE Asian Hardware Oriented Security and Trust Symposium (AsianHOST), Beijing, China; 2017. p. 37–42.

[44] Tokunaga C, Blaauw D, and Mudge T. True random number generator with a metastability-based quality control. IEEE Journal of Solid-State Circuits. 2008;43(1):78–85

[45] Holleman J, Bridges S, Otis BP, *et al.* A 3 μW CMOS true random number generator with adaptive floating-gate offset cancellation. IEEE Journal of Solid-State Circuits. 2008;43(5):1324–1336.

[46] Srinivasan S, Mathew S, Erraguntla V, *et al.* A 4Gbps 0.57pJ/bit process-voltage-temperature variation tolerant all-digital true random number generator in 45nm CMOS. In: IEEE Proc. 22nd International Conference on VLSI Design, New Delhi, India; 2009. p. 301–306.

[47] Tokunaga C, Blaauw D, and Mudge T. True random number generator with a metastability-based quality control. IEEE Journal of Solid-State Circuits. 2008;43(1):78–85.

[48] Turan MS, Barker E, Kelsey J, *et al.* Recommendation for the Entropy Sources Used for Random Bit Generation. Document SP 800-90B, Gaithersburg, MD, USA: NIST; 2018.

[49] Suresh VB and Burleson WP. Entropy and energy bounds for metastability based TRNG with lightweight post-processing. IEEE Transactions on Circuits and Systems I: Regular Papers. 2015;62(7):1785–1793.

[50] Rentería-Mejía CP and Velasco-Medina J. High-throughput ring-LWE cryptoprocessors. IEEE Transactions on Very Large Scale Integration (VLSI) Systems. 2017;25(8):2332–2345.

[51] Shaltiel R. How to get more mileage from randomness extractors. Random Structures & Algorithms. 2008;33:157–186.

[52] Li X, Cohen AB, Murphy TE, and Roy R. Scalable parallel physical random number generator based on a superluminescent LED. Optics Letters. 2011;36:1020–1022.

[53] Tkacik TE. A hardware random number generator. In: Cryptographic Hardware and Embedded Systems (CHES), San Francisco Bay, CA, USA; 2002. p. 450–453.

[54] von Neumann J. Various techniques used in connection with random digits. In: Monte Carlo Method, Appl Math Series. Washington, DC: National Bureau of Standards; 1951.

[55] Davies RB. Exclusive or (XOR) and hardware random number generators. In: Tech Rep; 2002.

[56] Zhou Q, Liao X, Wong Kw, *et al.* True random number generator based on mouse movement and chaotic hash function. Information Sciences. 2009;179:3442–3450.

[57] Omar V and Yaksic C. A Study on Hash Functions for Cryptography. GIAC Practical Repository, Sans Institute; 2003.

[58] Long HN and Roscoe AW. Short-Output Universal Hash Functions and Their Use in Fast and Secure Data Authentication. Berlin, Heidelberg: Springer; 2012.

[59] Machicao J and Bruno O. A cryptographic hash function based on chaotic network automata. Journal of Physics: Conference Series. 2017;936(1):012–058.

[60] ElRakaib M. Cryptographic hash function using cellular automata. International Journal of Computer Applications Technology and Research. 2016;5:238–240.

[61] McInnes JL and Pinkas B. On the impossibility of private key cryptography with weakly random keys. In: Menezes AJ, Vanstone SA, editors. Advances in Cryptology-CRYPTO'90. Berlin, Heidelberg: Springer Berlin Heidelberg; 1991. p. 421–435.

[62] Satpathy SK, Mathew SK, Kumar R, *et al.* An all-digital unified physically unclonable function and true random number generator featuring self-calibrating hierarchical Von Neumann extraction in 14-nm tri-gate CMOS. IEEE Journal of Solid-State Circuits. 2019;54(4):1074–1085.

[63] Dodis Y, Gennaro R, Hastad J, *et al.* Randomness Extraction and Key Derivation Using the CBC, Cascade and HMAC Modes. Berlin, Heidelberg: Springer. vol. 3152; 2004. p. 494–510.

[64] Sunar B, Martin W, and Stinson D. A provably secure true random number generator with built-in tolerance to active attacks. IEEE Transactions on Computers. 2007;56:109–119.

[65] Koc C. Cryptographic Engineering. Berlin, Heidelberg: Springer; 2009.

[66] Wang Q and Qu G. A silicon PUF based entropy pump. IEEE Transactions on Dependable and Secure Computing. 2019;16(3):402–414.

[67] Gassend B, Clarke D, van Dijk M, *et al.* Silicon Physical Random Functions. In: Proceedings of the 9th ACM Conference on Computer and Communications Security, Washington, DC, USA; 2002. p. 148–160.

[68] Suh G, O'Donnell C, Sachdev I, *et al.* Design and Implementation of the AEGIS Single-Chip Secure Processor Using Physical Random Functions. ACM Sigarch Computer Architecture News. vol. 33; 2005. p. 25– 36.

[69] William C, Crypt X Package Documentation, Information Security Research Centre and School of Mathematics, Queensland University of Technology.

[70] Walker J. ENT: A Pseudorandom Number Sequence Test Program; 1998. Available from: https://www.fourmilab.ch/random/.

[71] L'Ecuyer P and Simard R. TestU01: A C library for empirical testing of random number generators. ACM Transactions on Mathematical Software. 2007;33(4). Available from: https://doi.org/10.1145/1268776.1268777.

[72] Robert G, Brown DE, and Bauer D. Dieharder: A Random Number Test Suite; 2004. Available from: http://webhome.phy.duke.edu/~rgb/General/dieharder.php.

[73] Bassham LE, Rukhin AL, Soto J, *et al.* SP 800-22 Rev. 1a. A Statistical Test Suite for Random and Pseudorandom Number Generators for Cryptographic Applications. Gaithersburg, MD, USA; 2010.

[74] Institution BS. BS ISO/IEC 18031. Information technology. Security techniques. Random bit generation.

[75] Killmann W and Schindler W. A Proposal for: Functionality Classes for Random Number Generators. Available from: http://www.bsi.bund.de/SharedDocs/Downloads/DE/BSI/Zertifizierung/Interpretationen/AIS_31_Functionality_classes_for_random_number_generators_e.pdf?_blob=publicationFile.

[76] Fischer V and Lubicz D. Embedded evaluation of randomness in oscillator based elementary TRNG. In: Workshop on Cryptographic Hardware and Embedded Systems 2014 (CHES 2014). Busan, South Korea; 2014. p. 16. Available from: https://hal-ujm.archives-ouvertes.fr/ujm-01010404.

[77] Baudet M, Lubicz D, Micolod J, *et al.* On the security of oscillator-based random number generators. Journal of Cryptology. 2011;24:398–425.

[78] Yang J, Ma Y, Chen T, *et al.* Extracting More Entropy for TRNGs Based on Coherent Sampling. In: International Conference on Security and Privacy in Communication Systems, Guangzhou, China; 2017. p. 694–709.

[79] Lubicz D and Bochard N. Towards an oscillator based TRNG with a certified entropy rate. IEEE Transactions on Computers. 2015;64:1191–1200.

[80] Schneider T. Information Theory Primer with an Appendix on Logarithms PDF Version. National Cancer Institute; 2013.

[81] Bayon P, Bossuet L, Aubert A, and Fischer, V. Fault model of electromagnetic attacks targeting ring oscillator-based true random number generators. Journal of Cryptographic Engineering. Berlin, Heidelberg: Springer; 2015: 6.

[82] Bayon P, Pourcheret F, Bossuet L, *et al.* Contactless Electromagnetic Active Attack on Ring Oscillator Based True Random Number Generator. In: Anonymous. Berlin, Heidelberg: Springer; 2012. pp. 151–166.

[83] Markettos AT and Moore SW. The frequency injection attack on ring-oscillator-based true random number generators. In: Lecture Notes in Computer Science. Berlin, Heidelberg: Springer Berlin Heidelberg; 2009. p. 317–331.

[84] Bahadur V, Selvakumar D, Vijendran, and Sobha PM. Reconfigurable side channel attack resistant true random number generator. In: 2016 International Conference on VLSI Systems, Architectures, Technology and Applications (VLSI-SATA), Bangalore, India; 2016. pp. 1–6. doi: 10.1109/VLSI-SATA.2016.7593048.

[85] Yang B, Rožić V, Mentens N, *et al.* TOTAL: TRNG on-the-fly testing for attack detection using lightweight hardware. In: 2016 IEEE Design, Automation Test in Europe Conference Exhibition (DATE), Dresden, Germany; 2016. p. 127–132.

Chapter 7

Micro-architectural attacks and countermeasures on public-key implementations

Sarani Bhattacharya[1] and Debdeep Mukhopadhyay[2]

7.1 Introduction

Micro-architectural features leave footprint in the processor, which is often captured by side-channels. Side-channel attacks allow malicious users to gain access to sensitive data of the system under attack by monitoring power consumption, timing, or electromagnetic radiation of the microprocessor. In recent microprocessors, various architectural components are incorporated in the system to improve system performance; however, these are emerging as new sources of side-channel leakage. The recent micro-architectural attacks named Spectre and Meltdown have taken the world by storm by uncovering two processor vulnerabilities. These vulnerabilities were there in the processor design for decades and use architectural constructs such as the branch predictions and speculative execution optimizations to realize a practical breach of security. Most of the architectural design choices present in modern-day processor architecture have evolved with performance as its foremost criteria. The micro-architectural attacks are eye-openers to computer architecture community that secure implementations can prove to be insecure only because of how it is getting executed on underlying hardware. These architectural components when designed did not consider security evaluation on cross-platform or cloud-based insecure environments that could take the advantages of the component structure to realize a side-channel attack. The present chapter addresses this very important aspect of performance versus secure design choices in computer architecture with respect to several new attacks in this direction.

In this chapter, we will look closely into one of the most important artifacts of computer architecture, which is the branch predictors. This hardware component decides on which direction the instruction pipeline proceeds depending on the history of the previous branch instructions. An accurate branch predictor improves the performance of processor to a great extent. In this chapter, we would delve into the impact of these branch predictors and their history sequences when cryptographic

[1]COSIC, ESAT, KU Leuven, Belgium
[2]Department of Computer Science and Engineering, Indian Institute of Technology Kharagpur, Kharagpur, India

algorithms are run on them. When asymmetric-key cryptographic algorithms are naïvely implemented on computers with such branch predictors, the ciphers are subjected to side-channel attacks because of their key-dependent input sequences. In this chapter, we discuss the vulnerability of monitoring the Hardware Performance Counter (HPC) event branch misprediction in context of such public-key cipher implementations. HPCs are special-purpose registers or counters that are specific to each processor's core on a multiprocessor system and are responsible for providing the micro-architectural event statistics pertaining to each core. Branch misprediction is one of the many events that are monitored by the HPCs. Using HPC-based measurements, we show that unknown predictors even for computing systems from Intel, AMD can be approximated by classic text book 2-bit predictor models. We show that execution footprints in the HPC traces when analyzed accompanied with these models can leak secrets for different variants of Rivest Shamir Adleman (RSA) or Elliptic Curve Cryptography (ECC) implementations.

Subsequently we delve into the aspect of fault tolerance in conjunction with micro-architectural security. It is often conjectured that faults in computing systems are created by hardware injection methods such as voltage or clock glitches and electromagnetic pulses. However, in our work, we develop the first practical software-triggered fault analysis of a 2048-bit RSA implementation by fine controlling a DRAM vulnerability called rowhammer. While rowhammering works by accessing the targeted locations faster than refresh cycles, leading to bit flips in adjacent lines, the crux of the work is in the combination of cache timing analysis and virtual-to-physical address mapping with hammering to increase the probability of a bit flip in the stored key in DRAM. Motivated by the practicality of bit fault models, subsequently we explain a differential fault analysis of simple 2-bit branch predictors and show how well they correlate with that in actual systems. Such strong correlations can lead to attacks against cipher implementation with various forms of countermeasures against side-channel attacks. Next, we further show that popular countermeasures like scalar blinding and splitting can be attacked even without faults using a stronger model of contemporary Intel branch predictors, where the branch misses are observed using asynchronous HPC measurements.

7.2 Related works

Public-key cryptographic algorithms are indispensable in daily life due to authentication, key exchange, and digital signatures, which are required in almost every application we use while connecting to the Internet. In his pioneering work, Kocher [1] showed that the time to process different inputs can be used as a side-channel information to find the exponent bits of the secret keys for RSA, Diffie–Hellman, Digital Signature Standard, etc. Bhattacharya and Mukhopadhyay [2] first established that branch misses from HPCs can reveal the secret key in RSA. Though there have been several proposed countermeasures to prevent Simple Power Analysis (SPA), there exist more powerful attacks such as Differential Power Analysis (DPA), proving simple side-channel countermeasures to be ineffective. In particular, while the authors [2] targeted SPA countermeasures, their attack is ineffective on countermeasures blinding the secret. The most popular countermeasures against DPA in Elliptic Curve Cryptography (ECC) have been proposed by Coron [3] and Clavier and Joye [4], namely, scalar splitting, scalar randomization, and point blinding. All of these blinding countermeasures randomize either the secret scalar or the base point of the curve. They were considered to be effective countermeasures against Hardware Performance Counters (HPCs) side-channel attacks [2]. Such countermeasures may be observed in conjunction with many open-source library like RELIC [5] which are used for the development of cryptographic software and are promoted via corresponding literature [6].

The vulnerability of Branch Prediction Unit (BPU) was first demonstrated in [7] using timing side-channel information from a standard RSA implementation. The variations of the timing information was performed utilizing the underlying BPU, using both synchronous and asynchronous techniques. A more sophisticated form of the branch prediction attack was demonstrated on recent Intel processors in [2], which utilize the strong correlation between the branch mispredictions from the victim system BPU, observed using HPCs and that from a simplistic classic BPU model to retrieve the complete secret exponent of public-key cryptosystems, such as RSA. A differential attack using branch misprediction events has also been carried out elaborately in [8]. In addition to targeting cryptographic algorithms, side-channels on the BPU can be used to perform other types of side-channel attacks, such as deriving kernel and user-level ASLR (Address Space Layout Randomization) offset [9], or covert channels [10,11]. Evtyushkin *et al.* [10] introduced a new covert channel to perform secret communication between the Trojan and the spy processes that exploits the residual state of dynamic branch-predictor behavior of the system, while the authors in a separate paper [11] describe an implementation of contention-based covert channel. The authors also provide a comparison of both types of covert channel communication in both the single-threaded and simultaneously multithreading environment. An attack has been developed in [9] to derive kernel and user-level ASLR offset that exploits the BTB (Branch Target Buffer) collisions between the branch instructions. A recent work named BranchScope [12] from the same group of authors targets the directional branch predictor and shows a series of steps that can force the complex structure of hybrid predictors to collide to the same data structure.

BranchScope [12] aims to create an interference of two branch instructions, one from the victim that is targeted by another branch owned by the spy. This is handled by using a prime and probe attack strategy, where a spy thread primes the bimodal predictor to one of the required states and allows the victim branch to run. The execution footprint of the victim is revealed in the probe phase, where the spy comes and detects the nature of the branch misprediction for its own execution and gains an insight of the previous state. BranchScope shows that this could be achieved in two different ways: first is the collision in the Pattern History Table (PHT), which could be difficult to achieve for all existing processor architectures. The authors thus recommend a second way to create a prime+probe strategy to work. This begins with confusing the two-level branch predictor with crafted branching sequences such that the predictor starts using the prediction output from the first-level (bimodal) predictor instead of the one with indexed PHT as the final prediction output. Once the final prediction begins to get decided from the first-level prediction, the spy uses the collision in the first-level predictor to get the secret from the victim process.

7.2.1 Speculative execution

Speculative execution was first introduced by Intel in their Pentium Pro and Pentium II processors over two decades ago. Soon AMD and ARM followed suit with their K9 and Cortex A9 series processors. Before speculative execution became a must-have feature in almost all CPUs, the earlier processors used what is called "in-order" processing. The instructions were executed in the order they were given to the processor without any rearrangement. The major problem with this kind of approach is that the entire CPU stops in an event of a pipeline stall. To increase parallelism, multiple instructions are processed in a pipelined manner in modern systems. Also, the instructions are not executed in the order they are issued. Modern processors use "Out-of-Order" Execution (OoOE), where the instructions are rearranged and executed without changing the final state of the registers and memory. Speculative execution plays an important part in achieving OoOE along with branch prediction and data flow control mechanism.

7.2.2 Speculative execution attacks

Although speculative executions do not alter the architectural state of the processor, they often affect the micro-architectural states, such as information stored in Translation Lookaside Buffers and caches. These information create a side-channel leakage that has been recently exploited in Spectre [13] and Meltdown [14] attacks. While Meltdown exploits a bug Intel's OoOE engine, Spectre uses the branch predictors to mount practical attacks. In OoOE, the exceptions are not handled until the retirement of the instruction. Therefore, in the case of an exception, the instructions are executed speculatively, which creates changes in micro-architectural states before getting discarded. The side-channel leakage thus produced has been exploited in to read data from the faulty memory access before the changes are rolled back by the processor. Spectre [13], on the other hand, comes in different variants, where variant 1 mistrains the BPU and variant 2 injects BTB with attacker-chosen target addresses. These attacks can bypass process isolation imposed by the operating system and allow cross-process BTB in injection to force victim process to execute arbitrary speculative execution.

7.3 Branch-predictor security

Computer architecture has evolved over decades with performance improvisation being its sole motivation and objective. All of these constructs of micro-architecture are heavily explored research topics for significant long time with performance as the main concern, until recently when security engineers started evaluating these constructs in the presence of intelligent adversaries. In this chapter, we elaborate the security evaluation of one of the most important architectural components—the branch predictors. The role of being a security engineer is a difficult task, since given a decade-long history of branch-predictor evolution, it is indeed difficult to guess which particular design has been incorporated in a modern-day multipurpose commercial system. We started off with the difficult task of partially reverse-engineering the branch predictor and found a surprising similarity to that of the simple bimodal structure. The correlation between the unknown underlying predictor and the known predictor model was high enough to craft a secret-recovery attack on public-key ciphers. Using this result, multiple popular implementations of public-key ciphers were proven to be vulnerable.

The work motivates itself from the fact that there exist no security guidelines to sensitive cryptographic applications executing in multi-tenant server or cloud environment in context of micro-architectural attack. There still exists open-source legacy library implementation such as RELIC [5] and text books [6], which suggests these libraries to be efficient, but these libraries are highly vulnerable to micro-architectural attacks, which we elaborate later in this chapter. Thus, the objective of this chapter is to create a conscious understanding of the design choices from security engineering point of view and lead to inherent secure alternatives. The basic objectives can be articulated with the following:

- HPC-based evaluation of side-channel leakage of asymmetric key algorithms: targets the public-key framework and exploits the secret-dependent conditional structure of the most popular algorithms using the branch misprediction statistics from HPCs. The objective was solely to report the inherent vulnerability of the branch predictor unit in the context of the implementations of the public-key ciphers.
- HPC-based evaluation of side-channel leakage of asymmetric key designs with countermeasures: aims to evaluate the security of the DPA-secure countermeasures in the presence of granular observations of branch mispredictions from HPCs. The implementations of public-key cryptographic algorithm in the presence of blinding countermeasures are observed to be still vulnerable when branch mispredictions were observed over periodic intervals.
- Software-triggered fault analysis of asymmetric key ciphers using rowhammer: explores the DRAM reliability event rowhammer to induce bit flips in the cryptographic secret. The work also formally models the fault analysis of the branch predictors in the presence of the software-driven bit-flip model.
- The final objective is to explore an HPC-based side-channel online monitor and design of architectural artifacts, in particular BPU, with security as its design criteria.

Before exploring the security of branch predictors, the basic architecture designs of branch predictors are explained next.

7.3.1 Dynamic branch predictor

The 2-bit dynamic branch-predictor state machine is one of the various predictor algorithms that is most often used in practice [15] (Figure 7.1). This is a deterministic

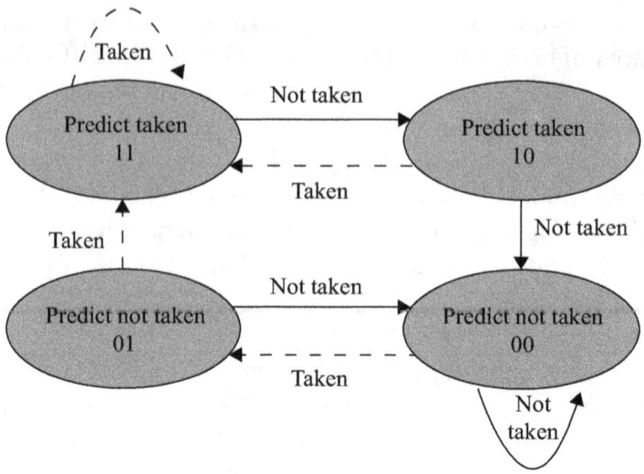

Figure 7.1 Dynamic 2-bit predictor state machine [15]

algorithm predicting next branch to be *taken* or *not taken* depending on the history of previously taken branches. In a 2-bit prediction scheme, the predictor must miss twice before the prediction changes.

But conditional branches that occur in a regular recurring pattern are not predicted well by this bimodal predictor.

In such cases, a two-level adaptive predictor [16] works better as the predictor remembers the last k occurrences of a branch instruction and uses an s-bit prediction function (such as an s-bit predictor) for each of the 2^k history of patterns. The first level of the two-level adaptive predictor uses a *branch history register* that is a shift register storing the history of the last k branches. The branch history register indices to a second level called *PHT*, which can hold 2^k entries, each of s bits. When a conditional branch B is getting predicted,

7.3.2 Branch predictors and branch mispredictions

A *branch* in a program can be caused by a conditional jump (e.g., an `if-else` block). When the condition is satisfied, the branch is *not-taken*, i.e., the next instruction to be executed is the one coming after the branch. When the condition is not satisfied, the branch is *taken*, i.e., the next instruction to be executed is the one at the address of the argument of a jump instruction. Compiler optimizations do not affect this.

Commonly, the implementations of public-key exponentiation algorithms and the scalar multiplications algorithms in ECC contain a conditional block, where the branches are conditionally dependent on secret key bits. This is the case in the double-and-add algorithm or the SPA resistant Montgomery ladder algorithm. Let the n-bit secret scalar in ECC be denoted as $(k_0, k_1, \ldots, k_i, \ldots, k_{n-1})$. The trace of taken and not-taken branches depending on scalar bits is expressed as $(b_0, b_1, \ldots, b_{n-1})$.

- If $k_j = 1$, then the conditional addition statement in the double-and-add algorithm gets executed. The branch is not-taken, i.e., $b_j = 0$.
- If $k_j = 0$, then the addition operation is skipped and the execution continues with the next squaring statement. The branch is taken, i.e., $b_j = 1$.

In order to avoid pipeline stalls, the predictor predicts whether the next branch is taken or not, based on the history of branches that have already been encountered. Predicted instructions are then fetched in the instruction pipeline. The condition is only evaluated during the "execute" stage. In the case of a mismatch between the predicted and the evaluated branches, the corresponding instruction is flushed from the instruction pipeline resulting in a pipeline stall. This is called a *branch misprediction* or a branch miss.

7.4 Branch misprediction attack

BPU was earlier shown to lead to attacks using timing as side-channel by authors in [17] under certain assumptions. In [2], the authors demonstrated an attack on the branch-predictors of contemporary Intel systems such as Intel Core i5, Core i3, and Core-2 Duo using the HPC event counts. It is to be noted that in the newer version of the Linux kernels, HPCs have been made unavailable for the user-level process access, as it conveys most precise information about the hardware performance events in a system. Branch misprediction penalty could also be measured from other side-channels such as execution time, latency, and power signals, but since these are affected by many more factors other than branch misprediction, the data retrieved would be high in noise. But this does not limit the attack with information obtained only through HPCs, and this could be alternatively applicable with timing side-channel with relatively more data samples. Surprisingly, being oblivious of the underlying branch predictor, the attacker approximates them by a very well-known primitive structure of predictor, which correlates well with the actual statistics of branch misses reported via HPCs. Using these approximations, the adversary modeled an iterative attack algorithm that can successfully recover the entire secret key of RSA.

Modular exponentiation of RSA has been implemented using both naïve square and multiply and Montgomery ladder, while the underlying multiplication and squaring have been implemented using Montgomery's method. The iterative attack as illustrated in [2] targets the key-bits of 1,024-bit RSA, where in offline phase, the system's underlying branch predictor is approximated by a theoretical predictor in literature. The authors perform subsimulation to classify the message-space into distinct partitions based on the event branch misprediction and the target key bit value. In online phase, the secret key bit is targeted using branch mispredictions obtained from the hardware performance monitors that reflect the information of branch miss due to the underlying predictor hardware. The attack works in two phases:

- An offline phase, during which the system branch predictor is approximated by a theoretical model (namely, 2-bit or 2-level adaptive predictor) and is used to classify the message-space into distinct in partitions based on the event of branch misprediction and assuming the value of the *i*th key bit. After this step, four different input sets are created based on guessed bit value and the event misprediction.

- In the online phase, the *i*th key bit is retrieved using the branch mispredictions obtained from the values of the performance monitors. In this step, a pair of input sets for a particular guess of the *i*th key bit from the partitioned set shows a correct separation in branch mispredictions from HPCs, while the other pair exhibits an incorrect separation. The guessed bit with the correct separation is shown to be exactly the same as the correct secret bit.

In this section, we illustrate that powerful side-channels exist through the HPCs that monitor the branch misses. Figure 7.2(a) illustrates that the 2-bit bimodal

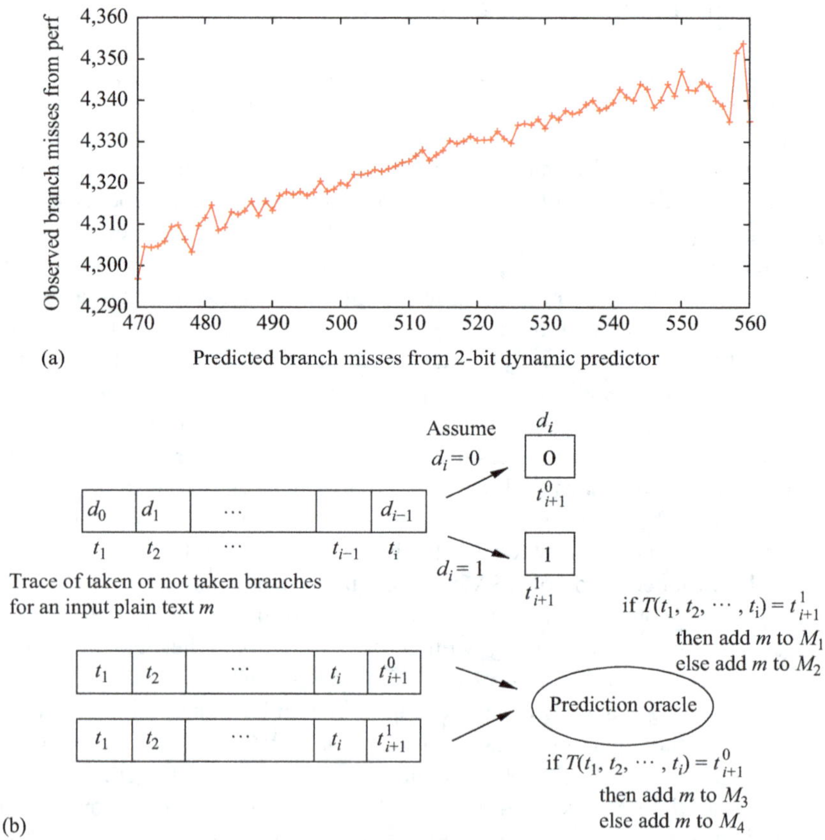

Figure 7.2 (a) Variation of branch misses from performance counters with increase in branch miss from 2-bit predictor algorithm [2] and (b) partitioning randomly generated ciphertexts set based on simulated branch miss modeling [2]

predictors bear a direct correlation to the underlying hardware predictor in Intel systems. Subsimulations are performed to classify the message-space into distinct partitions based on the event branch misprediction and the target key bit value as illustrated in Figure 7.2(b). In online phase, the secret key bit is ascertained using branch mispredictions obtained from the hardware performance monitors that reflect the information of branch miss due to the underlying predictor hardware. The authors also provide an analysis to justify that the probability of success to guess the *i*th bit correctly is equivalent to the extent of resemblance of theoretical predictor behavior to the underlying system predictor hardware. Figure 7.3 shows the correct and incorrect separations for all the four sets (separated by simulations over two-level adaptive predictor) for the randomly chosen 548th bit location of the target key-stream. Figure 7.3(a) plots average branch misses observed from performance counters for each elements in sets M_1 and M_2 (each set having $L = 1{,}000$ elements) and the experiment is repeated over $I = 1{,}000$ iterations in order to check the consistency of the output.

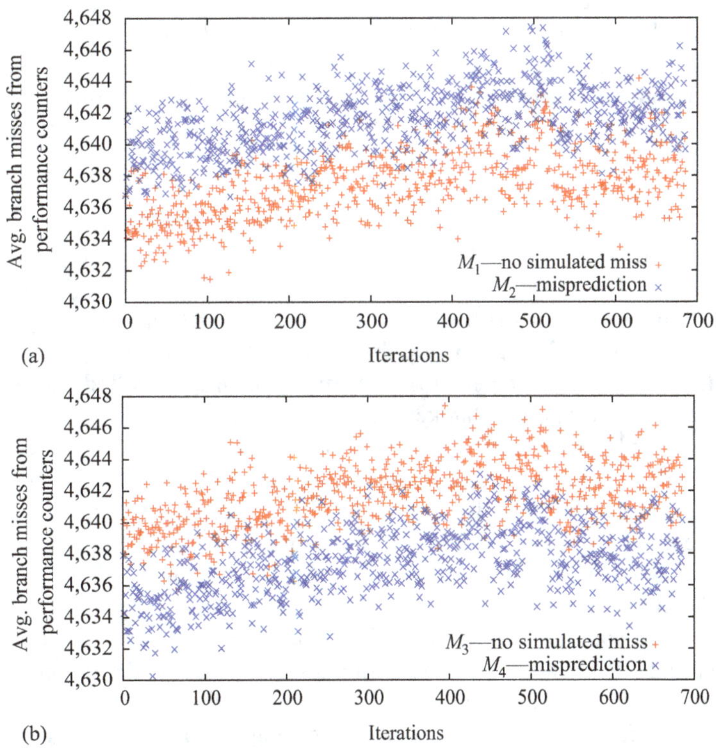

Figure 7.3 Branch misses from HPCs on square and multiply correctly identifies secret bit $d_i = 1$, ciphertext set partitioned by simulated misses of two-level adaptive predictor [2]: (a) correct assumption $d_i = 1$ and (b) incorrect assumption $d_i = 0$

It is evident from the figure that in most of the iterations the average branch miss for set M_2 is more than the branch misses for set M_1(as expected). On the contrary, Figure 7.3(b) plots average branch misses observed from performance counters for each elements in sets M_3 and M_4. But we observe an incorrect separation as in most of this case, ciphertexts in set M_4 is having lesser branch misses than in set M_3, which is incorrect, since theoretically it should be the reverse. Thus, from these two figures, the correct exponent can be easily identified showing correct difference in branch misses.

The authors also fine-tune this original attack strategy in [18] such that this alternative strategy can retrieve the secret bits much more efficiently and requires much lesser number of inputs. The experimental validations of both the attack strategies are illustrated on public-key cryptographic algorithms as RSA and ECC on several Intel platforms show significant success revealing the secret exponent and scalar value. The attacks have also been extended to the RSA-OAEP-randomized padding procedure where the adversary targets the decryption phase of the implementation and the branch miss side-channel information of the entire decoding procedure and can be successfully exploited to reveal the secret exponent. What makes these results more relevant is the fact that protections that fuzz the timing channels are not sufficient to thwart these attacks, and present performance counters as a distinct side-channel that attracts attention to further systematic research to develop systems that are inherently secure.

7.5 Inserting real-time faults in public-key secret using rowhammer

In this section, we explain software-induced fault induction in the cryptographic secrets as appears in [19]. In this work, the authors bring into practice a methodology to combine timing analysis to perform rowhammering in a controlled manner in order to create bit flips in cryptographic keys that are stored in memory. The attack would require only user-level privilege for Linux kernel versions before 4.0 and is unaware of the memory location of the key. An intelligent combination of timing prime + probe attack and row-buffer collision is shown to induce bit-flip faults in a 1,024-bit RSA key on modern processors using realistic number of hammering attempts. This demonstrates the feasibility of fault analysis of ciphers using purely software means on commercial x86 architectures.

The authors combine knowledge of reverse engineering of Last Level Cache (LLC) slice and DRAM addressing with timing side-channel to determine the bank in which secret resides. The overall idea of the attack involves three steps. The attacker successfully identifies an *eviction set* by observing the differences in variation of timing samples observed as illustrated in Figure 7.4(a) and (b). The eviction set comprises of a set of data elements which maps to the same cache set and slices as that of the secret exponent. Construction of such set comprises of timing analysis using *prime* + *probe* methodology. This set essentially behaves as an alternative to `clflush` statement of x86 instruction set.

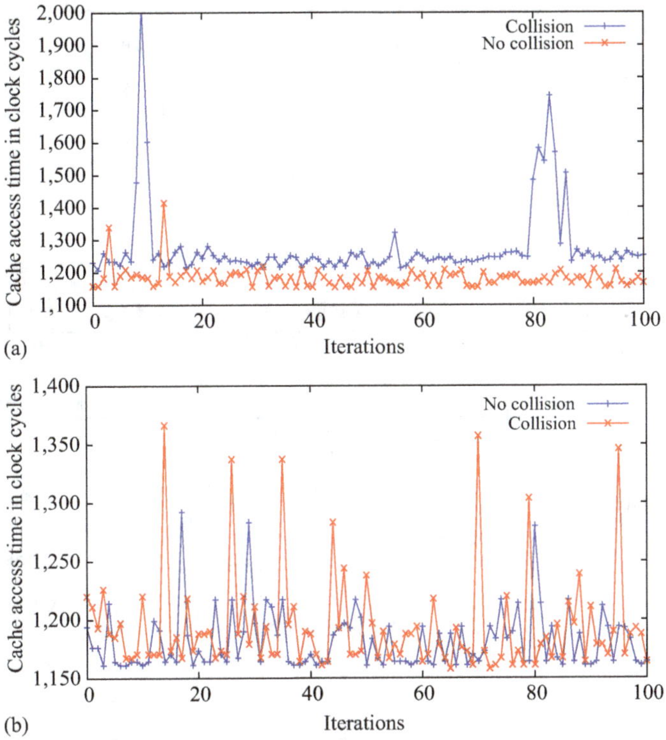

Figure 7.4 Timing observations of (a) cache set collision from optimal eviction set (LLC) and (b) cache slice collision [19] using equations in [20]

In the next phase, the attacker observes timing measurement to DRAM access by following the timing variation as in Figure 7.5(a) and (b) to determine the DRAM bank in which the secret resides in. The variation in timing is observed due to the row-buffer conflict forced by the adversary. This helps in precisely triggering rowhammer to addresses, which collides to the same bank as the secret. This increases probability of bit flip in the secret exponent and the work provides a series of steps to improve the controllability of fault induction.

A statistic over observations of bit flips in respective banks is reported in Figure 7.6. The bar graph shows the number of bit flip instances that were observed for respective banks of a single Dual In-line Memory Module (DIMM). A real-time software-driven fault attack on cryptographic secret was demonstrated in this work. This makes the software bit-flip fault models to be practical. It also explains a series of steps which takes the advantage of the DRAM reliability phenomenon named rowhammer and develops a workable environment that could induce bit flips in the cryptographic secret.

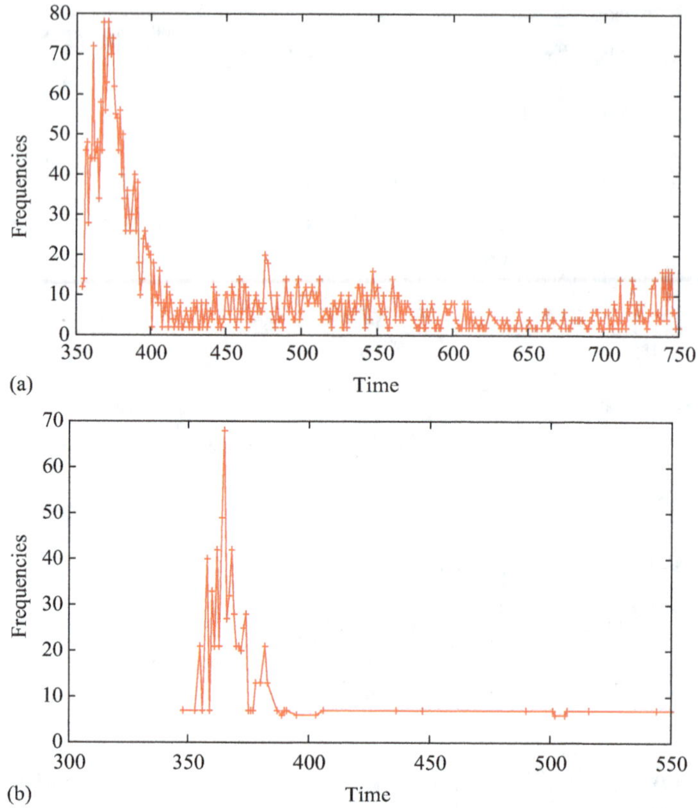

Figure 7.5 Timing observations for row-buffer collision in clock cycles (a) for DRAM bank collision and (b) of separate DRAM bank access [19]

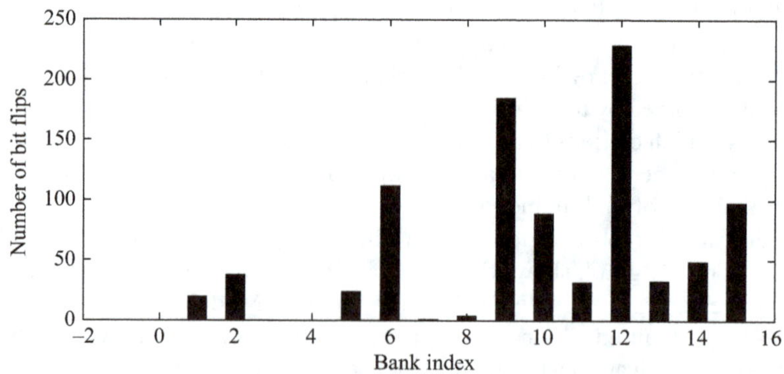

Figure 7.6 Number of bit flips observed in all banks of a single DIMM [19]

7.6 Fault attack revealing secret keys of exponentiation algorithms from branch prediction misses

In this section, we extend our study of the combination of branch-predictor algorithms and performance counters as in [21] to explore a fault attack on the popular square-and-multiply-based exponentiation algorithm used in RSA. We demonstrate the power of the attack exploiting branch misses from performance monitors by formalizing a fault attack model, where the adversary is capable of performing a bit flip at a desired bit position of the secret exponent.

In this work, the authors model the effect of the bimodal predictor to exploit the side-channel leakage of branch misses from the performance counters. The characterization of the differential of branch misses from correct and faulty branching sequences is made based on the behavior of 2-bit predictor. If there is a sequence of n branches denoted as $(b_0, b_1, \ldots, b_{n-1})$ generated from execution of the algorithm under attack, a fault at the ith execution of the algorithm changes the branching decision for the ith instance. The difference in branch misses (Δ_i) between the correct branching sequence $(b_0, b_1, \ldots, b_i, \ldots, b_{n-1})$ and the faulty sequence $(b_0, b_1, \ldots, \overline{b_i}, \ldots, b_{n-1})$ simulated theoretically over a 2-bit predictor algorithm can be at least -3 and at most 3. The simulated difference for any branch sequence will always take a value in the range $[-3, 3]$ and this depends on the subsequent branching sequences starting from the ith flipped branch. This effectively means that the branching decision for the ith instance has a great impact in determining the range of Δ_i. Based on the formal model of these differences in branch misses, some selected examples are illustrated in Figures 7.7(a), (b) and 7.8(a), (b). This characterization is exploited to develop an iterative attack algorithm where knowledge of the previously determined key-bits and the difference of branch misses (as gathered from the performance counters) are utilized to determine the next bit. This formal model takes the difference of branch mispredictions into consideration and characterizes based on the known bits and the future guesses, which eventually leads to an attack algorithm that demonstrates a full-key recovery.

7.7 Deduce and remove attack on blinded scalar multiplication with asynchronous perf ioctl calls

In this section, we explore branch prediction attack on the protected implementation of ECC algorithm as appears in [22]. In this work, the authors reverse engineer the undisclosed model of Intel's Broadwell and Sandybridge branch predictor and further utilize the largely unexplored perf ioctl calls in sampling mode to granularly monitor the branch prediction events asynchronously when a victim cipher is executing. With these artifacts in place, the scalar blinding and splitting countermeasures are targeted to develop a key retrieval process using a technique called deduce and remove. The deduce step uses template based on the number of branch misses as expected from the 3-bit model of the BPU to infer the matched candidate values. In the remove step, any

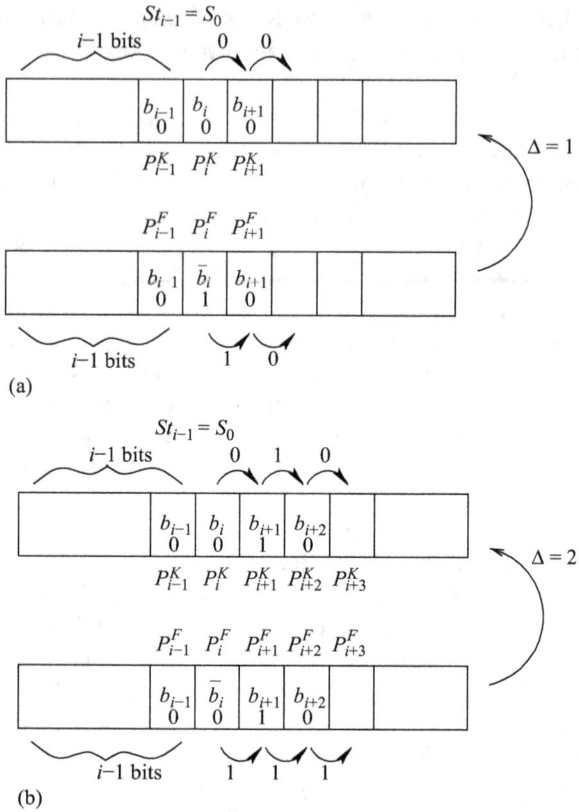

Figure 7.7 Variation of simulated branch misses on the ith branching decision having $St_{i-1} = S_0$ [21]: (a) $\Delta_i = 1$ and (b) $\Delta_i = 2$

erroneous conclusions are corrected by using the properties of the blinding technique under attack.

The success rates of the attack on scalar splitting are illustrated in Figure 7.9. The figure shows the percentage of sequences that has successfully revealed the blinded scalar for four consecutive bit positions. Individual template matching on each of the split scalars reveals the split scalar and the random mask efficiently. The experiments revealed 4 bits consecutively on Curve1174 long integer multiplication over split scalars and applied the final check using the *remove* step to prune out the wrongly deduced scalars. Similarly, this attack is highly efficient to retrieve blinded scalar bits in a scalar randomization countermeasure as illustrated in Figure 7.10. In this figure, success, failure, and inconclusive percentages of the blinded scalar are plotted which are retrieved. Since the template matching algorithm reveals 1 bit at a time, the *remove* step for scalar randomization cannot be applied to each bit. In this case, the

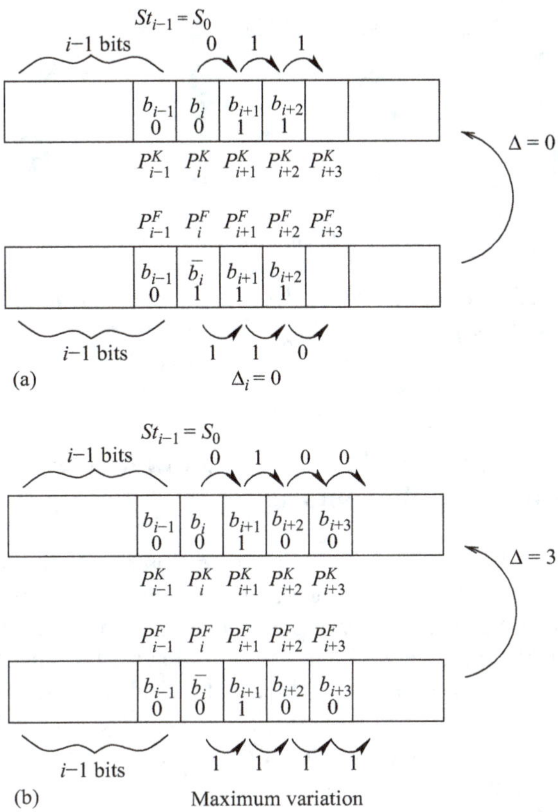

Figure 7.8 Variation of simulated branch misses on the ith branching decision having $St_{i-1} = S_0$ [21]: (a) $\Delta_i = 0$ and (b) maximum variation

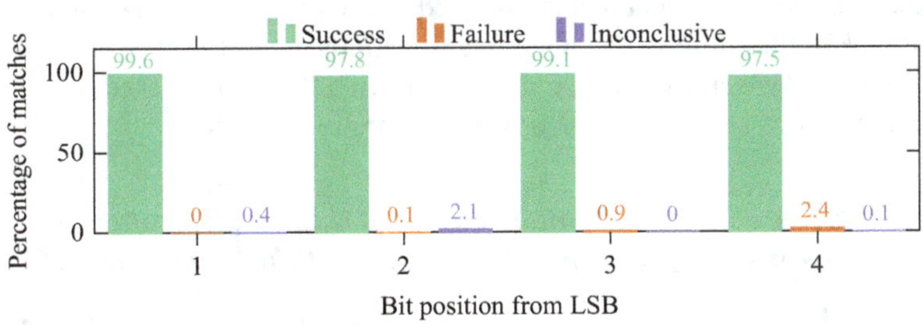

Figure 7.9 Probabilities of bit retrieval for scalar splitting in Curve1174 [22]

template matching step is performed on *t* multiple bits subsequently and the *remove* algorithm is applied on these multiple retrieved bits together. The experimental results illustrate the effectiveness and efficiency of the proposed attack for scalar splitting, scalar randomization, and point blinding countermeasures.

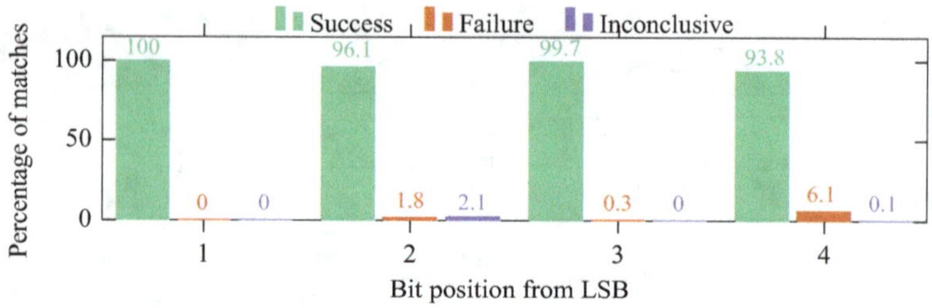

Figure 7.10 Probabilities of bit retrieval for scalar randomization in Curve1174 [22]

7.8 Extending *deduce* and *remove* to a publicly available cryptographic implementation

Continuing the discussion on Curve1174, in this section we will implement the attack from the previous section to RELIC, a publicly available cryptographic library. In this section, we illustrate the applicability of our branch-prediction-based template attack to a publicly available implementation of Curve1174 named RELIC-toolkit. RELIC-toolkit [5] is a modern cryptographic library that provides several ECC and pairing implementations, implemented with emphasis on efficiency and flexibility [6]. There are a number of algorithmic choices that a developer can use while executing the particular curve operations.

We target the elliptic curve scalar multiplication algorithm provided in the RELIC library as in Listing 7.2. The list for each of these operations is illustrated in Appendix 7.12. The elliptic curve addition operation is conditioned on the secret bits of the scalar. Thus, for every bit, the doubling step gets executed, but there is a conditional addition statement that is executed when the bit is 1. The sets of equations by which each of the addition and doubling operations for Curve1174 is implemented are different (see Appendix 7.12), unlike the unified doubling and addition formula used in the previous section for performing scalar multiplication.

7.8.1 Difference in branch misprediction due to difference in operations involved in Addition and Doubling in RELIC

We incorporate a template-based analysis targeting the branch mispredictions of the operations that follow the statement conditional on secret bits. The doubling operation will be executed regardless of the bit being 0 or 1. The operation following doubling is an addition if the secret bit is 1 else a doubling for next bit. So we target the operation following the unconditional doubling. In the setting of RELIC implementation, we took a sequence of masked scalars and targeted the particular operation that follows the unconditioned doubling. We plot the branch misprediction samples from HPCs for these two operations separately in Figure 7.11.

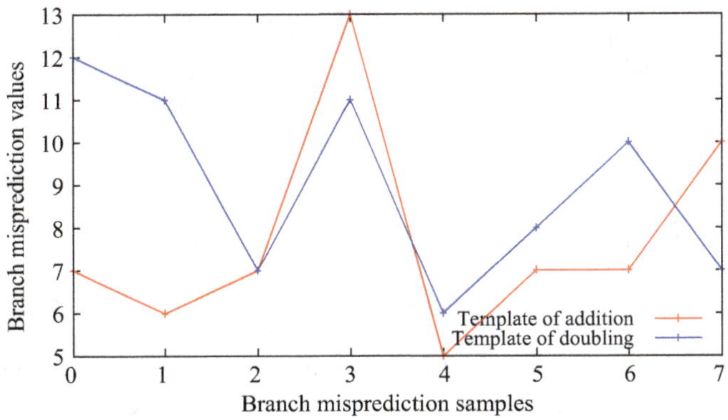

Figure 7.11 *Difference in templates of the conditioned addition and the subsequent doubling step executed in two different execution scenario for the target bit*

7.8.2 Template building and matching in RELIC

The RELIC library comes with a number of test codes along with the implementation. We target one such test code for Curve1174. The template building and matching steps are such that here we construct the branch misprediction templates for doubling and addition for every bit instead of looking into individual modular reductions involved in each multiplication operation. For each blinded scalar bit, we do the following steps:

- For each guess of the scalar bit: (a) we obtain the branch misprediction samples from perf, (b) simulate mispredictions using 3-bit predictor and predict that whether the execution of next operation (doubling if 0 and addition if 1) results in branch miss.
- Similar to the previous attack, at the end of the template building stage we gather corresponding template traces for four different bins, $bin\_miss_{(i=0)}$, $bin\_nomiss_{(i=0)}$, $bin\_miss_{(i=1)}$ and $bin\_nomiss_{(i=1)}$, i.e., misprediction and no misprediction bins for each of 2 bits.
- Like the previous attack methodology, we incorporate an iterative attack on each bit of the blinded scalar. The algorithm under analysis being an MSB first algorithm, Figure 7.12 illustrates the success rate of the scalar-blinded bits retrieved correctly using the template matching technique.

This work aimed a higher order attack on the blinding countermeasures of the asymmetric key cryptographic algorithms. In the course of this analysis, a partial reverse engineering of the underlying predictor is performed with the help of the granular observations of branch mispredictions. This process of granular observation was entirely asynchronous to the target executable thus justifies targeting a real-life crypto execution scenario. This work demonstrated a template-based deduce and

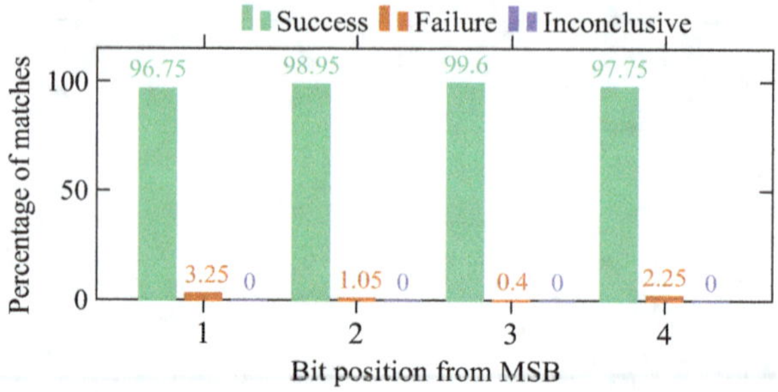

Figure 7.12 Probabilities of bit retrieval for scalar randomization in RELIC implemented Curve1174

remove algorithm to guess partial secret bits and finally remove wrong candidates using the properties of the implementation. This made it even more clear that even implementations that are inherently free from balanced branch-based structures could be put to threat by exploiting the data-dependent modular reduction implementations, where a generic implementation of constant time reduction operation is difficult to conceive and inefficient to use.

7.9 Online detection and reactive countermeasure for leakage from BPU using TVLA

Through our previous sections, we have established that a detection and solution to this leakage are not going to get achieved by patching the implementations here and there. It needs a serious detection framework, even in scenarios where the underlying algorithm is not much known to the user. We illustrate a branch misprediction leakage-detection framework that triggers an alarm on detecting leakages using Test Vector Leakage Analysis (TVLA). In this section, we explore an online leakage-detection tool Branch-Monitor [23] for branch misprediction traces that does an online detection of the leakage and raises an alarm if there exists a significant difference in the distribution of branch misses for selected inputs. Branch-Monitor is proposed as a watchdog to the information leakage from the HPCs. The information leakage for a shared user platform from the branch misprediction hardware is significant and our proposed detection tool can detect the leakage by applying a statistical test on the captured traces of the branch mispredictions using the perf "ioctl" calls. Branch-Monitor tool observes the branch misprediction traces for an executable and performs an online t-statistic test over time. For each set of points in the trace, the statistic is modified accordingly, and an alarm is raised if the value exceeds the threshold + or −4.5.

The authors have targeted the Edward-1174 curve implementation written with long integer multiplications compiled using GNU C compiler. Figure 7.13(a) and (b) represents two scenarios where the leakage is detected in as low as 30 traces on an Intel-i5-5200U processor running Ubuntu 16.04 LTS, and a particular part of the trace is observed to leak more as the number of traces is increased. The $|t|$ statistic is calculated over each point on the trace, and the figures clearly show that both of the traces leak information since the $|t|$ statistic is having a value above 4.5 and below −4.5. The x-axis in the figures represents the individual sampling point of a branch misprediction trace and the y-axis values represent their appropriate $|t|$-statistic value. The experiments on a comparatively older processor Intel-i5-3470 running Ubuntu-12.04 LTS, and the leakage-detection scenario is illustrated in Figure 7.13(a) and (b). The figures clearly show that there are multiple points of leakage where the $|t|$ statistic crossed the range (4.5, −4.5).

Randomizing the internal states of the branch prediction hardware is a practical countermeasure to thwart these forms of leakages. The BPU being shared by all processes of the same processor core, the effect of one process is predominant on the concurrently running execution on the same processor core. The t-statistic test is performed on the execution time of the Edward-1174 curve implementation when the

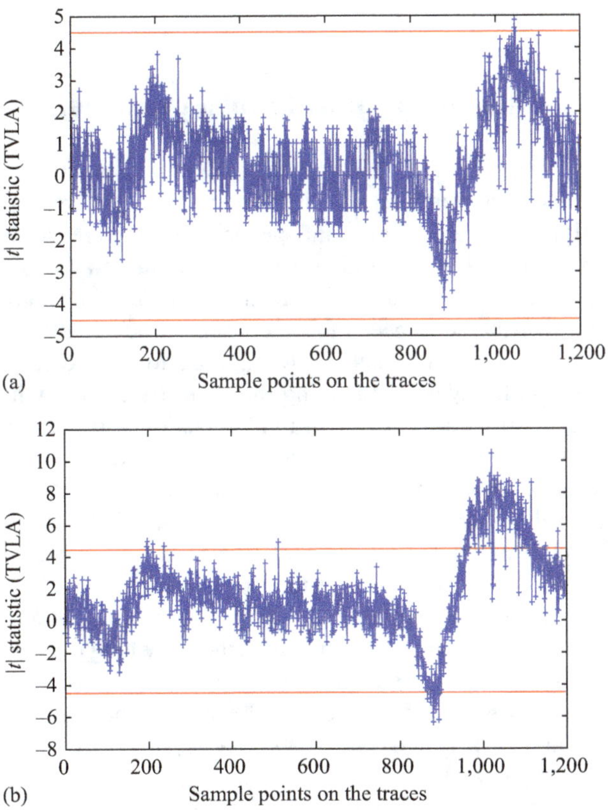

Figure 7.13 Test vector leakage analysis on Edward-1174 implementation on Ubuntu 16.04 in Intel Core i5-5200 [23]: (a) TVLA on 30 traces and (b) TVLA on 200 traces

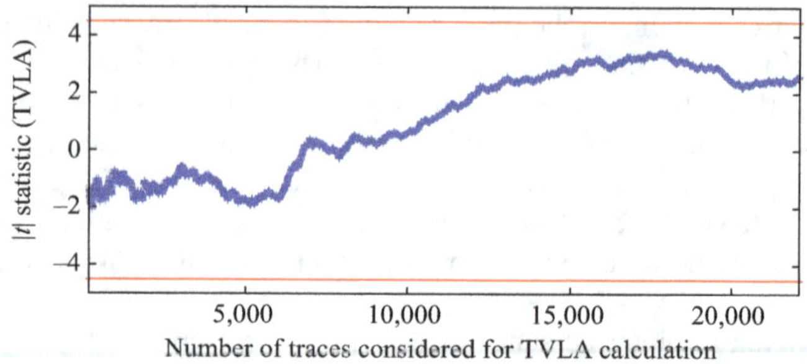

Figure 7.14 Variation of timing-TVLA with increasing number of traces [23]

randomization module is also getting executed in parallel to the cipher implementation. Figure 7.14 illustrates how the $|t|$-statistic is changing as the number of traces is increased, and even after observing for 25,000 traces the TVLA value indicates that there is no leakage through timing channels.

7.10 General mitigation against branch prediction attacks

At the *application level*, the first and the most obvious countermeasure suggested by the previous works in literature is to implement the scalar multiplications such that the control flow of the execution is independent of the secret scalar K. However, in this chapter we showed that the data-dependent branches can still be attacked by the adversary and thus are inefficient to thwart branch misprediction attacks. Thus, we suggest that the code needs to be constant-time such that it is free from both secret and data-dependent branching. Randomization of projective coordinates, like point blinding, only involves changing the input base point without changing the secret-dependent structure. Projective randomization and point blinding would be vulnerable if we target the naïve Montgomery ladder with branches depending on secret. However, when secret-dependent branches are removed, the present attack does not directly apply. Thus, the attack presented in Section 7.8 still finds relevance in the presence of such point randomization, since the control flow of execution depends directly on secret bits. However, the fact that the modulo operations have branches depending on the data can be exploited using cross-correlating branch miss traces. For doubling operation, there are many pairs with the same long integer multiplication inputs (and thus have the same branch pattern).

Another countermeasure is to randomize the state of the predictor intermediate to the execution. This can be achieved, e.g., at the *system level*, by introducing random branching executions in parallel to the execution of the victim library. The objective is to randomize the state of the predictor to inhibit the *deduce* step of the attack.

Randomization in several layers of algorithm and measurement can only make the attack difficult, as the adversary could be more powerful and have more fine-grained traces. This brings us again to the open challenge of rethinking the structure of the branch predictor at the *hardware level*, such that they are inherently secure against these attacks.

7.11 Existing countermeasures

The countermeasures proposed for all the micro-architectural attacks uncovered till date are very specific to the target implementation or target hardware. Depending on the proposed countermeasure, we divide them in the following types.

7.11.1 Altering the structure of the target implementation

Asymmetric-key cryptographic implementations are appropriate examples of this specific genre. These algorithms are often iterative and sequential in nature. The naive implementations of these algorithms are the most efficient implementations of such algorithms proposed but the unbalanced structure of the algorithms is vulnerable to timing side-channel attacks. As for example, the square and add algorithms of RSA or their equivalent double and add versions of ECC algorithm, though vulnerable to timing attacks but have faster execution time compared to the timing side-channel secure Montgomery ladder. Though the Montgomery ladder was thought to be secure against timing attacks, it was found to be vulnerable to branch prediction attacks. In the case of always double and add algorithms for ECC implementation, the nonconstant time conditional reductions in the underlying Montgomery multiplication subroutine makes such implementation insecure against timing vulnerability and branch misprediction attacks. In this context, Daniel Bernstein's Networking and Cryptographic library (NaCl) [24] is one efficient library that performs all cryptographic implementations in constant time using bit-wise operators. The library is highly efficient in terms of performance and is shown to have no leakage with respect to timing side-channel. However, this library supports popular algorithms and their implementations, some special curves are not included in the library.

On the other hand, there still exists open-source legacy library implementation such as RELIC [5] and text books [6] that suggest these libraries to be efficient. But these libraries are nonconstant time and are highly vulnerable to micro-architectural attacks. Thus, in most of such detected vulnerabilities, patching implementations and making them constant time for a very specific platform and/or specific algorithm is the most common countermeasure. Making an implementation constant time does not always make it side-channel secure [25]. This paper by Yarom *et al.* shows that even a constant time OpenSSL RSA implementation could be subjected to stronger forms of the prime and probe-based cache attacks. Also, it was reported by Reparaz in [26] that even the most constant time library implementations are not actually constant time, and they show leakages when large number of traces reveal the leakage points.

7.11.2 Patching architecture here and there

The attempt to completely do away with different forms of cache-based attacks with inherent modification to cache organization was proposed in [27,28]. Similarly, based on an information theoretic leakage model, the hardware prefetcher of the processor was shown to be vulnerable, and in this context the even–odd sequential prefetcher was shown to be a secure alternative as described in one of our papers [29]. Also in context to buffer overflow attacks, the common solutions that thwart buffer-overflow-based attack incorporate bound checking, while the accesses requested from the processor are addressed. Bound checking seems to be a reasonable solution to such problem, also addressed in one of our papers [30], which shows that intelligent bound checking could thwart all the forms of control flow vulnerabilities. But on commercial systems, there are even more optimization constructs that were purely incorporated with performance as its basic concern. As for example, recent vulnerabilities like Spectre [13], Meltdown [14], and Foreshadow [31] showed that even deferred bound checking could be detrimental in the presence of speculative execution and OoOE. This definitely brings us to the bigger problem of performing an overall security analysis of the target hardware and algorithm. Securing components separately would not be a wise design choice. Secure design of the branch predictor unit has not been explored and targeted in literature. In this chapter, we make an attempt to give an overview of the vulnerabilities of BPU and also discuss the secure design alternatives to the existing predictors.

7.11.3 Countermeasures and patches are expensive

Considering parameters such as area, performance penalty, and energy consumption, hardware countermeasures are faster and are considered to be more reliable. However, the issue with proposing a new hardware every time is that they are required to be implemented by hardware manufacturers to be operational. The countermeasures are not always as fast as the original implementation. Eventually, most academic countermeasures are never deployed in practice.

There exist a number of surveys on micro-architectural attacks and countermeasures where each of them provides a vivid description of each of these countermeasure along with their pitfalls and benefits [32–36]. The timing attack countermeasures typically try to make the number of instructions constant such that there is no dependency on timing variation in such cases. Further data dependency is also removed from such algorithms. Balancing statements may lead to leakages through branch misprediction events. To completely get rid of such leakages, conditional statements are suggested to be avoided, which leads to somewhat inefficient implementations than original cases. Fault attack countermeasures on the other hand are expensive in terms of the area requirements. Thus, more or less, efficient countermeasures are possible but not proposed in abundance.

7.12 Conclusion

Micro-architectural attacks have found tremendous relevance in cloud-based and server-based environments because many architectural components are found vulnerable when exploited by intelligent adversaries. These micro-architectural components have evolved over the years with performance improvement as its sole motivation. However, when cryptographic algorithms are implemented on these hardware architectures, the execution footprints leak the secret keys which the cipher tries to protect. This chapter attempts to make a detailed overview on the possible vulnerabilities of important micro-architectural features and their impacts on the security of executing cryptographic algorithms. This chapter starts with the security analysis of one of the most important architectural construct-branch predictors and exposes that commercial branch predictors existing even in well-known processors can be approximated by classic text book branch prediction methods. It further shows how a combination of such an approximated model, combined with the branch misses observed via side-channels such as HPCs or timing, can be utilized to reveal secret keys of standard RSA-like ciphers. The work subsequently shows the first reported software-driven fault analysis of such RSA-like ciphers, by exploiting rowhammer attacks on contemporary DRAM chips. The feasibility of such bit flips motivated us to perform a detailed fault analysis of the branch-predictor units, showing the importance of fault tolerance in architectures. The attacks are further extended to several protected designs using blinding a novel attack technique, named as deduce and remove, showing that secured implementations can be targeted due to the leakage from underlying architectures. The later part of the chapter engages itself in detecting leakages from branch predictors using statistical t-test. This chapter finally touches upon the very important aspect of performance versus secure design choices in designing branch predictors.

Appendix A: Perf handler Code

```
// Function to handle overflow interrupts
static void perf_event_handler(int signum, siginfo_t* info, void*
      ucontext) {
 // Only POLL_HUP should happen
 if(info->si_code != POLL_HUP) {
  exit(EXIT_FAILURE);
 }
 count++;
 // Read branch_misses value
 read(fm, &sample_counter, sizeof(long long));
 // Reset the branch_misses counter to zero
 ioctl(fm, PERF_EVENT_IOC_RESET, 0);
 fprintf(fp, "%lld  ", sample_counter);
 // Refresh the instruction count buffer
 ioctl(info->si_fd, PERF_EVENT_IOC_REFRESH, 1);
}

// Configure the signal handler
struct sigaction sa;
memset(&sa, 0, sizeof(struct sigaction));
// Specify function to execute on interrupt
```

```
sa.sa_sigaction = perf_event_handler;
sa.sa_flags = SA_SIGINFO;

// Configure perf_event_attr struct
struct perf_event_attr pe, pm;
memset(&pe, 0, sizeof(struct perf_event_attr));
pe.type = PERF_TYPE_HARDWARE;
pe.size = sizeof(struct perf_event_attr);
// Event to be monitored for sampling mode
pe.config = PERF_COUNT_HW_INSTRUCTIONS;
pe.disabled = 1;  //Event is initially disabled
pe.sample_type = PERF_SAMPLE_IP;
// Number of counts of the monitored event on which interrupt occurs
pe.sample_period = 1600;
pe.exclude_kernel = 1; // Exclude events happenning in kernel space
pe.exclude_hv = 1; // Exclude events happenning in hypervisor
```

Listing 7.1 Sampling mode of `perf ioctl` *calls.*

Appendix B: RELIC codes

```
void ep_mul_basic(ep_t r, const ep_t p, const bn_t k) {
        int i, l;
        ep_t t;
        ep_null(t);
        if (bn_is_zero(k)) {
                ep_set_infty(r);
                return;
        }
        TRY {
                ep_new(t);
                l = bn_bits(k);
                ep_copy(t, p);
                for (i = l - 2; i >= 0; i--) {
                    ep_dbl(t, t);
                    if (bn_get_bit(k, i)) {
                        ep_add(t, t, p);
                    }
                }
                ep_norm(r, t);
        }
        CATCH_ANY {
                THROW(ERR_CAUGHT);
        }
        FINALLY {
                ep_free(t);
        }
}
```

Listing 7.2 Basic multiplication call for Curve1174 as implemented in RELIC

```
/* dbl-2007-bl formulas: 1M + 8S + 1*a + 10add + 1*8 + 2*2 +
    1*3 */
/* http://www.hyperelliptic.org/EFD/g1p/auto-shortw-jacobian.
    html#doubling-dbl-2007-bl */

/* t0 = x1^2, t1 = y1^2, t2 = y1^4. */
fp_sqr(t0, p->x);
fp_sqr(t1, p->y);
fp_sqr(t2, t1);

if (!p->norm) {
 /* t3 = z1^2. */
 fp_sqr(t3, p->z);

 if (ep_curve_get_a() == OPT_ZERO) {
  /* z3 = 2 * y1 * z1. */
  fp_mul(r->z, p->y, p->z);
  fp_dbl(r->z, r->z);
 } else {
  /* z3 = (y1 + z1)^2 - y1^2 - z1^2. */
  fp_add(r->z, p->y, p->z);
  fp_sqr(r->z, r->z);
  fp_sub(r->z, r->z, t1);
  fp_sub(r->z, r->z, t3);
 }
} else {
 /* z3 = 2 * y1. */
 fp_dbl(r->z, p->y);
}

/* t4 = S = 2*((x1 + y1^2)^2 - x1^2 - y1^4). */
fp_add(t4, p->x, t1);
fp_sqr(t4, t4);
fp_sub(t4, t4, t0);
fp_sub(t4, t4, t2);
fp_dbl(t4, t4);

/* t5 = M = 3 * x1^2 + a * z1^4. */
fp_dbl(t5, t0);
fp_add(t5, t5, t0);
if (!p->norm) {
 fp_sqr(t3, t3);
 switch (ep_curve_opt_a()) {
  case OPT_ZERO:
   break;
  case OPT_ONE:
   fp_add(t5, t5, ep_curve_get_a());
   break;
  case OPT_DIGIT:
   fp_mul_dig(t1, t3, ep_curve_get_a()[0]);
   fp_add(t5, t5, t1);
   break;
  default:
   fp_mul(t1, ep_curve_get_a(), t3);
   fp_add(t5, t5, t1);
   break;
 }
```

Listing 7.3 Doubling implementation in RELIC implementation

```
	/* madd-2007-bl formulas: 7M + 4S + 9add + 1*4 + 3*2. */
	/* http://www.hyperelliptic.org/EFD/g1p/auto-shortw-jacobian-3.
	   html#addition-madd-2007-bl */

	if (!p->norm) {
	 /* t0 = z1^2. */
	 fp_sqr(t0, p->z);

	 /* t3 = U2 = x2 * z1^2. */
	 fp_mul(t3, q->x, t0);

	 /* t1 = S2 = y2 * z1^3. */
	 fp_mul(t1, t0, p->z);
	 fp_mul(t1, t1, q->y);

	 /* t3 = H = U2 - x1. */
	 fp_sub(t3, t3, p->x);

	 /* t1 = R = 2 * (S2 - y1). */
	 fp_sub(t1, t1, p->y);
	 fp_dbl(t1, t1);
	} else {
	 /* H = x2 - x1. */
	 fp_sub(t3, q->x, p->x);

	 /* t1 = R = 2 * (y2 - y1). */
	 fp_sub(t1, q->y, p->y);
	 fp_dbl(t1, t1);
	}

	/* t2 = HH = H^2. */
	fp_sqr(t2, t3);

	/* If E is zero. */
	if (fp_is_zero(t3)) {
	 if (fp_is_zero(t1)) {
	  /* If I is zero, p = q, should have doubled. */
	  ep_dbl_projc(r, p);
	 } else {
	  /* If I is not zero, q = -p, r = infinity. */
	  ep_set_infty(r);
	 }
	} else {
	 /* t4 = I = 4*HH. */
	 fp_dbl(t4, t2);
	 fp_dbl(t4, t4);

	 /* t5 = J = H * I. */
	 fp_mul(t5, t3, t4);

	 /* t4 = V = x1 * I. */
	 fp_mul(t4, p->x, t4);

	 /* x3 = R^2 - J - 2 * V. */
	 fp_sqr(r->x, t1);
	 fp_sub(r->x, r->x, t5);
	 fp_dbl(t6, t4);
	 fp_sub(r->x, r->x, t6);
```

```
        /* y3 = R * (V - x3) - 2 * Y1 * J. */
        fp_sub(t4, t4, r->x);
        fp_mul(t4, t4, t1);
        fp_mul(t1, p->y, t5);
        fp_dbl(t1, t1);
        fp_sub(r->y, t4, t1);

        if (!p->norm) {
            /* z3 = (z1 + H)^2 - z1^2 - HH. */
            fp_add(r->z, p->z, t3);
            fp_sqr(r->z, r->z);
            fp_sub(r->z, r->z, t0);
            fp_sub(r->z, r->z, t2);
        } else {
            /* z3 = 2 * H. */
            fp_dbl(r->z, t3);
        }
    }
    r->norm = 0;
```

Listing 7.4 Addition step in RELIC implementation

References

[1] P. C. Kocher. Timing attacks on implementations of Diffie-Hellman, RSA, DSS, and other systems. In N. Koblitz, editor, *CRYPTO '96: Proceedings of the 16th Annual International Cryptology Conference on Advances in Cryptology*, volume 1109 of *Lecture Notes in Computer Science*, pages 104–113, London, UK, Springer-Verlag, 1996.

[2] S. Bhattacharya and D. Mukhopadhyay. Who watches the watchmen?: Utilizing performance monitors for compromising keys of RSA on intel platforms. In T. Güneysu and H. Handschuh, editors, *Cryptographic Hardware and Embedded Systems – CHES 2015 – 17th International Workshop, Saint-Malo, France, September 13–16, 2015, Proceedings*, volume 9293 of *Lecture Notes in Computer Science*, pages 248–266. Saint-Malo: Springer, 2015.

[3] J. Coron. Resistance against differential power analysis for elliptic curve cryptosystems. In *Cryptographic Hardware and Embedded Systems, First International Workshop, CHES '99, Proceedings*, volume 1717 of *Lecture Notes in Computer Science*, pages 292–302. Worcester, MA: Springer-Verlag, 1999.

[4] C. Clavier and M. Joye. Universal exponentiation algorithm. In *Cryptographic Hardware and Embedded Systems – CHES 2001, Third International Workshop, Proceedings*, volume 2162 of *Lecture Notes in Computer Science*, pages 300–308. Paris: Springer-Verlag, 2001.

[5] D. F. Aranha and C. P. L. Gouvêa. RELIC is an Efficient Library for Cryptography, 2019. https://github.com/relic-toolkit/relic.

[6] N. E. Mrabet and M. Joye. *Guide to Pairing-Based Cryptography*. Chapman & Hall/CRC, 2016.

[7] O. Aciiçmez, Ç. K. Koç, and J.-P. Seifert. On the power of simple branch prediction analysis. *IACR Cryptology ePrint Archive*, 2006:351, 2006.
[8] S. Bhattacharya and D. Mukhopadhyay. Formal fault analysis of branch predictors: Attacking countermeasures of asymmetric key ciphers. *PROOFS*, 2016.
[9] D. Evtyushkin, D. V. Ponomarev, and N. B. Abu-Ghazaleh. Jump over ASLR: attacking branch predictors to bypass ASLR. In *49th Annual IEEE/ACM International Symposium on Microarchitecture, MICRO 2016, Taipei, Taiwan, October 15–19, 2016*, pages 1–13, 2016.
[10] D. Evtyushkin, D. V. Ponomarev, and N. B. Abu-Ghazaleh. Covert channels through branch predictors: a feasibility study. In *Proceedings of the Fourth Workshop on Hardware and Architectural Support for Security and Privacy, HASP@ISCA 2015, Portland, OR, USA, June 14, 2015*, pages 5:1–5:8, 2015.
[11] D. Evtyushkin, D. Ponomarev, and N. B. Abu-Ghazaleh. Understanding and mitigating covert channels through branch predictors. *TACO*, 13(1): 10:1–10:23, 2016.
[12] D. Evtyushkin, R. Riley, N. B. Abu-Ghazaleh, and D. Ponomarev. Branchscope: A new side-channel attack on directional branch predictor. In *ASPLOS*, pages 693–707. Williamsburg, VA: ACM, 2018.
[13] P. Kocher, D. Genkin, D. Gruss, *et al.* Spectre attacks: Exploiting speculative execution. *CoRR*, abs/1801.01203, 2018.
[14] M. Lipp, M. Schwarz, D. Gruss, *et al.* Meltdown: Reading kernel memory from user space. In *USENIX Security Symposium*, pages 973–990. Baltimore, MD: USENIX Association, 2018.
[15] J. L. Hennessy and D. A. Patterson. *Computer Architecture: A Quantitative Approach, 4th Edition*. Morgan Kaufmann, 2006.
[16] T.-Y. Yeh and Y. N. Patt. Two-level adaptive training branch prediction. In *MICRO*, pages 51–61. Albuquerque, NM: ACM/IEEE, 1991.
[17] O. Aciiçmez, Ç. K. Koç, and J. Seifert. On the power of simple branch prediction analysis. *IACR Cryptology ePrint Archive*, 2006:351, 2006.
[18] S. Bhattacharya and D. Mukhopadhyay. Utilizing performance counters for compromising public key ciphers. *ACM Transactions on Privacy and Security*, 21(1):5:1–5:31, 2018.
[19] S. Bhattacharya and D. Mukhopadhyay. Curious case of rowhammer: Flipping secret exponent bits using timing analysis. In *Cryptographic Hardware and Embedded Systems – CHES 2016 – 18th International Conference, Santa Barbara, CA, USA, August 17–19, 2016, Proceedings*, pages 602–624, 2016.
[20] C. Maurice, N. L. Scouarnec, C. Neumann, O. Heen, and A. Francillon. Reverse engineering Intel last-level cache complex addressing using performance counters. In *Research in Attacks, Intrusions, and Defenses – 18th International Symposium, RAID 2015, Kyoto, Japan, November 2–4, 2015, Proceedings*, pages 48–65, 2015.
[21] S. Bhattacharya and D. Mukhopadhyay. Formal fault analysis of branch predictors: attacking countermeasures of asymmetric key ciphers. *Journal of Cryptographic Engineering*:1–12, 2016.

[22] S. Bhattacharya, C. Maurice, S. Bhasin, and D. Mukhopadhyay. Branch prediction attack on blinded scalar multiplication. *IEEE Transaction of Computers*, 2019.

[23] S. Bhattacharya, S. Bhasin, and D. Mukhopadhyay. Online detection and reactive countermeasure for leakage from BPU using TVLA. In *31st International Conference on VLSI Design and 17th International Conference on Embedded Systems, VLSID 2018, Pune, India, January 6–10, 2018*, pages 155–160, 2018.

[24] D. J. Bernstein. Cryptography in NaCl.

[25] Y. Yarom, D. Genkin, and N. Heninger. Cachebleed: A timing attack on OpenSSL constant time RSA. In *Cryptographic Hardware and Embedded Systems – CHES 2016 – 18th International Conference, Santa Barbara, CA, USA, August 17–19, 2016, Proceedings*, pages 346–367, 2016.

[26] O. Reparaz, J. Balasch, and I. Verbauwhede. Dude, is my code constant time? In *Design, Automation & Test in Europe Conference & Exhibition, DATE 2017, Lausanne, Switzerland, March 27–31, 2017*, pages 1697–1702, 2017.

[27] F. Liu and R. B. Lee. Random fill cache architecture. In *47th Annual IEEE/ACM International Symposium on Microarchitecture, MICRO 2014, Cambridge, United Kingdom, December 13–17, 2014*, pages 203–215, 2014.

[28] F. Liu, H. Wu, K. Mai, and R. B. Lee. Newcache: Secure cache architecture thwarting cache side-channel attacks. *IEEE Micro*, 36(5):8–16, 2016.

[29] S. Bhattacharya, C. Rebeiro, and D. Mukhopadhyay. A formal security analysis of even-odd sequential prefetching in profiled cache-timing attacks. In *Proceedings of the Hardware and Architectural Support for Security and Privacy 2016, HASP@ICSA 2016, Seoul, Republic of Korea, June 18, 2016*, pages 6:1–6:8, 2016.

[30] D. B. Roy, M. Alam, S. Bhattacharya, *et al.* Customized instructions for protection against memory integrity attacks. *Embedded Systems Letters*, 10(3):91–94, 2018.

[31] J. V. Bulck, M. Minkin, O. Weisse, *et al.* Foreshadow: Extracting the keys to the intel SGX kingdom with transient out-of-order execution. In *USENIX Security Symposium*, pages 991–1008. Baltimore, MD: USENIX Association, 2018.

[32] O. Aciiçmez and Ç. K. Koç. Microarchitectural attacks and countermeasures. In Ç. K. Koç, editor, *Cryptographic Engineering*, pages 475–504. Springer, 2009.

[33] Q. Ge, Y. Yarom, D. Cock, and G. Heiser. A survey of microarchitectural timing attacks and countermeasures on contemporary hardware. *Journal of Cryptographic Engineering*, 8(1):1–27, 2018.

[34] J. Betz, D. Westhoff, and G. Müller. Survey on covert channels in virtual machines and cloud computing. *Transactions on Emerging Telecommunications Technologies*, 28(6), 2017.

[35] J. Szefer. Survey of microarchitectural side and covert channels, attacks, and defenses. *IACR Cryptology ePrint Archive*, 2016:479, 2016.

[36] G. Doychev and B. Köpf. Rigorous analysis of software countermeasures against cache attacks. *CoRR*, abs/1603.02187, 2016.

Chapter 8
Mitigating the CACHEKIT attack

Mauricio Gutierrez[1], Ziming Zhao[2], Adam Doupé[1], Yan Shoshitaishvili[1], and Gail-Joon Ahn[1]

8.1 Introduction

ARM processors are at the core of today's interconnected and ever-growing technological world. As of September 2016, ARM has shipped out more than 86 billion chips that are used to securely and efficiently power anything from "smart cities, smart phones, supercomputers, cars, spacecraft, networking base stations and servers" to Internet of Things devices [1]. As the demand for processors to power such technologies continues to increase, so does the need to secure them against malicious actors and attacks. At the forefront of ARM's defense mechanisms are a set of hardware security extensions known as TrustZone (TZ). In TZ, the processor is divided at the hardware level to create two independent units. One is used as a Trusted Execution Environment (TEE) known as Secure World (SW) that has privileged access to all system resources. The other is known as the Normal World (NW); this is where the typical rich Operating System (OS) and untrusted applications run, and it has access only to resources that are specifically allocated to it.

This means that anything executing on SW has the privilege to access NW memory and CPU registers, but NW cannot affect or access the TEE in SW. Therefore, SW can be used to load trusted system integrity checking and malware detection tools to securely monitor an NW that could potentially become compromised. Since the release of TZ, there have been many research efforts into determining the best way to leverage the TEE provided in hardware by these new security extensions to prevent and identify exploits, including the detection of rootkits in memory [2,3]. However, all of these solutions are predicated on the idea that the high privileged SW must have access over all system resources so as to detect malicious code being present in memory. For example, Trustdump is a forensic toolkit that has a small piece of memory acquisition and integrity checking code stored in SW [4]. From the TEE, it can attempt to verify physical memory regions of the NW, beyond the reach of the potentially corrupted system.

[1]School of Computing, Informatics, and Decision Systems Engineering, Arizona State University, Tempe, AZ, USA
[2]Department of Computing Security, Rochester Institute of Technology, Rochester, NY, USA

A gap or vulnerability in this fundamental idea would therefore pose a critical threat to security. As attackers continue to evolve, the next generation of rootkits has been identified that tries to avoid memory introspection tools running in the SW, by exploiting the fact that SW does not have access to the NW cache [5]. As it turns out, in the ARM TZ architecture all processor resources are tagged with an additional bit known as the nonsecure (NS) bit. This bit is used to identify which world they belong to. The presence of this additional NS bit means that the SW cannot inherently access the NW cache, which is referred to as a cache incoherence. This therefore identifies at least one blind spot for the SW. Coupled with cache-locking mechanisms offered by some processors, this vulnerability opens up a terrifying attack vector. CacheKit is a new type of rootkit that exploits cache incoherence and cache-locking mechanisms to hide itself from introspection tools running in the SW. By hiding the rootkit in NW cache, it avoids introspection from SW tools and therefore evades the primary method of detection used against rootkits. This provides an unprecedented level of stealth which could mean that a compromised system is never able to detect the malicious software running silently in the cache.

Consequently, many real-time and embedded systems that offer cache-locking mechanisms are left vulnerable and exposed to this next generation of rootkits. For processors to continue to provide these hardware facilities for timing and performance-sensitive processes, a new defense approach must be established to address these vulnerabilities. Throughout this chapter, various possible solutions are proposed and analyzed to determine the best approach to defend against CacheKit. Through thorough research and experimentation, we arrive at the optimal defense mechanism that we call CacheLight, a lightweight approach for preventing malicious abuse of cache-locking mechanisms. This novel solution leverages both the TZ and virtualization extensions in the ARM architecture to address the vulnerability identified by CacheKit. In this chapter, we cover the design, implementation, and implications of CacheLight and how it is used to allow legitimate users to continue to utilize cache locking while giving SW the power to ensure system security by controlling and verifying the use of said mechanisms.

8.2 Background: ARM, cache, and TrustZone

8.2.1 ARM architecture

ARM processors have nine modes of operation when the ARM TrustZone extensions are enabled. The modes, respective privileges of software executing in that mode, and security states in which that mode can operate are summarized in Table 8.1. The `usr` mode is where user space programs run and has the lowest privilege level of 0. The `svc` mode is where most parts of the kernel run, this mode has privilege level 1. It is also importation to note that `mon` mode has privilege level 1 but only ever runs in secure state.

At a system-level view, the AMRv7 architecture has 16 core registers when security extensions are implemented; 13 general-purpose 32-bit registers `R0–R12`, the

Table 8.1 ARM processor modes

Processor mode	Privilege level	Security state
User (usr)	PL0	Both
FIQ (fiq)	PL1	Both
IRQ (irq)	PL1	Both
Supervisor (svc)	PL1	Both
Monitor (mon)	PL1	Secure only
Abort (abt)	PL1	Both
Hypervisor (hyp)	PL2	Nonsecure only
Undefined (und)	PL1	Both
System (sys)	PL1	Both

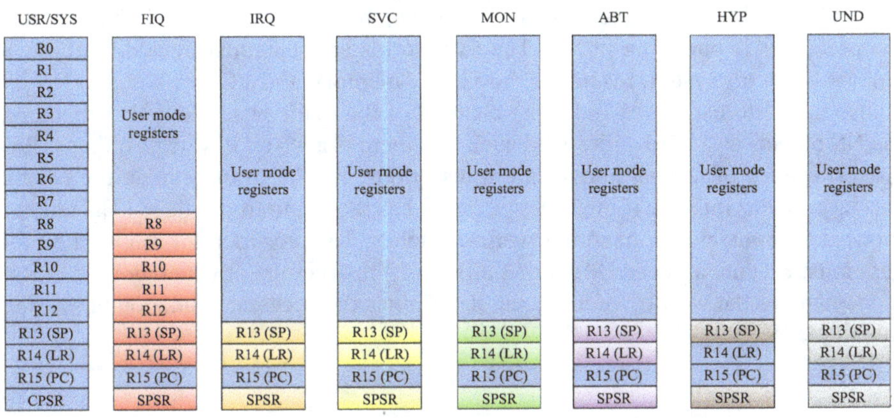

Figure 8.1 ARMv7 core registers

Stack Pointer (`SP`), Link Register (`LR`), and the Program Counter (`PC`). Additionally, the `Current Program Status Register CPSR` holds processor status and control information, including the current processor mode. The value of this register is copied and saved upon entry in the respective `Saved PSR` by all modes except user and system. This system-level view is shown in Figure 8.1.

To extend the functionality of the ARM processor, the ARM architecture supports coprocessors, each with its own set of registers. Coprocessor instructions provide access to 16 coprocessors in the ARMv7 architecture described as `CP0–CP15`. `CP15` is called the system control coprocessor and is reserved for the control and configuration of the ARM processor system, including architecture and feature identification [6].

8.2.2 ARM TrustZone

ARM processors power billions of networked embedded devices around the world. To ensure that processors could provide privacy and security in the quickly expanding Internet of Things, ARM developed TZ, a new technology to provide hardware-assisted security. TZ is a set of hardware security extensions on the processor, memory, and peripherals that ensure complete system isolation for running secure code. As stated, these isolated environments are often referred to as world. The SW is where a small Trusted Code Base (TCB) runs for security assurance, while the NW is the name given to the environment where the OS and other user programs run. The SW is often thought of as overseeing the NW as it has higher access privileges. For this to work, the SW must have access to resources that belong to the NW. Conversely, the NW must not have access to any of the resources allocated to the SW. Figure 8.2 shows the isolated environments provided by TZ.

To divide the two worlds, `CP15` has a `Security Configuration Register` (`SCR`) with an NS bit that determines the security context of the processor. When the NS bit is set, the processor is in the NW. When the NS bit is cleared, the processor is operating in SW. The two worlds are separated by adding this NS control bit to all system resources. To isolate memory and I/O devices, the NS bit is also added to each read and write channel of the main system bus. For memory, the NS bit acts as a 33rd address bit such that there is a 32-bit Physical Address (PA) space for each of normal and secure transactions. For I/O transactions, the NS bit acts as a flag to indicate if the access attempt has the required privilege [7]. Monitor mode is responsible for handling world-switches. To enter monitor mode, the process must execute a Secure Monitor Call (SMC) instruction, at which point monitor mode handles the world switching and determines the necessary function or handler to execute [8].

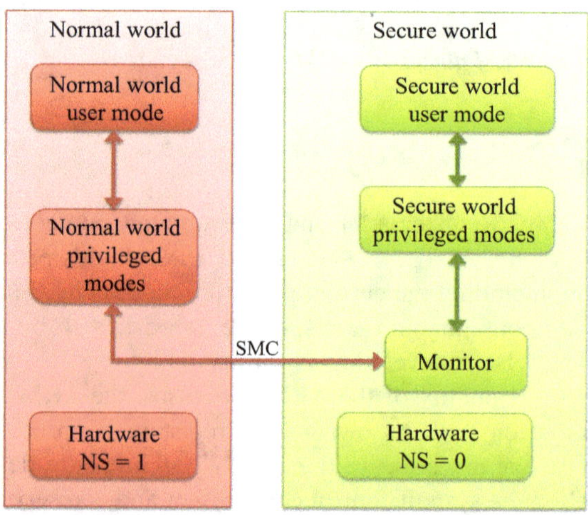

Figure 8.2 ARM TrustZone security extensions

8.2.3 ARM cache

Caches can be divided into types based on whether the index and tag bits correspond to physical or Virtual Addresses (VAs). Physically Indexed, Physically Tagged (PIPT) caches use the PA for both the index and the tag. This approach is slow as the PA must be looked up in the translation tables before that address can be looked up in the cache; however, it is also simple and avoids problems with aliasing. Virtually Indexed, Physically Tagged (VIPT) caches use the VA for the index. They are faster than PIPT caches as the cache line can be looked up in parallel with the Translation Lookaside Buffer (TLB); however, the tag cannot be compared until the PA is available. Though they still face aliasing problems, the physical tag bits make sharing more manageable than with Virtually Indexed, Virtually Tagged (VIVT) caches that use the VA for both the index and the tag. In general, the ARM L1 instruction cache is VIPT for greater speed and since instructions generally are not shared. However, the L1 data cache and any lower level caches are generally PIPT, with some options for implementing them as VIPT [9].

With the TZ extension, the processor cache is also tagged with an additional bit. This tag bit is used to record and identify whether a given memory access was secure or NS. With this additional bit, the processor can always know which lines of cache belong to which world, and cache flushing between world switches is not necessary. The two worlds can evict each other's lines in the cache as needed [6]. However, it is possible to prevent cache lines from being evicted by locking them. While this allows for performance optimization, it also gives rise to the cache incoherence, which CacheKit exploits to hide from memory-inspection tools. After the virtual and PA translations have occurred in the TLB, cache stores the memory data of the PA and keeps the NS bit to indicate the security state needed to access the entry. This NS bit is set by hardware and it is not directly accessible by system software [10].

8.3 The Genode operating system framework

In this chapter, Genode is used as the OS running in the SW. The Genode OS framework is the implementation of the Genode architecture. It is a tool kit for building highly secure special-purpose OSs. It scales from embedded systems to highly dynamic general-purpose workloads. The system is based on a recursive structure where each program is executed in a dedicated sandbox. Each sandbox is granted only those access rights and resources that it requires to fulfill its specific purpose. Hierarchies are formed by allowing programs to create and manage sub-sandboxes out of their own resources. Policies can then be applied at each level to establish various layers of security. Programs are only allowed to communicate and trade their resources with each other through strictly defined manners. This rigid regime reduces the attack surface of security-critical functions by orders of magnitude compared to contemporary OSs. Like Unix OSs, Genode is a collection of small building blocks, out of which sophisticated systems can be composed. However, unlike Unix, those building blocks include not only applications but also all classical OS functionalities, including kernels, device drivers, file systems, and protocol stacks [11].

8.4 Background: CacheKit attack

CacheKit is a new type of rootkit that exploits a cache incoherence and cache-locking mechanisms in the ARM TZ architecture [5]. Dividing the two worlds completely results in a cache incoherence where the contents of secure and NW are different even if they map to the same PA. CacheKit exploits the fact that even though the SW can access the memory of NW, the two worlds are separated such that they cannot gain access to the other's cache. There are three major steps in establishing CacheKit: loading, locking, and hiding. First, a technique known as Cache-as-RAM is used to ensure that the rootkit is loaded only into cache of the NW, where it can avoid detection from the SW [12,13]. Then, the ARM hardware support settings are exploited to keep the code persistent in cache as long as possible. Finally, the translation tables are modified such that the malicious code in cache maps to unused I/O addresses in physical memory so that if cache content is flushed to RAM for inspection, the data are simply lost. This ensures that even if the rootkit were to be flushed into memory for any reason, any trace of the malicious code would be lost [5].

8.4.1 Loading

In the ARM architecture, the only way to read or write a cache line is to have the processor read from or write to virtual memory. Therefore, to load the rootkit into cache, the CacheKit module must first enable caching on memory. This is done by setting the paging table memory attribute as WriteBack. The WriteBack configuration ensures that load register (LDR) instructions trigger a cache line fill. The key of cache loading is to ensure that data are loaded to cache and cache only, the rootkit should not be stored in RAM at any point as it would make it visible to memory inspection.

Figure 8.3 shows how, using the Cache-as-RAM technique, it may be possible to have code in cache that is not consistent with what is in main memory, thereby allowing an attacker to modify benign code in the cache with malicious code without it being written back to memory right away.

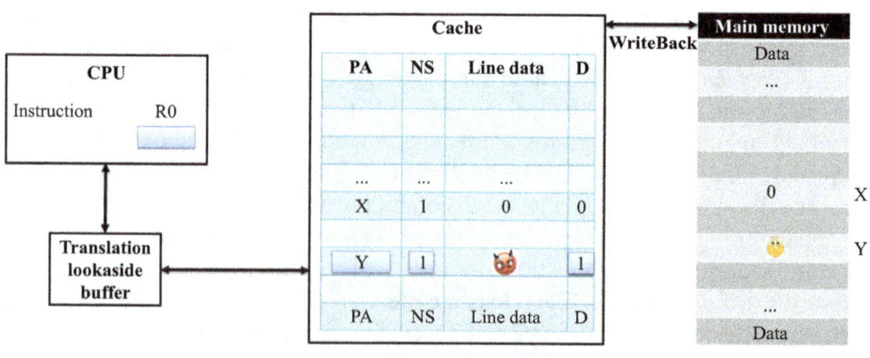

Figure 8.3 Exploiting the Cache-as-RAM technique

8.4.2 Locking

The Cortex-A8 processor's cache is divided into eight cache ways, of which up to seven are allowed to be locked at a time. A processor that offers cache-locking mechanisms is essential to this attack as without it the malicious code would not be able to persist in an infected system. With cache being very volatile memory, locking is essential to ensure a rootkit's purpose of remaining hidden within an infected system. While using Cache-as-RAM enables us to better hide the malware, cache locking allows the code to survive long enough to make it useful. For this to work effectively, the cache lines corresponding to all the memory addresses to be locked in cache must first be evicted. Second, the cache way to be used is unlocked and the other ways are locked, thereby ensuring that any cache fills made by the LDR and STR instructions will be made to the way that the attacker has designated for the rootkit. Finally, once the rootkit has been loaded into cache the way that has been reserved for the rootkit is locked and the rest of the cache is unlocked. Therefore, using the Cache-as-RAM technique to store the rootkit directly into the cache gives it exceptional stealth as it is never exposed to memory-inspection tools. However, there is a trade-off in storing the rootkit in cache. Cache is very volatile and even though cache lines can be locked with hardware control, they will still respond to cache-maintenance instructions if they are called. Therefore, such an instruction can still cause the cache to be written out to memory and become exposed to inspection tools.

8.4.3 Hiding

Therefore, once the rootkit has been successfully loaded and stored in cache, CACHEKIT must now overcome the two main obstacles with remaining concealed. First are the cache-maintenance instructions that even locked lines must respond to. As stated, ARMv7 provides various cache-maintenance instructions that can cause the contents of even locked cache to be written out to memory. This is dangerous because if the malicious code is flushed to memory, then it becomes detectable to memory-inspection techniques. Similarly, the second problem has to do with writing back to memory and introspection from the NW kernel. Detection methods that sequentially map each physical page into the kernel memory space would still be able to read the cache and detect the rootkit. To overcome these two obstacles, cache lines are mapped to unused physical I/O address space. The key is to find the memory allocation of the system in question and determine which part of the address space is used for I/O devices. In 32-bit ARM, the convention is to have the first 2 GB of memory mapped to I/O space. Furthermore, specific boards can have reserved, unused I/O space. Therefore, any access requests to this area are redirected to peripheral bus. CACHEKIT takes advantage of this and carves out this space for potential flushes by reconfiguring it as memory space using the Cache-as-RAM technique. Now, all read and write operations are redirected to the cache of the processor. This newly configured mapping is neither backed by RAM nor any real I/O device. Therefore, should the malicious code ever be evicted by a cache clean or flush there is no physical device that will respond to the memory write. This ensures that upon cache eviction, instead of being written to memory where it can be discovered, the malicious code is simply lost with no error, response, or detection, as though it had never existed. This allows the intentions and operations of the rootkit to remain unknown and untraced, perhaps for future reloading. While mapping the cache to I/O address space ensures that an attacker can maintain stealth by destroying the data since no actual backup memory exists, it also means that they lose the rootkit completely if it is ever flushed from cache. Since the effectiveness of a rootkit depends on how long it can remain hidden in the system, this is a major trade-off between stealth and persistence. Therefore, for CacheKit to be most effective, it is imperative to have an environment where data can persist in cache.

8.5 Defeating CacheKit attacks: naïve approaches

In this section, we discuss several possible solutions to defeat CacheKit attacks. We compare the benefits and limitations of each approach and propose that CacheLight provides the best solution.

8.5.1 Naïve prevention

First, we focus on preventing rootkits from hiding in NW cache. According to the ARM technical reference manual for the Cortex-A8 [8], the CL bit determines whether or not cache locking can be done in the NW. If CL is set (to 1), then cache locking is available in both the worlds. However, if the CL bit is cleared (to 0), then cache locking is only available in the SW. Attempting to use registers associated with cache locking, namely, the `L2 Lock Down Register`, results in an undefined instruction exception. Therefore, this provides a simple and direct solution to preventing CacheKit attacks by disabling the cache-locking capabilities in the NW and only allowing trusted applications running in privileged SW levels to lock cache entries. While this would prevent such attacks entirely because the rootkits would no longer be able to remain persistent in cache, it also limits a lot of what NW users can accomplish in their applications.

Therefore, disabling cache locking may not be an option. Many modern processors feature cache-locking mechanisms to allow for finer control of cache eviction policies and thereby improving the cache hit rate. This can have a considerable impact on the performance of a system if managed correctly [14]. For this reason, a wide variety of processors across different manufacturers and processor families offer this option. Some of ARM's Cortex and ARM11 processors allow for way locking in which locking is available at the granularity of ways of a set-associative cache. Line locking is a more fine-grained approach, where it is possible to have a different number of locked lines in different sets of the cache; Intel's XScale, the ARM9 family, and BlackFin 5xx family processors support this kind of locking mechanism [15].

Originally, cache locking was designed to offer timing predictability for real-time systems, where techniques focused mostly on improving worst case execution time. However, recent research has expanded this use to better performance of general embedded systems by improving average-case execution time. While the improvement can be significant, there is also the risk that an improper implementation will end up negatively impacting performance; therefore, careful management is essential to fully realize the potential of cache locking [16]. Several techniques have been developed to try to use cache locking to optimize performance as categorized by Mittal [16].

Many different mechanisms have been developed, each with their own benefits and limitations, to try to leverage cache locking to the fullest. These include both dynamic and static cache locking as well as partial and full cache locking [17]. There are also works that focus on guaranteeing a minimized predictable average worst case execution time for embedded systems by exploring different profiles [14,15]. In fact, the i.MX53 was chosen because it features an ARM Cortex-A8 processor that supports cache locking. Considering how important cache locking can be in both real-time and embedded systems across so many processors, we cannot assume that disabling the mechanism is an option.

8.5.2 Naïve detection

Therefore, we move to detecting rootkits hidden in the cache. Given that it is most likely a part of a real-time or general embedded system that depends on cache locking to give critical processes the timing predictability or the ability to meet stringent performance requirements, we cannot move this functionality to the SW. Not only would it affect the timing with the need for world switching but it would also greatly broaden the TCB, defeating the purpose of having a small, isolated, trusted environment. Therefore, it is necessary to enable cache locking in the NW and thus we must be able to detect malicious attempts by a potentially compromised NW OS.

Originally, we had considered that in order to detect an attack, we would need not only the attempt to lock the cache but also the remapping of a VA to a PA that does not correspond to system memory. That is, these two fundamental necessities of CACHEKIT would be enough to determine that a malicious user is attempting to hide code in the NW cache. CACHEKIT prevents its data from being retrieved for forensic purposes by mapping to the physical region of reserved unused I/O so that if the cache is ever flushed, the malicious code is simply lost and cannot be analyzed. However, these two functionalities together make for a very specific signature that can most definitely signal malicious activity. This is because while some time-critical programs may need instructions or data to be locked in the cache, we would never need anything outside the system memory region to be locked.

The first step is to detect if NW has attempted to lock the cache by reading the value of the `L2 Lock Down Register`. Anything other than 0 would indicate that one or more cache ways are locked. If we are able to then find a page-table entry that maps a VA to a PA outside of the system memory region, we would be able to detect a rootkit hiding in the cache. This may require a page-table walk, because we have to check all entries to ensure completeness. While a page-table walk is expensive, we would only need to execute it when a lock is detected. In this manner, we would then be able to force a cache flush or invalidation to trigger a write back to memory. To ensure that we are able to retrieve the malicious code, we also have to modify the VA that has an invalid mapping to a new, valid memory region that we have reserved from SW for memory inspection. However, we find that this approach will be ineffective

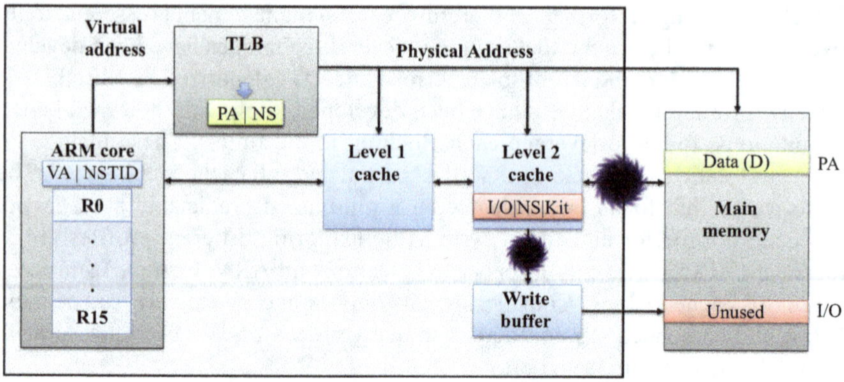

Figure 8.4 PIPT caches prevent the retrieval of hidden data

in retrieving the malicious data in the cache because even if we are able to remap the VA to a valid PA, the L2 cache in the ARM architecture is PIPT. This means that once the remapped address has been loaded into the cache, it is already too late to retrieve the data. This approach would work only with a processor that supports both cache locking and VIVT caches. In Cortex-A8, the L2 cache is PIPT [8]. This means that the tag and index bits stored in the cache to identify which memory address the cache line maps to are taken from the PA and not the VA.

As shown in Figure 8.4, even if we are successful in remapping the VA to a valid PA, once we flush the cache, the write to memory will be done according to the tag and index bits stored with the respective cache lines. Therefore, once the rootkit has been stored in the cache with the modified PA that points to reserved, unused I/O, any eviction will always be done to this location. CACHEKIT creates a "black-hole" situation where once it is in the cache, there is no retrieving it because there is no way to access the tag and index bits of the cache other than with perhaps a JTAG debugger. Therefore, the only solution left is to develop a defense mechanism to prevent rootkits, or any malicious code, from being loaded into the cache in the first place. This makes a detection solution considerably difficult. However, by leveraging TZ technology, prevention is still possible.

8.6 Defeating CACHEKIT attacks: CACHELIGHT

In this section, we present CACHELIGHT, a defense mechanism for preventing malicious software from inhabiting the cache for any significant period of time. We leverage the TZ security extensions to ensure that even if an attacker is able to briefly compromise the NW kernel, they will not be able to leave behind any malicious code in the NW cache. The main idea is to expand on the most simple solution of disabling the ability to lock the cache by clearing the CL bit. However, we still want user programs that have time-sensitive operations in real time and embedded systems to be able to make use of the cache-locking mechanism. Therefore, we implement CacheLight as an SW kernel function that can be called from NW kernel to have the secure OS perform the locking for the NS OS. This means that while users can still take advantage of the benefits of cache locking for legitimate purposes, the secure OS can have better control over what gets locked in the cache and perform integrity checking to prevent malicious code from being loaded in. This is implemented as an SMC that the NW can call and the SW can then handle.

8.6.1 Workflow

The first step is to determine what arguments to pass with the SMC to SW. In order to service the call, the SMC handler in SW must know:

1. OP code: The operation code to determine what the SMC is for and properly handle the request.
2. VA: The VA that corresponds to the data that the NW wants to lock in cache.
3. LockDownReg: The `L2 Lockdown Register` value to be used. This determines which cache ways the NW wishes to lock.
4. Size: The size or amount of data to be loaded into the cache from the base address (VA).

With these four arguments, we have all the information we need to load and lock the data into the cache on behalf of the NW. More importantly, the data must first be loaded into physical memory by the NW, which ensures that whatever is being loaded into the cache is consistent with what appears in RAM and therefore exposed to memory-inspection tools.

Once we have established the necessary parameters to pass to SW, we can detail the necessary work-flow, or steps, of the approach as shown in Figure 8.5. SW has all the necessary information to first verify that the request to lock memory is valid

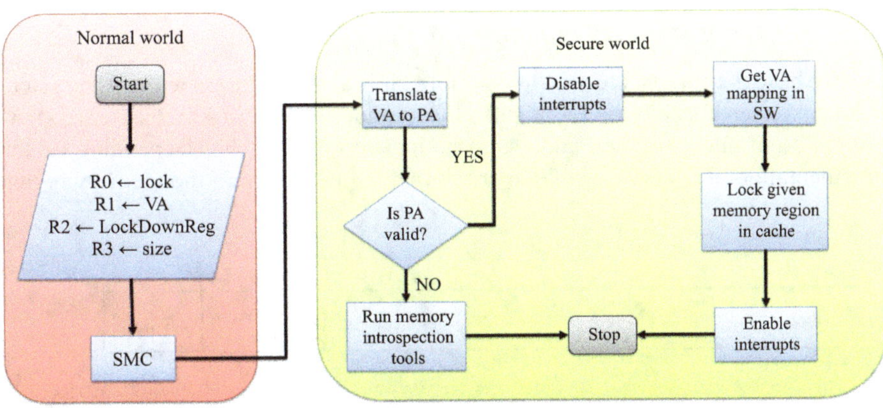

Figure 8.5 Overall workflow of CacheLight

by ensuring only memory regions allocated to NW are being requested. Furthermore, all the data must first be loaded into main memory by the NW, and are therefore available to inspection, should SW find an invalid PA is given. Otherwise, it can then go through the process of servicing legitimate requests and locking the requested memory in cache. The steps for this case are detailed in the following sections.

8.6.2 Virtual-to-physical address translation

VA translation is provided by way of the Memory Management Unit (MMU). In ARMv7, the MMU works with the L1 and L2 memory systems to translate VAs to PAs and controls accesses to and from external memory. Therefore, it is enhanced with security extensions and multiprocessor extensions to provide address translation and access permission checks. Virtual-to-PA mappings and memory attributes are defined in main memory as page tables; each world has its own set of page tables and TLB identifiers to remove the requirement for context switch TLB flushes [8].

Therefore, the first step in preventing rootkits from being loaded into cache is to translate the given VA to the PA. The ARMv7 architecture provides VA to PA translation operations using registers from `CP15`. There is a `Physical Address Register(PAR)` that holds the PA after a successful translation or the source of the abort after an unsuccessful translation. It is a read/write register banked in secure and NS states and accessible in privileged modes only [8]. Figure 8.6 shows the format of the `PAR` register. The `PAR` bits [31:12] contain the PA after a successful translation. Bits [11:0] are taken from the VA to obtain the complete PA. To make use of this register for our address translation, we do VA to PA translation in the other secure or NS state. The purpose of the VA to PA translation in the other secure or NS state is to translate the address with the current virtual mapping in the NS state, while the core is in the secure state [8]. To access the VA to PA translation in the other secure or NS state, we write CP15 c7 with Opcode2 set to 4 for privileged read permission. In this manner, the given VA will be translated using the NW's translation tables and we can retrieve a PA to work with.

8.6.3 Verifying memory contents

Once we have the PA, we are able to determine it is a valid address within the system memory. We can check if the PA maps to I/O memory region, in which case we know we should not allow such an address to be locked in the cache. Depending on the implementation, it is also possible to check that the PA is within the memory region

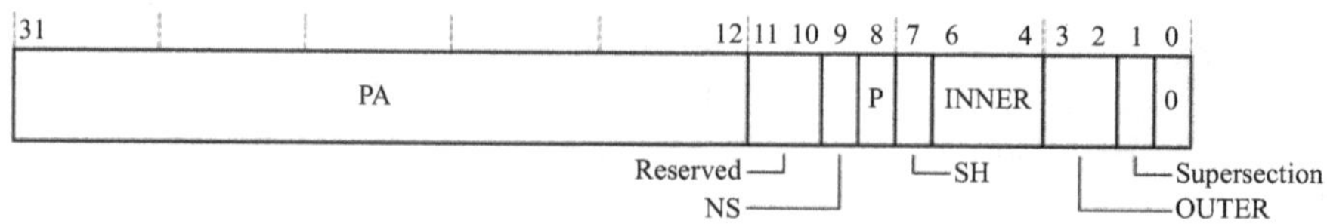

Figure 8.6 Physical address register in ARM [8]

allocated to the NW by the SW for a much better, tighter security check. At this point, we can flush the cache and run the memory-inspection tools of the SW to retrieve and analyze the malicious code that the attacker attempted to leave behind. Since in this scheme only the SW is allowed to perform cache locking, we are able to ensure that any memory that will remain in the cache does not differ to what is in RAM and thus we can be certain that there is no incoherence from what we are inspecting in the RAM and what the system is executing in the cache.

On the other hand, if the request is for a valid PA, then the SW must be able to perform the loading and locking of the requested memory region into the specified cache way.

8.6.3.1 Enabling and disabling interrupts

The first step is to disable interrupts so that we cannot be preempted. To lock the given memory, and only the given memory, in the given cache way(s) the locking process must not be interrupted. Therefore, it is imperative that this code be within a non-preemptible critical section. Once the locking is complete, SW can enable interrupts again.

8.6.4 Mapping normal-world memory to secure world

With the virtualization extensions active in both worlds, it is necessary to map a VA in the SW to the given PA from the NW. This means that the NW memory must be mapped in the SW as well. Memory that both worlds have access to is referred to as World-Shared Memory (WSM). WSM is designated as NS since NW can only make NS accesses, while SW can make both secure and NS accesses. Now that valid VA entries exist for NW PAs, SW can use these to load the requested memory region and thereby create the cache line entries on behalf of the NW. Given the PA that the NW would like to lock, a helper function is defined to build the offset within the physical NW memory and add that offset to the virtual start address of the mapping within the SW kernel, returning the SW VA for the specified PA. Since memory is shared, this VA can be used to create cache entries tagged as NS that will generate cache hits when the NW attempts to access said memory, effectively allowing the SW to create locked cache entries for the NW.

8.6.5 World-shared memory

As mentioned in Section 8.2, all addresses are tagged with an additional NS bit to indicate what mode the entry belongs to. NW is only able to access entries tagged with an NS bit equal to 1. Therefore, when we load the data into the cache, we have to ensure that it is tagged with NS = 1 so that NW will have a cache hit when it tries to do a look-up in the TLB. Figure 8.7 shows how the memory system of a theoretical ARM processor might handle the state associated with security extensions when accessing the memory system [10].

First, the core processing logic attempts a data load, a data store, or an instruction prefetch. The hardware passes the VA and the current world (NS Table Identifier or NSTID) to the TLB to enable it to perform address translation. The NSTID is passed by the hardware and depends on the current state of the processor, not the NS bit. In this case, the NSTID will represent an SW entry [10].

Then, the TLB loads the PA and the NS bit associated with the VA and NSTID was passed, performing a page-table walk and forcing NS = 1 if NSTID = 1 if necessary. The translation tables are responsible for keeping track of the NS bit and whether the VA resolves to a secure or NS PA [6]. The TLB then passes this information to the cache to perform the actual data or instruction access. If the SW has mapped the NS memory containing the data, the NW wants to lock in its translation tables, then the SW can directly access the NS cache lines. Therefore, an NW application can pass data to the SW through any level in the cache hierarchy. This enables a high-performance system in comparison to solutions that require cached data to be flushed out of the cache and in to external memory [10].

Therefore, when the loading and locking of memory is performed, the resulting cache entries will be tagged with an NS bit of 1 rather than 0. This means that when the NW attempts to access this address that it requested SW to load into the cache for it, it will generate a cache hit. NW will find a cache entry that matches the PA translation with an NS bit of 1, indicating that this memory is designated for NW use and will return the data from that cache line. This also means that SW can gain access to the NW cache if it creates the necessary translation table entries by implementing WSM.

8.6.6 Locking NW memory into cache from SW

Finally, we can now perform the loading and locking of the memory in the SW exactly as it would have been done in the NW. We pass the VA, the value to use for the `L2 Lockdown Register`, and the size of the data to load as arguments. Therefore, SW has all the information required to load and lock data to the requested cache ways. By implementing WSM, the SW can issue NS memory accesses that create NS cache entries for the NW. When control is returned to the NW, the legitimate process will

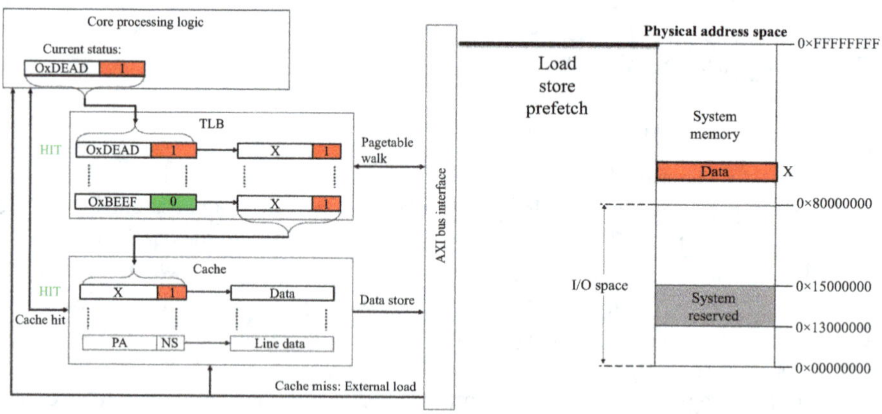

Figure 8.7 Memory system for an ARM core

Table 8.2 Comparison of different defense approaches

Approach	Category	Pros	Cons
CacheKit detection	Detection	Flag potential malware and flush the cache	• Cannot verify existence of rootkit • Cannot retrieve code for analysis or forensic purposes
Disabling cache locking in NW	Prevention	• Simple • Direct • No overhead • Done upon system initialization	• Cache-locking mechanisms not available to normal user • Processor might have been selected for those specific features
CacheLight	Prevention	• Lightweight • Overhead in setup, not execution time • Prevents loss of malicious code for analysis • Control over what is locked in cached • NW supports cache locking	• Additional world-switch overhead in set-up time • Can still load malicious code into cache

be able to access the time-critical section locked in the cache as intended. Therefore, this mechanism gives the SW a lot of power and control over what gets loaded into the cache, allowing it to ensure that there is no incoherency with what is stored in memory and what is stored in NW cache.

8.6.7 Comparing approaches

To the best of our knowledge, there have been no other defense mechanisms proposed against these kinds of attacks. Therefore, to evaluate the best solution, we provide a comparison of the three approaches introduced here in Table 8.2. We conclude that, while there is room for further work, the most robust and effective solution is the CacheLight approach. In the following chapters, we implement and evaluate CacheLight in terms of security and performance.

8.7 CacheLight implementation

8.7.1 Genode: a secure world OS

To defend against this new type of attack, we first replicate the original prototype scenario on the i.MX53 Development Board. Once functionality was verified, we moved to deploy the defense environment. The first step is to have a TEE in the

SW. After careful consideration, we decided to use the open-source Genode Project that supports the i.MX53. Genode provides support for various NW OSs, hardware platforms, and scenarios. The scenario we choose was designed for the i.MX53 with TZ hardware capabilities. The Linux kernel was modified such that certain kernel functions could handle world switches using an SMC call to the SW. The CacheKit attack module can then be cross-compiled for the new target kernel in the NW and deployed on the new Genode environment [18].

8.7.2 Building and deploying the environment

Using Genode 18.02, we build the Genode TZ Virtual Machine Monitor (TZ-VMM) scenario for the i.MX53 board. This build generates a bootable image for the target board as specified in the configuration file. The next step is to obtain the necessary bootloader provided by Genode Labs and generate the boot image. Finally, a bootable SD card is prepared for the board using both the boot image and the Genode image generated earlier. Now we have the Genode TZ-VMM running on the i.MX53 development board. The details of using and understanding Genode and all of the different scenarios can be found in their Foundations Book [11].

8.7.3 Deploying the CacheKit attack

We replicated CacheKit on the i.MX53 development board, which features a single ARM Cortex-A8 processor. The project is implemented as a kernel module that once installed on the board can use the cache manipulation tools in ARM to successfully load, lock, and conceal the rootkit in cache. Next, we must be able to cross-compile the CacheKit attack module against the new modified target kernel running in the NW. Unfortunately, the modified kernel is provided as a pre-compiled image in the Genode repository. Therefore, we obtain the source code of the modified kernel and create our own build. This then gave us a target to cross-compile the CacheKit attack module against to be able to deploy it on the new environment.

Now that we have our prototype setup, we deploy the attack module and find that it is unable to perform the hiding functions in the new environment. After examining the Genode source code for the SW, we find that upon initialization the CL bit of the `NS Access Control Register (NACR)` is cleared. Therefore, when the kernel module attempts to lock the cache, an undefined instruction exception is thrown by the system. When the attack module attempts to access the `L2 Lock Down Register` to lock down a cache way, an undefined instruction exception is thrown and the module terminates. Notice also that when we read the NACR, the value is given as initialized Genode where only access to coprocessors 10 and 11 is enabled [18]. This means that the CL bit is set to 0 and thus the attack module is rightfully denied the ability to lock down the cache. This is a very simple and direct method for preventing rootkits from hiding in the NW cache to avoid memory introspection. While having it disabled by default is a good idea, as discussed removing cache-locking capabilities for the NW is not always acceptable as they are necessary for certain applications. Therefore, it is enabled to replicate a real environment and the CacheKit module is successfully deployed to compromise it.

8.7.4 Deploying the CacheLight defense

The NW kernel was modified such that certain kernel functions could handle world switches using an SMC to the SW. According to their documentation, there were six different SMCs added to the Linux kernel [19]. However, upon closer inspection of the current release, Genode 18.02, there are actually only four SMCs that are implemented. Furthermore, they do not follow the ARM SMC Calling Convention [20], but rather implement their own, simpler version of handling SMC's. In Genode, an SMC is still used to generate a synchronous exception that is handled by Secure Monitor code running in Exception Level 3. The Secure Monitor Mode handles the world switching and then hands off the handling of the exception to the VMM running as an unprivileged user-level component in Genode. However, instead of using the ARM Calling Convention, since there are only four different calls implemented, the function to perform is simply passed as an argument in register `R0`. Registers `R1–R10` are then used to pass any necessary arguments to the SW. Upon a world-switch the monitor will save the state of the Virtual Machine, including these registers, so that the SW can have access to them. The VMM then simply checks the value in R0 to determine what function the NW needs it to perform for the SMC and expects any relevant arguments in the other registers [19]. This makes it fairly simple for us to add and define our own, fifth, SMC for a request to SW to lock memory in the cache. The following table shows the different SMC's that Genode had already implemented with the addition of our own "LOCK" SMC with code 4.

As shown in Table 8.3 our SMC has a function code of 4, and this is passed through register `R0` and is used to determine what function the SW needs to perform and the respective arguments in the remaining registers. In our case, we pass the following arguments:

1. R1 is the VA to be loaded and locked in the cache.
2. R2 contains the lock down register value to be used. This determines which cache ways the NW wishes to lock.
3. R3 is the size or amount of data to be loaded into the cache from the base address in R1.

With these three arguments, we have all the information we need to load and lock the data into the cache on behalf of the NW. More importantly, we have the location of the data before it get stored in the cache that allows us to regulate what can be locked

Table 8.3 SMC implementation in modified linux

SMC number	Function
0	Framebuffer
1	Inout
2	Serial
3	Block
4	Lock

into the NW cache. Once the request is handled by Genode, it restores the VM's state and switches back to NW operations.

After implementing an SMC for CacheLight in the Genode core, we need to define what happens once it is called by the Linux kernel. The SMC would be called whenever the NW has some code that needs to be locked in the cache. The NW would load the operation code, VA, value for the lock down register, and size of data to be loaded into the respective registers and perform an SMC. Once Genode reaches the exception handling for the call, we need to be able to create an entry in the cache with the data requested by the NW that is accessible by NW. Overall, the entire process is implemented as roughly 200 lines of C and assembly code in the SW kernel, which is a minimal increase in the TCB of the SW.

Genode utilizes the ARM virtualization extensions for the TZ-VMM scenario [21]. Therefore, to establish WSM, the Genode bootstrap process is modified to map the RAM region that is later allocated to the NW VM to a designated VA space in the SW. Once defined, the mapping can be added in the platform constructor of the bootstrap process to create the page-table entries in the SW for the NW RAM space, effectively creating WSM. The entries are created with the necessary flags, among them the NS bit that indicates the VAs resolves to NS physical space.

Once we deploy CacheLight, we are able to detect whenever the CacheKit module is attempting to lock malicious code into the cache. By doing security checks on the PA being locked into cache, CacheLight can detect anomalous behavior and if needed deploy inspection tools. If the attempt to lock the cache is malicious, it can then flush the caches and run memory introspection tools to determine the nature of the attack and retrieve any relevant data for forensic analysis. On the other hand, if the request is determined to be legitimate, CacheLight can service it by taking advantage of WSM.

8.8 Evaluation

In this section, we verify the ability of the two worlds to share data through any level of the cache hierarchy using WSM as well as the performance impact of CacheLight.

8.8.1 Effects of world-shared memory

For CacheLight to work, SW must be able to create cache entries tagged with an NS bit that NW can then access. While NW is forced by hardware to only make NS memory accesses, SW has the option to perform both secure and NS accesses. The type of access made is determined by the NS bit in the first-level page-table descriptor [6]. This NS bit defines the PA space, secure or NS, for all of the large and small pages of memory described by that table. For accesses from secure state, it determines whether the access is to secure or NS memory. However, it is ignored by accesses from NS state [10]. In a cache-able secure memory access, the line-fill is requested using secure access and the data are tagged in the cache as secure data. In a cache-able NS memory access, the line-fill is requested using NS access and if no security error is received, the data is tagged in cache as NS data. Therefore, the

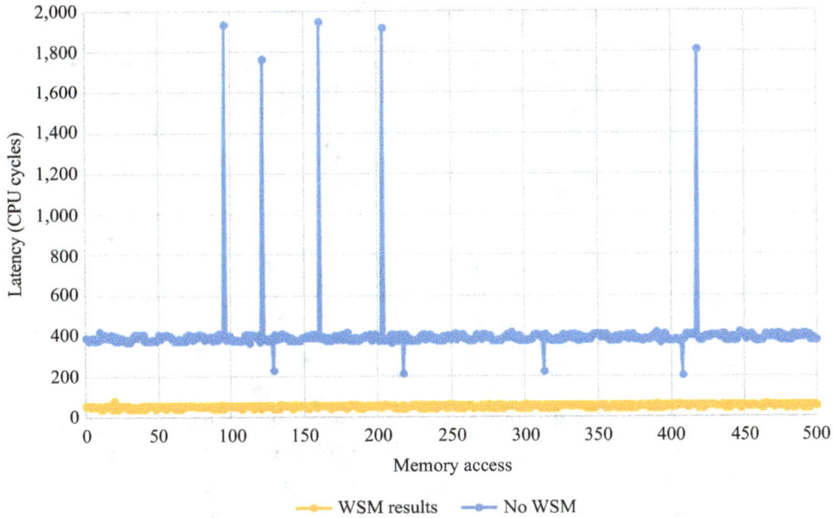

Figure 8.8 WSM allows SW to create non-secure cache entries for NW

NS bit that is used to tag the cache records the type of memory access that was made, allowing SW to cause allocation of NS lines into the cache.

Figure 8.8 shows the effects of implementing WSM to be able to create and lock NS cache lines from the SW. As shown, if there is no WSM then SW generates a secure memory access and the cache entries created and locked are tagged as secure. If the NW attempts to access that memory, it gets a miss in the cache and has to load that same address but with an NS tag which it would be unable to lock in the cache. However, with WSM, the SW can create and lock cache entries tagged as NS. Then, when the NW attempts to access that memory, it will get a hit in the cache and can use the locked entries. Therefore, it is possible to restrict cache-locking capabilities to the SW for security measures while still allowing legitimate NW programs to request locked NS cache entries from SW.

8.8.2 Performance evaluation

To defend against malware that attempts to evade memory introspection tools, CACHELIGHT, incurs some overhead in the cache-locking process. In a typical system, an NW-privileged program can directly perform the locking; however, with CACHELIGHT, there are additional security checks and world-switches that must occur. Therefore, to measure the performance impact of CACHELIGHT, the approach is split into three parts. First, there is the essential and necessary locking and loading of the cache which is code that runs regardless of whether CACHELIGHT is implemented or not. In addition, there is the overhead of the security checks and address translations

Table 8.4 Performance impact of CACHELIGHT when locking one cache way

	CPU cycles	Nanoseconds	Percentage of total
Total time	1,930,060	1,930.06	100
Locking time	654,733	654.73	33.92
World-switch overhead	1,272,500	1,272.50	65.93
CACHELIGHT overhead	2,827	2.83	0.15

done by CACHELIGHT. Finally, that leaves the overhead of switching to and from SW to lock the cache. Timing analysis of these three parts is performed using the available ARM Performance Monitoring Unit. As, the overhead portions are constant with respect to the size of the data being locked, analysis is performed for locking a single cache way of size 32 kB.

As shown in Table 8.4, the world-switch to and from SW is the main source of overhead for CACHELIGHT. In this experiment, with the i.MX53 running Genode, it makes up almost 66% of the time taken for CACHELIGHT to process a valid lock request for a single cache way. However, this is unavoidable as switching between worlds is necessary for the implementation. This can only be addressed by how the ARM architecture handles a world-switch or the efficiency of the Genode monitor code for switching between worlds. Furthermore, while it is considerable, it is a one-time cost of setting up the locked cache lines and therefore should occur scarcely. On the other hand, the actual overhead of the security measures of CACHELIGHT is minimal.

8.9 Related work

Rootkits and the methods for detecting and defending are always in a constant evolutionary race to outclass each other. Originally, persistent rootkits were needed to modify nonvolatile storage to survive system power cycles, which meant that file integrity checking tools could effectively detect them [22]. Therefore, rootkits switched to reside only in the OS kernel memory to defeat this storage-based detection. To detect this new type of rootkit, defenders acquire the system memory using a dedicated secure coprocessor [23] or physical hardware [24]. In their search to acquire higher root privileges, attackers have developed several different rootkits. Virtual machine-based rootkits insert a customized malicious hypervisor beneath the currently running OS [25]. Firmware-based rootkits infect the firmware on I/O devices [26] or the system BIOS [27].

As their counterpart, new hardware and software rootkit detectors with higher privilege are also proposed. Hypervisors are commonly used to introspect the untrusted OS [28]. However, vulnerabilities are frequently found in the hypervisors. This gave rise to the use of hardware features, such as security extensions in various processors. Throughout the chapter we focus on ARM TZ; however, each developer has their own set of hardware security extensions. For example, AMD has SVM [29] and Intel provides TXT [30]. These hardware features provide a TEE with guaranteed isolation and the highest privilege that defenders can claim as their own. CacheKit seeks to exploit weaknesses in these new hardware defense features to evade detection by the OS kernel. Another proposed design that has much the same goals is Shadow Walker, as it exploits the I-TLB and D-TLB coherency problem in the Intel architecture to hide the rootkits [31]. There is also an implementation called Cloaker that can hide its presence by locking the page translation it altered in the TLB [32]. However, none of these have ever been able to avoid the last-level defense of memory inspection. Thus, we note that CacheKit provides a new level of stealth as it is able to evade physical memory inspection by hiding in the NW cache.

CacheKit is the only one able to evade the physical memory check. Furthermore, by mapping the cache lines to unused I/O space we ensure that the malicious code cannot be examined. In the worst case, it is evicted and then destroyed before it can be inspected and analyzed. To the best of our knowledge, there have been no proposed defenses against this new kind of attack, leaving many ARM systems vulnerable and exposed. This prototype is implemented on the i.MX53 development board and thus the immediate, and complete application of the concept is limited to similar single Cortex-A8 processor systems. While this is only a subset of all ARM processors, it still poses a major threat as a nearly undetectable rootkit mechanism in embedded systems. To the best of our knowledge, there have been no defense mechanisms proposed for this kind of attack.

8.10 Future work

While this initial implementation of CacheLight shows great potential, there is still much work that can be done in this area. There still remain various applications and security checks for which CacheLight could be leveraged to increase security. CacheLight allows for control over what gets locked in NW cache but optimizing and evaluating to what extent this new control is used is yet to be explored, for example, limiting or regulating how much of the cache a request can lock. Additionally, there are instances where a rootkit could be loaded but not locked in the cache with a remapped address. This would result in a failed attempt so it was not of central focus on this chapter. Conversely, it would have been a malicious attempt nonetheless, one which would not be retrievable for analysis. Therefore, in such cases, valuable forensic data are lost about the attack and attacker. Perhaps, there could be work done to expand the robustness of the approach in such cases. Finally, there is still more work that can be done on the response after an attack is detected and verified. Looking deeper into what inspection tools to run in case CacheLight flags potential malicious code and the appropriate response mechanisms to employ if the tool does find malicious code in memory.

8.11 Conclusion

This chapter explored the CacheLight solution, a lightweight approach to preventing malicious use of cache-locking mechanisms while allowing time-critical applications to legitimately utilize them to ensure execution times in embedded and real-time systems. CacheLight allows the NW to perform cache-locking through requesting it as a service from the SW. Additionally, it only requires a minimal increase in the TCB to handle a new SMC, which the OS running in the TEE can then validate and verify to prevent any malicious code from being hidden in the cache to begin with.

The SW can now handle and verify the validity of any cache-locking request upon world switch. This allows the trusted code to ensure that any data to be locked in the cache not only map to a valid address in memory but are also consistent with what is present in main memory, effectively bringing the contents of the cache to light. Moreover, because the SW does not hand control back to NW after verifying the address, but rather performs the loading and locking on behalf of the NW, the attacker cannot bypass the security checks by passing different addresses in the arguments. Should CacheLight find that the attempt to lock the cache is malicious, it can then flush the caches and run memory introspection tools to determine the nature of the attack and retrieve any relevant data for forensic analysis. On the other hand, if the request is determined to be legitimate, CacheLight can service it by taking advantage of WSM.

Therefore, CacheLight can successfully prevent malicious code from hiding from SW introspection tools in the NW cache for any significant amount of time. Additionally, while we present a solution for the ARM architecture, the approach can be generalized to any architecture that employs the same execution separation idea. If the attack can be modified to a new architecture, then so can the defense. Furthermore, CacheLight incurs the overhead of a world-switch for the setup of the time-critical data. However, the initial setup of locking data in the cache is already expected to be expensive so that the performance and timing requirements can be met once the setup is done and the application is running. CacheLight makes additional overhead to the set-up process but not the execution of the time-critical process that requested the lock. Given that it provides security against an otherwise undetectable attack, the trade-off in set-up time is extremely worthwhile.

References

[1] ARM. 2016 ARM: Media Fact Sheet. ARM News. https://www.arm.com/-/media/arm-com/news/ARM-media-fact-sheet-2016.pdf?la=en. (2016).

[2] A. M. Azab, P. Ning, J. Shah, *et al.* 2014. Hypervision across worlds: Real-time kernel protection from the ARM TrustZone secure world. In *Proceedings of the 21st ACM Conference on Computer and Communications Security (CCS)*. ACM, Scottsdale, AZ, 90–102.

[3] F. Zhang, J. Wang, K. Sun, and A. Stavrou. 2014. HyperCheck: A Hardware-Assisted Integrity Monitor. *IEEE Transactions on Dependable and Secure Computing* 11, 4 (2014), 332–344.

[4] H. Sun, K. Sun, Y. Wang, J. Jing, and S. Jajodia. 2014. Trustdump: Reliable memory acquisition on smartphones. In *Proceedings of the 19th European Symposium on Research in Computer Security (ESORICS)*. Springer, Wroclaw, Poland, 202–218.

[5] N. Zhang, H. Sun, K. Sun, W. Lou, and Y. T. Hou. 2016. CacheKit: Evading memory introspection using cache incoherence. In *Proceedings of the 1st IEEE European Symposium on Security and Privacy*. IEEE, Saarbrücken, Germany, 337–352.

[6] ARM. 2012. ARM Architecture Reference Manual, ARMv7-A and ARMv7-R edition. http://infocenter.arm.com/help/index.jsp?topic=/com.arm.doc.ddi0406c/index.html. (2012).

[7] T. Alves and D. Felton. 2004. Trustzone: Integrated Hardware and Software Security. https://github.com/laginimaineb/cve-2016-2431. (2004).

[8] ARM. 2006–2010. ARM Cortex-A8 Technical Reference Manual. http://infocenter.arm.com/help/topic/com.arm.doc.ddi0344k/DDI0344K_cortex_a8_r3p2_trm.pdf. (2006–2010).

[9] ARM. 2015. ARM Cortex-A Series Programmer's Guide for ARMv8-A. http://infocenter.arm.com/help/topic/com.arm.doc.den0024a/index.html. (2015).

[10] ARM. 2009. ARM Security Technology Building a Secure System Using TrustZone Technology. http://infocenter.arm.com/help/index.jsp?topic=/com.arm.doc.prd29-genc-009492c/index.html. (2009).

[11] N. Fenske. 2017. Genode Operating System Framework Foundations. (2017). Available from: https://genode.org/documentation/genode-foundations/index.

[12] Y. Lu, L. Lo, G. R. Watson, and R. G. Minnich. 2006. CAR: Using Cache as RAM in LinuxBIOS. http://rere.qmqm.pl/mirq/cache as RAM lb 09142006.pdf. (2006).

[13] V. J. Zimmer, M. A. Rothman, and S. M. Datta. 2004. Using a processor cache as RAM during platform initialization. US Patent 20,040,103,272. (2004).

[14] T. Liu, M. Li, and C. J. Xue. 2012. Instruction Cache Locking for Embedded Systems Using Probability Profile. *Journal of Signal Processing Systems* 69, 2 (2012), 173–188.

[15] Y. Liang, T. Mitra, and L. Ju. 2015. Instruction Cache Locking Using Temporal Reuse Profile. *IEEE Transactions on Computer-Aided Design of Integrated Circuits and Systems* 34, 9 (2015), 1387–1400. https://doi.org/10.1109/TCAD.2015.2418320.

[16] S. Mittal. 2016. A Survey of Techniques for Cache Locking. *ACM Transactions on Design Automation of Electronic Systems* 21, 3, Article 49 (2016), 24 pages. https://doi.org/10.1145/2858792

[17] K. Anand and R. Barua. 2015. Instruction-Cache Locking for Improving Embedded Systems Performance. *ACM Transactions on Embedded Computing Systems* 14, 3, Article 53 (2015), 25 pages. https://doi.org/10.1145/2700100

[18] Genode Labs. 2017. Genode: Operating System Framework. https://github.com/genodelabs/genode. (2017).

[19] Genode Labs. 2017. An Exploration of ARM TrustZone Technology. Genode OS Documentation and Articles. https://genode.org/documentation/articles/trustzone. (2017).

[20] ARM. 2016. SMC CALLING CONVENTION System Software on ARM Platforms. http://infocenter.arm.com/help/topic/com.arm.doc.den0028b/ARM_DEN0028B_SMC_Calling_Convention.pdf. (2016).

[21] Genode Labs. 2017. An In-Depth Look into the ARM Virtualization Extensions. Genode OS Documentation and Articles. https://genode.org/documentation/articles/arm_virtualization. (2017).

[22] G. H. Kim and E. H. Spafford. 1994. The design and implementation of tripwire: A file system integrity checker. In *Proceedings of the 2nd ACM Conference on Computer and Communications Security (CCS)*. ACM, Fairfax, VA, 18–29.

[23] N. L. Petroni, T. Fraser, J. Molina, and W. A. Arbaugh. 2004. Copilot – A coprocessor-based kernel runtime integrity monitor. In *Proceedings of the 13th USENIX Security Symposium (Security)*. USENIX, San Diego, CA, 179–194.

[24] E. Chan, S. Venkataraman, F. David, A. Chaugule, and R. Campbell. 2010. Forenscope: A framework for live forensics. In *Proceedings of the Annual Computer Security Applications Conference (ACSAC)*. ACM, Austin, TX, 307–316.

[25] J. Rutkowska. 2006. Subverting Vistatm Kernel for Fun and Profit. Black Hat Briefings. (2006).

[26] J. Heasman. 2007. Implementing and Detecting a PCI Rootkit. Black Hat DC. (2007).

[27] S. Embleton, S. Sparks, and C. Zou. 2013. SMM Rootkit: A New Breed of OS Independent Malware. Security and Communication Networks. (2013).

[28] A. M. Azab, P. Ning, and E. C. Sezer. 2009. A hypervisor-based integrity measurement agent. In *Proceedings of the Annual Computer Security Applications Conference (ACSAC)*. ACM, Honolulu, Hawaii, 461–470.

[29] Advanced Micro Devices. 2013. Amd64 Architecture Programmer's Manual. AMD. (2013).

[30] Intel Trusted Execution Technology. 2016. *Intel Software Development Guide*. Intel. (2016).

[31] S. Sparks and J. Butler. 2005. Shadow Walker: Raising the Bar for Rootkit Detection. Black Hat Japan (2005), 504–533.

[32] F. M. David, E. M. Chan, J. C. Carlyle, and R. H. Campbell. 2008. Cloaker: Hardware supported rootkit concealment. In *Proceedings of the 29th IEEE Symposium on Security and Privacy (Oakland)*. IEEE, Oakland, CA, 296–310.

Chapter 9
Deep learning network security

Si Wang[1] and Chip-Hong Chang[1]

9.1 Introduction

The explosion of data usage has contributed to the requirement of processing extensive amount of data for most of the applications on smart devices and edge- and fog-computing nodes. Due to the scale and complexity of the tasks, decision support systems can greatly benefit from the use of machine learning (ML) techniques to correlate multimodal sensing to make accurate predictions and powerful inferences. Traditional ML algorithms have to be fed with previously extracted features. These features are usually identified in advance to reduce the complexity of the data and increase the visibility of the patterns to the learning algorithms [1]. Furthermore, in some circumstances, like multiple object detection, the task needs to be divided into parts and solved individually and the partial results are combined at the final stage. The required human intervention and discontinuity in the process of accomplishing the tasks contribute to the reduced efficiency of the conventional ML algorithms in the face of massive raw data and intricate tasks. Deep learning (DL), also referred to as deep neural network (DNN), has overcome the weakness of the need for human's participation on effective feature identification and hard-core feature extraction. It learns the high-level features from raw data in an incremental manner and solves the problems end-to-end. As a result, DL has now become a preferred option for handling majority of the challenging tasks in image classification [2], speech recognition and language processing [3].

In the early 2010s, DL has undergone a blossoming in applications with two breakthroughs: Microsoft's speech recognition system in 2011 [4] and Krizhevsky *et al.*'s demonstration of the outstanding performance of AlexNet system on the ImageNet Large Scale Visual Recognition Challenge task in 2012 [5]. These early successes inspired prodigious research activities directed toward algorithmic development of DNN. While superior DNN architectures and regulation mechanisms have been explored to achieve remarkable accuracy and generalization, efficient processing techniques that ease the high computational complexity of DNNs without compromising performance accuracy or increasing hardware cost have started to be the center of attention. Optimizing memory hierarchy improves the energy efficiency

[1]School of Electrical and Electronic Engineering, Nanyang Technological University, Singapore

by introducing low-cost data movement; quantization of the data values reduces the number of bits for the representation, resulting in reduced storage and relaxed computation requirements; compression of the networks reduces the number of operations and model size by exploiting the sparsity of activations and weights [1]. These techniques have proven to be significant in driving a wider deployment of DNNs in artificial intelligence (AI) systems. The rise of DNN accelerator research aiming at improving the throughput and energy consumption, e.g., Neuflow [6], DaDianNao [7] and Eyeriss [8], has alleviated the stringent computing requirements and resource limitations on the underlying hardware platform. With the progressive advancements of models, easy access to required hardware resources for training and lower performance bottleneck in hardware implementation, DL is rapidly growing and expanding into safety and security critical applications, such as self-driving cars, malware detection and surveillance. The recent adoption of face ID security on mobile devices is an indication of how fast DL solutions have infiltrated into our daily lives [9].

Despite the impressive outcomes demonstrated by DL in licit application scenarios, the dark side of not being able to fully understand and explain what have been done within these superintelligence, which can turn an inherently benevolent system to malevolent. Szegedy *et al.* [10] exploited an inherent vulnerability of DNNs in the context of image classification. They found that the networks are susceptible to adversarial samples obtained by maliciously adding imperceptibly small perturbations to the original inputs. Even with high generalization ability, the model could rarely refrain from misclassification in such an adversarial environment. In addition, these adversarial distortions are relatively robust across different classifiers. The intriguing findings raised security concern of DL and triggered a number of studies on efficient algorithms for adversarial image generation. One of the representatives is fast gradient sign method (FGSM) [11] that takes advantage of the gradient of loss function to determine the injection of perturbations. Apart from image-specific abnormalities, Moosavi-Dezfooli *et al.* [12] proposed universal perturbations that enable class alteration on any image. The later realization of black-box attack [13] against DNNs in 2017 compensates the shortcomings of requiring the knowledge of either DNNs model internals or training data inherited in the previously proposed algorithms. Moreover, Eykholt *et al.* [14] successfully brought adversarial attack to the real world by physically perturbing the objects with dedicatedly designed stickers.

Existing defend strategies to tackle with the abovementioned security breaches can be dichotomized into robustness enhancement and adversaries detection. The former approach aims to reduce the model sensitivity to malicious changes in inputs. Related works could be broadly grouped into three categories: adversarial training, gradient masking and input transformation. Adversarial training augments the original dataset with discovered adversarial examples for training. It has later shown to be nonadaptive to new attacks [11]. Gradient masking seeks to conceal gradient information by producing near-zero gradients but found to be feeble in mitigating black-box attack because of the transferability of adversarial samples [13,15]. Input transformation attempts to eliminate the effect of adversarial perturbations by introducing variations to the inputs. Previous detection techniques [16–18] could also be broadly categorized into three classes: sample statistics, detector training and prediction inconsistency. Sample statistics require large-scale datasets for the analysis and have shown to be less effective in filtering out the adversarial samples located near the legitimate distribution [19]. Training a detector adds additional subnetworks that are dedicated for distinguishing adversarial inputs from the clean ones [20]. Prediction inconsistency measures the degree of consensus among multiple models in predictions [21,22].

These attacks or countermeasures, mostly analyzed at software level, have not considered the feasibility and effectiveness when the trained DNN models are implemented on hardware platforms. On the other hand, the globalization of supply chain in integrated circuit design and manufacturing makes injection of malicious hardware components into the neural network circuit a potential and realistic threat [23,24]. Noninvasive side-channel attacks that leverage on information leakage or backdoor created unconsciously in the hardware optimization and implementation are also feasible on DNN accelerators. In addition to misprediction danger that maliciously alters prediction results, confidentiality of the model has also shown to be possibly compromised. Hua *et al.* [25] successfully conducted a reserve-engineering attack on the convolutional neural network (CNN) accelerator through side channels, including the off-chip memory access and timing. Since a commercially valuable model that charges others for queries is a critical piece of intellectual property, uncovering its architecture as well as parameters enables subversion of model and privacy intrusion.

The security weaknesses of DL are mainly rooted in the lack of explainability in DNN-based systems. The provided opaque insights into behaviors and thought processes of the models obscure detection and awareness of malicious manipulations. Furthermore, compared with black-box DNNs, transparent DNNs are beneficial for algorithmic reliability guarantee, potential bias recognition and performance commitment [26]. Thus, various approaches have been proposed for the interpretability of deep network processing, representation and self-explainable systems. One technique is layer-wise relevance propagation (LRP) that identifies the most influential features of an input on the classifications [27].

The remainder of this chapter is organized as follows. Section 9.2 reviews preliminaries of DNNs, including relevant definitions, typical structures and custom hardware acceleration techniques. Classification of emerging misprediction attacks is presented in Section 9.3, alongside with their vital characteristics, examples and the respective defenses. Section 9.4 highlights the confidentiality threat crucial for commercial practice and the related defensive mechanisms. The fundamental concepts of explainable DNNs and the taxonomy pertaining to current methodologies to achieve transparency are elucidated in Section 9.5. Finally, the chapter is concluded in Section 9.6.

9.2 Preliminaries

9.2.1 Artificial neural networks (ANNs) and DNNs

Artificial neural networks (ANNs) are a large subfield of ML. They are inspired from the working principal of brain that is currently the best "engine" for learning and solving problems [1]. The essential component of an ANN is neuron computation that imitates the way the brain works. As shown by Figure 9.1(a), a biological neuron consists of the dendrites, a cell body and an axon. The dendrites transmit received signals to the cell body. The signals are normally scaled by synapses that are connections between the dendrites of current neuron and axons of former neurons. Then, the cell body performs a computation on the incoming scaled signals and generates an output activation that is later transmitted to other neurons as input signal by axon. The whole process is mathematically reconstructed in an ANN illustrated in Figure 9.1(b), where x_i indicates the ith input signal to the neuron, and w_i indicates the scaling factor of the input signal x_i. w_i is called *weight* to mimic the function of synapse in the biological model. The neuron computes a weighted sum of the input values. Nonlinearity is introduced by applying an activation function f to the sum to obtain an output activation. With multiple neurons in each layer, Figure 9.2 shows the calculation of one layer, where $y_j = f(\sum_{i=1}^{3} w_{ij}x_i)$.

Weights of an ANN are derived in the training phase where the model acquires its learning ability. Using image classification as an example, the output of an ANN

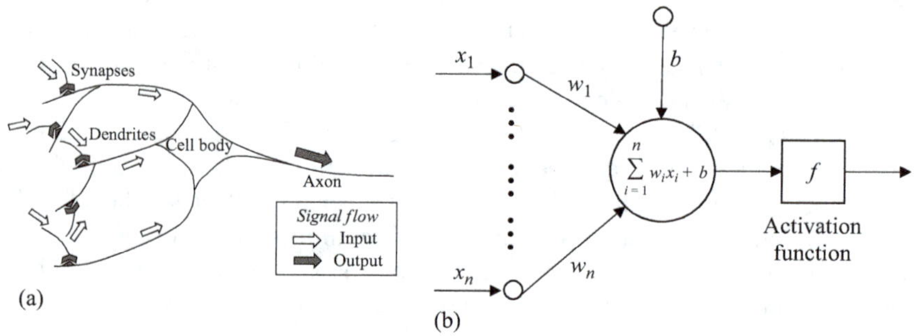

Figure 9.1 (a) A biological neuron and (b) mathematical model for a neuron in an ANN

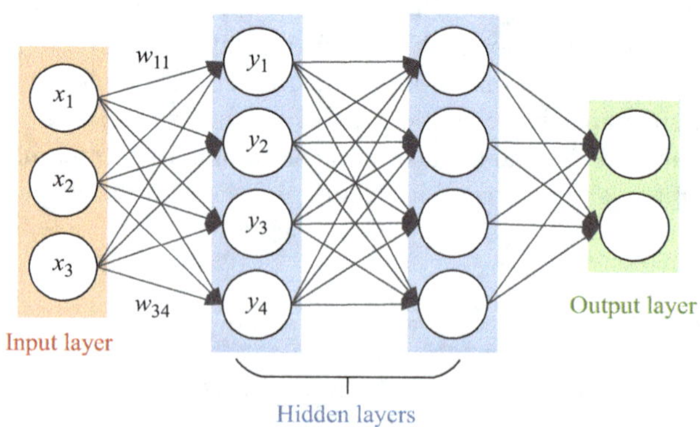

Figure 9.2 Output activation calculation at each layer of an ANN

will be a vector of scores that imply the possibilities for an image to be classified to each object class. The weights should be set at the values such that the output has the highest score in the correct class and that score is maximized. The deviation between the ideal score and the score calculated by an ANN based on its current weights is called the loss (L). Weights are updated through an iterative process until the loss is minimized. The optimization utilizes a gradient decent approach to find the local minimum of the loss by taking steps proportional to the negative of the gradient of loss function at the current point. The basic formula is presented as follows [1]:

$$w_{ij}^{t+1} = w_{ij}^{t} - \alpha \frac{\partial L}{\partial w_{ij}} \tag{9.1}$$

where α refers to the learning rate that has an influence on the convergence speed.

The gradient is often obtained by back-propagation method. It calculates the partial derivatives of the loss with respect to the weight backward through the network by applying the chain rule [5]. Apart from supervised learning mentioned previously where all the training samples are labeled, there are three other techniques to train the network. Unsupervised learning takes unlabeled data to explore the possible clusters; semi-supervised learning applies unknown examples to determine the decision boundaries and annotates a small portion for labeling groups; reinforcement learning is often applied in the policy network that outputs a decision and receives the corresponding reward as a feedback from the environment. It helps to make decisions that maximize the reward as well as balance the exploration and exploitation [3,28,29]. The determined weights are then adopted in the inference stage, where the model performs its task by forward propagating.

DNNs are one category of ANNs but with more layers in their architectures. Typically, a DNN can contain five to more than a thousand layers [1]. One interpretation for the inner processing of the deep networks in vision field is hierarchical feature extraction. When an image is fed into the first layer of the DNN in the form of pixels, the corresponding activations can be construed as miscellaneous subordinate features of the image-like lines. These features are then combined into more and more advanced ones at following layers. For example, lines are aggregated into corners, which are further merged into other shapes. Finally, the network yields a probability that the obtained high-level features compose a specific object.

9.2.2 Fundamental components of DNNs

DNNs can be broadly classified into recurrent (RDNN) and feedforward (FDNN) structures, as shown in Figure 9.3. A recurrent network allows the presence of feedback, thus enables the existing internal states to store the required intermediate activations for later processing. It is temporally dependent. This dynamic behavior facilitates the use of RDNNs in applications like unconstrained handwriting and speech recognition. In contrast, a feedforward network performs computations in sequence. All the operations are completed with the outputs of the previous layer as inputs. There is no loop in the information flow. One prevalent type of FDNNs are CNNs. It is widely used in image classification for its shared-weight architecture

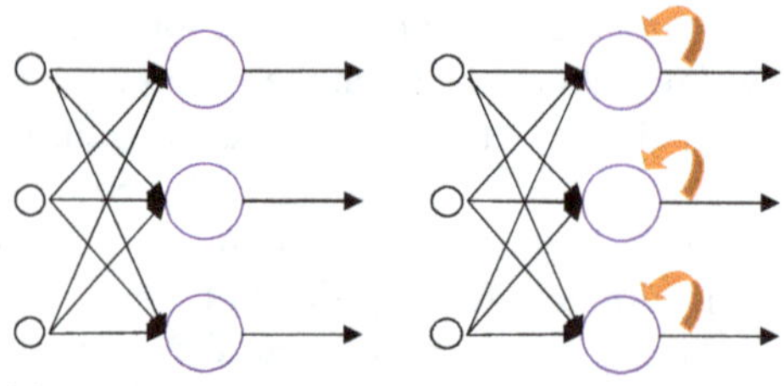

Figure 9.3 (a) Feedforward and (b) recurrent networks

and translation invariance properties [30,31]. A CNN has five fundamental building blocks: convolutional (Conv), pooling, rectified linear units (ReLUs), fully connected (FC) and softmax layers.

Conv layer is the essential element of a CNN. It exploits the feature abstraction of high-dimensional inputs such as images that are described in terms of height (H), width (W) and channel (C). To take spatial structure of data into consideration, each neuron in the Conv layer is connected to a small region of the input volume. This region is referred to as the receptive field of the neuron. Connections to the corresponding receptive fields are local in space but extend along all the channels of the input volume. This kind of local connectivity enhances the interpretation of the spatially local input patterns [32]. The weights of Conv layer are composed of a set of learnable 3D filters. Each filter has the same size of receptive field. As shown in Figure 9.4, each channel slice of an output feature map (ofmap) is obtained by convolving the width and height of an input feature map (ifmap) with the respective filter. The dot products between entries of each filter and windowed ifmap at each stride are computed and summed to produce the output value at each point of the channel slice of the ofmap. Spatial dimensions of ofmap are thereby determined by those of ifmap, filter and stride length (Str). The number of channels of the ofmap is equivalent to that of the filters. Figure 9.5 serves as a specific example to explain the operations of high-dimensional convolutions with a $4\times4\times3$ image as the input data and a $3\times3\times3$ filter. With $Str = 1$, the ofmap in this case is of dimensionality $2\times2\times1$. It can be seen that the same set of weights is used to obtain every output in the same channel slice of the ofmap. This weight-sharing scheme significantly alleviates the storage requirements of the weights.

Pooling layer is periodically introduced in-between the successive Conv layers to down-sample the processed data. It not only reduces the amount of computations and parameters but also contributes to the translation invariance of the network [33]. Pooling is applied to each channel separately. Only the spatial dimension will be changed after pooling. The length of the channel will remain unchanged. There are several pooling techniques such as max pooling and average pooling. As shown in Figure 9.6, pooling divides a channel slice of the ifmap into nonoverlapping blocks and outputs the maximum or average value of each block accordingly. With $Str = 2$

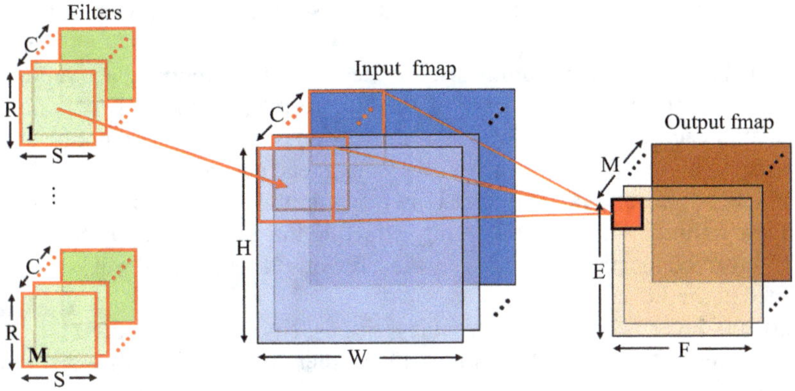

Figure 9.4 High-dimensional convolution operation

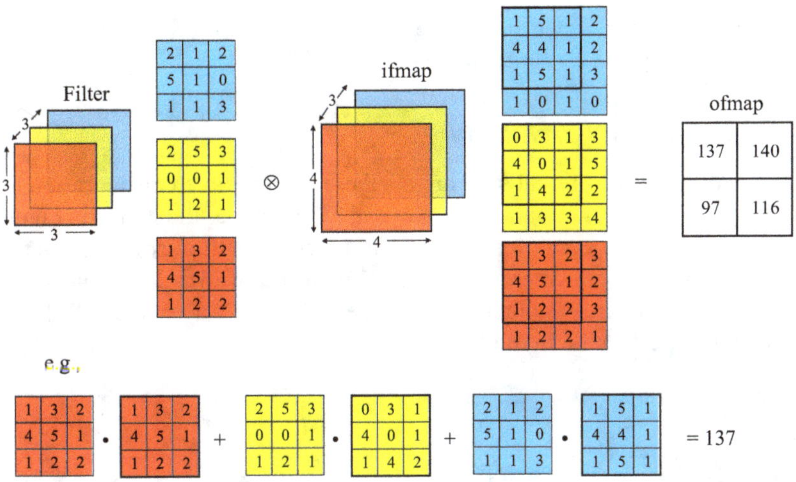

Figure 9.5 An example of high-dimensional convolutions

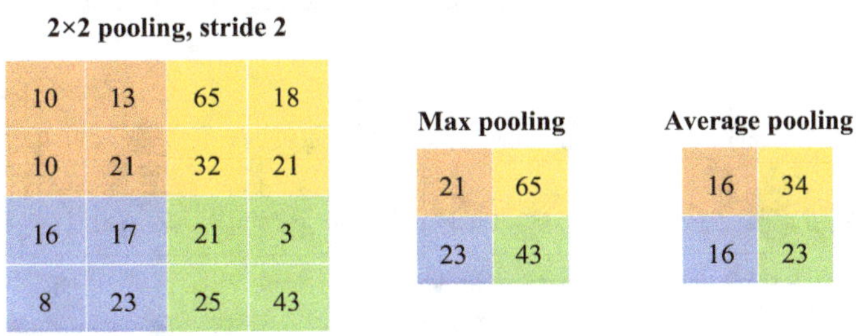

Figure 9.6 Pooling operations

and a receptive field of size 2×2, pooling decreases the input dimension from 4×4 to 2×2.

ReLU layer introduces nonlinearity to the network by applying activation function $f(x) = \max(0, x)$ to each entry of the feature map (fmap) [5]. It is normally inserted after each Conv or FC layer. The dimension of the fmap will not be changed afterward. Although there are many other nonlinear activation functions, such as sigmoid and hyperbolic tangent, ReLU is often preferred due to its simplicity and accelerated training without significantly affecting the generalization accuracy [5].

FC layer connects all neurons in the previous layer. High-level reasoning is performed in the FC layers.

Softmax layer is normally the last layer of a CNN that takes logits (the outputs of the previous layer) as inputs and normalizes them into a probability distribution. It is defined as follows [34]:

$$\sigma(z)_j = \frac{e^{z_j}}{\sum_{k=1}^{K} e^{z_k}} \quad \text{for } j = 1, \ldots, K \tag{9.2}$$

where K is the number of neurons in the softmax layer, z is a K-dimensional vector of the logits and σ is the K-dimensional output.

9.2.3 Popular DNN architectures

Most of the deep convolutional structures adhere to the prime configuration of five convolutional (Conv) sections and one to several FC blocks, as illustrated in Figure 9.7. Each Conv section can contain several Conv blocks. Feature extraction is accomplished by Conv sections, where the spatial dimension of ifmap is reduced and the depth is extended. Each FC block is composed of an FC layer and a ReLU layer. Classification is performed in the FC blocks.

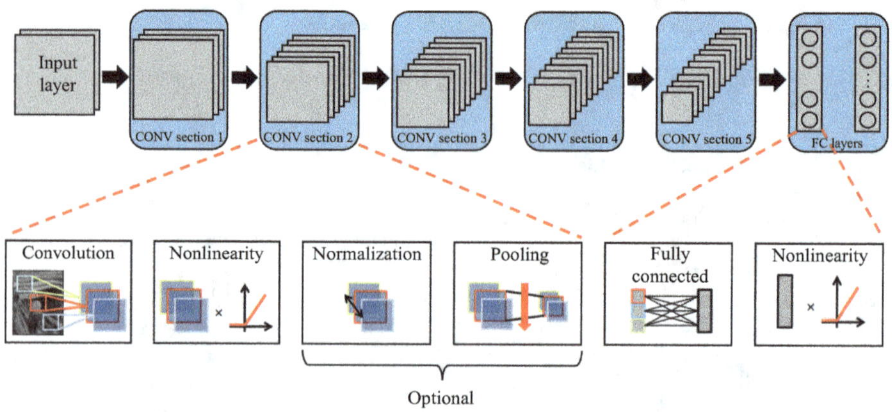

Figure 9.7 Conventional configuration of DNN

Conventionally, each Conv block consists of a Conv layer, a ReLU layer and optionally a normalization layer and a pooling layer. Examples are AlexNet [5] and VGGNets [35]. The former has only one Conv block for each Conv section and the first two blocks use 11×11 and 5×5 kernels, while the rest of the blocks use 3×3 kernel. The latter sequentially accumulates multiple Conv blocks in the same section and uses 3×3 kernel all the way.

Figure 9.8 shows some variants of Conv blocks. Inception [36] provides a distinct block formation. It executes multiple convolutions with diverse filter size in parallel and concatenates the resulting ofmaps along the channel dimension before going to the next block. An extra 1×1 kernel is often utilized before a larger convolution for dimension compression. The adoption of inception module enables the model to choose the best type of convolution for the Conv layer without human intervention. Furthermore, the model is capable of recovering both the local features via smaller convolutions and the high abstracted features via larger convolutions. ResNets [37] introduce shortcuts among Conv blocks to mitigate the gradient vanishing issue in the training. Since merely increasing the network depth will result in infinitesimal gradient when back-propagated to earlier layers, the performance of the model can easily be saturated or even start to demean abruptly. The presence of identity connections enables the layers to be skipped and a residual mapping to be learnt. A 1×1 convolution is also applied to reduce the number of weight parameters. DenseNets [38] follow this strategy by outspreading the identity connections to all the subsequent layers.

Unlike other networks that have five Conv sections, MobileNets [39] has six Conv sections. They are built based on depth-wise separable convolutions that factorize a standard convolution into a depth-wise and a point-wise. The former uses a single filter to convolve with each input channel. The results are then combined by the latter

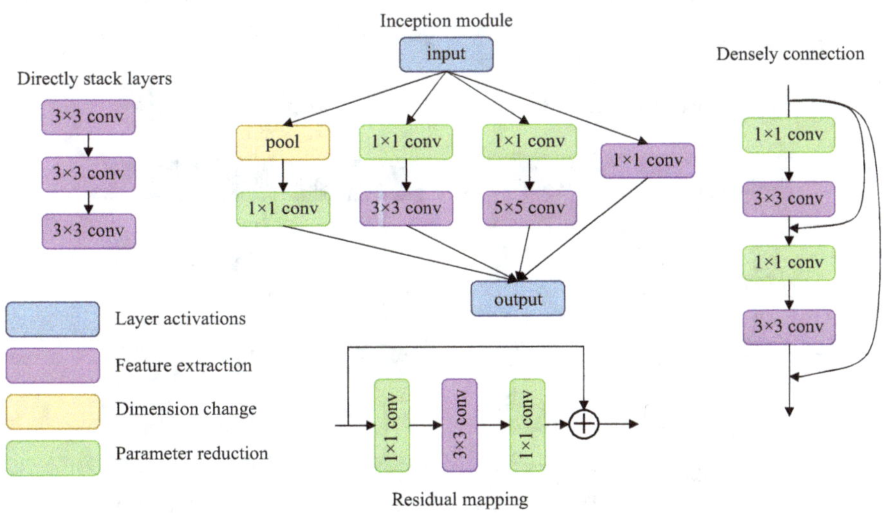

Figure 9.8 Different variants of convolutional block design

using a 1×1 convolution. MobileNets are dedicated for mobile and embedded vision applications with constrained resources.

9.2.4 Representative techniques for DNN hardware acceleration

Computational complexity increases the difficulty in deploying DL in portable and smart wearable devices. Since the essential computational component of DNNs is multiply-and-accumulate (MAC) operation, improving its efficiency has become an inevitable approach to produce an energy-efficient and low-latency accelerator. MAC operations can be sped up by parallel processing schemes. Both temporal and spatial architectures as shown in Figure 9.9 are regularly used. The temporal architecture, also known as single instruction multiple thread or single instruction multiple data, is usually adopted in CPUs or GPUs. Its control and memory are shared by numerous arithmetic logic units (ALUs), each of which works independently without intercommunication. In contrast, spatial architecture allows data transfer between adjacent ALUs and possesses private control logic and local memory (register file) for each ALU. Such ALU is known as processing element. Spatial architecture is commonly realized in application-specific integrated circuit (ASIC) designs.

To speed up the kernel computation on CPUs and GPU platforms, data transformation is applied to reduce the number of multiplications, often at the expense of increasing the number of additions with irregular data access patterns. A typical data transformation example is fast Fourier transform (FFT) [40,41]. By taking the FFT of the filter and ifmap, multiplications are carried out in the frequency domain. The inverse FFT of the result is finally calculated to obtain ofmap in spatial domain. On the other hand, dataflow processing leverages memory hierarchy to alleviate the memory wall problem. Multiple levels of local memory hierarchy with various energy costs are introduced to offer low-cost data movements [8]. The efficiency is further

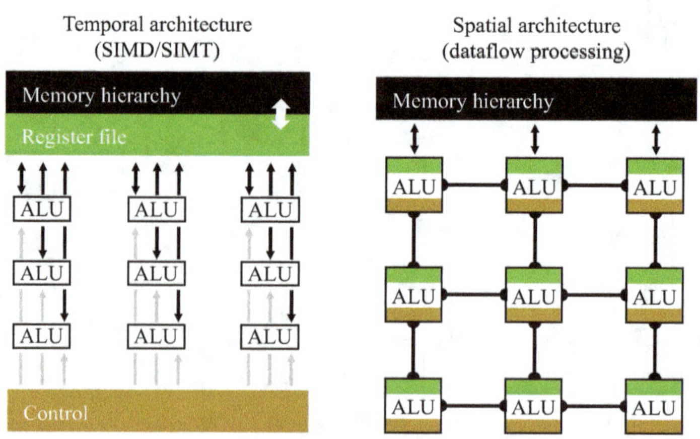

Figure 9.9 Parallel processing schemes [1]

enhanced by three forms of data reuse: convolution, feature map and filter. Convolution reuse refers to the use of the same ifmap and filter, fmap reuse refers to the use of the same ifmap across different filters and filter reuse represents multiple implementations of the same filter for diverse ifmaps (batch processing). As a result, relevant data can be stored locally and accessed several times to save off-chip memory visits. The corresponding dataflow strategies are categorized into weight stationary, output stationary, no local reuse and row stationary [1,6,7,42–44].

In addition to direct hardware optimization, codesign between the network model accuracy and hardware performance optimization takes implementation complexity into consideration to increase the likelihood of adoption into practical applications [1]. Precision reduction is one popular approach. Generally, 32-bit floating point is adopted on CPU and GPU platforms. It has a wide dynamic range that favors accuracy but consumes more energy and area. Recent works [45] have inspected the effect of reduced precision on energy and area costs and found that they scale nearly linearly and quadratically with the number of bits for a fixed-point adder and multiplier, respectively. Most research works focus on weight instead of activation quantization, since weights directly influence the storage capacity requirement. Several techniques have been proposed for weight quantization, which can be broadly categorized into linear and nonlinear quantizations.

Another approach is to decrease the number of operations and model size. This can be achieved via activation statistics utilization, network pruning, architecture enhancement or knowledge distillation. Since ReLU zeros negative values in the ofmaps, the resulting activation is sparsely distributed. By taking advantage of these intrinsic characteristics, the costly external memory accesses can be reduced by compression through run length coding [46]. The sparsity can be further increased by clipping insignificant activations. Moreover, memory readings and MAC operations on zero-valued activations can also be eliminated in hardware design. Besides, a large amount of weights are found to be redundant and can be pruned with negligible impact on accuracy. The accuracy loss after network pruning is commonly compensated by fine-tuning. The number of parameters and operations can also be constrained using a compact network architecture through convolution factorization [47]. As shown in Figure 9.10, a 5×5 filter can be replaced with either two 3×3 filters or a 1×5 filter and a 5×1 filter. The number of weights is thus reduced from 25 to 18 and 10, respectively. Last but not the least, establishing a simple student model based on the knowledge transferred by an ensemble of complex teacher nets could achieve good performance while shrinking the model size [48,49].

9.3 Misprediction attacks

9.3.1 Threat model

Barreno *et al.* [50] have provided the following attributes to define a space of misprediction attacks on ML systems. Several typical scenarios are then given to facilitate understanding.

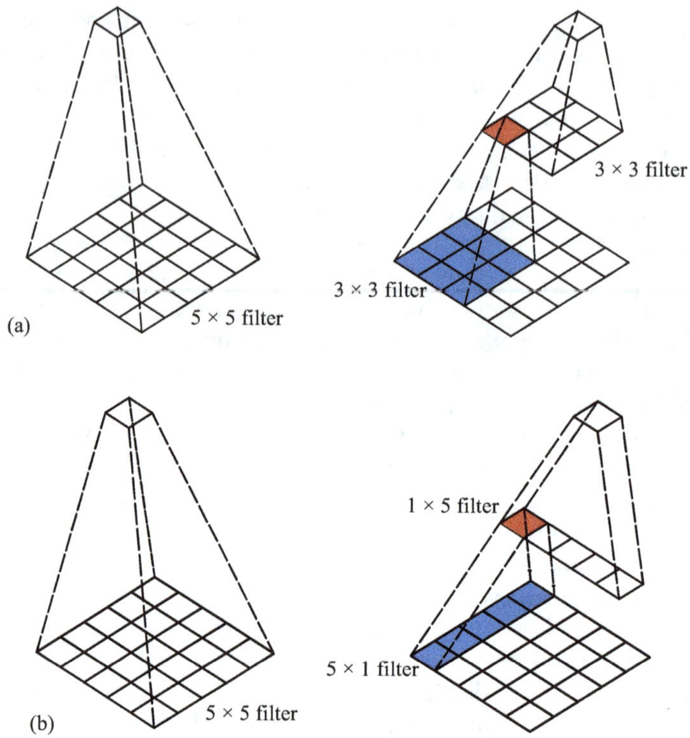

Figure 9.10 Decomposition of a 5×5 convolution to (a) two 3×3 convolutions or (b) 1×5 and 5×1 convolutions [1]

Attack taxonomy

Influence

Causative—Causative attacks alter the training process through influence over the training data.

Exploratory—Exploratory attacks do not alter the training process but use other techniques, such as probing the learner or offline analysis, to discover information.

Specificity

Targeted—The specificity of an attack is a continuous spectrum. At the targeted end, the focus of the attack is on a particular point or a small set of points.

Indiscriminate—At the indiscriminate end, the adversary has a more flexible goal that involves a very general class of points, such as "any false negative."

Security violation

Integrity—An integrity attack results in intrusion points being classified as normal (false negatives).
Availability—An availability attack is a broader class of attack than an integrity attack. An availability attack results in so many classification errors, both false negatives and false positives, that the system becomes effectively unusable.

Example 9.1 (Causative integrity attack). *In an intrusion detection system, the defender periodically retrains the model with new traffic. The adversary contaminates the information flow by carefully injecting some abnormalities that resemble legitimate instances to mislead the learner, resulting in no flagging for an intrusion.*

Example 9.2 (Causative availability attack). *In an email spam filter system, the attacker sends harmful messages, including both malicious words and benign parts, when the defender is collecting data for training. Since the legitimate components are mistakenly correlated with spam, the system is likely to block a similar normal email next time.*

Example 9.3 (Exploratory integrity attack). *In a face recognition system for authorization, the adversary perturbs the captured sensor data during the inference to grant accesses to unauthorized users.*

Example 9.4 (Exploratory availability attack). *In an image classification system, the attacker injects noise to the model during the inference to corrupt the classification ability of the model.*

In the following subsections, three existing popular misprediction attacks, namely evasion, poisoning and backdoor, aiming at altering the inference results of DL models are discussed in detail.

9.3.2 Evasion attacks

9.3.2.1 Data evasion

This kind of attacks distorts input data during the inference phase so that the DNN model fails to identify the input. This attack falls into the exploratory category of the taxonomy. The input modifications can be made either digitally or physically. One threat that is extensively studied in academia for digital perturbation is by introducing adversarial examples that are usually well crafted by adding imperceptibly small perturbations to the original benign inputs. It is first demonstrated by Szegedy *et al.* [10], who used box-constrained limited-memory-Broyden–Fletcher–Goldforb–Shanno algorithms to craft malicious samples, as shown in Figure 9.11. Five

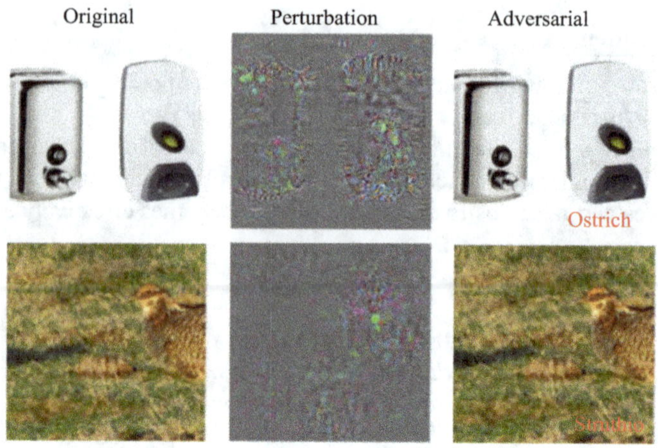

Figure 9.11 Illustration of adversarial examples generated using [10] for AlexNet [5]. The perturbations are magnified 10× for better visualization. The predicted classes for adversarial images are also shown.

representatives of adversary generation methodologies are elaborated subsequently. In the following, $x, \widehat{x}$, F and t are used to denote a clean input sample, an illegitimate image, a DNN model and a target class to be perturbed into, respectively.

FGSM: FGSM uses an L_∞-norm distance metric to quantify the similarity between the adversarial and the respective original samples. It is defined as follows [11]:

$$\widehat{x} = x - \varepsilon \cdot \operatorname{sign}\left(\nabla J_{F,t}(x)\right) \tag{9.3}$$

where J and ε denote the cost function used to train the model and an input variation parameter that controls the amplitude of distortion, respectively. ε should be sufficiently small to be unobtrusive.

FGSM computes the gradient direction of the loss function for each pixel of an input image to determine the direction of added changes that will potentially minimize the loss. It is a one-step-targeted attack that aims to be fast rather than optimal. Its variant, basic iterative method (BIM), applies FGSM multiple times with a smaller step size to improve the success rate.

JSMA: Jacobian-based saliency map attack (JSMA) [51] adopts an L_0-norm distance metric to find adversarial examples. Unlike FGSM that changes all the pixels, JSMA restricts perturbations to a limited number of input features. It utilizes the forward derivatives of the network with respect to the input to construct a saliency map that reflects the impact of each input component on the target classification. The most significant pixel is then selected and modified to increase the likelihood of target class until the threshold of being undetectable is reached or misclassification is achieved. Given

$$\begin{cases} a_i = \dfrac{\partial F_t(x)}{\partial x_i} \\ b_i = \sum\limits_{k \neq t} \dfrac{\partial F_k(x)}{\partial x_i} \end{cases}, \tag{9.4}$$

the saliency map constructed in each round is expressed as follows:

$$\text{Smap}(x,t)_i = \begin{cases} a_i \times |b_i| & \textit{if } a_i > 0 \text{ and } b_i < 0 \\ 0 & \text{otherwise} \end{cases}, \tag{9.5}$$

where i and k represent the input pixel and any object class other than t, respectively.

It is difficult to obtain the adversarial examples by selecting only one pixel at each iteration. Instead, a candidate pair of pixels (p_1, p_2) are collected each time using the following heuristic:

$$\arg\max_{(p_1,p_2)} \left(\frac{\partial F_t(x)}{\partial x_{p_1}} + \frac{\partial F_t(x)}{\partial x_{p_2}} \right) \times \left| \sum_{i=p_1,p_2} \sum_{k \neq t} \frac{\partial F_k(x)}{\partial x_i} \right| \tag{9.6}$$

Although JSMA is capable of generating adversarial examples with the minimum disturbance to fool a classifier, the search for a suitable pair of input features for each repetition is time consuming and requires a lot of memory. Due to this constraint, JSMA has been demonstrated on only small datasets like Modified National Institute of Standards and Technology (MNIST) database, which is a dataset of handwritten digits.

C&W: C&W attack [52], named after Nicholas Carlini and David Wagner, translates the initial formulation of the problem [10]:

$$\textit{minimize } D(x, \widehat{x}), \textit{ such that } F(\widehat{x}) = t \wedge \widehat{x} \in [0, 1]^n, \tag{9.7}$$

to an easier optimization problem:

$$\textit{minimize } \|\delta\|_p + c \cdot f(\widehat{x}, t), \textit{ such that } \widehat{x} \in [0, 1]^n, \tag{9.8}$$

where D is equivalent to the distance metric $\|\bullet\|_p$. δ and c represent the perturbation and a carefully chosen constant, respectively. $\widehat{x} = x + \delta$. f is defined as an objective function such that $F(\widehat{x}) = t$ if and only if $f(\widehat{x}, t) \leq 0$. Since the nonlinearity of the constraint $F(\widehat{x}) = t$ increases the difficulty of solving this optimization problem directly, the introduction of f eases the problem. The best f is found to be

$$f(\widehat{x}, t) = \max(\max\{G(\widehat{x})_k : k \neq t\} - G(\widehat{x})_t, 0), \tag{9.9}$$

where $G(\widehat{x})$ is the output probability vector of a classifier. To bypass the box constraint $\widehat{x} \in [0, 1]^n$ in expression (9.8), δ is replaced with a new variable r such that

$$\delta = 0.5(\tanh(r) + 1) - x. \tag{9.10}$$

Since $-1 \le \tanh(r) \le 1, \widehat{x}$ unconditionally lies in the boundaries. Therefore, the final problem definition is given by:

$$\arg\min_r \|0.5(\tanh(r)+1) - x\|_p + c \cdot \max(\max\{G(0.5(\tanh(r)+1))_k : k \neq t\} \\ -G(0.5(\tanh(r)+1))_t,\ 0). \tag{9.11}$$

All the L_2-norm, L_∞-norm and L_0-norm versions of C&W attacks have been proposed and validated to be more effective than earlier methods for generation of adversarial examples.

DeepFool: Unlike the three previously discussed algorithms, DeepFool [53] is an indiscriminate attack optimized for L_2-norm distance metric. It seeks to explore the minimum perturbations required to cross the decision boundary. It simplifies the DNN model to an entirely linear structure where the classes are separated by hyperplanes. The problem is then formulated into computing the distance between the input point and the closest hyperplane. To generalize the solution to the nonlinear classifier, DeepFool approximates the boundary of the initially correct class label of the point of interest iteratively by borrowing the idea of Newton's iterative algorithm.

Universal adversarial perturbations: Moosavi-Dezfooli *et al.* [53] extended DeepFool to a generalized image-agnostic adversarial perturbation. It can be mathematically expressed as follows [12]:

$$\delta := \left\{ \delta | Err(X_\delta) := \frac{1}{m} \sum_{i=1}^{m} 1_{F(x_i+\delta) \neq F(x_i)} \ge 1 - acc \right\},\ s.t.\ \|\delta\|_p \le \xi. \tag{9.12}$$

where X represents a set of natural images $\{x_1, \ldots, x_m\}$ collected from a population, δ is a fixed perturbation vector that leads to the misclassification of most samples in X, X_δ defines a set of perturbed data points $\{x_1 + \delta, \ldots, x_m + delta\}$, acc quantifies the accuracy of the tampered samples and ξ constrains the extent of distortion δ. This algorithm iteratively feeds the benign inputs by adding previously derived universal perturbation into the classifier. It calculates the minimum distortion that sends the current tampered input to the decision boundary in each iteration. The instance of universal aberration is updated by adding the computed distortion and then used it for the next round. As can be observed in Figure 9.12, these universal perturbations remain quasi-imperceptible to human vision systems.

All the previously discussed means rely on the assumptions that the adversary has full access to the target DNN model, including the architecture and parameters. Hence, they are all white-box techniques. The later method [13] exploits the transferability of adversarial examples among different classifiers for black-box attack. Although the hypothesis of such attack seems to be realistic, Su *et al.* [54] discovered that the strength of transferability varies largely across models and attacking algorithms. In a black-box setting, the output of the model corresponding to an input image is used to train a substitute model that intends to imitate the mapping of the source model. The adversarial generation process is then made with respect to the replacement. FGSM and JSMA crafting algorithms have been evaluated on small-scale MNIST and GTSRB (the German Traffic Sign Recognition Benchmark)

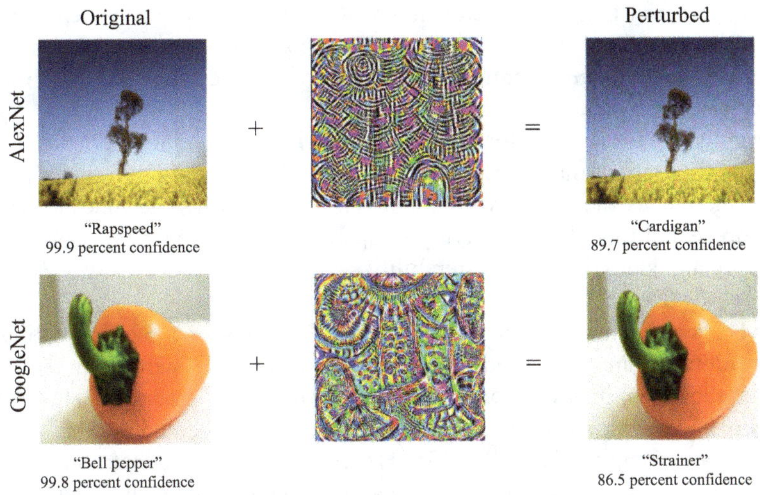

Figure 9.12 Illustration of adversarial examples generated using universal perturbations[12] for AlexNet [5] and GoogLeNet [36]

datasets for black-box attack in [13]. An average success rate of more than 80 percent is achieved. However, the performance is reversed when employing a more complicated dataset like ImageNet. In [54], the authors addressed the transferability through comprehensive ImageNet classification experiments conducted on every 2 of the 18 networks from 7 families: AlexNet [5], VGGNets [35], Inception nets [36], ResNets [37], DenseNets [38], MobileNets [39] and NASNets [55]. Proportion of adversarial images generated from one model that are wrongly perceived by another model in the same target class corresponding to the targeted transferability and the indiscriminate transferability without class constraints are separately analyzed. The former can be referred to as matching rate, whereas the latter is equivalent to the error rate. Several main findings on transferability by [54] are listed below.

- The adversarial images possess considerably higher indiscriminate transferability than targeted transferability.
- FGSM and BIM yield much higher untargeted transferability than C&W, while their adversarial generation success rates are worse than C&W. The generation and transfer success rates are ordered as follows:

$$\begin{aligned} \textit{Generation success rate}: \ & FGSM \leq BIM \leq C\&W \\ \textit{Transfer success rate}: \ & FGSM \geq BIM \geq C\&W \end{aligned}$$

  Furthermore, a larger ε for FGSM and lesser iteration for BIM are more beneficial to the indiscriminate transferability.
- Transferability is not necessarily symmetrical. In other words, good transferability of adversarial examples of one model to another model does not imply the same quality of transferability in the reverse.

- The adversarial images crafted using VGGNets are empirically better in transferability to other network families. This could be attributed to their basic architectures and simple convolution operations, which form the baseline of all other advanced models.
- The transfer rate is close to 100 percent when the two ends are from the same network family but much lower when their architectures are different. This could be exploited as the mechanism for reverse-engineering the inner structure of an unrevealed model by supplying tampered samples generated from a known network and studying the transferability.

For DL system implemented on CPU or GPU platforms, 32-bit floating point representation is commonly used. Therefore, the adversarial perturbation generated using the abovementioned attacks is calculated in 32-bit precision. However, the intensity of each component of a color image pixel is represented in 8-bit scale, with an integer value ranging from 0 to 255. Hence, precision loss that undermines the distortion efficiency is inevitable upon converting a 32-bit adversarial data array into 8-bit image format. Particularly for FGSM, BIM, C&W and DeepFool attacks that inflict infinitesimal changes for manipulation, failure rates have increased tremendously. Most derived deformations, which are smaller than the image precision, are eliminated during the transformation. Effectiveness might be further declined, while these attacks are imposed on a custom-integrated DNN system that normally utilizes quantization technique to save energy and area.

Countermeasures against digital data evasion attacks can be dichotomized into reactive and proactive categories. The former intends to improve model robustness, whereas the latter aims to detect adversaries. Reactive measures are carried out by three methods: adversarial training, gradient masking and input transformation. They are elucidated next.

Adversarial training: Adversarial training retrains the model with the addition of off-the-shelf adversarial examples into the original training dataset. The method indeed improves model resilience against used adversarial algorithms but remains vulnerable for unused ones. This nonadaptivity implies that the strength of this defense is determined by the number of the distinct attack algorithms involved in adversarial generation process. Apart from significantly high computational costs in crafting malicious images by most known techniques, it is also unrealistic to assume that the attacker is restricted to techniques that are known to the defender. The incurred doubled training cost further suppresses efficiency of adversarial training.

Gradient masking: Since many attacks are based on the extraction of gradient information, hiding gradient from adversaries seems to be a feasible defense. Papernot *et al.* [15] proposed a defensive distillation method to strengthen DNN models. This technique extracts probability vectors of a pretrained model as soft labels to build a distilled model with the same architecture. It is originated from the hypothesis that encoding probability vectors that contain correlation information among classes can improve the generalization ability of the model. Furthermore, a smoother softmax layer is designed by adjusting the temperature parameter for gradient concealment. The experiments showed that the success rates of some attacks on MNIST and CIFAR-10 datasets have been reduced significantly without compromising accuracies. However, it was later broken by C&W and black-box attacks.

Input transformation: A few studies have demonstrated the effectiveness of data transformation against evasion attacks by inhibiting the adversarial effect. Bhagoji *et al.* [56] introduced linear dimensionality reduction methods, such as principal component analysis, to project original data along the identified "principal axes" (i.e., directions that preserve the most crucial information of the data) to the training data. The same transformation procedure is repeated in inferencing before feeding the inputs to the model. Instead of building a fresh model specialized for the projected inputs, Meng and Chen [57] presented MagNet, a framework that neither modifies model internals nor requires adversarial creation process. It reconstructs the inputs beforehand using autoencoders, trained with sufficient clean examples, and is expected to move tampered images toward the distribution of legitimate ones. Since the output dimension of an autoencoder is the same as that of the input, spatial structure is not corrupted. MagNet is theoretically applicable to various classifiers. Besides, several feature squeezing strategies [16,58], such as binary filter, spatial smoothing and color depth reduction, have been suggested to shrink the feature space of the input to increase the difficulty of adversarial manipulation.

Compared to reactive defenses that passively increase model compatibility of malicious samples, proactive methods, with the capability of detecting attacks, have profound implications on follow-up actions toward adversaries. These methods can be divided into three main types: sample statistics, detector training and prediction inconsistency.

Sample statistics: Feinman *et al.* [19] discovered two features, density and Bayesian uncertainty estimates, to detect illegitimate points lying far from the natural data manifold and nearby. Density estimates, specifying the distance between a given input and submanifolds, are calculated using the activations of the last hidden layer, while Bayesian uncertainty estimates are extracted directly from the dropout layer. It can be observed that the computational cost increases substantially with the model complexity and the size of the dataset. Apart from the architecture or image features that are commonly explored for defense in literature, Nixon *et al.* [59] utilized the device feature, sensor pattern noise (SPN), for adversarial example detection dedicated to mobile applications. Dataflow of the DNN image recognition can be extended as shown in Figure 9.13 to introduce the role of SPN. Since digital perturbations are likely to be intervened if inserted before the third stage, they are assumed to be added somewhere between the third and fourth stages. SPN Dash system for adversarial perturbation detection is then introduced right before the classification phase. The detailed functionality is demonstrated in Figure 9.14. I_{enc}, SPN_{cur} and SPN_{dev} denote an output image of the third stage, the SPN of the current submitted I_{enc} and a reference SPN for a specific device, respectively. Although a 10× reduction in computational and memory overhead has been achieved empirically, the susceptibility of SPN detector as well as the device-dependent estimation accuracy are the main bottlenecks for the generalization of this defense.

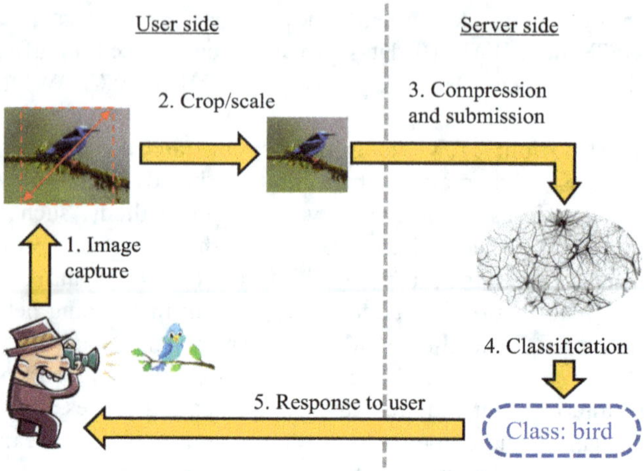

Figure 9.13 Extended dataflow of DNN image recognition [59]

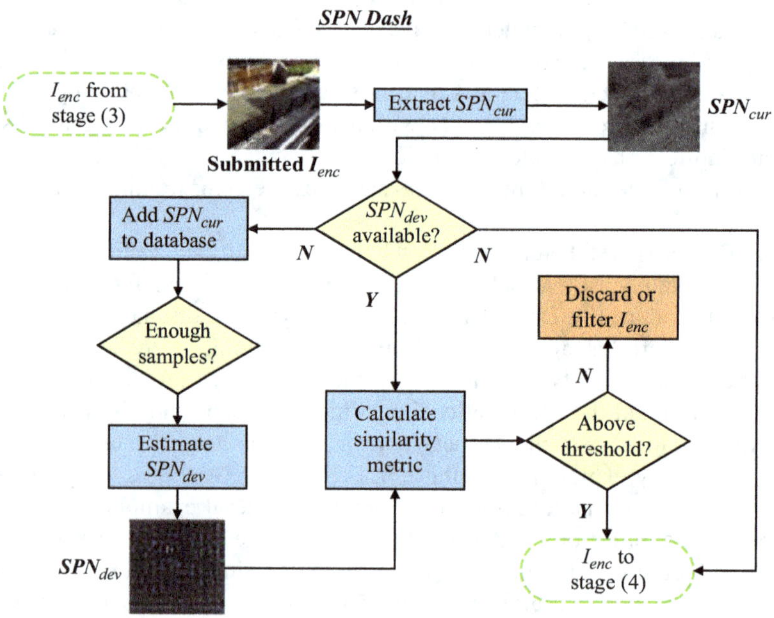

Figure 9.14 SPN Dash framework [59]

Detector training: Metzen *et al.* [17] devised augmentation of subnetworks as a detector to the deep net. This additional module is devoted to binary classification between clean and adversarial inputs. It is trained while freezing the parameters of the original model. Hence, the accuracy is not sacrificed. In contrast, Grosse *et al.*

[20] only revised the last layer with an extra class of "adversary", resulting in some accuracy loss on natural samples. Both defenses require massive adversarial examples for training, which is costly. Besides, they are also prone to over-generalization of adversarial attacks.

Prediction inconsistency: Prediction inconsistency uses the degree of consensus among multiple models in predictions as an indicator. Wang *et al.* [21] integrated mutation testing of DL systems [22] and statistic hypothesis testing into the detection algorithm. They found that adversarial images are more sensitive to the model mutants than clean images. Label change rate (LCR) is used to quantify this mutation sensitivity and is defined as follows:

$$\zeta(x) = \frac{|\{f_i|f_i \in F \cap f_i(x) \neq f(x)\}|}{|F|}, \tag{9.13}$$

where x denotes an input image, f is the original model, F is a set of mutated models $\{f_1, \ldots, f_m\}$ and $|F|$ is the number of elements of F. LCR is discovered to be significantly higher for a tampered input than that of a clean one against a number of DNN mutants. Therefore, statistic hypothesis testing is applied to determine the truthfulness of the input identity in terms of LCR. The strategy has only been implemented on small-scale datasets. Its feasibility and efficiency on complex database are untestified. Xu *et al.* [16] observed that unnecessarily large feature input space increases the degree of freedom for adversarial abuse. Hence, input compression could remove perturbations added to those eliminated unimportant features. The workflow is illustrate in Figure 9.15. It basically compares the classification results of the initial and squeezed inputs. If the difference is so great that it exceeds a predetermined threshold, it alerts that the input may be an adversary; otherwise, the system outputs a prediction result for this certified legitimate input.

Digital data evasion could occur via man-in-the-middle attack, but often with time constraint in order to be oblivious. Comparatively, physical perturbation is relieved from the harsh time condition and not required to be imperceptible since visible intrusion like a graffiti or a sticker is often not considered suspicious in real life. However, it suffers from varying environmental conditions, such as changing distances to the object, different angles and lighting, and preprocessing of the captured image like cropping and scaling. Eykholt *et al.* [14] proposed robust physical perturbations (RP_2)

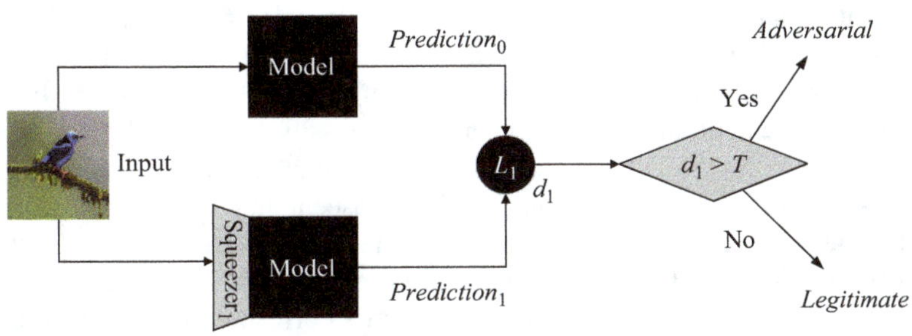

Figure 9.15 Workflow of feature squeezing method [16]

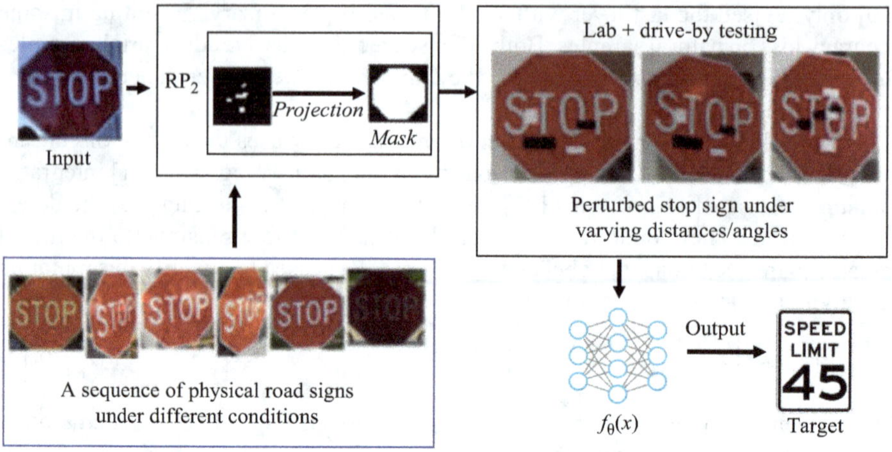

Figure 9.16 The pipeline of RP_2 attack [14]

algorithm to craft robust adversarial perturbations on real objects surviving from a variety of viewpoints, including significantly changing distances and angles. The pipeline of RP_2 attack is shown in Figure 9.16. Different image instances that encode both environmental variability and synthetic conversions are sampled to calculate an adversarial perturbation. The perturbation is then projected onto a mask to ensure that the position of the perturbation is within the surface area of the object. Finally, the derived mask is printed out and the perturbed area is cut out and stuck onto the target object. To account for the fabrication loss that occurred during printing, an additional term, based on non-printability score [60], is added to the objective function for perturbation optimization. LISA, a US traffic sign dataset, as well as GTSRB have been used in the experiments. An average targeted attack success rate of more than 60 percent and indiscriminate attack success rate of close to 100 percent have been reported, hinting an initial success of transferring data evasion attack from the digital to the physical paradigm.

9.3.2.2 Model evasion

As a variant of data evasion, model evasion injects noise to the model instead of the inputs. With the rise of hardware-based fault-injection attacks such as row hammering, laser beaming and clock glitch, tampering internal nodes of a DNN model deployed on dedicated hardware has been made possible. Liu *et al.* [61] proposed two approaches to conduct efficient model evasion attacks based on parameter modification: single bias attack (SBA) and gradient descent attack (GDA). SBA intends to distort one parameter to achieve the wrong decision. It has been found that the classification results are linearly dependent on the biases in the output layer. Therefore, moderately incrementing relevant bias for its corresponding adversarial class enables prediction alteration. SBA is also image-agnostic, thus is advantageous in run-time implementation. GDA aims at stealthy misclassification that only exerts an impact on the targeted inputs while maintaining the accuracy for others. It is realized by optimizing the following expression with gradient descent:

$$\arg\max_{\theta} \ F^{t}(\theta, x) - \lambda \left|\theta - \theta_b\right|, \tag{9.14}$$

where θ represents all the weights and biases in a DNN model, θ_b denotes the initial value of θ and $F^t(\theta, x)$ is the output probability of the targeted class t given an input x. Since gradient descent perturbs parameters according to their respective gradients, larger gradient parameters receive larger distortions. Furthermore, it is observed that the parameters in fewer parameter layers tend to have larger gradients. As a result, expression (9.14) tends to adjust the parameters in smaller layers in coarse granularity. Layer-wise instead of global search is adopted. To further fine-tune the solution, redundant parameters in the modification list are pruned to achieve better stealthiness. However, the work is only simulated by software, without demonstrating its physical attack viability. Breier *et al.* [62] proposed the first physical fault-injection attack to the DL model by introducing laser faults into the processing of activation function implemented on a low-cost microcontroller. MNIST is selected for evaluation, and the number of injected faults to produce a reasonable success rate ($\geq$50 percent) is recommended to be more than half or three-quarters of the neurons in the targeted layer using sigmoid and tanh, or ReLU activation functions, respectively. It can be observed that the feasibility of this attack in using large-scale database like ImageNet is not considered as usual, so is the usage of complex models. To replace invasive laser injection, clock glitch and row hammer are potential noninvasive attack to induce the same faulty behavior.

Clock glitch: Clock glitch introduces faults into a digital circuit through timing violation caused by irregular pulses in the clock signal. As shown in Figure 9.17, data d_{in} is transmitted from register A through a combinational logic to register B. The register should be triggered by the sampling edge of B to guarantee normal execution of the circuit. Therefore, the clock cycle should be larger than the maximum path delay τ of the combinational logic and the setup time of the register to ensure correct functionality. Figure 9.18 demonstrates a timing diagram comparison between the case of normal and glitched clock. It can be clearly seen that register B may sample the wrong data as the $(n-1)$th period is smaller than τ in Figure 9.18(b). By this means, an error can be introduced on register B. The generation of clock glitch can be done using two clock signals with different periods and a control signal *trigger*,

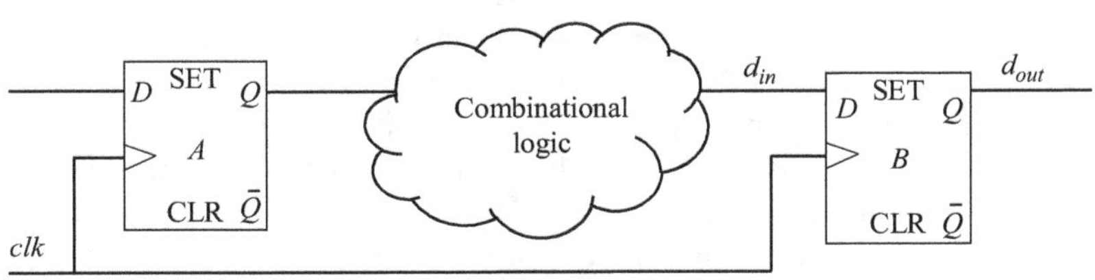

Figure 9.17 A typical sequential logic path

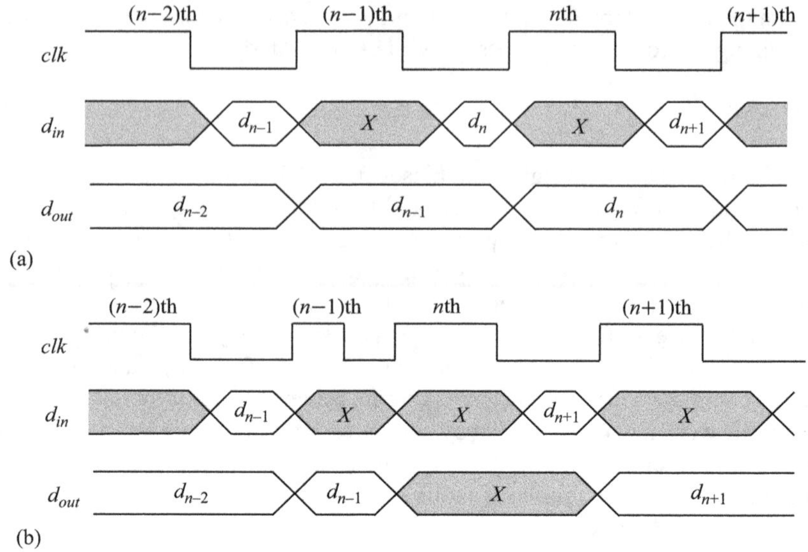

Figure 9.18 Timing diagrams under (a) a normal clock and (b) a glitched clock

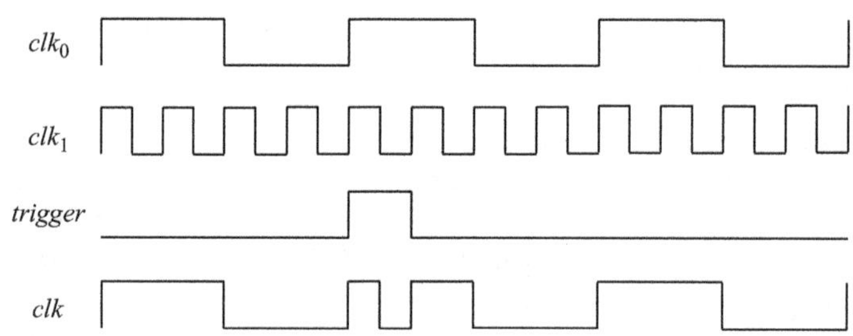

Figure 9.19 A simple design for generation of the clock with a glitch

as demonstrated in Figure 9.19. clk_0 is the normal clock. clk_1 is only selected when the *trigger* is enabled. *clk* represents the resultant clock signal with a glitch.

Row hammering: Since DNN accelerator relies heavily on the off-chip memory like dynamic random access memory (DRAM) to store its weights and intermediate results, the integrity of such memory can impact its accuracy. Row hammering is an attack that exploits the unfavorable side effect of DRAM induced by the poor separation of high-density memory cell array to enable electrical interaction in the neighborhood of a cell. This behavior could possibly corrupt data in nearby rows that are initially not included in memory access patterns. Kim *et al.* [63] successfully demonstrated a deliberate crafting of row hammering effect on Intel and advanced micro devices (AMD) systems by using a malicious program that activates the same row of DRAM intensively.

9.3.3 Poisoning attacks

Since DNNs require a large number of training samples to construct a mapping, the quality of collected samples is usually not guaranteed. By contaminating the training data with a set of deliberately crafted samples, attackers are able to make the trained model corrupted, thus affecting the normal functions of the DNN-based system. This is referred to as poisoning attack. It is a causative attack according to the taxonomy. Yang *et al.* [64] proposed both direct and generative gradient methods to produce tampered inputs with an assumption that the adversary has full knowledge of the targeted system. The latter is an acceleration of the former by bypassing expensive gradient computation using the concept of generative adversarial network (GAN) [10].

To defend against poisoning attacks, an activation clustering technique is proposed to filter out the poisoned training data points in [65]. It uses training data that include the abnormal data points to query the trained model during the inference and extract activations of the last hidden layer for each training sample. Since these activations are found to form two distinct regions for the class that has poisoned training dataset but form only one cluster for the clean one, they are grouped according to their label and each group is clustered separately.

9.3.4 Backdoor attacks

A backdoor attack implants a backdoor in the DNN model that is only triggered by the presence of a unique key. The process can be completed via influencing the training stage or an untrusted supply chain to modify the model before deployment. The former case is often in companion with poisoning attack, while the latter is also referred to as hardware Trojan attack. Gu *et al.* [66] devised a *BadNet* that incorporates an identifier for the backdoor trigger into a user-specified architecture under the scenario where users outsource pretrained models, that might be maliciously trained, to speed up the training procedures. The strategy has been demonstrated on a US street sign detector that recognizes stop signs as speed limits when a particular sticker is added to the stop sign while behaves normally on other cases. It has also been shown that the backdoor remains after the classifier is fine-tuned to perform a new task. Xinyun *et al.* [67] conducted backdoor poisoning attacks under a more harsh threat model, in which the adversary is only allowed to introduce limited volume of malicious samples in training but has neither controllability nor knowledge on the victim system. Two main strategies have been imposed on a face recognition system: input-instance-key and pattern-key. The first one uses an input instance, that belongs to the input space consisting of face photos, to be the key. It also takes variants of the selected input instance into consideration. The second one uses a pattern that can be any image to be the trigger. It offers more freedom in backdoor instances.

Globalized supply chain creates more opportunities for Trojan insertion into the hardware solution of DL. As shown in Figure 9.20, the lateral integrated circuit design and production flow involving a variety of vendors supplying electronic design automation tools, third-party intellectual properties, manufacture foundries, etc. resulting in increasing risks and challenges of deliberate hostility for supply chain security management. Any included party may have motivation to inject a hardware

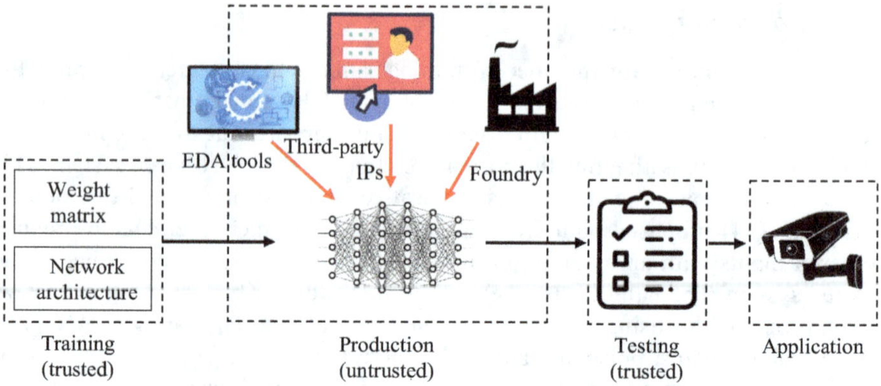

Figure 9.20 Modern integrated circuit design flow

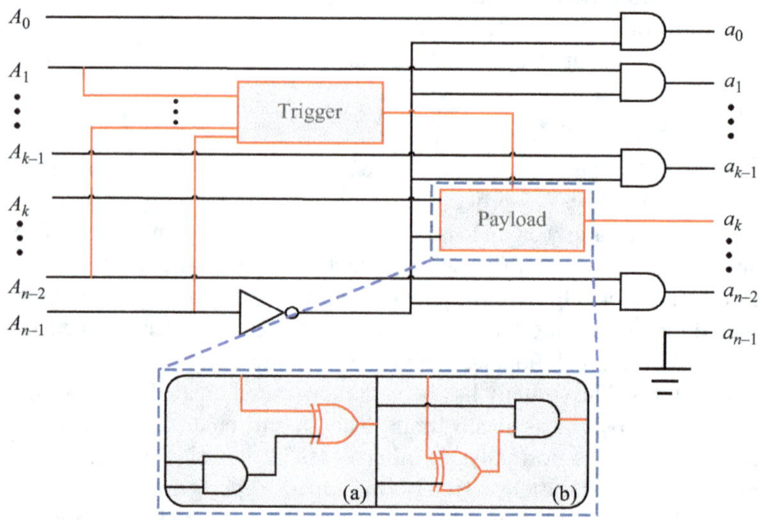

Figure 9.21 Hardware Trojan injection on ReLU implementation

Trojan to the design. Different from a backdoor poisoning attack, a hardware Trojan is a rancorous circuit containing a trigger and a payload. The payload circuitry will only be activated by the predetermined trigger condition. The attack has been mounted on a targeted operation in the DNN to add required perturbation in the payload that forces misclassification for a chosen input instance as the trigger [23,24]. The perturbation is computed through a modified adversarial sample generation algorithm. A simple example for hardware Trojan embedding in ReLU implementation is illustrated in Figure 9.21. Apart from computational Trojans, memory Trojans that monitor memory access patterns through the memory have also been proposed for a weaker adversarial setting [68]. It resets a random portion of the ofmaps written back to the DRAM upon detecting the specified input pattern to cause accuracy degradation. As shown in Figure 9.22, data communication between the DRAM and a DNN accelerator is supervised by the memory controller that contains information about the request types, memory addresses and request data [69]. The process can possibly be exploited given the fact that existing third-party memory controller IP is often used to build accelerators for faster design cycle [70,71]. With no model knowledge, a memory Trojan should be designed to enable identification of the input image data from the memory bus for later filtering of deliberately crafted trigger image. This is realized by investigating the number of write accesses that conveys information about the layer boundaries and read over write access (r/w) ratio that encodes knowledge regarding layer types. Since the input data are always fetched after the last-layer operation and the last layer is often the FC layer, extracting such layer information is beneficial to input identification. For the trigger logic, as the key is selected to be fractal and symmetric geometrical pattern, accurate and efficient detection of such pattern is the main concern. A similarity correlation analysis between a selected reference sub-image and other sub-images for an input image is taken for trigger recognition. The mechanism is shown to support certain extent of tolerance in translation variance.

Existing countermeasures against backdoor attacks either utilize input transformation to undermine the trigger identification or recover the integrity or confidentiality of the DNN to mute the malicious component or prevent its abuse from the beginning. An autoencoder can be introduced to preprocess the input for the first strategy [72]. However, the input transformation is only applicable to those backdoors that use illegitimate inputs as keys. It ignores scenarios where clean images or inner features are used. A fine-pruning technique has been devised for *BadNet* attack [66,72]. It prunes redundant neurons and fine-tunes the pruned model to restore the accuracy. This methodology is based on the finding that backdoored neurons are often dormant in the presence of benign inputs [66]. As for keeping confidentiality of DNNs from adversaries, cryptography could be adopted to increase security strength. Oblivious RAM (ORAM) [73] has been proposed to conceal memory access patterns by continuously shuffling and re-encrypting currently accessed data. In this way, there is

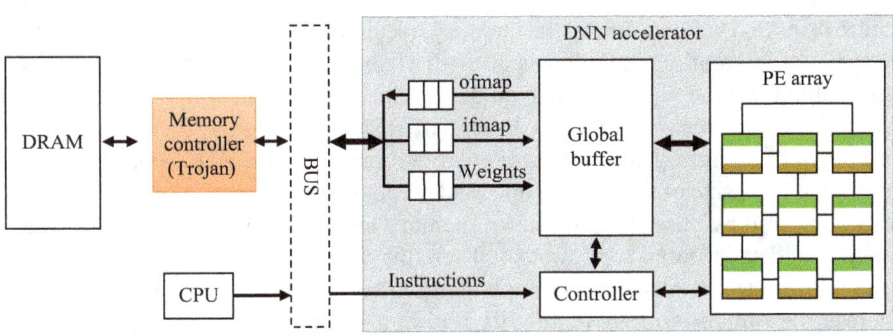

Figure 9.22 Hardware Trojan embedded in memory controller [68]

no chance for memory Trojan attack. Nevertheless, ORAM is designed for small storage. Its practicality on memory-intensive applications like DNN accelerators is questionable.

9.4 Confidentiality attacks

9.4.1 Incentive

To train a DNN model, enormous sample collection, human power and computational resources are required. While they may be adequately provided in large companies, they are more difficult to be satisfied for small corporations. For business competitiveness, the trained model is seldom released to the public. Therefore, it is likely for the business opponents to steal the model to build their own AI products or solutions to subvert the market dominance of the original model owner. Furthermore, the establishment of a DL model is inseparable from the data. In many cases, the data are sensitive and likely to be eavesdropped for malpractice.

9.4.2 Model confidentiality attacks

Model confidentiality attacks are also referred to as model extraction attacks. They analyze input, output and other extrinsic knowledge of a system to extract the model information, including the architecture and weights. Recent studies have shown that memory visits can be observed through physical exploitation of the data bus or side channels such as cache conflicts and page faults. Feature maps and parameters are typically stored in off-chip memory to ease the on-chip storage requirement, and they are visited when needed in the DNN accelerator implementation [74,75]. Hua *et al.* [25] leveraged this vulnerability of memory trace that leaks layer information as well as parameters to construct a similar DNN model that achieves comparable accuracy as the original one. The reverse-engineering process has been evaluated on an accelerator integrated with the modern processor's security capabilities, including prohibition of direct access and tampering to the inner operations, and encryption of feature maps and weights during the memory-accelerator transmission [76]. However, the types of instruction and address are still visible for each external memory visit when submitting the input. Layer boundaries can be determined by read-after-write dependency since the start of a new layer is often accompanied by the read operation on a memory address that has already been written to previously. Once the borders are resolved, the size of the filters, input and output feature maps for each layer can be further inferred by observing their own memory locations within a layer, since each category of data is stored contiguously. Then, the data type can be identified through inspection of the memory instruction. Weights are often read only during inferencing, whereas the ofmap is often write only within a layer. In addition, activation and pooling operations are often merged with convolution as a single layer to produce an ofmap in the DNN accelerator. Therefore, the in-between outputs are hidden from the view of an adversary. To fully uncover the structure, the problem is to solve the 11 variables listed in Table 9.1 for each layer using the following equations [25]:

$$Size_{ifmap} = W_{ifmap}^2 \times D_{ifmap} \tag{9.15}$$

$$Size_{ofmap} = W_{ofmap}^2 \times D_{ofmap} \tag{9.16}$$

$$Size_{Fil} = W_{Fil}^2 \times D_{ifmap} \times D_{ofmap} \tag{9.17}$$

$$W_{ofmap} = \frac{((W_{ifmap} - W_{Fil} + Pad_{conv})/Str_{conv}) + 1 + (Pad_{pool} - W_{pool}) \times P}{Str_{pool} \times P + \overline{P}} + P \tag{9.18}$$

$$Str_{conv} \leq W_{Fil} \leq \frac{W_{ifmap}}{2} \tag{9.19}$$

$$Str_{pool} \leq W_{pool} \leq \frac{W_{ifmap} - W_{Fil} + Pad_{conv}}{Str_{conv}} + 1 \tag{9.20}$$

$$Pad_{conv} \leq W_{Fil} \tag{9.21}$$

$$Pad_{pool} \leq W_{pool} \tag{9.22}$$

To approach the correct solution, the execution time for each layer is recorded and compared with the expected number of MAC operations for each possible configuration. Since the operating clock cycles should be approximately proportional to the number of MAC operations, candidate structures that violate this rule can be excluded. The authors also proposed reverse-engineering of weights by taking advantage of zero pruning technique that is often applied to reduce the computational complexity and memory access intensity in the DNN accelerator [46,77,78]. Zero pruning, that only captures nonzero values, leaks information about the number of zeros in the ofmap. By gradually changing the values of interested input pixels and observing the respective changes in ofmap, zero-crossing points can be determined, and a set of linear equations with respect to the weights and bias can be obtained. In other words, each weight can be expressed as a function of the bias. The exact values of parameters can be determined by using a tunable activation function instead of ReLU to remove smaller pixel values below a threshold [46,78]. The bias is thereby obtained by zeroing all the input pixels and adjusting the threshold. Weights can be subsequently derived

Table 9.1 Layer variables to construct a CNN architecture

Layer variable	Definition
$W_{ifmap/ofmap/Fil/pool}$	Width of the input/output feature map/filter/pooling operation
$D_{ifmap/ofmap}$	Depth of the input/out feature map
$Str_{conv/pool}$	Stride of the convolution/pooling operation
$Pad_{conv/pool}$	Number of pixels padded in the convolutional/pooling layer
P	Existence of the pooling layer

based on the previously extracted relationship with the bias. However, the process is founded on the assumption that the threshold can be acquired. One possible defense strategy against such side channel information leakage is the incorporation of ORAM algorithm [73] to obfuscate the memory trace, yet substantially increase the number of entries to hinder its usage on memory-intensive tasks, as discussed in Section 9.3.4.

9.4.3 Data confidentiality attacks

The potential data confidentiality threats of the DL-based system can be analyzed from two viewpoints: service providers and users. Provision of the AI services to the public could possibly result in undesirable disclosure of sensitive information encoded in the used training dataset. It has been demonstrated that the targeted data information, such as rough training data distribution and existence of an individual, could be extracted with merely the availability of the confidence scores for the corresponding queries to the model [79,80]. From the perspective of the users, the controllability of their own data is undermined with the permission given to cloud computing when users' private information are uploaded to the server. Although enabling cloud service gets rid of device constraints and allows the users to enjoy DL applications anywhere, personal information is at severe risk. They can be exploited during the data transmission, thus raising privacy concerns.

To preserve confidentiality of the concerned data, private aggregation of teacher ensembles approach has been proposed to protect the privacy of sensitive training data [81]. As shown in Figure 9.23, this method builds multiple independent DNN models, each of which is trained by a different disjoint set of the original training data. In the inference stage, the final prediction result will be the majority voting of these classifiers but with random noise added to the vote counts to introduce ambiguity. A student model is then trained using insensitive unlabeled datasets through semi-supervised learning where some of the unlabeled samples are clustered and some of them are labeled by querying the previous aggregated teacher model to label the clusters. This student model is the one to which the users make queries. Since the amount of information leakage related to the training data is dependent on the number of users' queries to the model trained by this training data, the privacy loss of the initial training data is effectively limited by the number of queries made to the teacher model

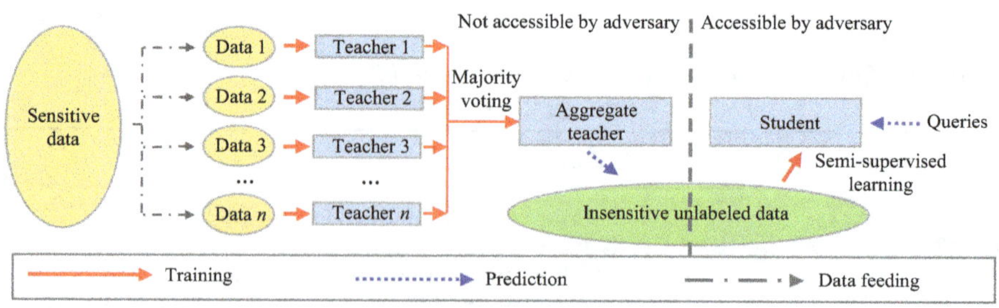

Figure 9.23 Private aggregation of teacher ensembles approach [81]

used for training the student model. Besides, homomorphic encryption (HE) can be adopted in DL to secure outsourced data in cloud processing. It is a cryptographic scheme that enables computation on ciphertexts without having to decrypt the inputs. It can be defined as follows:

$$Enc(a) \diamond Enc(b) = Enc(a * b) \tag{9.23}$$

where Enc defines an encryption method that maps a set of plain messages X to a set of ciphertexts Y: $X \rightarrow Y$, a and b are plaintexts in X, $\diamond$ and $*$ are linear operators defined in the domain of Y and X, correspondingly. Initially, semi-HE can only be applied to either addition or multiplication [82]. Attempts to design fully HE that allows both operations on ciphertexts were then made [83–85], but it will significantly increase the computational complexity and degrade the performance due to the intrinsic nonlinearity of the DL.

9.5 Explainability

As AI technologies have been rapidly developed and extensively applied for commercial use, the black-box nature of some AI algorithms has raised concerns about the reliability of the solutions. In 2016, a risk assessment tool, that has been commonly used in courtrooms for a defendant's future crime prediction to evaluate his/her rehabilitation needs, has proved to be biased against blacks [86]. Furthermore, as discussed earlier, DNNs can be easily obfuscated by inputs with small distortions even if they are highly generalized. The existence of such undesirable behaviors serves as an incentive for explainability study on AI methods. Explainability is defined as the ability to provide reasons for a specific decision derived by the AI. It can be evaluated by two means: interpretability and completeness. Interpretability aims at describing inner operations in a simpler way that is easy for the users to understand, while completeness measures the level of preciseness for an explanation. Perfect completeness can be achieved by illustrating all the mathematical functions and parameters in a system. It is difficult to preserve the advantages of both metrics concurrently, since a highly interpretable system is usually weak in prediction, whereas a precise description is often hardly understandable for human. An optimal extent of explainability should be figured out to strike a good balance between interpretability and completeness, instead of blindly pursuing interpretability to fawn over the users. Explainability of a DL-based system is sourced from two perspectives: the processing and representation of data. The former focuses on input–output mapping explanation, while the latter analyzes inner information flows. Representative explainable strategies that improve explainability of the DNN are highlighted subsequently.

9.5.1 Explainability of DNN processing

The tremendous computations involved in DL processing have significantly increased resistivity in providing explanations for the decision-making. Thus, the core for processing explainability enhancement is to reduce the operational complexity of the targeted DNN. It can be realized by either building a proxy model that has comparable performance with the original model but is more perceivable, or constructing a saliency map to underline the most influential operations.

Proxy models: Ribeiro *et al.* [87] presented a local interpretable model-agnostic explanations (LIME) approach to create a linear proxy model. It uses instances within the neighborhood of an interested input to provide explanations for each individual prediction. As shown in Figure 9.24, the blue and pink backgrounds indicate a learned complex decision function F, the bold red cross is the point of interest that needs to be explained and other underlined crosses or circles are adjacent instances whose relevance to the targeted input is represented by the size. LIME utilizes the outstanding weighted samples to learn a linear decision boundary denoted by the dashed line in Figure 9.24. It can be observed that the obtained explanation is locally reliable to the original model. The problem is formulated as follows:

$$\arg\min_{g} L(F, g, \pi_x) + \Omega(g) \tag{9.24}$$

given

$$\begin{cases} L(F, g, \pi_x) = \sum_{z, z' \in Z} \pi_x(z)(F(z) - g(z'))^2 \\ \pi_x(z) = \exp\left(-\frac{D(x, z)^2}{\sigma^2}\right) \end{cases} \tag{9.25}$$

where L is the measure of unfaithfulness, g is the linear proxy model that uses binary interpretable representation $\{0, 1\}^{d'}$ as its domain, x is the original representation of the targeted sample, $\pi_x(z)$ is the measure of the locality of the perturbed instance z around x defined on some distance function D, z' is the perturbed sample that uses binary interpretable representation, Z is the set of perturbed instances with their own labels obtained by performing F, and Ω is the measure of complexity for a given explanation. Since LIME provides the specific measurement for both fidelity and completeness, it becomes advantageous over other methodologies. Alternative options for proxy models can be decision trees [88] or falling rule lists [89–93].

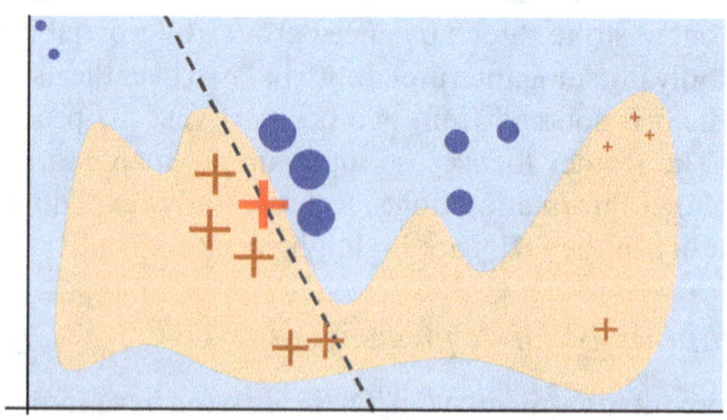

Figure 9.24 An example to present intuition for LIME [87]

Saliency mapping: Bach *et al.* [27] decomposed a prediction $F(x)$ into a set of pixel-wise relevances R to monitor the contribution of each input pixel in an image x to the classification result, as expressed in the following:

$$F(x) \approx \sum_{d=1}^{N} R_{x_d} \tag{9.26}$$

where x_d is the dth input pixel of x, N is the total number of input pixels and R_{x_d} quantifies the contribution of x_d to the output. This pixel-wise decomposition is achieved by LRP, with the assumption that F can be perceived as a sequence of layer computations. Then the problem is to find the relevance score R^l for each neuron in the lth layer, given R^{l+1}, such that

$$F(x) = \cdots = \sum_{d=1}^{N(l+1)} R_d^{l+1} = \sum_{d=1}^{N(l)} R_d^l = \cdots = \sum_{d=1}^{N(1)} R_d^1 \tag{9.27}$$

where $N(l)$ indicates the number of neurons in the lth layer and R_d^l denotes the relevance score for the dth neuron in the lth layer. In addition to the constraint of relevance preservation across layers, the node relevance should be equal to not only its total incoming contribution scores, but also to the sum of its outgoing relevance scores. Other than the LRP method, Deep Learning Important FeaTures (DeepLIFT) [94], class activation mapping [95] and SmoothGrad [96] techniques have also been proposed to efficiently extract the saliency map. A comprehensive summary of the state-of-the-art strategies can be found in [97].

9.5.2 Explainability of DNN representations

A deep network consists of multiple layers, each of which can be further decomposed into neuron or filter computations. Understanding the role of the corresponding bottleneck is the goal for enhancing the explainability of DNN representations.

Role of layers: The role of layers can be understood by testing their ability to tackle different problems. Razavian *et al.* [98] discovered that generic image representations produced by a CNN targeting ImageNet classification can be directly reused to solve other recognition tasks such as object localization, scene identification, fine-grained bird classification and attribute detection. By applying the deep feature vector, which is the output of an inner layer in the CNN, to the support vector machine algorithm, the resultant model is capable of achieving superior performance compared with modern custom architectures. This is known as transfer learning. It substantially reduces the time of model establishment for a new task.

Role of units: The role of individual units like neurons and filters can be explained by either estimating the input patterns as a result of unit-wise response maximization or examining their ability of transfer learning. The former can be accomplished by gradient-based methods [99–101] that compute the gradients of a given unit with respect to the input for visualization. Constructing a generative network that outputs images with the inputs of features is an alternative option [102]. The latter can be achieved using network dissection [103] technique that associates individual units with the corresponding interpretable visual concepts drawn from the broadly and densely labelled dataset (Broden). Moreover, network pruning [104] can also be an indirect method that identifies redundant neurons to help understanding the level of importance for each unit. It further implies the existence of a smaller model that is possibly more interpretable for the same problem.

9.5.3 Self-explainable systems

Self-explainable systems are designed to encode additional architectures that provide certain explanations for their behaviors to improve transparency. They can be trained to embed attention, learn disentangled representations or produce generative explanations.

Attention networks: Attention mechanism is a strategy of helping neural networks to explore the location of important information when a sequence of data is being processed. For example, several relevant questions are raised after the readers have read a long paragraph of texts. It is common that the readers do not remember the answers unless they read the texts again by paying attention to these questions. Attention networks work in a similar fashion. A representation of question is supplemented to generate an attention vector. It is used to learn the significance coefficients for patches of the input data by looping over the input data. Attention-based techniques have shown a great success in applications such as natural language processing [105], fine-grained image classification [106] and image caption generator [107].

Disentangled representations: Disentangled representations mean that the units in DL networks can learn approximately complete concepts alone. A promising solution for encoding disentanglement is auto-encoding variational Bayes. It utilizes the stochastic gradient variational Bayes estimator to approximate the posterior inference efficiently [108]. Chen *et al.* [109] developed a scalable unsupervised methodology to learn highly semantic and meaningful factors: information maximizing GANs. It extends GAN framework to an objective-driven maximization of mutual communication between a small portion of latent variables and the observations. Furthermore, β variational auto-encoding (β-VAE) augments a hyperparameter β to the original VAE technique to control the learning constraints imposed on the model. It is able to extract the most efficient latent factors of the data [110]. It also contributes to the quantitative measurement for the extent of disentanglement to benchmark different techniques against each other.

Generative explanations: Some AI tasks are designed to provide explanations for inputs. As shown in Figure 9.25, an image together with the image-related questions are submitted as the input. The model is required to generate the corresponding answer as the output. This visual question answering system is valuable in assisting visually impaired users and visual information analysis [111].

9.6 Conclusion

DL, as a distinct representative of emerging ML techniques, has been given close attention since its first practical application for digit recognition in the late 1980s. Thirty

Does it appear to be rainy?
Does the girl have 20/20 vision?

What ingredients does it have?
Is this a vegetarian dish?

Figure 9.25 Explanation-generation examples

years of development have bred various architectures for DL, which are different from the original feedforward FC multilayer perceptrons. Vast efficient processing technologies have also been proposed to deploy DNN models on hardware platforms such as CPU, GPU, FPGA and ASIC, without compromising the predictive power. The rapid growth of data being processed and tremendous improvements in computing power, boosted by open-source software and cloud service in providing frameworks or computing and storage resources to speed up the design cycles, have largely contributed to the current proliferation of DL in a broad range of fields, including transportation, healthcare and manufacturing [112]. Many big companies such as Amazon, Google and Microsoft have already integrated DL into commercial systems and continue charting their future road map on DL applications. For example, Google smart reply auto-magically answers messages that hit a user's inbox; Apple Siri interprets what the iPhone user asks of her; Google's self-driving cars show the unimaginable potential of how DNNs change human life. The prowess and prevalence of DL will also attract malevolence and warrant more research effort and close scrutiny of the security in DNN design and deployment.

This chapter attempts to provide a comprehensive review of the DL security. It focuses on two types of DL attacks and their related countermeasures. First, malicious distortions injected in either the inputs or models can affect the correct classification results of the system. This kind of attack is well studied at algorithmic level. Practical constraints such as limited accessibility and effect of actual deployment may impact the success rate of such attack. On the other hand, reverse engineering and potential exploitation of backdoor or flaws through deployed hardware accessibility and unreliable supply chains of integrated circuit design may pose greater threats to the DL system. Second, the sensitive data used for training, and the trained model can be stolen through queries or side channels in the context of privacy-preserving applications. Existing countermeasures mostly and mainly focus on software-level protection, vulnerabilities in the underlying hardware accelerator design are less studied. Modern primitives for securing hardware, such as physical unclonable functions [113] and logic locking [114], could be applied to the DL hardware to mitigate the general vulnerabilities. However, without considering the intrinsic weaknesses of DNN implementation, the overhead and performance penalty may be unacceptable. It is desirable to have a hardware-supported solution that takes attributes of DNNs into account at design time for system-level defense. Furthermore, the effectiveness of current protection measures and emerging approaches on large-scale datasets are seldom evaluated, so are the complex models. Scalability is thus another metric that needs more attention in the development of defense methodologies. The opaque nature of DL has further worsened the defender's situation and has become a big concern that affects the consumers' confidence and acceptance of DNN solutions, especially in safety and security critical tasks. This calls for a learning paradigm shift from data driven to knowledge driven for explainability enhancement.

Acknowledgment

This work is supported by the Singapore Ministry of Education Academic Research Fund Tier 2 Grant No. MOE2015-T2-2-013.

References

[1] V. Sze, Y.-H. Chen, T.-J. Yang, and J. S. Emer, "Efficient processing of deep neural networks: A tutorial and survey," *Proc. IEEE*, vol. 105, no. 12, pp. 2295–2329, 2017.

[2] Y. Bengio, "Learning deep architectures for AI," *Found. Trends Mach. Learn.*, vol. 2, no. 1, pp. 1–127, 2009. [Online]. Available: http://dx.doi.org/10.1561/2200000006.

[3] J. Schmidhuber, "Deep learning in neural networks: An overview," *Neural Netw.*, vol. 61, pp. 85–117, 2015.

[4] L. Deng, J. Li, J. T. Huang, *et al.*, "Recent advances in deep learning for speech research at Microsoft," in *2013 IEEE Int. Conf. Acoustics, Speech Signal Processing*, Vancouver, BC, Canada, May 2013, pp. 8604–8608.

[5] A. Krizhevsky, I. Sutskever, and G. E. Hinton, "ImageNet classification with deep convolutional neural networks," in *Proc. Advances Neural Info. Processing Syst.*, Harrah's & Harveys Lake Tahoe, Stateline, NV, USA, Dec. 2012, pp. 1097–1105.

[6] V. Gokhale, J. Jin, A. Dundar, B. Martini, and E. Culurciello, "A 240 G-ops/s mobile coprocessor for deep neural networks," in *Proc. IEEE Conf. Comput. Vision Pattern Recogn. Workshops*, Columbus, OH, USA, Jun. 2014, pp. 682–687.

[7] Y. Chen, T. Luo, S. Liu, *et al.*, "DaDianNao: A machine-learning supercomputer," in *Proc. 47th Annual IEEE/ACM Int. Symposium Microarchitecture*, Cambridge, UK, Dec. 2014, pp. 609–622.

[8] Y.-H. Chen, T. Krishna, J. S. Emer, and V. Sze, "Eyeriss: An energy-efficient reconfigurable accelerator for deep convolutional neural networks," *IEEE J. Solid-State Circuits*, vol. 52, no. 1, pp. 127–138, 2016.

[9] N. Akhtar and A. Mian, "Threat of adversarial attacks on deep learning in computer vision: A survey," *IEEE Access*, vol. 6, pp. 14 410–14 430, 2018.

[10] C. Szegedy, W. Zaremba, I. Sutskever, *et al.*, "Intriguing properties of neural networks," in *Int. Conf. Learning Representations (ICLR)*, Scottsdale, AZ, USA, May 2013. [Online]. Available: https://arxiv.org/abs/1312.6199.

[11] I. J. Goodfellow, J. Shlens, and C. Szegedy, "Explaining and harnessing adversarial examples," in *Int. Conf. Learning Representations (ICLR)*, San Diego, CA, USA, May 2014. [Online]. Available: http://arxiv.org/abs/1412.6572.

[12] S.-M. Moosavi-Dezfooli, A. Fawzi, O. Fawzi, and P. Frossard, "Universal adversarial perturbations," in *Proc. IEEE Conf. Comp. Vision Pattern Recogn.*, Honolulu, HI, USA, Jul. 2017, pp. 1765–1773.

[13] N. Papernot, P. McDaniel, I. Goodfellow, S. Jha, Z. B. Celik, and A. Swami, "Practical black-box attacks against machine learning," in *Proc. 2017 ACM Asia Conf. Comp. Comm. Security*, Abu Dhabi, United Arab Emirates, Apr. 2017, pp. 506–519.

[14] K. Eykholt, I. Evtimov, E. Fernandes, *et al.*, "Robust physical-world attacks on deep learning visual classification," in *Proc. IEEE Conf. Computer Vision Pattern Recogn.*, Salt Lake City, UT, USA, Jun. 2018, pp. 1625–1634.

[15] N. Papernot, P. McDaniel, X. Wu, S. Jha, and A. Swami, "Distillation as a defense to adversarial perturbations against deep neural networks," in *Proc. 2016 IEEE Symp. Security Privacy (SP)*, San Jose, CA, USA, May 2016, pp. 582–597.

[16] W. Xu, D. Evans, and Y. Qi, "Feature squeezing: Detecting adversarial examples in deep neural networks," in *2018 Net. Distributed Syst. Security Symp. (NDSS)*, San Diego, CA, USA, Feb. 2018. [Online]. Available: https://arxiv.org/abs/1704.01155.

[17] J. H. Metzen, T. Genewein, V. Fischer, and B. Bischoff, "On detecting adversarial perturbations," in *5th Int. Conf. Learning Representations (ICLR)*, Toulon, France, Apr. 2017. [Online]. Available: https://arxiv.org/abs/1702.04267.

[18] J. Wang, J. Sun, P. Zhang, and X. Wang, "Detecting adversarial samples for deep neural networks through mutation testing," 2018. [Online]. Available: http://arxiv.org/abs/1805.05010.

[19] R. Feinman, R. R. Curtin, S. Shintre, and A. B. Gardner, "Detecting adversarial samples from artifacts," 2017. [Online]. Available: https://arxiv.org/abs/1703.00410.

[20] K. Grosse, P. Manoharan, N. Papernot, M. Backes, and P. McDaniel, "On the (statistical) detection of adversarial examples," 2017. [Online]. Available: https://arxiv.org/abs/1702.06280.

[21] J. Wang, G. Dong, J. Sun, X. Wang, and P. Zhang, "Adversarial sample detection for deep neural network through model mutation testing," in *Proc. 41st Int. Conf. Software Engineering*, Montreal, QC, Canada, May 2019, pp. 1245–1256.

[22] L. Ma, F. Zhang, J. Sun, *et al.*, "DeepMutation: Mutation testing of deep learning systems," in *2018 IEEE 29th Int. Symp. Software Reliability Engineering (ISSRE)*, Memphis, TN, USA, Oct. 2018, pp. 100–111.

[23] C. Joseph and L. Yingjie, "Hardware Trojan attacks on neural networks," 2018. [Online]. Available: https://arxiv.org/abs/1806.05768.

[24] C. Joseph and L. Yingjie, "Hardware Trojan design on neural networks," in *2019 IEEE Int. Symp. Cir. Syst. (ISCAS)*, Sapporo, Hokkaido, Japan, May 2019, pp. 1–5.

[25] W. Hua, Z. Zhang, and G. E. Suh, "Reverse engineering convolutional neural networks through side-channel information leaks," in *2018 55th ACM/ESDA/IEEE Design Automation Conf. (DAC)*, San Francisco, CA, USA, Jun. 2018, pp. 1–6.

[26] L. H. Gilpin, D. Bau, B. Z. Yuan, A. Bajwa, M. Specter, and L. Kagal, "Explaining explanations: An overview of interpretability of machine learning," in *2018 IEEE 5th Int. Conf. Data Sci. Advanced Analytics (DSAA)*, Turin, Italy, Oct. 2018, pp. 80–89.

[27] S. Bach, A. Binder, G. Montavon, F. Klauschen, K. R. Müller, and W. Samek, "On pixel-wise explanations for non-linear classifier decisions by layer-wise relevance propagation," *PLoS One*, vol. 10, no. 7, p. e0130140, 2015.

[28] Y. Bengio, A. Courville, and P. Vincent, "Representation learning: A review and new perspectives," *IEEE Trans. Pattern Anal. Mach. Intell.*, vol. 35, no. 8, pp. 1798–1828, 2013. [Online]. Available: https://doi.org/10.1109/TPAMI.2013.50.

[29] Y. LeCun, Y. Bengio, and G. Hinton, "Deep learning," *Nature*, vol. 521, no. 7553, p. 436, 2015.

[30] W. Zhang, "Shift-invariant pattern recognition neural network and its optical architecture," in *Proc. Annual Conf. Japan Society Applied Physics*, Japan, May 1988.

[31] W. Zhang, K. Itoh, J. Tanida, and Y. Ichioka, "Parallel distributed processing model with local space-invariant interconnections and its optical architecture," *Appl. Opt.*, vol. 29, no. 32, pp. 4790–4797, 1990.

[32] T. N. Sainath, O. Vinyals, A. Senior, and H. Sak, "Convolutional, long short-term memory, fully connected deep neural networks," in *Proc. 2015 IEEE Int. Conf. Acoustics, Speech Signal Processing (ICASSP)*, Brisbane, Queensland, Australia, Apr. 2015, pp. 4580–4584.

[33] D. Scherer, A. Müller, and S. Behnke, "Evaluation of pooling operations in convolutional architectures for object recognition," in *Proc. 20th Int. Conf. Artificial Neural Networks: Part III*, ser. ICANN'10, Berlin, Heidelberg, Germany, 2010, pp. 92–101. [Online]. Available: http://dl.acm.org/citation.cfm?id=1886436.1886447.

[34] N. M. Nasrabadi, "Pattern recognition and machine learning," *J. Electron. Imaging*, vol. 16, no. 4, p. 049901, 2007.

[35] K. Simonyan and A. Zisserman, "Very deep convolutional networks for large-scale image recognition," 2014. [Online]. Available: https://arxiv.org/abs/1409.1556.

[36] C. Szegedy, W. Liu, Y. Jia, *et al.*, "Going deeper with convolutions," in *Proc. IEEE Conf. Comp. Vis. Pattern Recogn.*, Boston, MA, USA, Jun. 2015, pp. 1–9.

[37] K. He, X. Zhang, S. Ren, and J. Sun, "Deep residual learning for image recognition," in *Proc. IEEE Conf. Comp. Vis. Pattern Recogn.*, Las Vegas, NV, USA, Jun. 2016, pp. 770–778.

[38] G. Huang, Z. Liu, L. Van Der Maaten, and K. Q. Weinberger, "Densely connected convolutional networks," in *Proc. IEEE Conf. Comp. Vis. Pattern Recogn.*, Honolulu, HI, USA, Jul. 2017, pp. 4700–4708.

[39] A. G. Howard, M. Zhu, B. Chen, *et al.*, "MobileNets: Efficient convolutional neural networks for mobile vision applications," 2017. [Online]. Available: https://arxiv.org/ abs/1704.04861.

[40] M. Mathieu, M. Henaff, and Y. LeCun, "Fast training of convolutional networks through FFTs," 2013. [Online]. Available: https://arxiv.org/ abs/1312.5851.

[41] C. Dubout and F. Fleuret, "Exact acceleration of linear object detectors," in *Euro. Conf. Comput. Vision*, Firenze, Italy, Oct. 2012, pp. 301–311.

[42] M. Sankaradas, V. Jakkula, S. Cadambi, *et al.*, "A massively parallel coprocessor for convolutional neural networks," in *2009 20th IEEE Int. Conf. Application-specific Syst., Architectures Processors*, Boston, MA, USA, Jul. 2009, pp. 53–60.

[43] V. Sriram, D. Cox, K. H. Tsoi, and W. Luk, "Towards an embedded biologically-inspired machine vision processor," in *2010 Int. Conf. Field-Programmable Technology*, Beijing, China, Dec. 2010, pp. 273–278.

[44] S. Chakradhar, M. Sankaradas, V. Jakkula, and S. Cadambi, "A dynamically configurable coprocessor for convolutional neural networks," *SIGARCH Comput. Archit. News*, vol. 38, no. 3, pp. 247–257, 2010.

[45] M. Horowitz, "1.1 computing's energy problem (and what we can do about it)," in *2014 IEEE Int. Solid-State Circuits Conf. Digest Technical Papers (ISSCC)*, San Francisco, CA, USA, Feb. 2014, pp. 10–14.

[46] J. Albericio , P. Judd, T. Hetherington, T. Aamodt, N. E. Jerger, and A. Moshovos, "Cnvlutin: Ineffectual-neuron-free deep neural network computing," *ACM SIGARCH Comput. Archit. News*, vol. 44, no. 3, pp. 1–13, 2016.

[47] F. Chollet, "Xception: Deep learning with depthwise separable convolutions," in *Proc. IEEE Conf. Comput. Vision Pattern Recogn.*, Honolulu, HI, USA, Jul. 2017, pp. 1251–1258.

[48] G. Hinton, O. Vinyals, and J. Dean, "Distilling the knowledge in a neural network," 2015. [Online]. Available: https://arxiv.org/abs/1503.02531.

[49] C. Bucilu, R. Caruana, and A. Niculescu-Mizil, "Model compression," in *Proc. 12th ACM SIGKDD Int. Conf. Knowledge Discovery Data Mining*, Philadelphia, PA, USA, Aug. 2006, pp. 535–541.

[50] M. Barreno, B. Nelson, R. Sears, A. D. Joseph, and J. D. Tygar, "Can machine learning be secure?" in *Proc. 2006 ACM Symposium Info., Comput. and Commun. Security*, Taipei, Taiwan, Mar. 2006, pp. 16–25.

[51] N. Papernot, P. McDaniel, S. Jha, M. Fredrikson, Z. B. Celik, and A. Swami, "The limitations of deep learning in adversarial settings," in *Proc. 2016 IEEE Euro. Symp. Security Privacy (EuroS&P)*, SaarbrÃ¼cken, Germany, Mar. 2016, pp. 372–387.

[52] N. Carlini and D. Wagner, "Towards evaluating the robustness of neural networks," in *Proc. 2017 IEEE Symp. Security Privacy (SP)*, San Jose, CA, USA, May 2017, pp. 39–57.

[53] S.-M. Moosavi-Dezfooli, A. Fawzi, and P. Frossard, "DeepFool: a simple and accurate method to fool deep neural networks," in *Proc. IEEE Conf. Computer Vision Pattern Recogn.*, Las Vegas, NV, USA, Jun. 2016, pp. 2574–2582.

[54] D. Su, H. Zhang, H. Chen, J. Yi, P.-Y. Chen, and Y. Gao, "Is robustness the cost of accuracy? – A comprehensive study on the robustness of 18 deep image classification models," in *Proc. Euro. Conf. Comp. Vision (ECCV)*, Munich, Germany, Sep. 2018, pp. 631–648.

[55] B. Zoph, V. Vasudevan, J. Shlens, and Q. V. Le, "Learning transferable architectures for scalable image recognition," in *Proc. IEEE Conf. Comp. Vis. Pattern Recogn.*, Salt Lake City, UT, USA, Jun. 2018, pp. 8697–8710.

[56] A. N. Bhagoji, D. Cullina, C. Sitawarin, and P. Mittal, "Enhancing robustness of machine learning systems via data transformations," in *2018 52nd Annual Conf. Info. Sci. Syst. (CISS)*, Princeton, NJ, USA, Mar. 2018, pp. 1–5.

[57] D. Meng and H. Chen, "MagNet: A two-pronged defense against adversarial examples," in *Proc. 2017 ACM SIGSAC Conf. Comp. Comm. Security*, Dallas, TX, USA, Oct. 2017, pp. 135–147.

[58] M. Osadchy, J. Hernandez-Castro, S. Gibson, O. Dunkelman, and D. Pérez-Cabo, "No bot expects the DeepCAPTCHA! introducing immutable adversarial examples, with applications to CAPTCHA generation," *IEEE Trans. Inf. Forensics Secur.*, vol. 12, no. 11, pp. 2640–2653, 2017.

[59] K. W. Nixon, J. Mao, J. Shen, H. Yang, H. H. Li, and Y. Chen, "SPN dash-fast detection of adversarial attacks on mobile via sensor pattern noise fingerprinting," in *2018 IEEE/ACM Int. Conf. Computer-Aided Design (ICCAD)*, San Diego, Canada, Nov. 2018, pp. 1–6.

[60] M. Sharif, S. Bhagavatula, L. Bauer, and M. K. Reiter, "Accessorize to a crime: Real and stealthy attacks on state-of-the-art face recognition," in *Proc. 2016 ACM SIGSAC Conf. Comp. Comm. Security*, Vienna, Austria, Oct. 2016, pp. 1528–1540.

[61] Y. Liu, L. Wei, B. Luo, and Q. Xu, "Fault injection attack on deep neural network," in *2017 IEEE/ACM Int. Conf. Computer-Aided Design (ICCAD)*, Irvine, CA, USA, Nov. 2017, pp. 131–138.

[62] J. Breier, X. Hou, D. Jap, L. Ma, S. Bhasin, and Y. Liu, "Practical fault attack on deep neural networks," in *Proc. 2018 ACM SIGSAC Conf. Comp. and Comm. Security*, New York, USA, Oct. 2018, pp. 2204–2206.

[63] Y. Kim, R. Daly, J. Kim, *et al.*, "Flipping bits in memory without accessing them: An experimental study of DRAM disturbance errors," *SIGARCH Comput. Archit. News*, vol. 42, no. 3, pp. 361–372, 2014. [Online]. Available: http://doi.acm.org/ 10.1145/2678373.2665726.

[64] C. Yang, Q. Wu, H. Li, and Y. Chen, "Generative poisoning attack method against neural networks," 2017. [Online]. Available: http://arxiv.org/abs/1703.01340.

[65] C. Bryant, W. Carvalho, N. Baracaldo, *et al.*, "Detecting backdoor attacks on deep neural networks by activation clustering," 2018. [Online]. Available: http://arxiv.org/ abs/1811.03728.

[66] T. Gu, B. Dolan-Gavitt, and S. Garg, "BadNets: Identifying vulnerabilities in the machine learning model supply chain," 2017. [Online]. Available: http://arxiv.org/abs/1708.06733.

[67] C. Xinyun, L. Chang, L. Bo, L. Kimberly, and S. Dawn, "Targeted backdoor attacks on deep learning systems using data poisoning," 2017. [Online]. Available: https://arxiv.org/abs/1712.05526.

[68] Y. Zhao, X. Hu, S. Li, *et al.*, "Memory Trojan attack on neural network accelerators," in *2019 Design, Automation & Test Euro. Conf. & Exhibition (DATE)*, Mar., pp. 1415–1420.

[69] B. Jacob, "The memory system: You can't avoid it, you can't ignore it, you can't fake it," *Synthesis Lectures on Computer Architecture*, vol. 4, no. 1, pp. 1–77, 2009, San Rafael, CA: Morgan & Claypool.

[70] M. Tehranipoor and F. Koushanfar, "A survey of hardware Trojan taxonomy and detection," *IEEE Des. Test. Comput.*, vol. 27, no. 1, pp. 10–25, 2010.

[71] S. Bhunia, M. S. Hsiao, M. Banga, and S. Narasimhan, "Hardware Trojan attacks: threat analysis and countermeasures," *Proc. IEEE*, vol. 102, no. 8, pp. 1229–1247, 2014.

[72] K. Liu, B. Dolan-Gavitt, and S. Garg, "Fine-pruning: Defending against backdooring attacks on deep neural networks," 2018. [Online]. Available: http://arxiv.org/abs/1805.12185.

[73] E. Stefanov, M. Van Dijk, E. Shi, *et al.*, "Path ORAM: An extremely simple oblivious ram protocol," *J. ACM (JACM)*, vol. 65, no. 4, p. 18, 2018.

[74] L. Li and A. Datta, "Write-only oblivious RAM-based privacy-preserved access of outsourced data," *Int. J. Inf. Secur.*, vol. 16, no. 1, pp. 23–42, 2017.

[75] Y. Xu, W. Cui, and M. Peinado, "Controlled-channel attacks: Deterministic side channels for untrusted operating systems," in *2015 IEEE Symposium Security Privacy*, San Jose, CA, USA, Mar. 2015, pp. 640–656.

[76] V. Costan and S. Devadas, "Intel SGX explained." *IACR Cryptol. ePrint Arch.*, vol. 2016, no. 086, pp. 1–118, 2016.

[77] A. Parashar, M. Rhu, A. Mukkara, *et al.*, "SCNN: An accelerator for compressed-sparse convolutional neural networks," in *2017 ACM/IEEE 44th Annual Int. Symposium Comp. Architecture (ISCA)*, Toronto, ON, Canada, Jun. 2017, pp. 27–40.

[78] B. Reagen, P. Whatmough, R. Adolf *et al.*, "Minerva: Enabling low-power, highly-accurate deep neural network accelerators," in *2016 ACM/IEEE 43rd Annual Int. Symposium Comp. Architecture (ISCA)*, Seoul, South Korea, Jun. 2016, pp. 267–278.

[79] F. Tramér, F. Zhang, A. Juels, M. K. Reiter, and T. Ristenpart, "Stealing machine learning models via prediction APIs," in *USENIX Security Symp.*, Vancouver, BC, Canada, Aug. 2016, pp. 601–618.

[80] M. Fredrikson, S. Jha, and T. Ristenpart, "Model inversion attacks that exploit confidence information and basic countermeasures," in *Proc. 22nd ACM SIGSAC Conf. Comput. Commun. Security*, Denver, CO, USA, Oct. 2015, pp. 1322–1333.

[81] N. Papernot, M. Abadi, U. Erlingsson, I. Goodfellow, and K. Talwar, "Semi-supervised knowledge transfer for deep learning from private training data," in *5th Int. Conf. Learning Representations (ICLR)*, Toulon, France, Apr. 2017. [Online]. Available: https://arxiv.org/ abs/1610.05755.

[82] C. Gentry, "A fully homomorphic encryption scheme," Ph.D. dissertation, Stanford, CA, USA, 2009, aAI3382729.

[83] M. Van Dijk, C. Gentry, S. Halevi, and V. Vaikuntanathan, "Fully homomorphic encryption over the integers," in *Annual Int. Conf. Theory Appl. Crypto. Techniques*, Monaco and Nice, France, May. 2010, pp. 24–43.

[84] Z. Brakerski and V. Vaikuntanathan, "Efficient fully homomorphic encryption from (standard) LWE," *SIAM J. Comput.*, vol. 43, no. 2, pp. 831–871, 2014.

[85] M. Yagisawa, "Fully homomorphic encryption without bootstrapping," *IACR Cryptol. ePrint Arch.*, vol. 2015, p. 474, 2015.

[86] J. Angwin, J. Larson, S. Mattu, and L. Kirchner, "Machine bias," *ProPublica*, vol. 23, p. 2016, 2016.

[87] M. T. Ribeiro, S. Singh, and C. Guestrin, "Why should i trust you?: Explaining the predictions of any classifier," in *Proc. 22nd ACM SIGKDD Int. Conf. Knowledge Discovery Data Mining*, San Francisco, CA, USA, Aug. 2016, pp. 1135–1144.

[88] J. R. Zilke, E. L. Mencía, and F. Janssen, "DeepRED – Rule extraction from deep neural networks," in *Int. Conf. Discovery Sci.*, Bari, Italy, Oct. 2016, pp. 457–473.

[89] F. Wang and C. Rudin, "Falling rule lists," in *Artificial Intell. Stat.*, San Diego, CA, USA, May 2015, pp. 1013–1022.

[90] L. Fu, "Rule generation from neural networks," *IEEE Trans. Syst., Man Cybern.*, vol. 24, no. 8, pp. 1114–1124, 1994.

[91] M. W. Craven, "Extracting comprehensible models from trained neural networks," Ph.D. dissertation, 1996, aAI9700774.

[92] H. Tsukimoto, "Extracting rules from trained neural networks," *IEEE Trans. Neural Netw.*, vol. 11, no. 2, pp. 377–389, 2000.

[93] T. Hailesilassie, "Rule extraction algorithm for deep neural networks: A review," 2016. [Online]. Available: https://arxiv.org/abs/1610.05267.

[94] A. Shrikumar, P. Greenside, and A. Kundaje, "Learning important features through propagating activation differences," in *Proc. 34th Int. Conf. Machine Learning-Volume 70*, Sydney, Australia, Aug. 2017, pp. 3145–3153.

[95] B. Zhou, A. Khosla, A. Lapedriza, A. Oliva, and A. Torralba, "Learning deep features for discriminative localization," in *Proc. IEEE Conf. Comput. Vision Pattern Recogn.*, Las Vegas, NV, USA, Jun. 2016, pp. 2921–2929.

[96] D. Smilkov, N. Thorat, B. Kim, F. Viégas, and M. Wattenberg, "SmoothGrad: removing noise by adding noise," 2017. [Online]. Available: https://arxiv.org/abs/1706.03825.

[97] M. Ancona, E. Ceolini, C. Öztireli, and M. Gross, "A unified view of gradient-based attribution methods for deep neural networks," in *NIPS 2017-Workshop Interpreting, Explaining Visualizing Deep Learning*, Long Beach, CA, USA, Dec. 2017.

[98] A. S. Razavian, H. Azizpour, J. Sullivan, and S. Carlsson, "CNN features off-the-shelf: An astounding baseline for recognition," in *Proc. IEEE Conf. Comput. Vision Pattern Recogn. Workshops*, Columbus, OH, USA, Jun. 2014, pp. 806–813.

[99] S. Karen, V. Andrea, and Z. Andrew, "Deep inside convolutional networks: Visualising image classification models and saliency maps," 2013. [Online]. Available: https://arxiv.org/abs/1312.6034.

[100] M. D. Zeiler and R. Fergus, "Visualizing and understanding convolutional networks," in *Euro. Conf. Comput. Vision*, Zurich, Switzerland, Sep. 2014, pp. 818–833.

[101] J. T. Springenberg, A. Dosovitskiy, T. Brox, and M. Riedmiller, "Striving for simplicity: The all convolutional net," 2014. [Online]. Available: https://arxiv.org/abs/1412.6806.

[102] A. Nguyen, A. Dosovitskiy, J. Yosinski, T. Brox, and J. Clune, "Synthesizing the preferred inputs for neurons in neural networks via deep generator networks," in *Advances Neural Inf. Processing Syst.*, Barcelona, Spain, Dec. 2016, pp. 3387–3395.

[103] D. Bau, B. Zhou, A. Khosla, A. Oliva, and A. Torralba, "Network dissection: Quantifying interpretability of deep visual representations," in *Proc. IEEE Conf. Comput. Vision Pattern Recogn.*, Honolulu, HI, USA, Jul. 2017, pp. 6541–6549.

[104] J. Frankle and M. Carbin, "The lottery ticket hypothesis: Finding sparse, trainable neural networks." in *Proc. Int. Conf. Learning Representations (ICLR)*, New Orleans, LA, USA, May 2019. [Online]. Available: http://dblp.uni-trier.de/db/conf/iclr/iclr2019.html#FrankleC19.

[105] A. Vaswani, N. Shazeer, N. Parmar, *et al.*, "Attention is all you need," in *Advances Neural Info. Processing Syst.*, CA, USA, Dec. 2017, pp. 5998–6008.

[106] T. Xiao, Y. Xu, K. Yang, J. Zhang, Y. Peng, and Z. Zhang, "The application of two-level attention models in deep convolutional neural network for fine-grained image classification," in *Proc. IEEE Conf. Comput. Vision Pattern Recogn.*, Boston, MA, USA, Jun. 2015, pp. 842–850.

[107] O. Vinyals, A. Toshev, S. Bengio, and D. Erhan, "Show and tell: A neural image caption generator," in *Proc. IEEE Conf. Comput. Vision Pattern Recogn.*, Boston, MA, USA, Jun. 2015, pp. 3156–3164.

[108] P. K. Diederik and W. Max, "Auto-encoding variational Bayes," 2013. [Online]. Available: https://arxiv.org/abs/1312.6114.

[109] X. Chen, Y. Duan, R. Houthooft, J. Schulman, I. Sutskever, and P. Abbeel, "InfoGAN: Interpretable representation learning by information maximizing generative adversarial nets," in *Advances Neural Info. Processing Syst.*, Barcelona, Spain, Dec. 2016, pp. 2172–2180.

[110] I. Higgins, L. Matthey, A. Pal, *et al.*, "beta-VAE: Learning basic visual concepts with a constrained variational framework." *ICLR*, vol. 2, no. 5, p. 6, 2017.

[111] S. Antol, A. Agrawal, J. Lu, *et al.*, "VQA: Visual question answering," in *Proc. IEEE Int. Conf. Comput. Vision*, Santiago, Chile, Dec. 2015, pp. 2425–2433.

[112] I. Stoica, D. Song, R. A. Popa, *et al.*, "A berkeley view of systems challenges for AI," 2017. [Online]. Available: http://arxiv.org/abs/1712.05855.

[113] C.-H. Chang, Y. Zheng, and L. Zhang, "A retrospective and a look forward: Fifteen years of physical unclonable function advancement," *IEEE Circuits Syst. Mag.*, vol. 17, no. 3, pp. 32–62, 2017.

[114] S. Dupuis and M.-L. Flottes, "Logic locking: A survey of proposed methods and evaluation metrics," *J. Electron. Test.*, vol. 35, no. 3, pp. 273–291, 2019. [Online]. Available: https://doi.org/10.1007/s10836-019-05800-4.

Chapter 10
Security implications of non-digital components

Xiaoxi Ren[1], Zhe Zhou[2], Di Tang[3], and Kehuan Zhang[3]

10.1 Introduction

Computers and computing systems are playing increasingly important roles not only in industry but also in people's daily life, and this has inevitably caused lots of problems, such as information security. Nowadays, adversaries are able to develop malicious code to take down a power grid, steal money from victim's account and build fine-grained profile for a victim user to launch accurate social engineering attacks.

Since digital computers can only handle data in zeros and ones, a normal computing system generally contains some non-digital components in order to interact with external non-digital world. Common examples of such non-digital components include things such as LCD screens, sensors and speakers. LCD screen and speaker modules are used to convert digital values into non-digital domain (like lights and sound), while sensors are used to capture physical values and process measurements in digital forms.

There are already lots of studies on the system security in digital domain, such as encryption algorithms and secure protocols, but there are few works on the security implications of those non-digital components. According to our studies, non-digital components will have some unique features that could be exploited to either enhance or undermine the security of a given system. For example, some non-digital components will demonstrate nonlinearity when processing the data, which had been proved in the DolphinAttack paper [1] that adversaries could take advantage of this feature to control voice-assistant by stealthily injecting voice command at frequency of ultrasound.

In this chapter, we will present two case studies that have utilized the feature of non-digital components to improve the security of a given system. We hope these works can inspire more following up works toward this direction and deepen our understandings over the security implications of non-digital components in computing systems.

[1]College of Computer Science and Electronic Engineering, Hunan University, Changsha, China
[2]School of Computer Science, Fudan University, Shanghai, China
[3]Department of Information Engineering, The Chinese University of Hong Kong, Hong Kong, China

10.2 Case study 1: Face Flashing—using light reflections to secure liveness detections

Face authentication systems are becoming increasingly prevalent, especially with the rapid development of deep learning technologies. However, human facial information is easy to be captured and reproduced, which makes face authentication systems vulnerable to various attacks. *Liveness detection* is an important defense technique to prevent such attacks, but existing solutions did not provide clear and strong security guarantees, especially in terms of time. In this case study, we will show how the light reflections on human faces can be used to construct a new challenge–response protocol called *Face Flashing* for liveness detection in face authentication systems. Light reflection is a natural process outside of digital world, but it can be used to significantly increase the bar for launching successful attacks on face authentication systems. By randomly flashing well-designed pictures on a screen and analyzing the reflected light, our protocol has leveraged physical characteristics of human faces: reflection processing at the speed of light, unique textual features and uneven three-dimensional (3D) shapes. By cooperating with working mechanism of the screen and digital cameras, our protocol is able to detect subtle traces left by an attacking process.

10.2.1 Architecture of face authentication systems

A typical architecture of face authentication system is illustrated in Figure 10.1. It is divided into two parts: front-end devices and the back-end server. The front-end devices comprise camera and auxiliary sensors such as flash lamps and microphones. The back-end server contains two main modules: a liveness detection module and a face recognition module. When the user commences the authentication process, the liveness detection module is initiated and sends generated parameters to front-end devices (Step 1). Subsequently, the front-end devices synthesize challenges according to the received parameters and deliver them to the user (Step 2). After receiving the challenges, the user makes expressions, such as smiling blinking, as responses. The sensors in the front-end devices capture such responses and encode them (Step 3). Either in real time or in postprocessing, the front-end devices send the captured responses to the liveness detection module in the back-end server (Step 4). The liveness

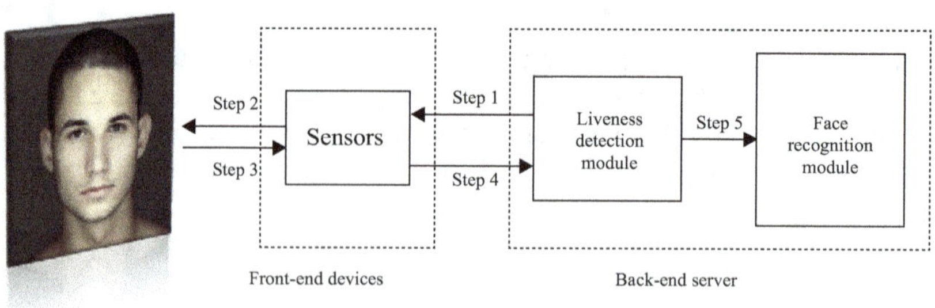

Figure 10.1 A typical face authentication system

detection module gathers all decoded data and checks whether the user is an actual human being. If so, the liveness detection module selects some faces among all the responses and delivers them to the face recognition module to determine the identity of the user (Step 5).

10.2.2 Attacks and solutions on liveness detection

In recent years, plenty of attacks have been developed to exploit the flaw that face recognition algorithms cannot determine whether a photograph taken by the front-end camera is captured from a real face, even if the recognition accuracy of some has exceeded human beings. In this study, we divide attacks into four categories and organize them as a tree, known as the attack tree, which is displayed in Figure 10.2. We first separate attacks into two categories: *static* and *dynamic*. *Static* attacks refer to the use of static objects, such as photographs, plastic masks and paper, as well as transformations of these objects (e.g., folding, creating holes through them and assembling them into certain shapes). Attacks using dynamic or partially dynamic objects are categorized into dynamic branch. Subsequently, we separate attacks into four subcategories: two-dimensional (2D) static, 3D static, 2D dynamic, and 3D dynamic. The 3D branches refer to attacks that use stereo objects, including sculptures, silicone masks and robots. More precisely, these objects must have notable stereo characters of human faces, such as a prominent nose, concave eye sockets and salient cheekbones; otherwise, the attacks are categorized into *2D* branches. Organized by this attacking tree, a brief review of relative attacks and solutions is presented next.

In the *2D static* branch, photograph-based attacks are the predominant form of attack. They are easily launched and effective at compromising primary recognition algorithms. Tracking eye movement was the first method proposed to counter such attacks. Jee *et al.* [2] proposed a method for detecting eyes in a sequence of facial images. Variations around the eyes are calculated, and whether the face is real is determined. Their basic assumption is that blinking and the uncontrolled movements of pupils are human nature behaviors. However, this method cloud be compromised by an adversary wearing a mask with eyeholes. A similar idea exploiting Conditional Random Fields was proposed by Sun *et al.* [3]. Its limitation is the same. Subsequently, lip movement and lip reading methods were developed by Kollreider *et al.* [4] for liveness detection. However, their method can also be fooled using carefully transformed photographs.

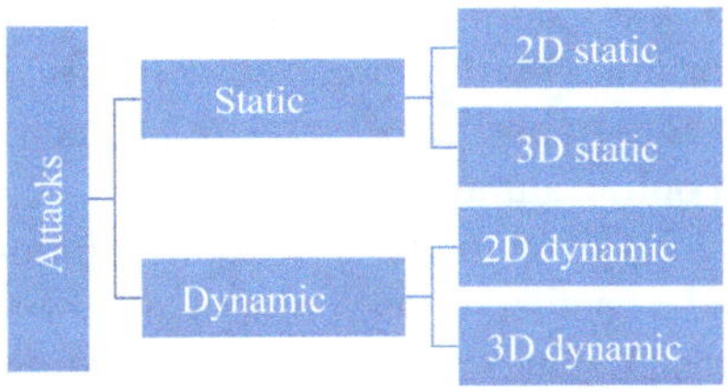

Figure 10.2 The attacking tree

To distinguish faces from photographs, Li *et al.* [5] leveraged the fact that faces and photographs have different appearances in frequency space. They conducted 2D Fourier spectral analysis to extract the high-frequency components of input images; faces contain more components in the high-frequency region than do photographs. However, adversaries can print a high-resolution photograph to bypass this method. Määttä *et al.* [6] observed that printed photographs are pixelized. That is, a face has more details than a photograph. Thus, they used a Support Vector Machine to extract microtextures from the input image. Later, Kim *et al.* [7] leveraged a more powerful texture descriptor, Local Binary Pattern to enhance performance. They additionally analyzed the information residing in both the low-frequency and high-frequency regions. However, all these types of solutions have a common drawback: low robustness. Motion blur or noise from the environment impairs their performance. Moreover, these methods are useless against screen-based attacks [8].

Because of strategies designed to protect against 2D attacks, adversaries have attempted to exploit 3D objects in 3D static attacks. Thus, researchers have developed novel methods to defend against these attacks also. Lagorio *et al.* [9] proposed a method to detect 3D facial features. They employed a two-camera device to ascertain the surface curvature of the subject seeking authentication. If the surface curvature is low, the subject is judged to be malicious. Although the accuracy is almost 100%, the sophisticated device required is expensive and the computational cost is unacceptable. Furthermore, Wang *et al.* [10] leveraged a one-camera, face alignment algorithm to ascertain 3D shape on the basis that forged faces are usually flatter than real faces. However, this method performed unsatisfactorily when applied to spectacle-wearers because of the limited capability of the face alignment algorithm.

In response to technological developments, adversaries must, in turn, develop more sophisticated attacks. One method is pasting stereo materials onto photographs. In contrast, researchers have developed practical and efficient methods to counter these threats, with the help from developments in computer vision. The fundamental idea behind these methods is that adversaries cannot manipulate static objects to simulate instructed expressions, even if these objects are similar to the human face. Thus, a common challenge–response protocol has been adopted, whereby users are asked to make expressions as the instructions, including happiness, despair and surprise. Such systems subsequently compare the captured video with stored data.

However, more powerful *2D dynamic* attacks have been developed, in which adversaries have exploited advanced deep learning models and personal computers with powerful processors. These attacks work by merging a victim's facial characteristics with photographs of the victim and using these forged photographs to bypass the face recognition algorithm. Furthermore, even if this operation requires time, adversaries can prepare photographs beforehand and launch offline attacks, sending forged photographs to an online authentication system.

To counter these new 2D dynamic threats, some solutions have been proposed. Bao *et al.* [11] introduced a method based on optical flow technology. The authors found that the motions of 2D planes and 3D objects in an optical flow field are the same in translation, rotation and moving, but not in swing. They used this difference to identify fake faces. Unfortunately, two drawbacks undermine this method: first, an uneven object will fail it; second, the method does not consider variation in illumination. Kollreider *et al.* [12] developed a method for detecting the positions and velocities of facial components by using model-based Gabor feature classification and optical flow pattern matching. However, its performance was impaired when keen edges were present on the face (e.g., spectacles or a beard). The authors admitted that the system is only error free if the data contain only horizontal movements. Findling and Mayrhofer [13] achieved liveness detection by combining camera images and movement sensor data. They captured multiple views of a face while a smartphone was moved. Li *et al.* [14] measured the consistency in device movement detected using inertial sensors and changes in the perspective of a head captured on video. However, both methods were demonstrated to be compromised by Xu *et al.* [15], who constructed virtual models of victims on the basis of publicly available photographs of them.

Less adversaries have attempted to launch *3D dynamic* attacks. They can reconstruct a 3D model of a victim's face [15] in virtual settings but hardly fabricate them in real scenes. We illustrate the difficulties in launching 3D dynamic attacks using the following three examples: first, building a flexible screen that can be molded into the shape of a face is expensive and may fail because the reflectance from a screen differs from that of a face. Second, 3D printing a "soft" mask is impractical, being limited by the printing materials available (see Section 10.2.5.4 for a fuller explanation). Third, building an android is infeasible and intricate and would involve face animation, precision control and skin fabrication. Additionally, building an android is costly, particularly a delicate android "face."

On the basis of the previous discussion, we observe that the current threats are principally 2D dynamic attacks because static attacks have been effectively neutralized and 3D dynamic attacks are hard to launch.

10.2.3 Design of Face Flashing protocol

The proposed protocol contains seven components that are illustrated in Figure 10.3, and eight steps are required to complete once challenge–response procedure where the challenge is flashing light and the response is the light reflected from the face.

- **Step 1: Generation of parameters.** Parameters are produced by the *generator* of the *liveness detection module* running on the back-end server, which works closely with the front-end devices. Such parameters include seed and N. Seed controls how random challenges are generated, and N determines the total number of challenges to be used. Communication of parameters between the back-end server and front-end devices is protected by standard protocols such as HTTP and TLS.
- **Step 2: Initialization of front-end devices.** After receiving the required parameters, the front-end devices initialize their internal data structures and start to capture videos from the subject being authenticated by turning on the *Camera*.
- **Step 3: Presentation of a challenge.** Once initialized, the front-end devices begin to generate challenges according to the received parameters. Essentially, a challenge is a picture displayed on a screen during one refresh period; light

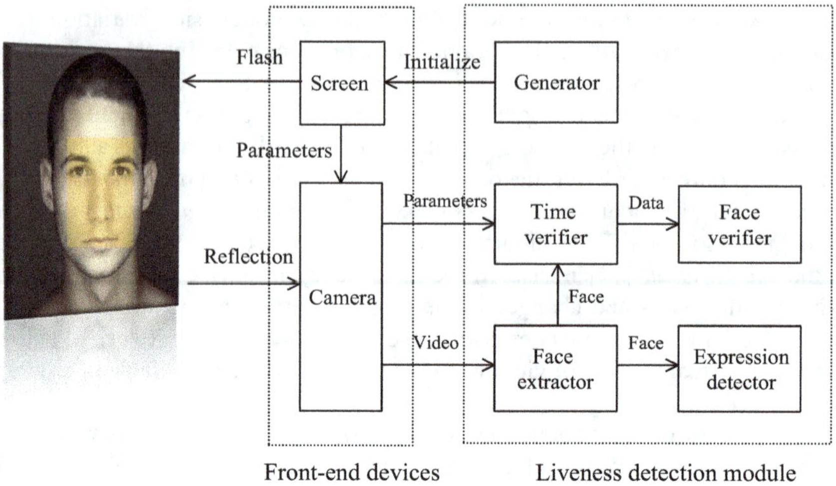

Figure 10.3 Architecture of Face Flashing

is emitted by the screen onto the subject's face. The challenge can be of two types: a background challenge that displays a pure color, and a lighting challenge that displays a lit area over the background color. More details are specified in subsequent sections.

- **Step 4: Collection of response.** The response is the light that is reflected immediately by the subject's face. We collect the response through the camera that has already been activated (in Step 2).
- **Step 5: Repetition of challenge–response.** Our protocol repeats Steps 3 and 4 for N times. This repetition is designed to collect enough responses to ensure high robustness and security so that legitimate users always pass, whereas adversaries are blocked even if they accidentally guess out the correct challenges beforehand.
- **Step 6: Timing verification.** Timing is the most crucial security guarantee provided by our protocol and is the fundamental distinction between genuine and fake responses. Genuine responses are the light reflected from a human face and are generated through a physical process that occurs simultaneously over all points and at the speed of light (i.e., zero delay). Counterfeit responses, however, would be calculated and presented sequentially, pixel-by-pixel (or row-by-row), through one or more pipelines. Thus, counterfeit responses would result in detectable delays. We detect delays among all the responses to verify their integrity.
- **Step 7: Face verification.** The legality of the face is verified by leveraging neural network that incorporates with both the shape and textual characters extracted from the face. This verification is necessary because without it our protocol is insufficiently strong to prevent from Media-based Facial Forgery (MFF) attacks, and face verification prolongs the time required by adversaries to forge a response, which makes the difference from benign response more obvious. The details are provided in Section 10.2.4.

- **Step 8: Expression verification.** The ability to make expressions indicates liveness. We verify this ability by ascertaining whether the detected expression is the one requested. Specifically, technology from [16] is embedded in our prototype for detecting and recognizing human expressions.

Details of Step 8 are omitted in this chapter so that we can focus on our two crucial steps: timing and face verifications. However, expression detection has been satisfactorily developed and is critical to our focus. Additionally, Step 8 is indispensable because it integrates our security boundary, which is elucidated in Section 10.2.5. The face extraction detailed in the next section is designed so that our two verification techniques are compatible with this expression detection.

10.2.4 Key techniques

10.2.4.1 Model of light reflection

Consider an image $I_{rgb} = \{I_r, I_g, I_b\}$ that is taken from a linear RGB color camera with black level corrected and saturated pixels removed. The value of $I_c, c \in \{r, g, b\}$ for a Lambertian surface at pixel position x is equal to the integral of the product of the illuminant spectral power distribution $E(x, \lambda)$, the reflectance $R(x, \lambda)$ and the sensor response function $S_c(\lambda)$:

$$I_c(x) = \int_{\Omega} E(x,\lambda)R(x,\lambda)S_c(\lambda)d\lambda, \ c \in \{r,g,b\} \tag{10.1}$$

where λ is the wavelength and Ω is the wavelength range of all visible spectrum supported by camera sensor. From the Von Kries coefficient law [17], a simplified diagonal model is given by:

$$I_c = E_c \times R_c, \ c \in \{r,g,b\} \tag{10.2}$$

Exploiting this model, by controlling the outside illuminant E, we can get the reflectance of the object. Specifically, when E_c for x and y are the same, then

$$\frac{I_c(x)}{I_c(y)} = \frac{R_c(x)}{R_c(y)}, \ c \in \{r,g,b\} \tag{10.3}$$

This means the lights captured by camera sensor at two different pixels x and y are proportional to the reflectance of that two pixels.

Similarly, for the same pixel point x, if applying two different illuminant lights E_{c1} and E_{c2}, then:

$$\frac{I_{c1}(x)}{I_{c2}(x)} = \frac{E_{c1}(x)}{E_{c2}(x)}, \ c1, c2 \in \{r,g,b\} \tag{10.4}$$

In other words, the reflected light captured by the camera in a certain pixel is proportional to the incoming light of the same pixel.

Implications of previous equations. Equations (10.3) and (10.4) are simple but powerful. They are the foundations of our liveness detection protocols. Equation (10.3) allows us to derive relative reflectance for two different pixels from the proportion of captured light from these two pixels. The reflectance is determined by the characteristics of the human face, including its texture and 3D shape. Leveraging (10.3), we can extract these characteristics from the captured pixels and further feed them to a neural network to determine how similar the subject's face is to a real human face.

Equation (10.4) states that for a given position, when the incoming light changes, the reflected light captured by the camera changes proportionally, and crucially, such changes can be regarded as "simultaneously" to the emission of the incoming light because light reflection occurs at the speed of light. Leveraging (10.4), we can infer the challenge from the current received response and detect whether a delay occurs between the response and the challenge.

10.2.4.2 Face extraction

To do our verifications, we need to locate the face and extract it. Furthermore, our verifications must be performed on regularized faces where pixels in different frames with the same coordinate represent the same point on the face. Concretely, when a user's face is performing expressions as instructed, it produces head movements and hand tremors. Thus, using only face detection technology is insufficient; we must also employ a face alignment algorithm that ascertains the location of every landmark on the face and neutralizes the impacts from movements. By using the alignment results, we can regularize the frames as we desired, and the regularized frames also ensure that our verifications are compatible with the expression detector.

First, we designed Algorithm 1 to quickly extract the face rectangle from every frame. In Algorithm 1, *track*(·) is our face tracking algorithm [18]. It uses the current frame as the input and employs previously stored frames and face rectangles to estimate the location of the face rectangle in the current frame. The algorithm outputs the estimated rectangle and a confidence degree, ρ. When it is small ($\rho < 0.6$), we regard the estimated rectangle as unreliable and subsequently use *detect*(·), our face detection algorithm [19], to redetect the face and ascertain its location. We employ this iterative process because the face detection algorithm is precise but slow, whereas the face tracking algorithm is fast but may lose track of the face.

Algorithm 1: Algorithm to extract the face

INPUT: *Video*
OUTPUT: $\{F_j\}$

for *frame* in *Video* **do**
    $Rect, \rho$ = track(*frame*)
    if $Rect = \emptyset$ or $\rho < 0.6$ **then**
        $Rect$ = detect(*frame*)
        $Rect \rightarrow$ track(·)
    end if
    $F_j = frame(Rect)$
end for

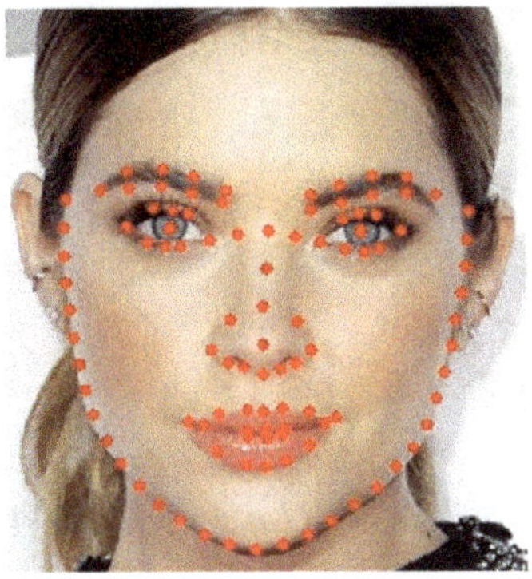

Figure 10.4 The 106 landmarks

After obtaining face rectangles, $\{F_j\}$, we exploit face alignment algorithm to estimate the location of 106 facial landmarks [20] on every rectangle. The locations of these landmarks are shown in Figure 10.4.

Further, we use alignment results to regularize every rectangle. Particularly, we formalize the landmarks on jth face as $L_j = (l_1, l_2, \ldots, l_{106})$, where l_i denotes $(x_i, y_i)^\top$, the coordinates of ith landmark. And, we calculate the transformation matrix T_j by

$$\begin{aligned} T_j &= \text{argmin}_T ||T\tilde{L}_j - L_{mean}||^2 \\ \text{where } L_{mean} &= \frac{\sum_j L_j}{\sum_j 1} \\ \tilde{L}_j &= \begin{bmatrix} x_1 & x_2 & \cdots & x_{106} \\ y_1 & y_2 & \cdots & y_{106} \\ 1 & 1 & \cdots & 1 \end{bmatrix} \end{aligned} \tag{10.5}$$

where T is a 3×3 matrix that contains rotation and shifting coefficients. We select the best T as T_j that minimizes the L_2 distance between the regularization target T_{mean} and $\tilde{L}_j$, the homogeneous matrix of the coordinate matrix. After that, we regularize the jth frame by applying the transformation matrix T_j to every pair of coordinates and extract the centering $1{,}280 \times 720$ rectangle containing the face. For the sake of simplicity, we use "frame" to represent these regularized frames containing only the face.*

10.2.4.3 Timing verification

Our timing verification is built on the nature of how camera and screen works. Basically, both of them follow the same scheme: refreshing pixel-by-pixel. Detailedly, after finishing refreshing one line or column, they move to the beginning of next line or column and perform the scanning repeatedly. We can simply suppose that an image is displayed on the screen line-by-line and captured by camera column-by-column,

*Since we just implemented those existing algorithms on face tracking, detection and alignment, we will not provide further details about them and interested readers can refer to original papers.

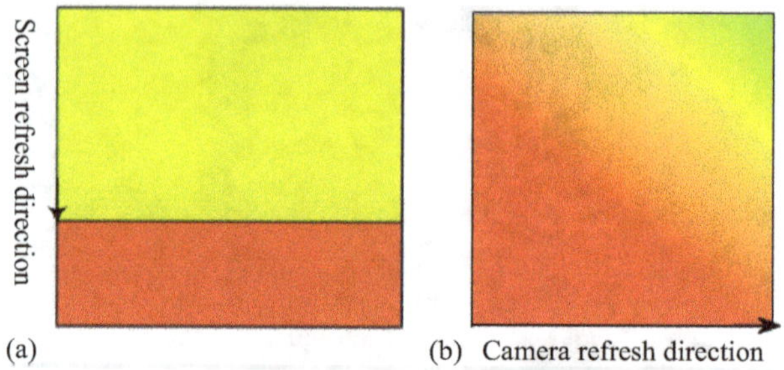

Figure 10.5 Working schemes of (a) screen and (b) camera refreshing

ignoring the time gap between refreshing adjacent pixels within one line or column that is much smaller than the time needed to jump to the next line or column. In other words, as to update any specific line on the screen, it has to wait for a complete frame cycle until all other lines have been scanned. Similarly, when a camera is capturing an image, it also has to wait for a frame cycle to refresh a certain column.

One example is given in Figure 10.5 to better explain the interesting phenomenon that is leveraged for our timing verification. Figure 10.5(a) shows a screen that is just changing the displaying color from red to green. Since it is scanning horizontally from top to bottom, the upper part is now updated to green but the lower part is still red. The captured image of a camera with column-scanning pattern from left to right is shown in Figure 10.5(b), which shows an obvious color gradient from red to green.†

To transform this unique feature into a strong security guarantee, the appropriate challenges must be constructed and verified to ensure the consistency of responses. In practice, we construct two types of challenges to be presented on front-end screen: one is the background challenge displaying a single color, and the other is the lighting challenge displaying a belt of different color on the background color. The belt of the different color from background is called the *lighting area*, and one example is shown in Figure 10.6(a), where the background color is red, while the lighting area is green.

To verify the consistency in responses, we defined another concept called Region of Interest (ROI), which is the region that the camera is scanning when the front-end screen is displaying the lighting area. The location of ROI is calculated as the following:

- Calculate t_u, the start time to show the lighting area:

$$t_u = t_{begin} + \frac{u}{rows} \times t_{frame} \tag{10.6}$$

where u is the upper bound of lighting area, *rows* is the number of rows contained in one frame, t_{begin} is the start time to show the current frame and t_{frame} is the during time of one frame.

†Similar but a little bit different color patterns can also be observed on cameras with row-scanning mode. Column-scanning mode that is used here is easier to understand.

- Find the captured frame, recording period of which covers t_u. Say the kth captured frame.
- Calculate the shift, l, against the first column of kth captured frame.

$$l = cols \times \frac{t_u - ct_k}{ct_{frame}} \tag{10.7}$$

where $cols$ is the number of columns contained in one captured frame, ct_k is the start time to expose the first column of kth capture frame and ct_{frame} is the exposure time of one captured frame.

After finding the location of ROI, we distill it by applying (10.4) on every pixel between the response of lighting challenge and background challenge. Two applied results are demonstrated in Figure 10.7. Now, the consistence can be verified. We check whether the lighting area can be correctly inferred from the distilled ROI. If it cannot, the delay exists and this response is counterfeit.

To infer the lighting area, we build four linear regression models handling different parts of captured frame (Figure 10.6(b)). Each model is fed with an average vector which is reduced from corresponding part of ROI, then estimates the location of $(u + d)/2$ independently. Next, we gather estimated results according to the size of each part. An example is shown in Figure 10.6(b), where the ROI is separated into two parts: the left part contains a columns and the right part contains b columns. The gathered result, $\hat{y}$, is calculated as following:

$$\hat{y} = \frac{a \times m_2 + b \times m_3}{a + b} \tag{10.8}$$

where m_2 and m_3 denote the estimated result made by model 2 and model 3, respectively.

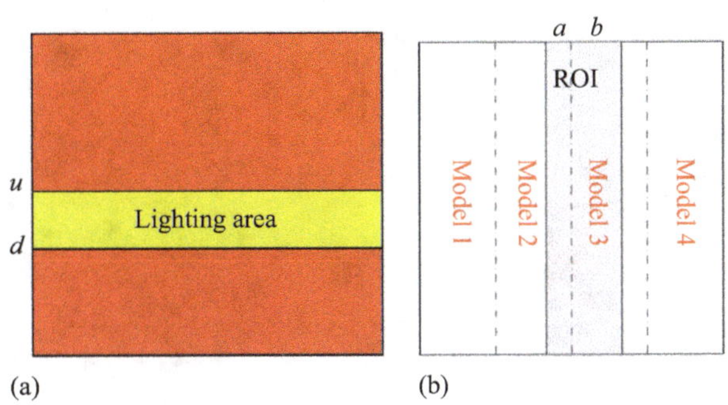

Figure 10.6 Example of lighting area and calculation of ROI: (a) lighting challenge and (b) captured frame

The final criteria of consistence is accumulated from $\hat{y}_i$, the gathered result of ith captured frame, as following:

$$\begin{aligned} d_i &= \hat{y}_i - \tfrac{u_i+d_i}{2} \\ mean_d &= \tfrac{\sum_{i=1}^{n} d_i}{n} \\ std_d^2 &= \tfrac{\sum_{i=1}^{n} (d_i - mean_d)^2}{n-1} \end{aligned} \tag{10.9}$$

We finally check whether $mean_d \times std_d$ is smaller than $exp(Th)$, where Th is a predefined threshold.

Note that legitimate responses are consistent with our challenges and will produce both small $mean_d$ and std_d. Adversarial responses will be detected by checking our final criteria. An additional demo is illustrated on Figure 10.7 to explain visually how the lighting area affects the captured frame.

10.2.4.4 Face verification

After preprocessing, we get a sequence frame with vibration removed, size unified and color synchronized. Further, we use (10.3) to generate the midterm result from the responses of a background challenge: First, we randomly choose a pixel on the face as the anchor point; then, we divide all the pixels by the value of that anchor point. Some midterm results are shown on Figure 10.8.

Without any difficulty, we can quickly differentiate results of real human faces from fake ones. This is because real human faces have uneven geometry and textures, while other materials, such as monitor, paper or iPad's screen, do not have. Based on this observation, we developed our face verification techniques, as described next.

- **Step 1: Abstract.** We vertically divide the face into 20 regions. In every region, we further reduce the image to a vector by taking the average value. Next, we

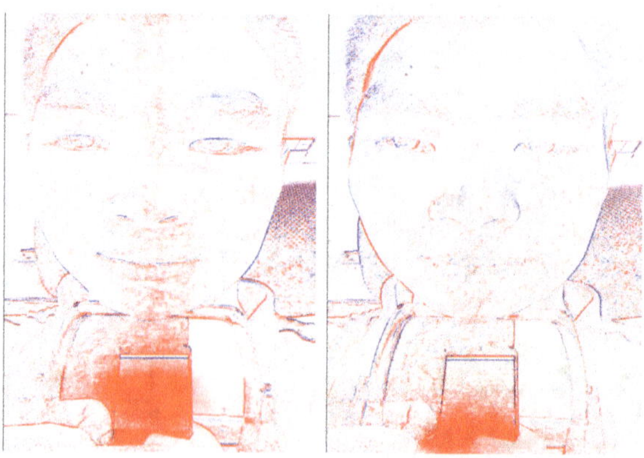

(a) (b)

Figure 10.7 Effect of (a) middle and (b) bottom lighting area. The mirrors in the bottom of both pictures will show the location of corresponding lighting area

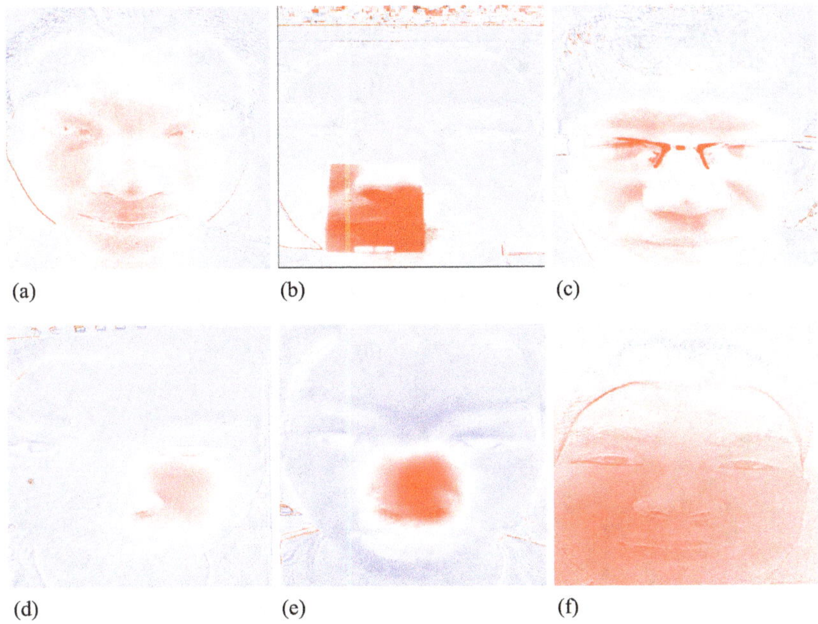

Figure 10.8 Examples of midterm results: (a) and (c) are captured from real human faces, (b) is captured from an iPad's screen, (d) and (e) are captured from an LCD monitor and (f) is captured from a paper

smooth every vector by performing polynomial fitting of 20 degrees with minimal 2-norm deviation. After that, we will derive images like Figure 10.9(c).

- **Step 2: Resize.** We pick out facial region and resize it to a 20 × 20 image by bicubic interpolation. An example is shown on Figure 10.9(d).
- **Step 3: Verify.** We feed the resized image to a well-trained neural network, and the decision will be made then.

The neural network we used contains three convolution layers with a pyramid structure, whose effectiveness was sufficiently proved in Cifar-10, a dataset used to train the neural network to classify ten different objects. In Table 10.1, we show the architecture of our network and the parameters of every layer.

10.2.5 Security analysis

In this section, we present the security analysis of Face Flashing. First, we abstract the mechanisms behind Face Flashing as a challenge–response protocol. Second, we analyze the security of two main parts in our protocol: timing verifications and face verification. Finally, we demonstrate how Face Flashing defeats three typical advanced attacks.

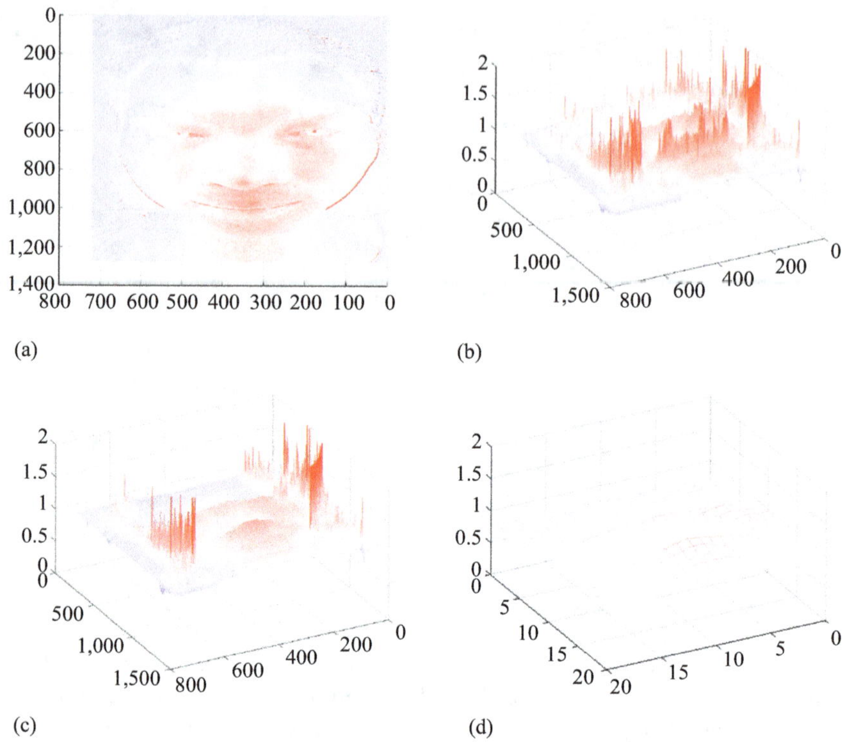

Figure 10.9 Face verification: (a) midterm result, (b) midterm result, (c) abstract result and (d) resize result

Table 10.1 Architecture of neural network

Input size	**Layer type**	**Stride**	**Padding size**
20 × 20 × 3	Conv 5 × 5	1	0
16 × 16 × 16	Conv 3 × 3	1	1
16 × 16 × 16	Pool 2 × 2	1	0
8 × 8 × 16	Conv 3 × 3	1	1
8 × 8 × 32	Pool 2 × 2	1	0
1 × 512	Inner product	0	0

It is certain that Face Flashing can defeat static attacks, as the expression detector, one component of our system, is sufficient to defeat them. Specifically, static materials cannot make expressions according to our instructions in time (e.g., 1 s) and attacks using them will be failed by expression detector. Therefore, the main task of our security analysis is to show that Face Flashing can defeat dynamic attacks.

10.2.5.1 Challenge–response elements

Face Flashing is a challenge–response protocol whose security guarantees are built upon three elements: unpredictable random challenges, hardly forged responses and the effective response verifications.

The challenges. Our challenge is a sequence of carefully crafted images that are generated at random. Since the front-end devices are assumed to be well protected, adversaries could not learn the random values. Besides, a verification session consists of tens of challenges. Even if the adversary can respond a right challenge by chance, it is unlikely for him to respond to a sequence of challenges correctly.

The responses. There are two important requirements for the responses: first, the response must be easily generated by the legitimate users, otherwise it may lead to usability problems or even undermine the security guarantee (e.g., if adversaries can generate fake responses faster than legitimate users). Second, the responses should include essential characteristics of the user, which are hard to be forged.

Face Flashing satisfies both the requirements. The response is the reflected light from the human face, and the user needs to do nothing besides placing her face against the camera. More importantly, such responses, in principle, are generated at the speed of light, which is faster than any computational process. Besides, the response carries unique characteristics of the subject, such as the reflectance features of her face and uneven geometry shapes, which are physical characteristics of human faces that are inherently different from other media, e.g., screens (major sources of security threats).

Response verification. We use an in-depth defense strategy to verify the responses and detect possible attacks.

- First, timing verification is used to prevent forged responses (including replay attacks).
- Second, face verification is used to check if the subject under authentication has a specific shape similar to a real human face.
- Third, this face-like object must be regarded as the same person with the victim by the face recognition module (orthogonal to liveness detection).

Considering the pre-excluded static object, it is very hard for adversaries to fabricate such a thing satisfying previous three rules simultaneously. In general, Face Flashing builds a high bar in front of adversaries who want to impersonate the victim.

10.2.5.2 Security of timing verification

The goal of the timing verification is to detect the delay in the response time caused by adversaries. Before further analysis, we emphasize that two points should be considered.

- First, according to the design of modern screens, the adversary cannot update the picture that is being displayed on the screen at the arbitrary time. In other words, the adversary cannot interrupt the screen and let it show the latest generated response before the start of the next refreshing period.

- Second, the camera is an integral device that accumulates the light during his exposure period. And, at any time, within an initialized camera, there always exist some optical sensors that are collecting the light.

For the sake of clarity, we assume that the front-end devices contain a 60-fps camera and 60-Hz screen. On the other side, the adversary has a more powerful camera with 240 fps and screen with 240 Hz. Under these settings, we construct a typical scenario security analysis which is illustrated in Figure 10.10.

In this scenario, the screen of the front-end device is displaying the ith challenge, and the adversary aims to forge the response to this challenge. The adversary may instantly learn the location of lighting area of the challenge after t_u, while she cannot present the forged response on her screen until v_k, due to the nature of how the screen works. Hence, there is a gap between t_u and v_k. Recalling our method described in Section 10.2.4.3, during the gap, some columns in the ROI have already completed the refreshing process. In other words, these columns' image will not be affected by the forged response displaying on the adversary's screen during v_k to v_{k+1}. We name this phenomenon as *delay*.

When *delay* happens, our camera will get an undesired response inducing four linear regression models to do deviated estimation about the location of lighting area. Besides, the standard deviation of these estimations will increase, for two reasons:

- The adversary's screen can hardly be synchronized with our screen. Particularly, it is different even the length of adjacent refreshing periods. Hence, the *delay* is unstable, so as the estimations.
- The precision of forging will be affected by the internal error of adversaries' measurement about time. This imprecision will be amplified again by our camera, which fluctuates the estimations.

In other words, if the adversary reduces $mean_d$ by displaying the carefully forged response, she will simultaneously increase std_d. On the other hand, if the adversary does nothing to reduce std_d, she will significantly enlarge $mean_d$. While for a benign user, the *delay* will not happen, the discordance between our camera and screen can

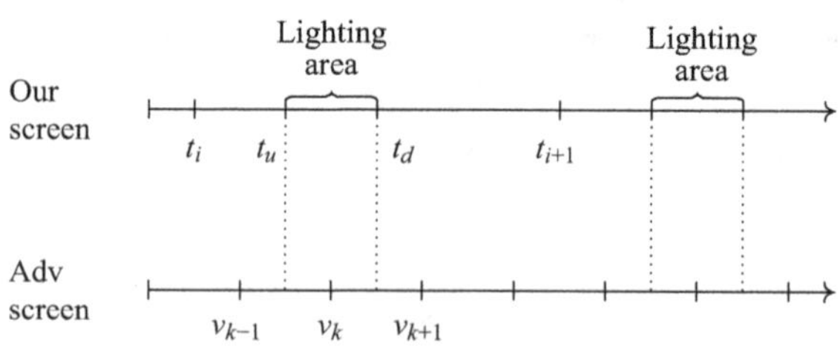

Figure 10.10 Security analysis on time

be solved by checking the timestamps afterward and both the accumulated $mean_d$ and std_d will be small, according to our verification algorithm.

In summary, we detect the delay by estimating the deviation. And the effectiveness of our algorithm provides a strong security guarantee on the timing verification.

10.2.5.3 Security of face verification

Our face verification abstracts the intrinsic information of shape through a series of purification. And we feed this information to a well-designed neural network.

If the adversary aims to bypass the face verification, there are two conundrums that need to be resolved. First, the adversary needs to conceal the specular reflection of the plain screen. Particularly, during the authentication procedure, we require the user to hold the phone so their face can occupy the entire screen. The distance, as we measured, is about 20 cm. In this short distance, the specular reflection is severe. In Figure 10.8(b), we demonstrate the result captured from a screen without any covering sheet. Even covered by a scrub film (Figure 10.8(d)), the screen's specular reflection is still intense.

Second, the forged object must have similar geometry shape with human faces. More precisely, its abstract result should like a transpose of "H" (Figure 10.9(c)). And this stereo object needs to make expression according to our instructions. Even if the adversary can achieve these, there is no promise that they can deceive our strong neural network modal every time. And there is no chance for the adversary to generate a response with low quality. The high recall of our model will be demonstrated in next section.

Previous two conundrums provide the security guarantee on face verification.

10.2.5.4 Security against typical attacks

Obviously, Face Flashing can defeat traditional attacks such as photo-based attacks. Here, we discuss its defenses against three typical advanced attacks.

Offline attacks. An offline attack is to record responses of previous authentications, and replay them to attack the current authentication. However, this attack is impossible to fool our protocol. First, the hitting possibility is small, as we require responses matched all the challenges. Concretely, if we use eight different colors and present ten lighting challenges, the hitting possibility will be less than 10^{-9}. Second, even if adversaries have successfully guessed the correct challenge sequence, displaying responses legitimately is difficult. Because displaying on screens will produce the intensely specular reflection that is easily detectable and displayed by projecting responses onto a forged object that leads to high std_d that also can be detected, as adversaries cannot precisely predict the length of every refreshing period of the screen.

MFF attacks. An MFF attack is to forge the response by merging victim's facial information and the currently received challenge. However, this attack is also useless, because it is hard to deceive our timing and face verifications simultaneously. First, to deceive our face verification needs forging high-quality responses, which is difficult and time-consuming. Particularly, high-quality forgery requires reconstructing the 3D model of victim's face and simulating the reflection process. Second, to deceive our timing verification needs to complete the previous forgery quickly. Actually, the available time is (1/240)/2 s for attacking a 60-Hz screen. Third, even if adversaries can quickly produce a perfectly forged response, displaying the response is not allowed (see the preceding paragraph).

3D-mask attacks. A 3D-mask attack is to wear a 3D mask to impersonate the victim. However, this attack is impractical. First, this attack needs to build an accurate mask that can fool our face recognition module, which is difficult.‡ Second, the legitimate mask is hard to be 3D printed. As the printed mask needs to have the similar reflectance of human skin and be so flexible that adversaries can wear it to make instructed expressions. However, the available 3D-printed materials are non-flexible under the requirement of Fused Deposition Modeling, so the prevalent 3D print technology cannot be used to build a 3D mask. Besides, the smallest diameter of available nozzles is 0.35 mm, which will produce coarse surfaces, and coarse surfaces can be distinguished from human skin.

In sum, Face Flashing is powerful to defeat advanced attacks, especially attacks similar to the ones mentioned previously.

10.3 Case study 2: Secure mobile payment via imperfection of LCD screens

Mobile payment, especially QR-code-based mobile payment, becomes increasingly popular, being offered by major banks (e.g., ICBC) and payment service providers (e.g., PayPal). Unlike mobile payment solutions provided by hardware vendors (e.g., Apple Pay and Samsung Pay), QR-code payment schemes do not rely on any hardware support and can therefore be easily deployed. However, the security guarantee of the new scheme is less clear: in the absence of hardware protection, users' digital wallet can be vulnerable to an OS-level adversary, who could steal his or her secret for generating payment tokens.

In this section, we will report our findings that the physical features of a phone's screen can enhance the security protection of this QR-code payment, serving as a second-factor authentication. Due to the manufacturing imperfection, the luminance levels of the pixels on the screen vary across the screen's display area, which can be used to uniquely characterize the screen. This physical fingerprint cannot be stolen even when the OS is fully compromised, since the adversary cannot observe the physical features of the screen.

However, screen fingerprinting could also undermine the mobile payer's privacy, as the less trusted merchants could use it to track customers and infer their purchase history and preferences. In this chapter, we propose a new authentication solution that *anonymously fingerprints mobile screens*. The approach, called *AnonPrint*, generates a random one-time mask to obfuscate a screen that hides its fingerprint from the merchants. In the meantime, the payment provider, who shares a secret with the payer,

‡Even though there is an existing study implying it is possible [21], performing it in real is not easy.

is able to reconstruct the mask and authenticate the payer through her obfuscated fingerprint.

10.3.1 Physical feature of screens

Brightness unevenness. Key to our screen fingerprinting technique is the observation that there are subtle differences in the maximum luminance of the pixels at different locations of the screen. This problem comes from the imperfection in the screen manufacturing, which could bring defects to the components in Figure 10.11: the light from the back light unit can be spread unevenly over the screen, with the broader darker than the center, causing the Mura effect; the transparency of the polarizer varies at different parts of the screen and so can be the default angle for bending the polarized lights when no voltage is applied. Although on a qualified product, all such subtle defects are barely observable to human eyes, they can however be reliably identified by a camera, as observed in our study.

Non-LCD screens. Not all smartphones have LCD screens. As a prominent example, Samsung products typically use AMOLED panels that have no back-end light source, and instead, they are built upon organic light emitting diode, with every pixel emitting lights of different strengths. Those panels also have luminance unevenness, which can be picked up by the same technique we developed for analyzing screens, as shown by our evaluations.

10.3.2 Off-line QR payment

As mentioned earlier, the QR-code payment is gaining popularity recently, due to its ease of use. Specifically, when a buyer checks out her purchases, she only needs to run a *wallet* app and click on its "pay" button to show a one-time QR code on the screen with a maximum luminance level and presents the screen under the camera

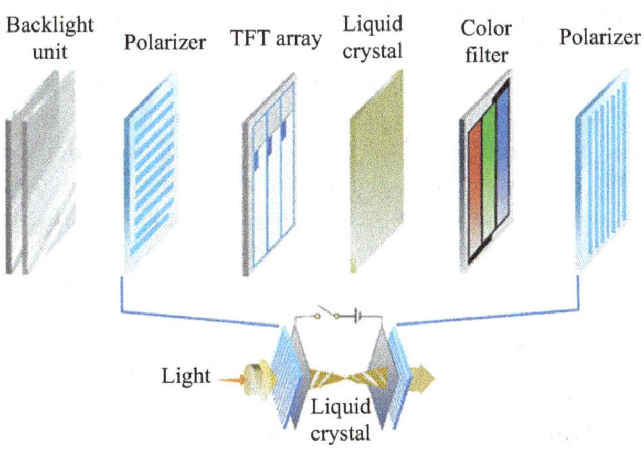

Figure 10.11 Typical structure of an LCD screen

Figure 10.12 A POS-supporting QR-code payment

of the vendor's Point of Sale (POS) system for scanning. After the vendor enters the right amount the buyer needs to pay, the scan starts and the POS will produce different sounds to notify both the parties whether the transactions get through with the payment service provider (see Figure 10.12, [22]).

Payment transaction. To use the payment scheme, the payer needs to open an account with the payment provider, install its wallet app and log in her account through the app when she is ready to pay. The app contains a piece of secret downloaded from the provider for generating one-time tokens.

When the payer initiates a payment transaction (by pressing the payment button of the wallet app), the app proceeds as illustrated in Figure 10.13 to complete the transaction. Specifically, it uses the current time, the user ID and the stored secret as inputs to generate a random token (usually using HMAC), which is then encoded into a QR code to be shown on the payer's screen. After being scanned into the POS system, the token is extracted, together with other transaction information such as the total amount and currency type, and is sent to the payment provider.

The provider, upon receiving the token and transaction information, recovers from the token the user ID and uses the information to retrieve the stored secret for the payer for verifying the token. When the token is valid and still fresh (within a given time interval), the provider moves forward on the transaction and notifies the vendor of the acceptance of the payment if a sufficient amount of fund is left in the payer's account.

Threat of the OS-level attack. However, compared to other mobile payment schemes, the QR-code payment is limited in its capability of defending against OS-level adversary. All off-line payment schemes rely on the secret stored on the mobile device, which is often protected through hardware means. For example, Apple Pay uses the secure element to store payment secret and runs the payment process within

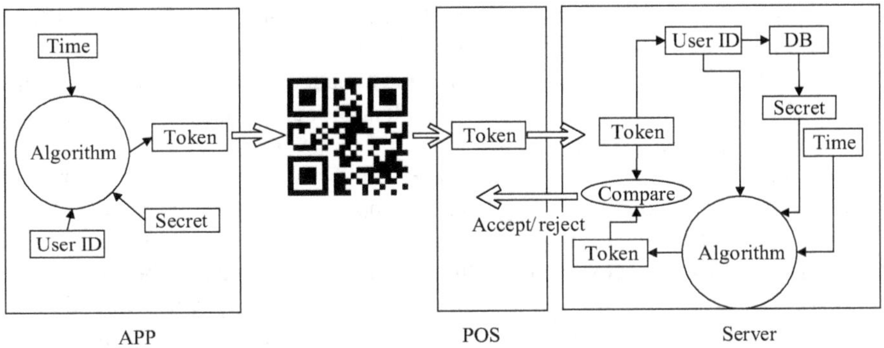

Figure 10.13 Off-line QR-code payment flow chart

its secure enclave. Samsung Pay leverages its KNOX (built upon top of ARM trustzone) to protect the payment information in its secure element. All of these techniques can mitigate the threats from the OS-level attackers, since the payment secret stays within the hardware component where the OS cannot access directly. For the payment schemes not provided by the device manufacturers, such as PayPal and AliPay, the hardware-level protection is not in place and can therefore only trust the OS to protect the payment process. Once the OS is compromised (particularly on rooted or jail-broken devices), the adversary can then get access to the payment secret to generate legitimate payment token to steal from the victim.

Actually, in our research, we analyzed AliPay on a jail-broken device and successfully extracted the secret.

10.3.3 Adversary model

An off-line payment transaction involves four actors: the payer, the OS, the vendor and the payment provider. In our research, we assume the provider to be trusted and collaborative, willing to protect the payer's account and her transaction privacy, while the OS and the vendor are considered to be less trusted, each with different capabilities. However, we do not assume that they collude.

Curious vendor. More specifically, we consider that the vendor is curious but honest. He can save a photo of the screen of payer's phone, during the QR code scan, by reprogramming[§] or customizing his scanner. However, we do not assume that he can acquire the secret within the payer's phone, including the payment secret and other information for generating the protection for the screen (Section 10.3.5). We neither assume that he can get or set the internal status of the POS (scanner is outside the POS and is controlled by the vendor). In the meantime, we assume that he does not attempt

[§]Some scanner manufacturers provide programming guide to help vendor modify the logic of the scanner [23].

to directly steal from the payer, an attempt that can be detected by the provider (e.g., overcharging the payers for the item they purchase).

Instead, the malicious intent of the vendor is to benefit from tracking or helping a third party track the payer's purchase activities. It is undoubted that purchasing activity logs are valuable to merchants who are eager to know what their customers bought, returned and what they might be interested. Starbucks, a well-known Apple-friendly brand, hesitated to support Apple pay because Apple pay is too anonymous [24].

Note that there are situations where the payment is done overtly, without hiding the payer's identity. Prominent examples include purchasing through membership card. In these cases, the identity information proactively provided by the payer enables the vendor to not only link her transactions together across different stores within the same organization but also connect her behaviors **across organizations**, should the vendor decide to sell the identity with purchase information to a third party such as the advertiser, which results in unexpected privacy leakage (imagine a restaurant learns what you would like to order according to your Walmart's history). However, the off-line scheme such as the QR pay is designed to provide the payer *option* to remain anonymous.

Untrusted OS. The operating system in our model is untrusted, which can be completely under the control of the adversary who not only can get access to the secret of the victim's wallet but can also learn other information, for example, the secrets for creating screen masks (Section 10.3.5). Further, the adversary also tries to acquire the fingerprint of the device, though he cannot directly observe it from the software of the device.

As mentioned before, the QR code payments like the one used by PayPal do not have a secure container for the payment secret in the phone and only ask the OS to protect the secret, so when the OS is compromised, everything is exposed to the adversary, except the hardware features of the screen that cannot be directly "seen" from the software stack of the system. With the wallet secret, the adversary can generate a valid token and QR code (Section 10.3.2), by using the same algorithm (which we assume is public) and time information, for off-line payment.

10.3.4 Generate screen fingerprint using brightness unevenness

Our research shows that unique physical features of smartphone screens can be used to identify them. Extraction of these features, however, is nontrivial, due to the impact of projective distortions introduced when the screen is scanned by the camera. In this section, we present a technique that addresses the issue and fingerprints a screen using the image that can be conveniently taken during a payment scan.

10.3.4.1 Photo extraction and correction

Figure 10.14 visualizes the luminance fingerprint of a screen, which is illustrated through distortion correction, blurring and contour drawing. This fingerprint can be picked up by the POS camera when it is positioned at a right distance and angle toward the screen. What is expected is that the screen is in parallel with the camera lens, with its center right under the camera. In practice, hardly can this be done

Figure 10.14 Fingerprint visualization

Figure 10.15 Illustration for distorted fingerprints

perfectly, as illustrated in Figure 10.12, even when a POS system is calibrated to identify the QR code shown on the screen (with the code boxed by a square visualized to the party who scans). As a result, the images snapped by the POS camera are often distorted, as illustrated in Figure 10.15, which brings in the trouble to effective extraction and comparison of fingerprints (distortion may make one phone similar to another). Following we elaborate how this challenge is addressed in our research through proper corrections.

Distortion correction. The picture of a screen can be viewed as a projection from the screen plane to the camera perspective plane (as they are not parallel), with an unknown angle, see Figure 10.16. Most important to the distortion correction is to identify the angle. For this purpose, we come up with a new approach that leverages the *QR code on the screen as a reference* to realign the whole image through a projective transformation, which turns out to be very effective in recovering the distorted screen fingerprint.

Specifically, we consider that the screen sits on a *display plane* (D) when its picture is taken, which is essentially a projection from D to a *shooting plane* (S). To recover full information from the distorted image, we need to project the image from S to a *correction plane* C parallel to D, through a transformation (T). This

Figure 10.16 Projective distortion illustration

Figure 10.17 Bullseyes locating algorithm

transformation maps each point with coordinate (x, y) on S to a point (x', y') on C as follows:

$$T\begin{bmatrix} x \\ y \\ 1 \end{bmatrix} = \begin{bmatrix} a & b & c \\ d & e & f \\ g & h & 1 \end{bmatrix}\begin{bmatrix} x \\ y \\ 1 \end{bmatrix} = \lambda\begin{bmatrix} x' \\ y' \\ 1 \end{bmatrix}$$

As we can see here, T links any points on the shooting plane S to those on the correction plane C. To find the transformation, we need four points (s_1, s_2, s_3, s_4) on the shooting plane and their counterparts (c_1, c_2, c_3, c_4) on the correction plane, as T has an eight-degree freedom (the variables $a, \ldots, h$) and every point pair yields two equations. We call the four points on C *fixed points* and the four points on S *moving points*.

Moving point locating. The QR code displayed on the screen has four "bullseyes" (see Figure 10.17, original) for the decoding program to locate the code, including three positioning markings to identify the direction of the code, and an alignment marking to the help with orientation. The coordinations of these bullseyes are used in our approach to compute the transformation. Specifically, on the correction plane, all we need are just the relative positions of these markings, which can be easily determined by generating a standard QR code on the plane, right in the middle of a rectangle $(0, 0)$ to (1,800, 1,080), which is an area where the screen picture will be projected to. A little more complicated is to get the moving points, their positions on the shooting plane S.

Actually, when a QR code is successfully scanned, the coordinates of all the four points should be correctly detected by the QR-code decoding. However, QR decoding algorithm cannot report these positions when decoding failed. In those cases where the QR code decoding failed but fingerprint is needed (multiple scan cases that will be introduced later), he needs to run our proposed algorithm over the picture of the screen to calculate the coordinates. The algorithm is described as follows:

1. Our approach recognizes rectangles using the MATLAB® image processing toolbox `regionprops`, which automatically identifies all quadrilaterals from the picture and then we remove those quadrilaterals totally unlikely to be a foursquare.
2. From the rest, our approach further selects the triples with their centroids forming an isosceles right triangle (approximately, due to the distortion). Oftentimes, this process will discover a triple that consists of the top-left, the top-right and the bottom-left bulleyes, as illustrated in Figure 10.17 middle.
3. Finally, we look for a quadrilateral (the alignment mark) whose centroid is approximately on the mid-perpendicular of the right-angle side of the isosceles we just found and is reasonably away from the right angle vertex, which is shown in Figure 10.17 right.

Figure 10.17 illustrates how this approach works. The four points identified in this way or reported by the decoding algorithm are then used as the moving points for restoring the image for the screen.

Screen image recovery. Using the positions of the four bullseyes on both S and C, we can compute T by simply solving an equation set consisting eight equations resulted from the four point pairs (Equation (10.10)).

$$\begin{bmatrix} s_{1x} & s_{1y} & 1 & 0 & 0 & 0 & -s_{1x}c_{1x} & -s_{1y}c_{1x} \\ s_{2x} & s_{2y} & 1 & 0 & 0 & 0 & -s_{2x}c_{2x} & -s_{2y}c_{2x} \\ s_{3x} & s_{3y} & 1 & 0 & 0 & 0 & -s_{3x}c_{3x} & -s_{3y}c_{3x} \\ s_{4x} & s_{4y} & 1 & 0 & 0 & 0 & -s_{4x}c_{4x} & -s_{4y}c_{4x} \\ 0 & 0 & 0 & s_{1x} & s_{1y} & 1 & -s_{1x}c_{1y} & -s_{1y}c_{1y} \\ 0 & 0 & 0 & s_{2x} & s_{2y} & 1 & -s_{2x}c_{2y} & -s_{2y}c_{2y} \\ 0 & 0 & 0 & s_{3x} & s_{3y} & 1 & -s_{3x}c_{3y} & -s_{3y}c_{3y} \\ 0 & 0 & 0 & s_{4x} & s_{4y} & 1 & -s_{4x}c_{4y} & -s_{4y}c_{4y} \end{bmatrix} \begin{bmatrix} a \\ b \\ c \\ d \\ e \\ f \\ g \\ h \end{bmatrix} = \begin{bmatrix} c_{1x} \\ c_{2x} \\ c_{3x} \\ c_{4x} \\ c_{1y} \\ c_{2y} \\ c_{3y} \\ c_{4y} \end{bmatrix} \tag{10.10}$$

The transformation enables us to restore the distorted image by determining the coordinates of every point on C using $T[s_x, s_y, 1]^T = \lambda[c_x, c_y, 1]^T$.

In our implementation, we computed the brightness values for all 1,800 × 1,080 points on plane C from the image on S. Specifically, for every point (c_x, c_y) on C, our approach uses the previous equation to identify its corresponding coordinates (s_x, s_y)

on S. In the case that these coordinates are not integers, we perform interpolation to correct them and find the point on S to assign its brightness level to c.

In addition to the distortion caused by a shooting angle, in reality, we also need to handle other artifacts introduced during picture taking before the fingerprint of the screen can be extracted. A prominent example is the spherical distortion resulted from the camera lens, which is amplified at the border of the projected image. To avoid the noise that could be brought in by this distortion, we simply remove a 60-pixel band from the peripheral of the image, which typically is wide enough to cover the border. Also, our approach drops the QR code and a few lines at the top and bottom of the image, should the gray scales of these lines be very different from those of their contiguous lines, which indicates the presence of floating status bars or virtual buttons.

10.3.4.2 Fingerprint extraction and comparison

Fingerprint extraction. The fingerprint of a screen should be built upon a set of *stable* features observed when each pixel on the screen reaches its highest brightness level in the white color, as what happens when a QR code is scanned. Such features are not supposed to change significantly across different observations (pictures taken) under various light conditions, as long as the camera is sufficiently close to the screen. So, we analyzed the photos for screens to seek a kind of stable representation to the brightness unevenness feature. We first found that the photo is interfered by relatively high-frequency noises, so we applied a radius 2 Gaussian filter to the photo to remove the high-frequency noise. Then after having analyzed the characteristics of different-color models such as HSV, HSL and HSI, we found the luminance level calculated by grayscaling the photo is the most stable representation to the screen feature, because this process eliminates saturation and hue information that turned to be unstable at all. More specifically, hue and saturation are mainly decided by the differences between the R, G and B values, while when displaying a pure white background, the differences between the three values are nearly zero, which means that they are easily affected by the ambient light (noise).

At last, a fingerprint of a screen is represented as a matrix of the *luminance* for each pixel (0–255). To visualize the screen features, we use contours to connect the pixels with similar luminance levels and then run Gaussian Blurs to smooth out the images. A screen image in the left side of Figure 10.14 is described by the fingerprint on the right.

Fingerprint comparison. Once the fingerprints are extracted, we need to have a way to compare them, so as to determine whether a screen has been seen before. For this purpose, we utilize the Pearson product–moment correlation coefficient to measure the similarity of two fingerprints. The correlation coefficient is chosen since it measures the distance between two vectors based upon the similarity of the *relations* among the elements within each individual vector. This allows us to overcome the differences in the absolute brightness values for the same pixel observed under various light conditions. Specifically, given two fingerprints, which are two matrices with each element being either the brightness level of a pixel (100–255) or of 0 if the value of the element is removed during the distortion correction (border or virtual button), we first convert these matrices into two vectors v and v', through concatenating their rows together, sequentially. Then we remove the elements from individual positions within v and v' if at least one of them in that position is 0 (as illustrated in Figure 10.18). After that, our approach compares v and v' by calculating the Pearson product-moment correlation coefficient. The result needs to be adjusted by the difference in the numbers of "effective elements" in these two vectors, that is, those not being removed (nonzero in value). This is important because the difference in vector lengths should also be taken into consideration in the similarity measurement, but our previous steps dropped this information. Let d be such a difference between the two matrices and e be the total number of nonzero elements copied to vectors. The *similarity* between two fingerprints, Δ, is calculated as follows:

$$\Delta = \left(1 - \frac{d}{e}\right) CorrCoef(v, v')$$

Figure 10.19 illustrates the distributions of Δ over 40 images taken from 20 phone screens, 2 images per screen. The blue solid line here represents the similarity between the images of the same phone, and the red dots are for those between two different phones. As we can see from the figure, rarely does the cross-phone similarity go above 0.9, while self-similarity stays above 0.95 most of the time. So we choose 0.90–0.95 as the threshold range for determining whether two fingerprints indeed belong to the same phone.

For this purpose, we further collect 160 images from 80 additional phones (still 2 images per phone). For the images, again we calculate the similarity and the self-similarity for each of them and further link two images together if their similarity goes above the threshold *th*, in an attempt to understand the feasibility of using the fingerprints to track users. This result is then compared with the ground truth (i.e., whether two images indeed come from the same phone), to identify the percentage of the images that have been correctly linked to their counterparts, which we call *tracking coverage*. In the meantime, we also inspect each image to find out how many other images it is similar to (with a similarity above the threshold).

The study shows that, when *th* is set to be 0.93, the coverage reaches 96.3%, but on average an image is linked to 2.195 images, indicating that at least one another device is confused with each phone, on average. This result shows that vendors can correlate 96.3% of their customers' transactions together, though may also incorrectly

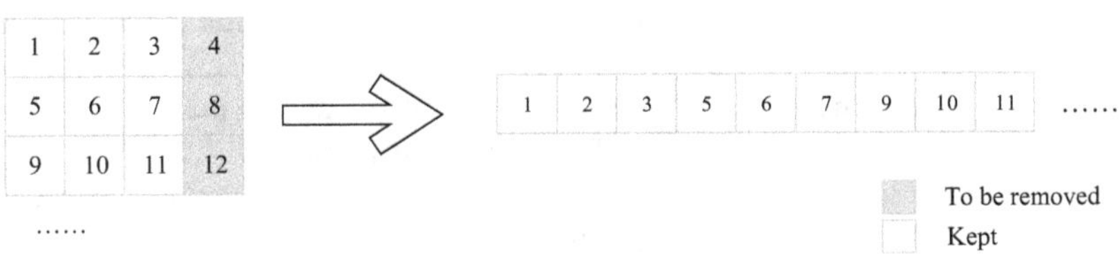

Figure 10.18 Converting fingerprint matrix to vector

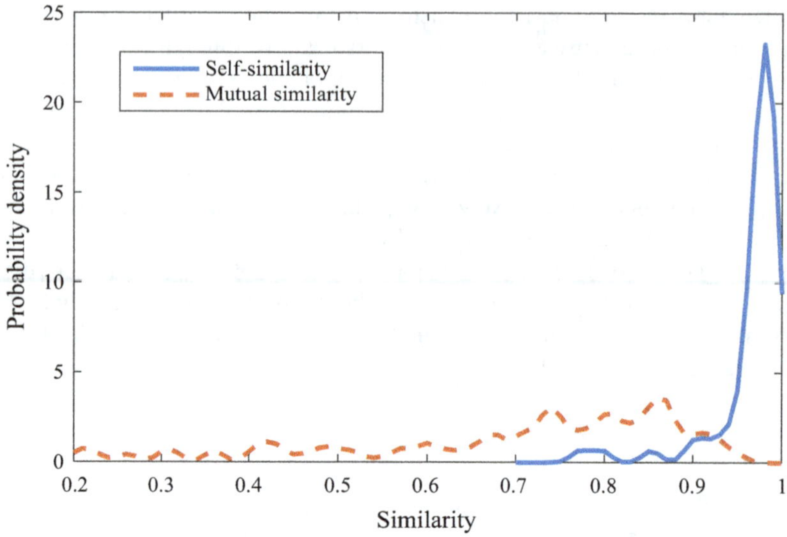

Figure 10.19 PDFs of similarities for vendor's tracking

bring another customer's transaction to one's profile. This could still be acceptable if the vendors use the linked transactions for purchasing preference mining and targeted advertisements. We can further raise *th* to 0.945. In this case, the coverage descends to 88.75%, while incorrect links per image drop to 0.070. This indicates that for 88.75% of transactions, the vendors can accurately identify other transactions from the same customer, by simply looking at the features of her screens.

10.3.5 Extension: anonymous screen authentication

To leverage the screen fingerprinting for better user protection but avoid its privacy leakage risk, here we describe our design of AnonPrint, which obfuscates a screen with a digital brightness mask to prevent the linking across payment transactions while still enabling an authorized party to authenticate the owner of the screen.

10.3.5.1 Framework overview

Figure 10.20 illustrates the framework of AnonPrint and the payment process that supports this anonymous screen authentication. The framework is built upon the existing off-line payment system, with only moderate changes made to the wallet app, POS scanner and the service on the payment provider side. More specifically, the payer first needs to submit the original screen fingerprint of her device to the payment provider when she opens an account. The wallet app is modified to synchronize a secret random seed with the provider, which is generated and updated through hashing the time for the payment (as encoded in the QR code) together with a shared secret using a cryptographic hash function (e.g., SHA-256). This seed is passed to a pseudorandom number generator each time when the wallet app needs to generate

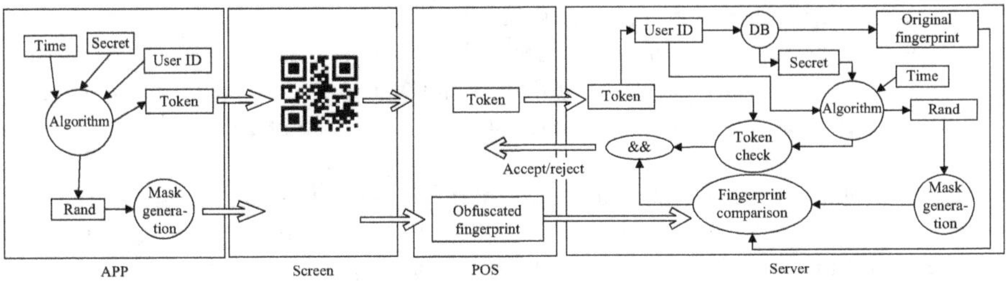

Figure 10.20 Overview of AnonPrint

a mask for obfuscation and when the provider needs to generate the same mask for authenticating the payer. This mask is displayed as the background for displaying the QR payment token, from which the POS scanner extracts the obfuscated screen fingerprint in addition to decode the QR code and passes the information to the payment provider. Upon receiving the fingerprint and the payer's token, the provider retrieves the shared secret and the original screen fingerprint using her claimed ID. Then, the same mask used by the payer is reconstructed and together with the original fingerprint serves as inputs for synthesizing a new obfuscated fingerprint. This fingerprint is compared with the one from the payer's screen and the transaction can be approved when their similarity is above a threshold and other security checks are made. Following we elaborate how the screen is masked, how the fingerprints are compared and our analysis of the security properties.

10.3.5.2 Screen obfuscation

Our idea to obfuscate a screen is to create a digital luminance pattern, called *mask*, to hide the screen's hardware fingerprint for each payment transaction. Such a mask is automatically generated by a digital wallet app, based upon a pseudorandom number generator seeded with a random number synchronized with the payment service provider, so the provider can also generate the same mask to authenticate the payer. The mask needs to be realistic, similar to a real fingerprint in terms of the distribution of brightness levels. Further, it should work on not only a physical screen but also its image, since the latter is all the provider has about the payer.

Mask generation. A screen fingerprint is characterized by a smooth luminance change observed across neighboring pixels: the luminous intensities of these pixels, when their inputs have all been set to the maximum (255), can be slightly different, due to their physical features; such a difference is minor between the pixels close to each other, since any large, abrupt change is likely to disqualify the whole product. To obfuscate this hardware fingerprint but maintain a screen's realistic looking, it is important that the mask generated will fully preserve this property. Here is how we do that in our research:

1. *Random zone selection:* Our approach first produces a 180×108 pure white (with all pixels set to 255) image as the background and randomly selects from the image 20 mutually disjoint zones, each with a size of 16×16.

2. *Dot darkening:* From each zone, we randomly choose 3 pixels and set their pixel value to a random number between 0 and 100.
3. *Smoothing:* For each zone, AnonPrint *blurs* it using *Gaussian Smoothing*, an imagine processing technique that, intuitively, "smoothes out" the dark color of the selected pixels to its neighboring pixels (i.e., elevating their values by proportionally reducing the values of their neighboring pixels, based upon their distances to these pixels, according to the Gaussian distribution). The radius of the Gaussian smooth filter was set to 10 in our implementation.
4. *Resizing:* The mask image is resized and scaled to a 1,800 × 1,080 matrix whose values range from 220 to 255. The size is identical to the original fingerprint, so that it can mask the whole area used to extract fingerprint. Besides, the resizing also implies smoothing.

Mask application. To obfuscate a screen, we only need to let the device display the mask as a background picture (may be scaled). Figure 10.21 shows two images of the same screen with and without a random mask, which look very similar. After that, the luminance level for each pixel on the screen is attenuated according to the mask, rather than its maximum value. In the meantime, the presence of such a mask helps one to mitigate the tracking risk without undermining the screen's authentication capability (Section 10.3.5.3): our experiment shows that the masked screen can no longer be linked to the original one by the fingerprint extracted from their images, except for the party who knows the mask, such as the service provider.

Applying the technique to the QR-code payment, each user first needs to register to the payment provider with an image of her unprotected screen when all pixels are set to the maximum gray scale. This can be done by taking a close-in picture of the screen using a second device (another phone or a laptop), or pay without mask at a trusted vendor. Also, the wallet app running on the user's device needs to share an additional pair of secrets for the mask generation, which are used together with the timing information to generate a seed that is the input to a pseudorandom number

Figure 10.21 Before mask, masked, masked by another mask

generator for computing the locations of the zones and the pixels within selected zones for generating the mask image.

10.3.5.3 AnonPrint verification

During the payment, an image of a masked screen is taken for authenticating the payer. This is done on the payment service provider's side by reconstructing the mask using the shared secret, and then obfuscate the fingerprint for comparing it with the image from the vendor. Most challenging issue here is how to build the obfuscated screen fingerprint, which, due to the vendor's lack of access to the physical device, can only be *synthesized* from the original fingerprint and the mask.

Screen attenuation model. To synthesize an obfuscated fingerprint, we need to understand how a pixel's brightness is attenuated when displaying the content with various pixel grayscale value. For this purpose, we selected 10 areas from a screen, each of 10×10 in size, and continuously reduced the value of each pixel in the area in a controlled environment, by one at a time, from 255 to 220. Each time we took a picture of the screen and measured each pixel's observed luminance value.

The observed brightness of a pixel attenuates approximately linearly with its decreased pixel value (0 for totally dark and 255 for pure white), which can be modeled by the attenuation function: $y = 1 - k(255 - x)/255$, where k is the contrast factor, x is the input pixel value and y is the normalized luminance level observed from the photo. A challenge here is how to determine k, which actually varies for different phones and even different shots of the same screen, due to the contrast level set for individual phones and most importantly the scanning camera's automatic adjustment of its ISO sensitivity according to the ambient light. Following we describe how to synthesize an obfuscated fingerprint using the model, particularly how to find k.

Fingerprint synthesization. Our attenuation experiment shows that there is a linear relation between observed luminance intensity and pixel grayscale value. This allows the payment service provider to synthesize the obfuscated fingerprint using the original screen's fingerprint and the reconstructed mask. Let the mask matrix be X and the original fingerprint matrix be f. The synthesized fingerprint f' can be calculated by $f' = f \circ (1 - k(255 - X)/255)$. The question here is how to determine k that varies across different scans.

We calculate k in our research based on the observation that the fingerprint synthesized using a right k should have a higher similarity with the fingerprint f_o extracted from an authorized screen, compared with that based upon an incorrect k. So, finding k can be modeled as the following optimization problem:

$$k = \arg\max_{\kappa \in D} \Delta(f \circ (1 - \kappa(255 - X)/255), f_o)$$

$$s = \Delta(f \circ (1 - k(255 - X)/255), f_o)$$

where Δ is the correlation coefficient mentioned in Section 10.3.4 and D is the range of k. To find out D, we recorded the optimized k using a large range [0, 10] over 100 images taken from 50 phones in common environment. The experiment shows that under normal ambient light situations (as under the POS machine), k falls between 0.25 and 6. Further, given an image of an authorized screen, a mask the screen wears and its original fingerprint, the similarity s between a synthesized obfuscated fingerprint and that extracted from the image is found to be approximately unimodal across possible k values, i.e., increase first and then decrease. Therefore, the provider can run a ternary search to find the k within [0.25, 6] that maximizes the similarity s. If s goes above the threshold, the device is authenticated, otherwise, it fails.

In a rare case when a legitimate screen fails to be recognized by the provider, service provider can ask the payer to enter a strong password (directly into the POS system) for the authentication.

Multi-scan. Actually, recognizing a QR-code is more complicated than it appears to be. A single scan on today's POS system typically yields a low recognition rate, typically below 50%. What happens in real world is multi-round, consecutive scans until a frame passes the decoder, and the message decoded from QR code passes the error-detecting code checks. This property is used in our design of AnonPrint, for the purpose of achieving a high security guarantee for the second-factor authentication without undermining the utility of the technique. To explicate the approach, we first need to define False Rejection Rate (FRR) and False Accept Rate (FAR), as follows:

$$FAR = \frac{number\ of\ accepted\ unauthorized\ fingerprints}{number\ of\ attack\ attempts}$$

$$FRR = \frac{number\ of\ failed\ authorized\ fingerprints}{number\ of\ authentication\ attempts}$$

Given an FRR of α and an FAR of $(1-\beta)$ for a single scan (i.e., showing a single mask), after N independent rounds, with each round a different mask shown on the screen, we are expecting a final FRR of α^N (i.e., rejecting a fingerprint when all N rounds fail), and FAR of $(1-\beta^N)$ (i.e., accepting a fingerprint when at least one round succeeds). AnonPrint is designed to utilize such a multi-round scan to strike a balance between the overall FAR and FRR. Specifically, during a QR scan, the wallet application continuously generates N different masks and display them as a background for the same QR code one after another, each lasting a fixed time interval. To help the scanner differentiate these masks, each mask is labeled by a mark—a small dark square with 10×10 pixels—displayed at a given position on the screen. The position of the square signals the scanner that a different mask is in use. In this way, totally N images will be sent to the payment provider, which accepts the transaction if at least one of them is considered to match the payer's screen fingerprint.

10.4 Conclusion

In this chapter, we presented two case studies that had leveraged the unique features of non-digital components to enhance the security of a given system. In the first case study, we proposed a novel challenge–response protocol, Face Flashing, to defeat the main threats against face authentication system—the *2D dynamic* attacks. We have systematically analyzed our method and illustrated that our method has strong security guarantees. In the second case study, we present a new technique that enhances the security protection to popular QR-based payment, without undermining the payer's privacy. Our technique leverages the features of the payer's screen, which is characterized by its unique luminance unevenness introduced by the imperfect manufacture process, ensuring that even when the payer's digital wallet has been fully compromised, an unauthorized payment still cannot succeed. The screen features, however, would raise a privacy concern, were it naively deployed to authenticate the payer, since it could be abused by the vendors to link one's different purchases together. To address this concern, we present AnonPrint that obfuscates the phone screen during each payment transaction. This new approach defeats the attempt to correlate the payer's purchase activities and also enables the payment provider, who can reconstruct the obfuscation mask, to authenticate the payer.

However, we believe that there should be more interesting examples where feature of non-digital components can be explored to enhance system security, or even be exploited by adversaries for hacking purpose. We hope that the case studies presented in this chapter could raise the awareness and research interest of our security community, so that we can have better understanding of potential security implications of those non-digital components.

References

[1] Zhang G, Yan C, Ji X, *et al.* DolphinAttack: Inaudible voice commands. In: Proceedings of the 2017 ACM SIGSAC Conference on Computer and Communications Security, CCS 2017, Dallas, TX, USA, October 30–November 03, 2017; 2017. p. 103–117. Available from: https://doi.org/10.1145/3133956.3134052.

[2] Jee HK, Jung SU, and Yoo JH. Liveness detection for embedded face recognition system. International Journal of Biological and Medical Sciences. 2006;1(4):235–238.

[3] Sun L, Pan G, Wu Z, *et al.* Blinking-based live face detection using conditional random fields. In: International Conference on Biometrics. Berlin: Springer; 2007. p. 252–260.

[4] Kollreider K, Fronthaler H, Faraj MI, *et al.* Real-time face detection and motion analysis with application in "liveness" assessment. IEEE Transactions on Information Forensics and Security. 2007;2(3):548–558.

[5] Li J, Wang Y, Tan T, *et al.* Live face detection based on the analysis of Fourier spectra. In: Defense and Security. Orlando, FL: International Society for Optics and Photonics; 2004. p. 296–303.

[6] Määttä J, Hadid A, and Pietikainen M. Face spoofing detection from single images using micro-texture analysis. In: Biometrics (IJCB), 2011 international joint conference on. Washington, DC: IEEE; 2011. p. 1–7.

[7] Kim G, Eum S, Suhr JK, *et al.* Face liveness detection based on texture and frequency analyses. In: 2012 5th IAPR International Conference on Biometrics (ICB). New Delhi: IEEE; 2012. p. 67–72.

[8] Chakraborty S and Das D. An overview of face liveness detection. International Journal on Information Theory. 2014;3(2).

[9] Lagorio A, Tistarelli M, Cadoni M, *et al.* Liveness detection based on 3D face shape analysis. In: Biometrics and Forensics (IWBF), 2013 International Workshop on. Lisbon: IEEE; 2013. p. 1–4.

[10] Wang T, Yang J, Lei Z, *et al.* Face liveness detection using 3d structure recovered from a single camera. In: Biometrics (ICB), 2013 International Conference on. Madrid: IEEE; 2013. p. 1–6.

[11] Bao W, Li H, Li N, *et al.* A liveness detection method for face recognition based on optical flow field. In: Image Analysis and Signal Processing, 2009. IASP 2009. International Conference on. Taizhou: IEEE; 2009. p. 233–236.

[12] Kollreider K, Fronthaler H, and Bigun J. Non-intrusive liveness detection by face images. Image and Vision Computing. 2009;27(3):233–244.

[13] Findling RD and Mayrhofer R. Towards face unlock: On the difficulty of reliably detecting faces on mobile phones. In: Proceedings of the 10th International Conference on Advances in Mobile Computing & Multimedia. Bali: ACM; 2012. p. 275–280.

[14] Li Y, Li Y, Yan Q, *et al.* Seeing your face is not enough: An inertial sensor-based liveness detection for face authentication. In: Proceedings of the 22nd ACM SIGSAC Conference on Computer and Communications Security. Denver, CO:ACM; 2015. p. 1558–1569.

[15] Xu Y, Price T, Frahm JM, *et al.* Virtual U: Defeating face liveness detection by building virtual models from your public photos. In: 25th USENIX Security Symposium (USENIX Security 16). Austin, TX: USENIX Association; 2016. p. 497–512.

[16] Smith RS and Windeatt T. Facial expression detection using filtered local binary pattern features with ECOC classifiers and platt scaling. JMLR: Workshop and Conference Proceedings. 2010;11:111–118.

[17] Brainard DH and Wandell BA. Analysis of the retinex theory of color vision. J. Opt. Soc. Am. A. 1986;3(10):1651–1661.

[18] Fernando W, Udawatta L, and Pathirana P. Identification of moving obstacles with Pyramidal Lucas Kanade optical flow and k means clustering. In: 2007 Third International Conference on Information and Automation for Sustainability. Melbourne: IEEE; 2007. p. 111–117.

[19] Yang B, Yan J, Lei Z, *et al.* Convolutional channel features. In: Proceedings of the IEEE International Conference on Computer Vision. Washington, DC: IEEE Computer Society; 2015. p. 82–90.

[20] Zhu S, Li C, Loy CC, *et al.* Unconstrained Face Alignment via Cascaded Compositional Learning. Las Vegas, NV: IEEE; 2016.

[21] Raghavendra R and Busch C. Robust 2D/3D face mask presentation attack detection scheme by exploring multiple features and comparison score level

fusion. In: Information Fusion (FUSION), 2014 17th International Conference on. IEEE; 2014. p. 1–7.

[22] Focus Technology Co., Ltd. POS Supporting QR Code; 2016. [Online; accessed 08 Aug 2020]. https://jepower.en.made-in-china.com/product/fNnQmIEPmxkd/China-IC-Card-Reader-POS-Device-Support-Four-Means-of-E-Payment-Functions.html.

[23] Wasp Barcode Technologies. Spec of a QR Code Scanner; 2016. [Online; accessed 08 Aug 2020]. https://www.waspbarcode.com/~/media/pdfs/waspbarcode/products/scanner-pdfs/wdi4600-2d-barcode-scanner.ashx.

[24] Brustein J. Apple Pay Is Too Anonymous for Some Retailers; 2016. [Online; accessed 08 Aug 2020]. https://www.bloomberg.com/news/articles/2014-10-20/apple-pay-is-too-anonymous-for-panera-starbucks-and-other-retailers.

Chapter 11
Accelerating homomorphic encryption in hardware: a review

Truong Phu Truan Ho[1] *and Chip-Hong Chang*[1]

The concept of homomorphic encryption has been long regarded as the Holy Grail in the cryptography community. In the era of cloud-centric and data-driven computing where individual privacy is often sacrificed for performance, homomorphic encryption can be considered as an important driver to outsourcing of computations. By allowing arbitrary evaluation and analyses to be performed directly on outsourced encrypted data, homomorphic encryption technology allows cloud operators to provide computation services to users by retaining the confidentiality of their data. Despite its conceptual simplicity and the amount of research effort to improve this technology over years, the practicality of homomorphic encryption is still considered unacceptable for most applications except several specific uses with the help of dedicated hardware accelerators. This chapter presents a detailed discussion on the most recent constructions of homomorphic encryption, as well as the algorithms and architectures for efficient implementation of performance bottleneck operations in hardware platforms. A recent construction of homomorphic encryption is first provided to identify the crux of the problem. Thereafter algorithms for efficient hardware accelerators of homomorphic encryption, namely, very large degree polynomial multiplication and very large number arithmetic, are discussed. The aim of this chapter is to provide a sufficiently comprehensive and self-contained overview so that engineers and researchers who are new to this emerging field can quickly grasp the underlying concepts, and further develop efficient hardware implementation of homomorphic encryption for their applications.

11.1 Introduction

The concept of homomorphic encryption was discovered and published in 1978 by Rivest *et al.* [1], right after they spotted the multiplicative homomorphism expressed in RSA scheme [2], i.e., the encryption of the product of two messages can be directly computed from the encryptions of the messages without requiring access to a secret

[1]School of Electrical and Electronic Engineering, Nanyang Technological University, Singapore

key. Since then, homomorphic encryption—the concept of computing on encrypted data—quickly became the Holy Grail in the cryptography community and is being touted as an ideal solution to privacy-preserving applications. Lacking further breakthrough and commercial support for more compelling use scenarios over a prolonged period, the problem of constructing cryptosystems expressing homomorphism gradually became forgotten. Although there were several constructions following the RSA scheme such as Goldwasser—S. Micali cryptosystem [3] in 1984, ElGamal cryptosystem [4] in 1985, Benaloh cryptosystem [5] in 1994 and Paillier cryptosystem [6] in 1999, those schemes were designed for the purpose of public-key encryption, and the homomorphisms are merely a byproduct spawned from and limited by their application requirements. As a consequence, and partly also because of privacy-preservation was not a dire concern during that time, homomorphic encryption did not receive its expected and deserved attention and their potential for consumerization has been played down [7].

Despite the scaling barriers, the last dozen years have witnessed the continuous increases in software to hardware integration and growth of semiconductor industry before Moore's law begins to slow down in many market segments. This results in an explosive growth of digital devices with various computing capabilities and power consumptions, from high-performance supercomputers to low-power microcontrollers. The emergence and convergence of new technologies, such as cloud computing, Internet of Things, big data analytics, artificial intelligence and 5G, have generated unprecedented massive amount of digital data. In every moment, new data are generated and processed, which led to boosted efficiency, high level of productivity and lowered cost. As the world is moving to cloud-centric and data-driven era, where more and more computations and storages are outsourced to clouds [8,9], data are easier to access and share than ever. A wide range of software tools are also available to greatly assist the analysis and utilization of this new-oil resource. However, this convenience is increasingly met with the threats of privacy infringement, which can be confirmed by rampant and severe data breaches happened recently. Typical examples include Facebook and SingHealth. In the early 2018, the Facebook–Cambridge Analytica data misuse was revealed and surprised the world. In this scandal, Facebook shared the online social behaviors of its users to Cambridge Analytica for political purposes [10]. In the middle of 2018, a cyber-attack targeting Singapore SingHealth compromised personal information of millions of patients, including the high-ranking government officials [11]. To act against the threat of data breaches, several governmental projects, such as HEAT funded by European Union [12], PROCEED funded by DARPA [13] and N-CRiPT and SCRIPTS funded by Singapore government [14], are currently being carried out, where the ultimate goal is to develop advanced homomorphic encryption technologies to efficiently process encrypted sensitive information.

In 2009, the first homomorphic encryption scheme supporting arbitrary computation was constructed by Craig Gentry [15]. Unlike previous constructions where limited homomorphism is supported [2–6], the scheme designed by Gentry marks the binary message with a small noise term to achieve homomorphism with respect to addition and multiplication modulo two. Based on this approach, an arbitrary computation can be carried out on encrypted data by representing the computation as a binary circuit consisting of AND gates (equivalent to multiplication modulo two) and XOR gates (equivalent to addition modulo two). This scheme [15], unfortunately, has its unique computational constraints. The added noise term grows quadratically for every homomorphic multiplication of two binary messages. Moreover, translating an arbitrary function to its equivalent binary form can result in a circuit with lots of AND gates, which further accelerates the noise growth. These deficiencies impose severe practical issues, as confirmed by several implementations of this scheme. For example, a single encryption takes 82 s on CPU [16] and 4.1 s on Nvidia GPU [17]. Driven by privacy preservation of outsourcing data computing to cloud, efforts have been continuously expended to address the crux of this performance bottleneck of Gentry's construction. Major advancements have been made to eventually reduce noise growth from quadratic to linear with extra computation after a homomorphic multiplication [18,19]. The support homomorphism for nonbinary messages further reduces the effect of noise growth [20]. Unfortunately, the reduced noise growth is still not good enough in practice to meet the desirable user experience today due to the algorithmic complexity and large size of parameters [21–23]. In a recent software implementation, it takes 36.7, 1.3 and 70 min to perform key generation, encryption and homomorphic function evaluation, respectively, of a lightweight block cipher SIMON-64/128 on a 4-core Intel Core i7 processor [24]. The computation time can be reduced significantly by using hardware coprocessor to handle the compute-intensive and performance-critical operations in homomorphic encryption. Specifically, the time taken for the homomorphic evaluation of SIMON-64/128 can be reduced significantly from 70 to 2.6 min with parallel processing in field programmable gate array (FPGA) implementation [25]. It is evident that hardware acceleration is a promising solution to closing in the performance gap of homomorphic encryption.

The remainder of this chapter is organized as follows. Section 11.2 describes a recent construction to give an appreciation of homomorphic encryption. While Section 11.3 focuses on algorithms for fast polynomial multiplication, Section 11.4 introduces the residue number system (RNS) for large number arithmetic. In Section 11.5, the techniques presented in Section 11.3 and 11.4 are employed for the design of hardware accelerators. The chapter is concluded in Section 11.6.

11.2 Fan–Vercauteren (FV) homomorphic encryption scheme

11.2.1 Ring learning-with-error assumption

Let R denote the polynomial ring $Z[x]/(x^n + 1)$. Each element in the ring R can be considered as an integer-coefficient polynomial with degree less than n. Operations in the ring R include addition and multiplication of polynomials. Addition of two polynomials in R is carried out by usual polynomial addition. To multiply two polynomials in R, the usual polynomial multiplication is carried out first. If the polynomial resulted from this multiplication has a degree greater than n, it is modulo reduced by $(x^n + 1)$.

Besides the polynomial degree n, two other parameters are important for the ring learning-with-error. They are an integer modulus q and an error distribution χ. The latter is a statistical distribution over the ring R. $R_q = Z_q[x]/(x^n + 1)$ is defined as a finite ring of polynomials with degree less than n and the integer coefficients are taken from the set $(-q/2, q/2]$. Let s and a be a secret and a plaintext polynomial uniformly chosen from the ring R_q. Then the ciphertext polynomial is computed by $b = as + e$, where e is a polynomial drawn from the error distribution χ. The computation of b is performed over the ring R_q by definition. The ring learning-with-error problem assumes that given the pair of polynomials (a, b), it is computationally infeasible to determine s. Please see Figure 11.1 for an illustration of the ring learning-with-error problem.

11.2.2 The encryption scheme

In this scheme, the message space is chosen as R_t for some integer t greater than one. The ciphertext space is taken as R_q. Let $\Delta = \lfloor q/t \rfloor$, where $\lfloor \ \rfloor$ means rounding down to the nearest integer. The Fan–Vercauteren (FV) public-key encryption scheme is defined as follows.

The secret key s is a polynomial sampled from the error distribution χ. Normally, s should be uniformly chosen from R_q. For efficiency, s can be sampled from χ without any reduction in security strength [26,27].

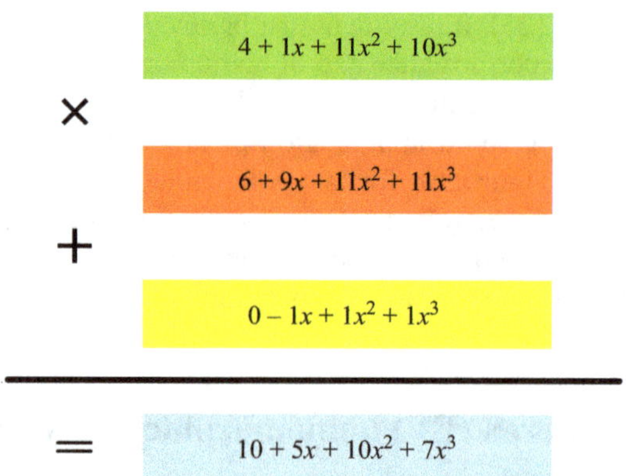

Figure 11.1 An example of ring learning-with-errors problem in polynomial ring $Z_{13}[x]/(x^4 + 1)$. The bottom-most polynomial (in blue-shaded text box) is a result of the product of the top-most input polynomial (in green-shaded text box) and the secret polynomial (in orange-shaded text box), marked with a small noise polynomial (in yellow-shaded text box). Ring learning-with-errors states that it is computationally infeasible to find the secret polynomial, given the input polynomial and the result polynomial (source: www.nxp.com)

To generate the public key, a polynomial a is uniformly chosen from R_q and a polynomial e is sampled from the error distribution χ. The public key pk is a pair of polynomials belongs to R_q and computed as

$$pk = (-as + e \mod q, a) \in R_q \times R_q \tag{11.1}$$

where "mod" denotes the modulo reduction operation.

The evaluation keys are generated by first selecting an integer $w \geq 2$. For efficiency, w can be chosen as a power of two. Let $l = \lfloor \log_w q \rfloor$, then the set, evk, of evaluation keys is given by

$$\begin{aligned} evk &= \{evk_0, evk_1, \ldots, evk_l\} \\ &= \left\{ \left(-a_i s + e_i + w^i s^2, a_i\right) \in R_q \times R_q : i = 0, 1, 2, \ldots, l \right\} \end{aligned} \tag{11.2}$$

where a_i is uniformly chosen from R_q and e_i is sampled from the error distribution χ.

Let $pk = (p_0, p_1)$. To encrypt a message $m \in R_t$, three polynomials, u, e_1 and e_2, are sampled from the error distribution χ. The encrypted message is a pair of polynomials belonging to R_q and is computed as

$$ct = (p_0 u + e_1 + \Delta m \mod q, p_1 u + e_2 \mod q) \in R_q \times R_q \tag{11.3}$$

Let the ciphertext $ct = (c_0, c_1)$. ct can be decrypted to m by

$$\left[\frac{t}{q} (c_0 + c_1 s \mod q) \right] \mod t \in R_t \tag{11.4}$$

where [ ] denotes rounding to the nearest integer.

Let $ct_0 = (c_0, c_1)$ and $ct_1 = (d_0, d_1)$ be the two valid ciphertexts corresponding to two messages, m_1 and m_2, respectively. Without decrypting these two ciphertexts, the ciphertext corresponding to the summation of m_1 and m_2 in R_t is computed by a homomorphic addition defined by

$$(c_0 + d_0, c_1 + d_1) \in R_q \times R_q \tag{11.5}$$

Homomorphic multiplication is more complicated. By treating the elements of a ciphertext ct as the coefficients of the polynomial $ct(s)$, we have

$$\begin{aligned} &(c_0 + c_1 s)(d_0 + d_1 s) \\ &\quad = c_0 d_0 + (c_0 d_1 + c_1 d_0)s + c_1 d_1 s^2 \\ &\quad = \Delta^2 m_1 m_2 + \cdots \end{aligned} \tag{11.6}$$

Indeed, $m_1 m_2$ can be recovered by using a different equation for decryption with the secret key s and its square s^2. However, it is necessary to keep a consistent decrypting equation for every ciphertext. The procedure to perform a homomorphic multiplication consists of two parts and involves the evaluation key evk that contains the information of s^2. In the first part of this procedure, the following intermediary is computed:

$$
\begin{aligned}
ct'_{mul} &= \left(c'_0, c'_1, c'_2\right) \\
&= \begin{pmatrix} \left[\frac{t}{q} c_0 d_0\right] \bmod q, \\ \left[\frac{t}{q}\left(c_0 d_1 + c_1 d_0\right)\right] \bmod q, \\ \left[\frac{t}{q} c_1 d_1\right] \bmod q \end{pmatrix} \in R_q \times R_q \times R_q
\end{aligned} \tag{11.7}
$$

In the second part, c'_2 is decomposed in base w such that

$$
c'_2 = \sum_{i=0}^{l} c'^{(}_2 i) w^i \tag{11.8}
$$

where each $c'^{(i)}_2$ is a polynomial with its coefficients taken from the set $\{0, 1, 2, \ldots, w-1\}$

Finally, the product is computed by

$$
f_0 = c'_0 + \sum_{i=0}^{l} evk_{i,0} c'^{(i)}_2 \tag{11.9}
$$

$$
f_1 = c'_1 + \sum_{i=0}^{l} evk_{i,1} c'^{(i)}_2 \tag{11.10}
$$

where $evk_{i,0}$ and $evk_{i,1}$ are the first and second elements of evk_i, respectively. The output $(f_0, f_1) \in R_q \times R_q$.

11.2.3 FV noise growth

In practice, the error distribution χ is fixed as a discrete Gaussian distribution with mean zero and standard deviation σ over the interval $[-B, B]$, where B can be selected to be an integer approximating 10σ [28]. The decryption equation can be expanded to

$$
\begin{aligned}
&\frac{t}{q}\left(c_0 + c_1 s \bmod q\right) \\
&= \frac{t}{q}\left(\Delta m + eu + e_1 + e_2 s + Kq\right) \\
&= m + v + Kt
\end{aligned} \tag{11.11}
$$

where K is a random integer and $v = (t/q)\left(eu + e_1 + e_2 s - (q/t - \Delta)\, m\right)$ is a polynomial with degree less than n and possesses rational number coefficients. The message m can be recovered correctly by rounding it to the nearest integer and modulo reduced t if and only if the absolute maximum value of the coefficients of v is less than 0.5. Here, v is referred to as the noise component of an FV ciphertext.

Homomorphic addition and multiplication cause noise growth. The analysis from [29] is used to provide an illustration of the effect of noise growth on the size of the ciphertext. In [29], the author relied on the infinity norm in the canonical embedding [30] to derive a strict upper bound estimation for the noise growth caused by the homomorphic operations. For a fresh ciphertext ct of a plaintext message $m \in R_t$, with very high probability, the canonical norm of the noise component v of ct satisfies

$$\|v\|^{can}_{initial} \leq \frac{42\sigma tn}{q} \tag{11.12}$$

where N_m is the number of nonzero coefficients in the polynomial m.

Let ct_1 and ct_2 be the two valid ciphertexts. Let ct_{add} and ct_{mul} be the ciphertexts resulting from the homomorphic addition and multiplication of ct_1 and ct_2, respectively. With very high probability, the canonical norms of the noise components v_{add} of ct_{add} and v_{mul} of ct_{mul} satisfy

$$\|v_{add}\|^{can} \leq \|v_1\|^{can} + \|v_2\|^{can} \tag{11.13}$$

$$\|v_{mul}\|^{can} \leq 14tn \max\left\{\|v_1\|^{can}, \|v_2\|^{can}\right\} \tag{11.14}$$

To analyze the noise growth caused by the homomorphic function evaluation, the authors of [29] use a benchmark to compute A level of additions, followed by one level of multiplication, and the operations are iterated D times. From (11.12)–(11.14), with very high probability, the noise component of the final ciphertext satisfies

$$\left\|v_{final}\right\|^{can} \leq \left(14tn2^A\right)^D \frac{42\sigma tn}{q} \tag{11.15}$$

For correct decryption, the norm of the noise component of the final ciphertext must be less than 0.5, which means

$$q > \left(14tn2^A\right)^D 84\sigma tn. \tag{11.16}$$

11.2.4 Parameter selection

For parameter standardization, the standard deviation σ is fixed as 3.2 [28]. With a fixed σ, the security of FV scheme is primarily determined by the polynomial degree n and the integer modulus q. For a fixed n, smaller q means higher security and vice versa. However, smaller q is not suitable for tasks with large multiplicative depth D, as indicated in (11.16). Table 11.1 provides several recommended values of the modulus q for different values of degree n and security levels in terms of number of bits [28].

To evaluate a specific function with homomorphic encryption, both security and correctness need to be assured. We assume that all inputs, outputs, as well as intermediate results of the target function can be encoded as polynomials of R_t. For the state-of-the-art encoding methods, we refer the readers to [29,31]. Through these methods, the degree n and the plaintext modulus t are determined. In the next step, (11.16) and Table 11.1 are employed to determine the smallest modulus q to ensure the correctness and security of homomorphic evaluation. As security level is at odd with the multiplicative depth of targeted function, in some cases, security level is sacrificed to allow for a higher multiplicative depth [31,32].

11.3 Polynomial multiplication

As presented in Section 11.2, homomorphic operations in FV scheme involve polynomial addition and multiplication in the ring $R_q = Z_q[x]/(x^n + 1)$. Homomorphic addition is easy to perform, because it is nothing but polynomial addition in the ring $R_q = Z_q[x]/(x^n + 1)$. All necessary computations are merely coefficient-by-coefficient modular q addition. On the other hand, homomorphic multiplication is more involved, as it consists of polynomial multiplications. The problem is escalated by the large degree n, as observed in Table 11.1. With such large value of n, naïve school-book method for polynomial multiplication is clearly inefficient due to its quadratic computational complexity.

11.3.1 Karatsuba algorithm

The first attempt to reduce the complexity of multiplication was due to Karatsuba [33]. His method is a recursive divide-and-conquer algorithm and its derivation is

Table 11.1 Recommended parameters of the polynomial degree n and ciphertext modulus q for different security levels with the standard deviation fixed as 3.2

n	Security level	Bit-length of q
1,024	128	29
	192	21
	256	17
2,048	128	55
	192	39
	256	51
4,096	128	106
	192	74
	256	58
8,192	128	208
	192	146
	256	114
16,384	128	415
	192	289
	256	226
32,768	128	831
	192	575
	256	449

given as follows. Let $a(x)$ and $b(x)$ be the two polynomials to be multiplied. Assume that the degree n of the polynomials is even. The product of $a(x)$ and $b(x)$ can be expressed as

$$\begin{aligned} & a(x) \times b(x) \\ & = \left(a_1(x)x^{n/2} + a_0(x)\right)\left(b_1(x)x^{n/2} + b_0(x)\right) \\ & = a_1(x)b_1(x)x^n + a_0(x)b_0(x) + (a_1(x)b_0(x) + a_0(x)b_1(x))x^{n/2} \end{aligned} \tag{11.17}$$

notice that

$$\begin{aligned} & a_1(x)b_0(x) + a_0(x)b_1(x) \\ & = (a_1(x) + a_0(x))(b_1(x) + b_0(x)) - a_1(x)b_1(x) - a_0(x)b_0(x) \end{aligned} \tag{11.18}$$

Substituting (11.18) into (11.17) gives

$$\begin{aligned} & a(x) \times b(x) \\ & = a_1(x)b_1(x)x^n + a_0(x)b_0(x) \\ & + ((a_1(x) + a_0(x))(b_1(x) + b_0(x)) - a_1(x)b_1(x) - a_0(x)b_0(x))x^{n/2} \end{aligned} \tag{11.19}$$

Equation (11.17) can be calculated with four multiplications and three additions of polynomials of degree $n/2$. Equation (11.19) can be calculated with only three multiplications but six additions of polynomials of degree $n/2$. Karatsuba's method reduces one multiplication at the cost of introducing three extra additions. This will still lead to hardware saving and a speedup in computation, as the cost and speed of polynomial addition is negligible compared to that of polynomial multiplication. It should be noted that Karatsuba's method can be applied recursively to further reduce the number of multiplications, as illustrated in Figure 11.2.

11.3.2 Number theoretic transform algorithm

The complexity of polynomial multiplication can be further minimized to quasilinear using number theoretic transform (NTT) algorithm. NTT algorithm is similar to fast Fourier transform (FFT) algorithm that is commonly used in the implementation of convolution for digital signal processing. Similar to FFT algorithm, NTT algorithm requires the existence of a root of unity. As FFT algorithm operates with the infinite ring of complex numbers, the root of unity always exists and is defined as $e^{i2\pi/N}$, where N is the number of FFT points. Indeed, FFT algorithm can be used for polynomial multiplication by choosing N in such a way that it is a power of two and greater than twice the maximum degree of the input polynomials. However, this approach is not suitable for polynomial multiplication in the ring $R_q = Z_q[x]/(x^n + 1)$, as the large values of n can lead to complicated round-off error analysis due to the floating-point arithmetic.

NTT algorithm avoids the complexity of floating-point arithmetic by replacing the complex root of unity $e^{i2\pi/N}$ by an integer root of unity in the integer ring Z_q, i.e., the set of q integers $\{0, 1, 2, 3, \ldots, q-2, q-1\}$. The arithmetic operations are performed modulo integer q. However, the number of NTT points N and the integer q must satisfy the constraint $q = 1 \mod N$ to guarantee the existence of the Nth integer root of unity w_N such that $w_N^N = 1 \mod q$. For example, with $q = 2^{64} - 2^{32} + 1$, 4,096 is the 16th root of unity in Z_q since $(4{,}096)^{16} = 1 \mod q$. The process of performing

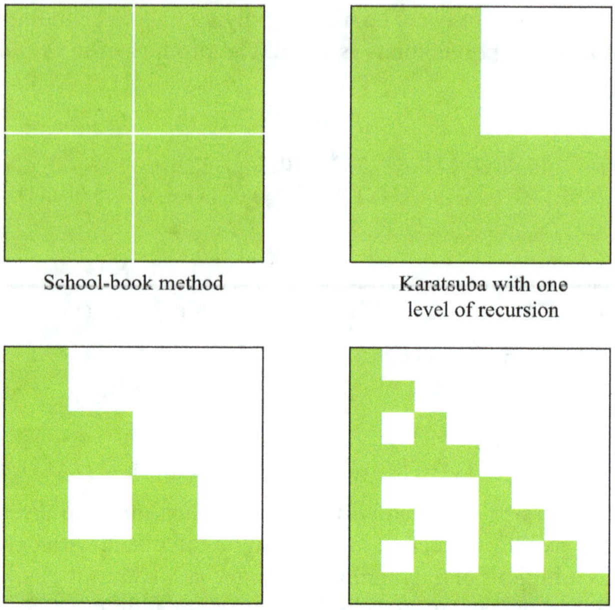

Figure 11.2 Cost (number of green squares) of Karatsuba multiplication algorithm for growing depth of recursion d [34]. The number of green squares is $(2^d)^{\log_2 3}$, where d is the recursion depth. Thus, the complexity of Karatsuba algorithm is estimated to be $O(n^{\log_2 3})$

NTT is similar to FFT in the complex domain. All it needs is to replace the root of unity in the complex domain $e^{i2\pi/N}$ by the integer root of unity in the finite ring w_N. Thus, the forward NTT is computed as

$$X_i = \sum_{j=0}^{N-1} x_j (w_N)^{ij} \pmod{q} \tag{11.20}$$

and the inverse NTT, or INTT is computed as

$$x_i = \frac{1}{N} \sum_{j=0}^{N-1} X_j (w_N)^{-ij} \pmod{q} \tag{11.21}$$

Given two polynomials, $\sum_i a_i x^i$ and $\sum_i b_i x^i$, belonging to $R_q = Z_q[x]/(x^n + 1)$, Figure 11.3 illustrates the procedure to perform their polynomial multiplication via the NTT algorithm. In order for NTT to work, N should be chosen as a power of two and greater than twice the maximum degree of the input polynomials. In practice, as the recommended values of n are all powers of two (see Table 11.1), we can simply set the number of NTT points $N = 2n$.

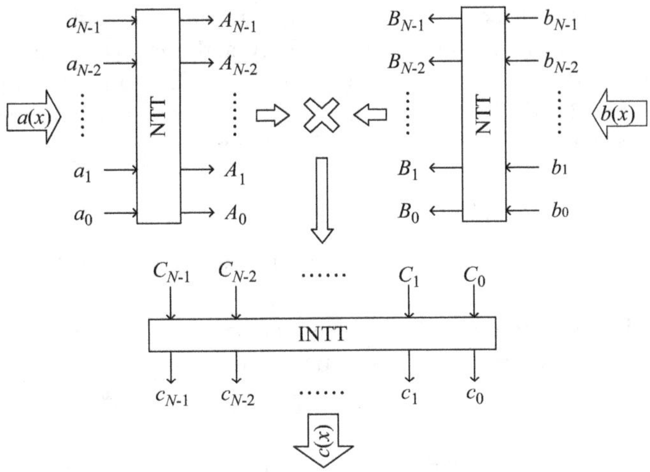

Figure 11.3 *N-point NTT-based polynomial multiplication. To compute $c(x) = a(x) \times b(x)$, (11.20) is employed to transform a(x) and b(x) to a vector $(A, A_1, A_2 \ldots)$ and a vector $(B, B_1, B_2, \ldots)$, respectively. The vector $(C, C_1, C_2, \ldots)$ is determined by point-wise multiplication of vectors $(A, A_1, A_2 \ldots)$ and $(B, B_1, B_2, \ldots)$. i.e., $C_0 = A_0 \times B_0$, $C_1 = A_1 \times B_1$, $C_2 = A_2 \times B_2$, etc. From vector $(C, C_1, C_2, \ldots)$, (11.21) is applied to recover the result c(x). Since NTT is defined over the ring $R_q = Z_q[x]/(x^n + 1)$, all arithmetic operations involved are modulo q, and c(x) must be modulo reduced by $(x^n + 1)$ if its degree is greater than or equal to n*

11.4 Residue number system

An RNS is defined by a set of K coprime integer numbers $\{m_1, m_2, m_3, \ldots, m_K\}$, i.e., the greatest common divisor of any pair of integers belonging to this set is equal to 1. The dynamic range M of an RNS so defined is the product of all integers in the set $\{m_1, m_2, m_3, \ldots, m_K\}$. An integer X in the range $[0, M)$ can be uniquely represented by a K-tuple $(x_1, x_2, x_3, \ldots, x_K)$, where each residue x_i can be determined by $X \mod m_i$, i.e., the remainder of dividing X by the modulus m_i. The K-tuple $(x_1, x_2, x_3, \ldots, x_K)$ is called the residue representation of X with respect to the base $\{m_1, m_2, m_3, \ldots, m_K\}$.

Let Z be the result of a modulo M arithmetic operation $\diamond$ on X and Y, where $\diamond$ can be an integer addition, an integer subtraction or an integer multiplication. The residue representations of X, Y and Z satisfy the following set of equations:

$$\begin{aligned} z_1 &= x_1 \diamond y_1 \mod m_1 \\ z_2 &= x_2 \diamond y_2 \mod m_2 \\ z_3 &= x_3 \diamond y_3 \mod m_3 \\ &\vdots \\ z_{K-1} &= x_{K-1} \diamond y_{K-1} \mod m_{K-1} \\ z_K &= x_K \diamond y_K \mod m_K \end{aligned} \tag{11.22}$$

Thus, RNS can be used to break very large integer arithmetic into many small residue operations to be executed independently and concurrently to improve the system's latency and throughput. For example, consider the moduli set $\{m_1, m_2, m_3\} = \{63, 64, 65\}$ for K=3. Let $X = 12{,}345$ and $Y = 54{,}321$. The residue representations of X and Y with respect to the base $\{63, 64, 65\}$ are computed as

$$\begin{aligned}(x_1, x_2, x_3) &= (X \text{ mod } 63,\ X \text{ mod } 64,\ X \text{ mod } 65) = (60, 57, 60)\\ (y_1, y_2, y_3) &= (Y \text{ mod } 63,\ Y \text{ mod } 64,\ Y \text{ mod } 65) = (15, 49, 46)\end{aligned}$$

Let $\diamond$ be the integer addition. The residue representation of $Z = X + Y = 66{,}666$ with respect to the base $\{63, 64, 65\}$ is computed from the residue representation of X and Y as follows:

$$\begin{aligned}(z_1, z_2, z_3) &= (x_1 + y_1 \text{ mod } 63,\ x_2 + y_2 \text{ mod } 64,\ x_3 + y_3 \text{ mod } 65)\\ &= (12, 42, 41)\end{aligned}$$

It can be easily verified that (z_1, z_2, z_3) is the correct representation of Z with respect to $\{63, 64, 65\}$ since:

$$(66666 \text{ mod } 63,\ 66666 \text{ mod } 64,\ 66666 \text{ mod } 65) = (12, 42,\ 41)$$

Given an arbitrary RNS base $\{m_1, m_2, m_3, \ldots, m_K\}$ and a residue representation $(x_1, x_2, x_3, \ldots, x_K)$, the binary representation of the integer X corresponding to this base can be computed from the Chinese remainder theorem (CRT) [35] by one of the following equations:

$$X = \left| \sum_{i=1}^{K} M_i \left| M_i^{-1} \right|_{m_i} x_i \right|_M \tag{11.23}$$

$$X = \left| \sum_{i=1}^{K} M_i \left| M_i^{-1} x_i \right|_{m_i} \right|_M \tag{11.24}$$

where $|a|_m$ denotes a mod m. $M_i = M/m_i$ and $\left|M_i^{-1}\right|_{m_i}$ is the multiplicative inverse of M_i mod m_i.

The difference between (11.23) and (11.24) lies in the number of subtractions of M from the weighted sum to carry out the modulo M operation. In (11.24), since $\left|M_i^{-1}x_i\right|_{m_i} < m_i$ and $M_i\left|M_i^{-1}x_i\right|_{m_i} < M$, the weighted sum is about K times greater than M. As a consequence, its modulo M reduction requires at most K subtractions of M. Equation (11.23), in contrast, requires many more subtractions of M from the weighted sum to complete the modulo M operation in the worst case. However, unlike (11.24), it requires no modulo multiplication to evaluate the weighted sum.

Consider the base $\{m_1, m_2, m_3\} = \{63, 64, 65\}$ as an example. Let $(x_1, x_2, x_3) = (23, 37, 41)$. Thus, M=262,080, M_1=4,160, M_2=4,095, M_3=4,032, $|M_1^{-1}|_{63} = 32$, $|M_2^{-1}|_{64} = 63$, $|M_3^{-1}|_{65} = 33$. Applying (11.23) gives

$$\begin{aligned} X &= |4{,}160 \times 32 \times 23 + 4{,}095 \times 63 \times 37 + 4{,}032 \times 33 \times 41|_{262{,}080} \\ &= |18{,}062{,}501|_{262{,}080} \\ &= 241{,}061 \end{aligned}$$

Applying (11.24) gives

$$\begin{aligned} (x_1, x_2, x_3) &= (23, 37, 41) \\ X &= |4{,}160 \times |32 \times 23|_{63} + 4{,}095 \times |63 \times 37|_{64} + 4{,}032 \\ &\quad \times |33 \times 41|_{65}|_{262{,}080} \\ &= |4{,}160 \times 43 + 4{,}095 \times 27 + 4{,}032 \times 53|_{262{,}080} \\ &= |503{,}141|_{262{,}080} \\ &= 241{,}061 \end{aligned}$$

11.5 Hardware accelerators

From Section 11.2, it is clear that homomorphic multiplication is the most time-consuming operation. Accelerating performance-critical cryptographic computations by exploiting parallel computations on hardware platform has been a common practice throughout the history of cryptology. Before discussing the state-of-the-art hardware architectures for accelerating homomorphic multiplication, a brief description of the homomorphic multiplication in YASHE [36]—a homomorphic encryption scheme that is very similar to FV, is provided. Like FV, YASHE also operates in the polynomial ring $R_q = Z_q[x]/(x^n + 1)$. YASHE used to be a favorite scheme in the cryptography community. However, after a recent so-called subfield attack that significantly reduces its security [37], YASHE started to lose its popularity to FV.

A YASHE ciphertext consists of only one element in R_q while an FV ciphertext contains two R_q elements. Given two YASHE ciphertexts, c_1 and c_2, their homomorphic multiplication is performed in two steps. The first step computes

$$c'_{mult} = \left[\frac{t}{q} c_1 c_2 \right] \bmod q \tag{11.25}$$

In the second step, c'_{mult} is decomposed in base w such that:

$$c'_{mult} = \sum_{i=0}^{l} c'^{(i)}_{mult} w^i \tag{11.26}$$

where each $c'^{(i)}_{mult}$ is a polynomial with coefficients taken from the set $\{0, 1, 2, \ldots, w-1\}$. The resulting product is given by

$$c_{mult} = \sum_{i=0}^{l} evk_i c'^{(i)}_{mult} \in R_q \tag{11.27}$$

where the evaluation key, $evk = \{evk_0, evk_1, \ldots, evk_l\}$, is a set of l elements in R_q. Comparing (11.7)–(11.10) with (11.25)–(11.27), it can be seen that the homomorphic multiplication procedure of YASHE involves the same type but a smaller number of operations than FV. This explains why YASHE has been a target for the design of several hardware accelerators before the subfield attack. The efficient implementations of FV as well as YASHE homomorphic multiplications depend heavily on the polynomial degree n and the modulus q. Different combinations of n and q will lead to different trade-offs in their hardware implementations.

11.5.1 Accelerating homomorphic encryption with number theoretic transform and Solinas prime

In [38], the authors implemented the homomorphic multiplication procedure of YASHE in an FPGA-based data center accelerator originally used for accelerating the Bing search engine [39]. The strategy they adopted is a combination of NTT algorithm and Solinas prime number [40]. While NTT algorithm helps to minimize the complexity for polynomial algorithm, setting the ciphertext q as a Solinas prime number of the form $2^y - 2^z + 1$ eases the fulfillment of the existence of an integer root of unity for NTT and supports efficient modulo reduction. As an example, let $q = 2^{64} - 2^{32} + 1$ and z be an arbitrary 128-bit number. z can be expressed as

$$z = 2^{96}a + 2^{64}b + 2^{32}c + d \tag{11.28}$$

where a, b, c and d are 32-bit numbers. By using the special number theoretic property of q, z mod q can be computed rapidly by

$$\begin{aligned} z &= 2^{96}a + 2^{64}b + 2^{32}c + d \\ &= (-1)a + (2^{32} - 1)b + 2^{32}c + d \\ &= 2^{32}(b + c) - a - b + d \pmod{q} \end{aligned} \tag{11.29}$$

This special value of q enables an efficient architecture for modulo q multiplier as shown in Figure 11.4.

Equations (11.25)–(11.27) are referenced to implement homomorphic multiplication in YASHE. Equation (11.26) can be implemented by a simple bit extractor since the base w is usually chosen as a power of two. Equation (11.27) can be executed by the NTT algorithm with the ciphertext modulus q. Special attention needs to be paid to (11.25), as it is more difficult to implement. In (11.25), c_1c_2 should be first computed without the reduction modulo q. The result is then scaled by t/q and rounded before it is reduced modulo q. To prevent overflow, the NTT for this equation should be used with a modulus Q such that $Q > n^2q$. Then, c_1c_2 is computed as

$$c_1c_2 = INTT_Q(NTT_Q(c_1) \odot NTT_Q(c_2)) \tag{11.30}$$

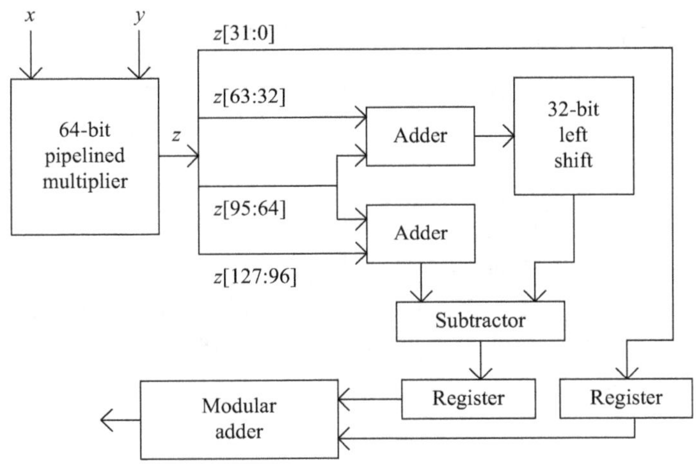

Figure 11.4 Efficient modulo multiplier design for Solinas number $q = 2^{64} - 2^{32} + 1$ [41]

Table 11.2 Parameter sets of YASHE with supported multiplicative depth

Set	***n***	***q***	***Q***	***l***	***t***
I	4,096	$2^{124}-2^{64}+1$	$2^{262}-2^{56}+1$	2	2^{10}
II	16,384	$2^{512}-2^{32}+1$	$2^{1,040}-2^{32}+1$	8	2^{10}

where $\odot$ denotes the point-wise multiplication. Table 11.2 provides the parameter sets for the implementation of [38].

Figure 11.5 shows the block diagram of the hardware architecture used to implement YASHE. The Stratix V GS D5 (5GSMD5) FPGA (dotted line box) is connected to two 4 GB DDR3-1333 SO-DIMM memory modules and communicates with the host computer via the PCIe (Peripheral Component Interconnect Express) interface. One DDR3 memory module stores the input/output as well as the intermediate polynomial coefficients, while the other stores the constants, including the twiddle factors, i.e., the powers of root of unity, and the NTT form of the evaluation key. As the evaluation key stays the same for every homomorphic multiplication, storing it in NTT form will save a considerable number of forward and INTTs. Taking the linearity of NTT into consideration, (11.27) can be expressed as

$$c_{mult} = INTT_q\left(\sum_{i=0}^{l} NTT_q\left(c'^{(i)}_{mult}\right) \odot EVK_i\right) \in R_q \tag{11.31}$$

where EVK_i is the forward NTT of evk_i in (11.27).

Two data movers are responsible for moving the data between the external memory modules and the FPGA memories. The NTT butterfly is composed of modulo

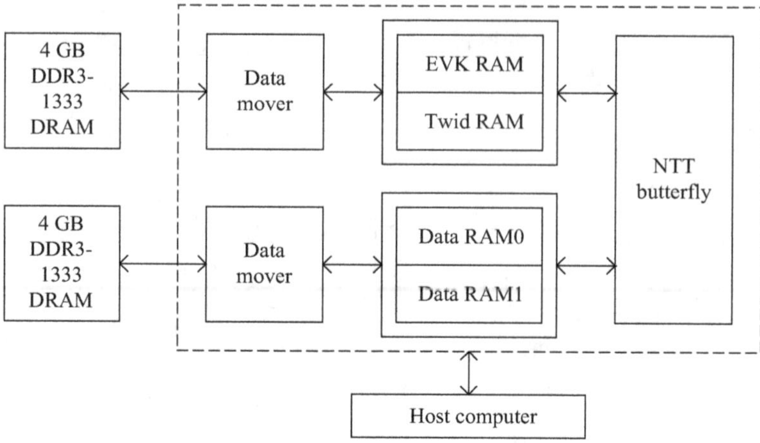

Figure 11.5 The block diagram of the hardware accelerator for homomorphic multiplication in YASHE

Table 11.3 Runtime of different operations in YASHE homomorphic multiplication procedure

Implementation	**Frequency (MHz)**	**Equation (11.25) (ms)**	**Equation (11.27) (ms)**
Set I	100	6.75	4.79
Set II	66	48.67	20.08

arithmetic units for two moduli, q and Q. As the FPGA memory capacity is limited, a cached NTT algorithm is adopted [42]. The idea is to process as many coefficients as possible before moving them from the onboard memories to the external modules.

The performance of the accelerator in [38] is shown in Table 11.3. Notice that with Set I, the number of polynomial multiplications for (11.27) is two, while with Set II, this number increases to eight. While the difference between the time taken for (11.25) and (11.27) is not obvious in Set I, the time taken for (11.25) is more than twice the time taken for (11.27) in Set II. The reason behind this is the arithmetic operations with multi-precision operands in Set II. To be specific, the implementation of Set II requires thousand bits wide adders and multipliers. For such large arithmetic units, it is very difficult to optimize them manually or automatically by FPGA compilers, let alone their added constraints to the backend placement and routing. The work in the following sections circumvents these performance stumbling blocks by exploiting RNS to reduce the cost of multi-precision arithmetic.

11.5.2 Accelerating homomorphic encryption with number theoretic transform and residue number system

In [25], the authors targeted a larger parameter set for YASHE than those presented in [38]. Their goal is to achieve a multiplicative depth large enough to support the evaluation of a lightweight block cipher on encrypted data. In particular, for this parameter set, the plaintext modulus t is 2, the polynomial degree n is 32,768 and the ciphertext modulus q is a 1,228-bit number. Consequently, the modulus Q for (11.25) to be realized by the NTT is chosen to be a 2,517-bit number to prevent integer overflow. Such value of Q leads to significant challenges in the design of an efficient multiplier. A bit-parallel 2,517-bit by 2,517-bit multiplier will incur huge silicon area and large delay. On the other hand, a word-serial multiplier will be even slower than a bit-parallel one.

This problem can be fixed by adopting RNS. By breaking large operand size arithmetic into many small residue operations to be executed in parallel and independently, the long computation time of multi-precision arithmetic can be overcome. To satisfy the requirement for the existence of an integer root of unity, q should be chosen as a product of K prime numbers q_i, i.e., $q = \prod_{i=1}^{K} q_i$, and each prime q_i must satisfy the condition $q_i = 1 \mod 2n$. In [25], the authors set q and Q as the product of forty-one and eighty-four 30-bit primes starting from 1008795649. The NTT–RNS polynomial multiplication algorithm is illustrated in Figure 11.6.

Consider two very high degree and very large coefficient polynomials, $a(x)$ and $b(x)$. Their product $c(x)$ can be calculated by the NTT–RNS algorithm as follows. First, $a(x)$ and $b(x)$ are converted into their equivalent RNS form. For each of them, a set of K high degree polynomials $a_i(x)$ and $b_i(x)$ with relatively small coefficients is generated. Second, each $a_i(x)$ is multiplied by the corresponding $b_i(x)$ to obtain $c_i(x)$

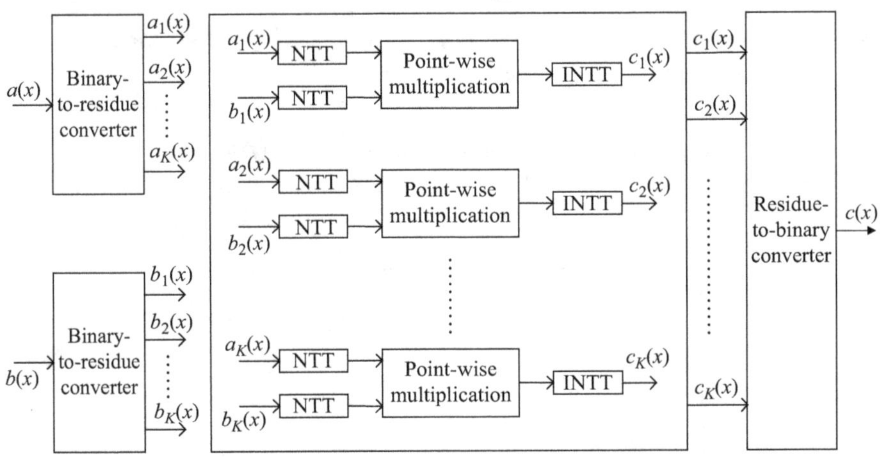

Figure 11.6 NTT–RNS polynomial multiplication algorithm

by the NTT algorithm. Lastly, $c(x)$ is recovered from the set of $c_i(x)$ by residue-to-binary conversion. Figure 11.7 shows the overall hardware architecture [25] to realize YASHE homomorphic multiplication procedure where each processor is dedicated to handle a single multiplication $c_i(x) = a_i(x) \times b_i(x)$. The performance result of this architecture is summarized in Table 11.4.

The use of RNS to decompose large coefficient polynomials opens out large design space exploration for the implementation of independent parallel processors to ease the complexity of placement-and-routing and performance tuning. However, RNS also introduces the expensive CRT operations as can be seen in Table 11.4. Recall that in (11.25), since the input polynomials, $c_1(x)$ and $c_2(x)$, are in R_q, small CRT is required to change their residue representations from one that corresponds to q to one that corresponds to Q. After the multiplication of $c_1(x)$ and $c_2(x)$, the

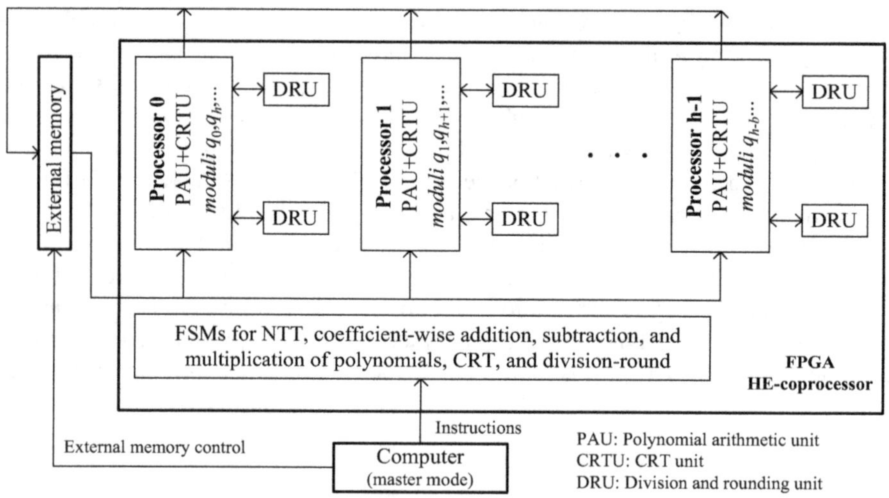

Figure 11.7 Parallel hardware architecture for in-place NTT–RNS polynomial multiplication algorithm

Table 11.4 Performance results at 143 MHz of NTT–RNS hardware architecture after placement and routing

Operation	**Time (ms)**
NTT	0.334
INTT	0.363
Polynomial addition/subtraction	0.029
Small CRT	5.743
Large CRT	19.248
Divide-and-round	19.678

result is a set of 84 residue polynomials, and a large CRT is required to convert these residue polynomials into a polynomial with the coefficients modulo Q ready to be scaled and rounded by t/q. Taking into consideration the fact that each Q is a product of 84 primes of 30 bits wide, and the large CRT that involves arithmetic operations on operands of size up to nearly 2,500 bits from (11.24), there is a limit on reaping the benefit of RNS computations. In particular, the cost of reduction modulo Q is not trivial when Q possesses no friendly arithmetic property.

Besides large CRT, there is another expensive trade-off when working with large parameter set that supports NTT. After performing the large CRT, each of the coefficients of the resulting polynomial should be scaled and rounded by t/q. The strategy adopted in [25] is to precompute t/q and multiply this precomputed term with the coefficient. To guarantee correctness for rounding, t/q should be precomputed with very high precision, since q is a 1,228-bit number, and each of the coefficients to be scaled can be 2,500 bits wide. According to [25], t/q should be precomputed up to a precision of 3,721 bits. This precomputation of scaling and rounding operation requires a 2,500-bit by 3,721-bit multiplier, which is not only costly in terms of silicon area but also adds to the critical path delay and becomes the performance bottleneck.

11.5.3 Accelerating homomorphic encryption with Karatsuba algorithm

In [43], the authors proposed a hardware/software codesign approach to accelerate the homomorphic multiplication in FV defined by (11.7)–(11.10). Instead of employing the NTT algorithm that offers the lowest asymptotic complexity for polynomial algorithm, the authors of [43] employed the Karatsuba algorithm. The NTT algorithm requires the ciphertext modulus q to satisfy a specific constraint, which is the main cause of the inefficient scale-and-round operation. The advantage of Karatsuba algorithm is that there is no specific constraint that q must satisfy. Thus, the plaintext modulus t and ciphertext modulus q can be chosen as power-of-two numbers to simplify the scale-and-round operation into bit extraction operation.

With the Karatsuba algorithm, the hardware/software codesign in [43] to accelerate FV homomorphic multiplication works as follows: the software module recursively applies (11.19) to create as many sub-polynomial products as possible and send them to the hardware module. The hardware module, which is typically an FPGA chip, is configured in such a way that it can support many sub-polynomial multipliers in parallel. The intermediate results from the hardware module are then sent back to the software module to reconstruct the final result. Note that the computation assigned to the software module involves only simple polynomial addition. The comparison between the Karatsuba-based accelerator [43] and NTT-based accelerator [38] is shown in Table 11.5.

The results from Table 11.5 show that for small parameter set, the Karatsuba-based accelerator is clearly preferred to the NTT-based accelerator. In particular, the Karatsuba-based accelerator consumes about half of the FPGA resources and achieves more than 20% improvement in performance compared to the NTT-based accelerator. However, the Karatsuba-based accelerator starts to lose its competitiveness for large

Table 11.5 Comparison between Karatsuba-based and NTT-based accelerators for different setup (n, $\log_2 q$)

Design	Karatsuba	FFT	Karatsuba	FFT
n, $\log_2 q$	2,560, 125	4,096, 125	6,144, 512	16,384, 512
ALM	29,534	69,058	153,642	141,090
Register	76,823	144,747	329,235	391,773
Memory bits	4,131	8,031,568	301,645	17,626,400
DSPs	100	144	456	577
Frequency (MHz)	331.13	100	262.47	66
YASHE (ms)	4.73	6.75	52.7	47.54
FV (ms)	11.9	15.46	124.36	122.30

parameter set due its higher asymptotic complexity. The statistics in the two rightmost columns of Table 11.5 indicate that both Karatsuba-based and NTT-based accelerators have equivalent hardware resources consumption and comparable computing time.

11.6 Conclusion

This chapter has discussed a recent popular construction of homomorphic encryption. Due to the effect of noise growth, the performance of homomorphic encryption is limited by the very high degree and very large coefficient polynomial multiplication. While the Karatsuba and NTT algorithms reduce the complexity of very high-degree polynomial multiplication, RNS helps mitigating the cost of very large coefficient arithmetic. Different hardware accelerators employing various algorithms have their own advantages and disadvantages depending on the choice of parameters. In several recent works, to apply homomorphic encryption for heart rate attack prediction based on encrypted medical data, it requires very large parameter sets [31,32]. This implies that the NTT and NTT–RNS-based accelerators are more suitable for real applications. There is also an extensive research effort to reduce the size of the recommended parameter sets for homomorphic encryption in the future [44], which may pave the way for a comeback of the Karatsuba-based accelerator.

References

[1] R. L. Rivest, L. Adleman, and M. L. Dertouzos, "On data banks and privacy homomorphisms", in *Foundations of Secure Computation*, R. A. Demillo *et al.*, Eds. New York, NY: Academic Press, 1978, pp. 169–179.

[2] R. Rivest, A. Shamir, and L. Adleman, "A method for obtaining digital signatures and public-key cryptosystems", *Communications of the ACM*, vol. 21, pp. 120–126, 1978.

[3] S. Goldwasser and S. Micali, "Probabilistic encryption", *Journal of Computer and System Sciences*, vol. 28, pp. 270–299, 1984.

[4] T. ElGamal, "A public key cryptosystem and a signature scheme based on discrete logarithms", *IEEE Transactions on Information Theory*, vol. 31, pp. 469–472, 1985.

[5] J. Benaloh, "Dense probabilistic encryption", in *Proceedings of Workshop on Selected Areas of Cryptography*, Kingston, Canada, pp. 120–128, 1994.

[6] P. Paillier, "Public-key cryptosystems based on composite degree residuosity classes", in *Proceedings of EUROCRYPT 1999*, Prague, Czech Republic, pp. 223–238, May 1999.

[7] O. Baudron, P-A. Fouque, D. Pointcheval, J. Stern, and G. Poupard, "Practical multi-candidate election system", in *Proceedings of the 20th Annual ACM Symposium on Principles of Distributed Computing*, RI, USA, pp. 274–283, Aug. 2001.

[8] Synergy Research Group, "2015 review shows $110 billion Cloud market growing at 28% annually | Synergy Research Group", *Srgresearch.com*, 2019. [Online]. Available: https://www.srgresearch.com/articles/2015-review-shows-110-billion-cloud-market-growing-28-annually.

[9] Cisco.com. [Online]. Available: https://www.cisco.com/c/dam/m/en_us/service-provider/ciscoknowledgenetwork/files/622_11_15-16-Cisco_GCI_CKN_2015-2020_AMER_EMEAR_NOV2016.pdf.

[10] J. Wong, "The Cambridge Analytica scandal changed the world – but it didn't change Facebook", *The Guardian*, 2019. [Online]. Available: https://www.theguardian.com/technology/2019/mar/17/the-cambridge-analytica-scandal-changed-the-world-but-it-didnt-change-facebook.

[11] I. Tham, "Singapore's privacy watchdog fines IHiS $750,000 and SingHealth $250,000 for data breach" *The Straits Times*, 15-Jan-2019. [Online]. Available: https://www.straitstimes.com/singapore/singapores-privacy-watchdog-fines-ihis-750000-singhealth-250000-for-data-breach.

[12] "Homomorphic Encryption Applications and Technology (HEAT)", *Heat-project.eu* [Online]. Available: https://heat-project.eu.

[13] "PROgramming Computation on EncryptEd Data (PROCEED)", *Darpa.mil*. [Online]. Available: https://www.darpa.mil/program/programming-computation-on-encrypted-data.

[14] Cherry Teh TL, "Two research centres at NUS, NTU set up to develop data protection tech in Smart Nation era", *The Straits Times*, 2019. [Online]. Available: https://www.straitstimes.com/singapore/strengthening-cyber-security-defences.

[15] C. Gentry, "Fully homomorphic encryption using ideal lattices" in *41st ACM Symposium on Theory of Computing (STOC)*, ACM, New York, NY, USA, pp. 169–178, 2009.

[16] C. Gentry and S. Halevi, "Implementing Gentry's fully homomorphic encryption scheme", in *Proc. Advances in Cryptology – EUROCRYPT 2011*, Tallinn, Estonia, pp. 128–149, May 2011.

[17] W. Wang, Y. Hu, L. Chen, X. Huang, and B. Sunar, "Exploring the feasibility of fully homomorphic encryption", *IEEE Transactions on Computers*, vol. 64, no. 3, pp. 698–706, 2015.

[18] Z. Brakerski, C. Gentry, and V. Vaikuntanathan, "(Leveled) fully homomorphic encryption without bootstrapping" in *Proc. 3rd Innov. Theor. Comput. Sci. Conf.*, New York, NY, USA, pp. 309–325, Jan. 2012.

[19] J. Fan and F. Vercauteren, "Somewhat practical fully homomorphic encryption" *IACR Cryptology ePrint Archive*, vol. 2012, p. 144, 2012.

[20] K. Nuida and K. Kurosawa, "(Batch) fully homomorphic encryption over integers for non-binary message spaces" in *Proc. 34th Annu. Int. Conf. Theory Appl. Cryptographic Tech.*, Sofia, Bulgaria, pp. 537–555, Apr. 2015.

[21] V. Vaikuntanathan, "Computing blindfolded: New developments in fully homomorphic encryption", in *Proceedings of FOCS 2011*, CA, USA, pp. 5–16, Oct. 2011.

[22] C. Aguilar-Melchor, S. Fau, C. Fontaine, G. Gogniat, and R. Sirdey, "Recent advances in homomorphic encryption: A possible future for signal processing in the encrypted domain", *IEEE Signal Processing Magazine*, vol. 30, no. 2, pp. 108–117, 2013.

[23] C. Moore, M. O'Neill, E. O'Sullivan, Y. Doroz, and B. Sunar, "Practical homomorphic encryption: A survey", in *Proceedings of 2014 IEEE International Symposium on Circuits and Systems*, Melbourne, Australia, pp. 2792–2795, Jun. 2014.

[24] T. Lepoint and M. Naehrig, "A comparison of the homomorphic encryption scheme FV and YASHE", in *Proceedings of AFRICACRYPT 2014*, Marrakesh, Morocco, pp. 318–335, May 2014.

[25] S. S. Roy, K. Jarvinen, F. Vercauteren, V. Dimitroz, and I. Verbauwhede, "Modular hardware architecture for somewhat homomorphic function evaluation", in *Proceedings of Cryptographic Hardware and Embedded Systems-CHES 2015*, Saint-Malo, France, pp. 164–184, Sep. 2015.

[26] B. Applebaum, D. Cash, C. Peikert, and A. Sahai, "Fast cryptographic primitives and circular-secure encryption based on hard learning problems", in *Proceedings of CRYPTO 2009*, Santa Barbara, CA, USA, pp. 595–618, Aug. 2009.

[27] V. Lyubashevsky, C. Peikert, and O. Regev, "On ideal lattices and learning with errors over rings", in *Proceedings of EUROCRYPT 2010*, French Riviera, Monaco, pp. 1–23, May 2010.

[28] "Standard – Homomorphic Encryption Standardization", *Homomorphicencryption.org*, 2019. [Online]. Available: https://homomorphicencryption.org/standard/.

[29] H. Chen, K. Laine, R. Player, and Y. Xia, "High-precision arithmetic in homomorphic encryption", in *Proc. CT-RSA 2018*, San Francisco, CA, USA, pp. 116–136, Apr. 2018.

[30] A. Costache and N. P. Smart, "Which ring-based somewhat homomorphic encryption scheme is best?", in *Topics in Cryptology – CT-RSA 2016*, Cham: Springer, vol. 9610, pp. 325–340, 2016.

[31] J. H. Cheon, J. Jeong, J. Lee, and K. Lee, "Privacy-preserving computations of predictive medical models with minimax approximation and non-adjacent form" in *Proc. FC 2017*, Sliema, Malta, pp. 53–74, Apr. 2017.

[32] J. Bos, K. Lauter, and M. Naehrig, "Private predictive analysis on encrypted medical data" *Journal of Biomedical Informatics*, vol. 50, pp. 234–243, 2014.

[33] A. Karatsuba and Y. Ofman, "Multiplication of multidigit numbers on automata", in *Doklady Akademii Nauk SSSR*, vol. 145, no. 2, pp. 293–294, 1962 (in Russian). English translation in *Soviet Physics-Doklady 7*, pp. 595–596, 1963.

[34] J. von zur Gathen and J. Gerhard, Modern Computer Algebra, Cambridge, UK: Cambridge University Press, 2003.

[35] P. V. A. Mohan, Residue Number Systems: Algorithms and Architectures, Norwell, MA: Kluwer, 2002.

[36] J. W. Bos, K. Lauter, J. Loftus, and M. Naehrig, "Improved security for a ring-based fully homomorphic encryption scheme", in *Proceedings of IMACC 2013*, Oxford, UK, pp. 45–64, Dec. 2013.

[37] M. Albrecht, S. Bai, and L. Ducas, "A subfield lattice attack on overstretched NTRU assumptions: Cryptanalysis of some FHE and graded encoding schemes" in *Proc. 36th Annu. Int. Cryptology Conf. Advances Cryptology*, pp. 153–178, 2016.

[38] T. Popplemann, M. Naehrig, A. Putnam, and A. Macias, "Accelerating homomorphic evaluation on reconfigurable hardware", in *Proceedings of Cryptographic Hardware and Embedded Systems-CHES 2015*, Saint-Malo, France, pp. 143–163, Sep. 2015.

[39] A. Putnam, A. M. Caulfield, E. S. Chung, *et al.*, "A reconfigurable fabric for accelerating large-scale datacenter services", *Communications of the ACM*, vol. 59, no. 11, 2016.

[40] J. A. Solinas, "Generalized Mersenne numbers" *National Security Agency (NSA)*, Tech. Rep., Sep. 1999.

[41] W. Wang, X. Huang, N. Emmart, and C. Weems, "VLSI design of a large number multiplier for fully homomorphic encryption", *IEEE Transactions on Very Large Scale (VLSI) Systems*, vol. 22, no. 9, pp. 1879–1887, 2014.

[42] B. Baas, "A generalized cached-FFT algorithm", in *Proc. IEEE Int. Conf. Acoustics Speech Signal Processing (ICASSP'05)*, Philadelphia, PA, USA, pp. 89–92, May 2005.

[43] V. Migliore, M. M. Real, V. Lapotre, A. Tisserand, C. Fontaine, and G. Gogniat, "Hardware/software co-design of an accelerator for FV homomorphic encryption scheme using Karatsuba algorithm", *IEEE Transactions on Computers*, vol. 67, no. 3, pp. 335–347, 2018.

[44] I. Chillotti, N. Gama, M. Georgieva, and M. Izabachene, "Faster fully homomorphic encryption: Bootstrapping in less than 0.1 seconds" in *Proc. 22nd Int. Conf. Theory Appl. Cryptology Inf. Secur. 2016*, Hanoi, Vietnam, pp. 3–33, Dec. 2016.

Chapter 12

Information leakage from robust codes protecting cryptographic primitives

Osnat Keren[1] and Ilia Polian[2]

Cryptographic hardware primitives are subject to fault injection attacks. Security-oriented codes are considered the most powerful technique for detecting such attacks. They provide (probabilistic) guarantees for the detection of maliciously injected faults even under the assumption of a sophisticated attacker with access to powerful equipment. In this chapter, we revisit the earlier finding that error-detection infrastructure may increase the undesired information leakage. It is shown that the probability of an attack to be undetected by the best robust codes known today is exponentially smaller than the entropy loss due to information leak from the checker. This means that an attack will be detected far before the attacker can acquire significant information. The analysis is constrictive; given a bound for acceptable information leakage, it allows the designer to easily choose the number of redundant bits required to detect the attack before this bound is reached.

12.1 Introduction

Fault injection attacks on cipher circuits aim to extract the secret key by analyzing correct and faulty ciphertext pairs. Such attacks grow in importance as novel cyber-physical applications and autonomous systems are physically exposed to and accessible by their users and can be manipulated. A large number of popular cryptosystems have been successfully attacked, ranging from simple block ciphers (e.g., AES [1], PRESENT [2] and LED [3]) to elaborate asymmetric [4] and postquantum schemes [5,6].

It is impossible to prevent an adversary from trying to mount an attack, but it is possible to make it more difficult. Typical countermeasures against fault attacks are divided into two groups that involve hardening either the hardware or the computations. Hardening hardware method involves better concealment of the sensitive parts of the chip (e.g., by bus scrambling, memory encryption, additional metal layers for

[1]Faculty of Engineering, Bar-Ilan University, Ramat Gan, Israel
[2]Institute of Computer Engineering and Computer Architecture, University of Stuttgart, Stuttgart, Germany

passive shielding) and by adding filters to the power and clock signal and/or by embedding security devices such as light or temperature sensors. Hardening the resistance of computation techniques includes information redundancy infective computations, parallel execution or inverse execution and branchless implementations. One of the most effective ways to mitigate fault injection attacks is by embedding security-oriented codes in hardware. In particular, these codes are essential for mitigating faulty ciphertext-only attacks [7].

Any coding scheme is effective in detecting and responding to the errors for which it was designed. This constitutes the fundamental difference between reliability-oriented and security-oriented codes. Reliability-oriented codes are designed to detect/correct a small number of erroneous symbols (bits, nibbles or bytes). Therefore, their effectiveness is determined by the minimum distance between code words, which defines the number of errors that can be detected. By contrast, codes that are designed to cope with an arbitrary number of erroneous symbols are said to be security oriented. Security-oriented codes are designed to protect against sophisticated attackers that can select the error vector, as well as bypasser attackers that can induce an arbitrary number of errors by manipulating the voltage or clock frequency. For example, Figure 12.1 shows the proportion of high-multiplicity and low-multiplicity errors due to clock duration manipulations. It shows that it is not enough to detect low-multiplicity errors alone, since a variety of error multiplicity can occur. Therefore, a code distance alone does not guarantee the detection of malicious faults in instances where the adversary is free to use high fault multiplicity.

Security-oriented codes can utilize deterministic or random encoders. The effectiveness of codes with random-encoding, e.g., the codes in [9–12], depends on the entropy of the random portion. In practice, it is difficult and expensive to implement a true (i.e., maximal entropy) random number generator that cannot be neutralized by fault injection. This makes codes with deterministic encoding an attractive

Figure 12.1 Error multiplicity: the number of faults that caused i erroneous bytes in a small-scale (full-scale) AES circuit [8]

alternative [13]. In this chapter, we focus on information leakage from robust codes with deterministic encoding.

Recently, several authors have observed that fault-tolerant digital design techniques affect side-channel attack (SCA) resistance [14,15]. In [14], the authors proposed a combined single-fault location attack on redundancy-based countermeasures assisted by supplementary side-channel information on the Hamming weight of the manifested error. However, this attack strategy applies to software implementations of AES and PRESENT ciphers that are protected by time/space redundancy. Note that these attacks were mounted against specific redundancy schemes that were not robust; their success does not automatically imply weaknesses in other detection methods and in particular in robust code-based solutions considered in this work.

In this chapter, we review known constructions of security-oriented codes and examine whether the use of robust codes with a deterministic encoder can degrade security; that is, whether the checker's response can narrow the key's search space. The (average) reduction in the size of the search space is evaluated in terms of the mutual information (MI) between the secret key and the checker's response. The MI is analyzed as a function of the injected error; it represents the (average) number of bits that can be learned about the key from a single injection of a specific error vector. It will be shown that when robust codes are used, the entropy loss due to information leak from the checker is small and increases linearly in the number of redundant bits r, whereas the probability that the attack cannot be detected decreases exponentially. The same relationship holds for the dependency of the entropy loss and probability of an undetected attack on the number of fault injections. In other words, we can expect that providing strong fault detection only has a limited and exponentially decreasing impact on information leakage and thus on security.

For example, assume that a 64-bit secret key is protected with a security-oriented code with 9 redundant bits. Table 12.3 provides one suitable code for this purpose. From our analysis (Section 12.5), it follows that a standard checker leaks 0.25562 bits per fault injection and that the probability that a fault injection will go undetected is 3.91×10^{-3}. Table 12.1 shows the total number of bits leaked after d fault injections and the probability that the attack will be detected. It can be seen that the number of attacks to extract a non-negligible amount of information is such that its detection is near to certain (e.g., masking probability of 10^{-30} for 16 fault injections needed for the extraction of 4 bits).

Table 12.1 Information leakage (bits) and cumulative probability of an undetected (masked) fault attack for a 64-bit key protected by 9 redundant bits

Number of faults d	**1**	**2**	**3**	**4**	**16**	**251**
Information leakage	0.25562	0.51124	0.76686	1.02248	4.08992	64
Masking probability	3.91E−03	1.53E−05	5.98E−08	2.34E−10	2.98E−39	<1E−604

12.2 Fault injection attacks

It has been noticed in [16] that faults that occur during a cryptographic operation can compromise its security. Fault injection attacks utilize this fact by deliberately inducing a physical disturbance during the execution of a cryptographic function. A plethora of fault injection techniques has been demonstrated, including inducing glitches on the circuit's inputs or clock lines; overheating; underpowering; applying electromagnetic pulses and illuminating the circuit with a laser [17,18]. The best known attacks, which need a single-fault injection to recover the complete secret key of a cipher [1,3], require very high temporal and spatial resolution, i.e., a fault injection in a precisely known clock cycle at a precisely known portion of the circuit. Even the most sophisticated fault injection techniques [19] currently do not provide such resolution, and as a consequence, an attack must be attempted several times. Robust codes used in this chapter make a conservative assumption of the attacker, namely, that the attacker can precisely select an arbitrary number of specific bits that will be flipped as a consequence of the fault injection. This is captured by defining the error as a binary vector that is added (XORed) to the circuit's outputs and errors that the attacker can choose.

Once a fault has been injected, the attacker can analyze the circuit's outputs (ciphertexts in the case of encryption) obtained in the presence and in the absence of the disturbance. Applying differential cryptanalysis techniques, secret key bits can be inferred. It is also possible to collect a larger number of the circuit's responses for faults of different intensities and to infer the key from a statistical analysis [20,21]. In this chapter, we make no assumption about which type of fault analysis the attacker will perform. Any injected fault that goes undetected by the robust code is considered critical and the codes aim at increasing the probability of detection. The question considered is whether knowing the check bits gives the attacker additional information about the secret key and how much such information leakage can occur.

Note that SCAs such as power analysis attacks are based on hypothesis testing. Thus, the computational complexity involved in conducting these attacks increases exponentially with the number of key-bits on which the hypothesis is made. Thus, although it is simpler to perform an SCA on sub-keys (nibbles/bytes), the power consumed while processing these sub-keys is smaller. This, in turn, decreases the SNR (i.e., the ratio between the energy of the signal that carries the information and the energy of the thermal and switching noises). The hardware implementation of codes for detecting fault injection attacks consumes power and hence alters the SNR.

It is often argued that deterministic code-based countermeasures against fault injection attacks can increase the information leakage via the side channels (such as power consumption and electromagnetic radiation). This claim is true to some extent; redundancy of any kind, as it is, carries information on the secret data it aims to protect. The question is how much exploitable information is carried by the redundant portion. In general, *duplication*-based redundancy or *repeated computation* increase the SNR. It also increases the correlation between the secret information and the side information. In some cases, the security level can still degrade even if more complex codes are employed. In [22], the authors studied the effect of *low-rate* error detection circuits on resistance to power analysis attacks. The evaluation that was carried on a single S-box circuit showed that an adversary who knows the coding scheme can take advantage of this additional information. An analysis of the MI between the secret 8-bit key and the power trace showed that this attacker could learn at most 0.5 bits (depending on the noise level), whereas an attacker who was not aware of the implemented code had a lower success rate. The observations presented in [22] apply to cases in which each S-box has its own redundant bits, i.e., when the code rate is low. In this chapter, we consider a different architecture in which the redundant bits are computed as *a function of the entire 64/128 state bits*. With a proper design, exploiting the information that leaks from the redundant bits via the side channels may become infeasible for a computationally bounded attacker.

12.3 Robust code-based architectures

There is a fundamental difference between protecting the output of combinational logic (in our case, a cipher) and protecting data stored in memory arrays or registers. In the latter case, the information portion is given explicitly, whereas in the first case, it is given implicitly. In other words, information about the secret key exists in the circuits' output as a function of functionality of the circuit, the user-controlled inputs and the internal state variables. In protected memory arrays, the code word is stored in memory, whereas in protected cryptographic modules a (systematic) code word is formed on the inputs of the checker.

When security-oriented codes are applied to hardware implementations of cryptographic primitives (e.g., block ciphers), they can interpret the entire state of the cipher (128 bits for AES-128, 64 bits for LED-64) as the information portion and add redundant bits according to the code structure. Detailed security-oriented code-based architectures are discussed in [8] for robust codes with and without error *correction* capabilities. Example 12.2 further illustrates the application for the reduced example of a 12-bit cipher state being protected by 2 redundant bits.

Formally, denote by k_b the number of state bits and by r_b the number of redundant bits. Then the number of wires that enter the checker is $n_b = k_b + r_b$. We refer to this n_b-bit binary vector as a word of length n symbols over a mixed alphabet, $n \leq n_b$. In a fault-free circuit, this vector is a code word c that consists of k information symbols that represent the output of the *original component*, and r redundant symbols generated by the predictor. That is, c can be read as $c = (c_1, c_2, \ldots c_{n_b})$ or as the vector

$$c = (\underbrace{\zeta_1, \zeta_2, \ldots, \zeta_k}_{\substack{k \\ \text{information} \\ \text{symbols}}}, \underbrace{\zeta_{k+1}, \ldots, \zeta_n}_{\substack{r \\ \text{redundant} \\ \text{symbols}}}) = (\xi_1, \ldots, \xi_k, \eta_1, \ldots \eta_r)$$

where $\zeta_i \in \mathbb{F}_2^{m_i}$ is an m_i-bit symbol from a finite field of characteristics 2. In addition, we have $\sum_{i=1}^{k} m_i = k_b$ and $\sum_{i=k+1}^{n} m_i = r_b$. If the code symbols belong to fields of different size, the code is said to be defined over a *mixed alphabet*. Usually, codes are defined over a single alphabet (e.g., BCH, parity and Quadratic-Sum (QS)). One exception is the compact protection code (CPC) [23] that will be discussed later. To simplify the writing and emphasize the role of each symbol, we denote an information symbol as ξ and a redundant symbol as η; that is, $\xi_i = \zeta_i$ and $\eta_i = \zeta_{i+k}$.

A schematic description of a circuit that implements a protected AES cipher is shown in Figure 12.2. The inputs to the circuit are the plaintext s and a key y. We assume that s is known to the attacker and that the random variable Y that represents the key has maximal entropy, i.e., $H(Y) = k_b$. In a fault-free scenario, the original component computes $x = f(s + y) \in \mathbb{F}_2^{k_b}$ in the first cycle and stores it in the register. From the point of view of the attacker, x is a uniformly distributed random variable.

The plaintext s and the random key y also enter a predictor block that computes r redundant symbols. Denote by w the binary vector at the output of the predictor; w is a deterministic function of $s + y$. Since f is a bijective function, w.l.o.g. we can refer to w as a function of x. That is, $w(x) = w(f(s + y)) \in \mathbb{F}_2^{r_b}$. The pair $c = (x, w(x))$ is stored in the register. This pair can be referred to as a code word of length $n = k + r$ symbols in a systematic code $\mathscr{C}$.

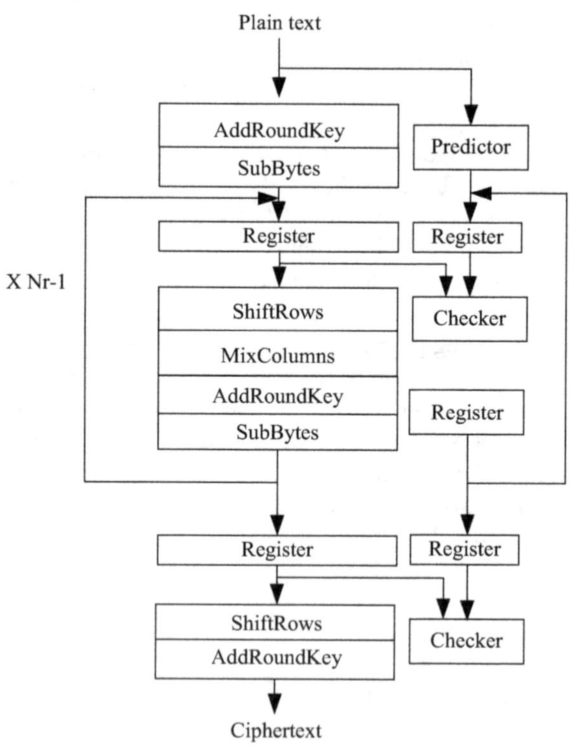

Figure 12.2 Protected AES architecture. The registers are located after the SubBytes module to enable on-line prediction and checking

An attacker may tamper with the original circuit and/or the predictor. Denote by $\hat{c}$ the word on the checker's inputs. In a fault-free circuit, $\hat{c} = c$. Experimental results reported in [24] show that some fault injection techniques, e.g., by jittering the clock, cause arbitrary numbers of bit flips. These bit flips are modeled as additive errors over a finite field of characteristic 2. We denote the difference between $\hat{c}$ and the correct code word c by e. That is, $e = \hat{c} - c \in \mathbb{F}_2^{n_b}$ is an additive error vector.

The checker examines the word $\hat{c}$ it sees on its inputs. If it decides that the word is not a legal code word or that it cannot be corrected into a legal one, it outputs the value $\perp$. Depending on the design, a $\perp$ may halt the processor or cause the system to replace the output by a random one. If the checker decides that the word is legal or that it is within the correction capability of the code, it decodes it into a legal code word. The checker's output is denoted by v, $v \in \mathscr{C} \cup \{\perp\}$. If $v \in \mathscr{C}$, it is of the form $v = (v_x, v_w)$.

The same takes place in the following cycles; the register is loaded with a code word $c \in \mathscr{C}$ computed by the predictor. However, here s is not the plaintext, but rather a deterministic function of the (key dependent) state bits of the register.

As the checker is implemented in hardware on the same circuit as the logic which it protects, the checker itself can be subject to fault injection as well. Usually, the checker is composed of two parts: an encoder and a comparator. An attack on the encoder logic is equivalent to an attack on the register and hence will be detected. Detection of an attack on the comparator requires a security-aware implementation [25].

Note that the data processed by the last two blocks in Figure 12.2 are not protected. An effective attack on these blocks will aim to stick the bits and not flip them. Thus, codes detecting stuck-at faults will be more effective. This, however, is beyond the scope of this chapter. A detailed analysis of the security provided by unidirectional error-detecting codes can be found in [26].

12.4 Security-oriented codes

In general, the effectiveness of a code as a countermeasure against fault injection attacks is measured in terms of its error detection capability. For a given error e, denote by $Q(e)$ the probability that e will be masked (undetected) by the code words of $\mathscr{C}$; that is,

$$Q(e) = 2^{-k_b}|\{c \ : \ c, c + e \in \mathscr{C}\}| \leq 1.$$

The set of errors that are masked by all code words (i.e., $Q(e) = 1$) forms the detection kernel of the code. The kernel is denoted by K_d. The error-masking probability (EMP) of a uniformly distributed code, $\bar{Q}$, is defined as

$$\bar{Q} = \max_{e \notin K_d} Q(e).$$

All errors can be divided into three sets (see Figure 12.3); errors that are always detected (e.g., e_1), those that are never detected (e.g., e_2) and those that are detected by some code words and are masked by others (e.g., e_3). Note that in linear codes, there are no errors of the third type. A code whose kernel has only one word (the

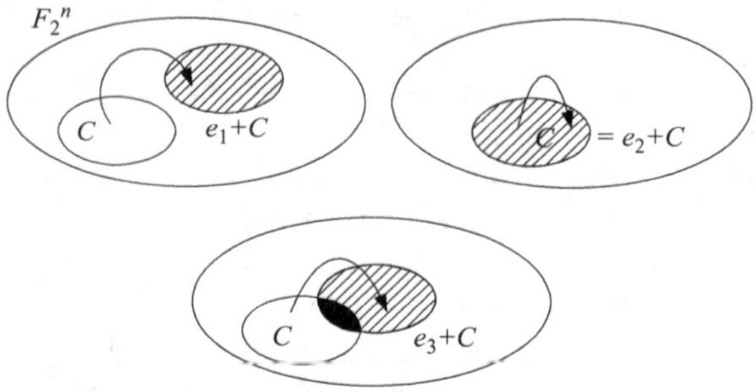

Figure 12.3 Codes with deterministic encoder—error types

all-zero vector) is called *robust*. The value of $\bar{Q}$ is lower bounded by the average EMP; namely, $\bar{Q} \geq 2^{-r_b}$. Codes that fulfill this bound on equality are called optimum robust codes. There are very few high rate optimum or close to optimum robust error-detecting codes; most are variants of two basic structures: the QS code [27] and Punctured-Square/Punctured-Cubic (PC) codes [28,29]. Punctured codes exist for any (k_b, r_b) pair but their implementation cost is high, whereas the QS-based codes have low implementation costs but impose restrictions on the relationship between k_b and r_b. The best robust code known so far is the CPC, which provides a high rate and low implementation cost. Table 12.2 presents the main robust and partially robust codes known to date. This table is an expansion of the table from [30]. The Vasil'ev code and its generalizations, the Phelps code, the One Switching code and the Multiple Switching code presented in the table are based on (not necessarily linear) Hamming codes; to simplify the description, the parameters of this Hamming code are denoted by $[n_V = k_V + r_V, k_V, 3]$. Note that most of the partially robust codes have error correction capability. For example, the Generalized Vasil'ev code [31] is constructed from two linear codes $[n_1, k_1, 4]$ and $[n_2, k_2, 3]$ and the minimum distance of the combined code is 3. Another example is the generalized Phelps code [30] that is constructed from two linear codes $[n_1, k_1, 3]$ and $[n_2, k_2, 3]$; the minimum distance of the combined code is 3 for the generalized code and 4 for the extended generalized code. In contrast, all the robust codes that have low EMP (except the RK code [32]) do not have error correction capability.

A CPC is defined over a mixed alphabet and hence exists for any k_b and r_b [23]. A CPC can protect any number of state bits with the probability of an undetected fault being at most 2^{-r_b+1}. A CPC is built upon a set of systematic robust (ground) codes $\{\mathscr{C}_j(n_j, k_j, m_j)\}_{1\leq j\leq l}$ of length n_j bits, size 2^{k_j} and $n_j - k_j$ redundant bits; the ground code itself is defined over a finite field $\mathbb{F}_{2^{m_j}}$. The dimension of a CPC is $\sum_{i=1}^{l} k_i = k_b$ and it has $r_b = \max_{1\leq j\leq l} r_j$ redundant bits. The k_js are chosen to minimize the hardware complexity. Formally, a CPC is constructed as follows [23]:

Construction 12.1 (Compact protection code (CPC)). Let

$$\mathscr{C}_j(n_j, k_j, m_j) = \{(x, w) : x \in \mathbb{F}_q^{k_i}, w = w(x) \in \mathbb{F}_2^{r_j}\}, \ j = 1, \ldots, l$$

Table 12.2 Binary robust and partially robust codes

Partially robust codes	n	k	$\log_2(K_d)$	$\overline{Q}$	d	**Complexity**
Vasil'ev [33]	$2^{r_V+1}-1$	$2^{r_V+1}-2-r_V$	$2^{r_V}-1$	≤ 0.5	3	Medium
Extended Vasil'ev [34]	$a+2^{r_V}+1$	$a+k_V$	a	≤ 0.5	4	Medium
General. Vasil'ev [31]	n_1+n_2	k_1+k_2	k_2	≤ 0.5	3	Medium
Phelps [35]	$2^{r_V+1}-1$	$2^{r_V+1}-2-r_V$	$2^{r_V+1}-2r_V-2$	$\geq 2^{-r_V+1}$	3	High
General. Phelps [30]	n_1+n_2+2	n_1+k_2	k_1+k_2	$\geq 2^{-r_V+1}$	3, 4	High
One Switching [36]	$2^{r_V}-1$	$2^{r_V}-1-r_V$	$2^{r_V-1}-1$	$1-2^{-2^{r_V-1}+1+r_V}$	3	High
Multiple Switching [37]	$2^{r_V}-1$	$2^{r_V}-1-r_V$	1	Close to 1	3	Very high
Cubic [38,39]	$k+r$	k	$k-r$	2^{-r+1}	1, 2	Medium
Inversion [40]	$k+r$	k	$k-r$	$2^{-2\lfloor\frac{r+1}{2}\rfloor+2}$	1, 2	Medium
Robust codes	n	k	$\log_2(K_d)$	$\overline{Q}$	d	**Complexity**
Quadratic Sum [27]	$(2s+1)r$	$2sr$	0	2^{-r}	1	Low
Punctured Cubic [29]	$k+r$	k	0	$\leq 2^{-r+1}$	1, 2	High
Robust Hamming [41]	2^r	2^r-1-r	0	0.5	3	Low
CPC [23]	$k+r$	$k=\sum_{i=1}^{l} k_i$	0	$\max_{1\leq i\leq l}\overline{Q}_i$	1, 2	Low
Rabii–Keren [32]	$k+r$	k	0	$\leq 2/q, 4/q$	d	Medium

be a set of l robust codes with EMP $\overline{Q}_j$. Denote by $\mathscr{C}_j'(n_j' = k_j + r_b, k_j, m_j)$ the code whose code words are derived from a $\mathscr{C}_j$ by appending $r_b - r_j$ zeros to each code word, i.e.,

$$\mathscr{C}_j' = \{(x, w') : w' = (w, \mathbf{0}_{n_j'-n_j}), (x, w) \in \mathscr{C}\}.$$

The CPC code $\mathscr{C}(n_b = k_b + r_b, k_b)$ consists of the code words (see Figure 12.4)

$$\mathscr{C} = \left\{ \begin{array}{l} (x, w) : x = (x_1, \ldots, x_l) \in \mathbb{F}_2^{k_b}, \\ \qquad w = \bigoplus_{j=1}^{l} w_j' \in \mathbb{F}_2^{r_b}, \\ (x_j, w_j') \in \mathscr{C}_j' \end{array} \right\}.$$

The EMP of the CPC is $\bar{Q} = \max_{1 \le i \le l} (\overline{Q}_i)$ [23].

For a given set of parameters, there are several ways to construct a CPC. Guidelines how to construct a CPC with minimal EMP and *low implementation cost* are given in [23]. The following example presents two (out of many) ways to construct a CPC of dimension 17 and 4 redundant bits.

Example 12.1. A $C(17 + 4, 17)$ code can be constructed by using $l = 2$ ground codes $QS(12 + 3, 12, 3) + PC(5 + 4, 5, 5)$ or using $l = 3$ ground codes, e.g., $PC(5 + 4, 5, 5) + PC(6 + 4, 6, 6) + PC(6 + 4, 6, 6)$. The first CPC consists of code words that satisfy

$$\mathscr{C} = \left\{ \begin{array}{l} (x, w) : x = (x_1 \in \mathbb{F}_{2^3}^4, x_2 \in \mathbb{F}_{2^5}) \in \mathbb{F}_2^{17}, \\ \qquad w_1 = x_{1,1} \cdot x_{1,2} \oplus x_{1,3} \cdot x_{1,4} \in \mathbb{F}_{2^3}, \\ \qquad w_2 = (x_2)^3 P_{5\times4} \in \mathbb{F}_2^4, \\ \qquad w = (w_1, 0) \oplus w_2 \in \mathbb{F}_2^4 \end{array} \right\}$$

where $P_{5\times4}$ is a binary 5×4 matrix of maximal rank 4. The first CPC has an EMP of $\bar{Q} = 2^{-3} = \max(2^{-3}, 2^{-4+1})$. Note that since the fields $\mathbb{F}_2^m$ and $\mathbb{F}_{2^m}$ are isomorphic, we associate a binary vector to the field in which the computation is performed to

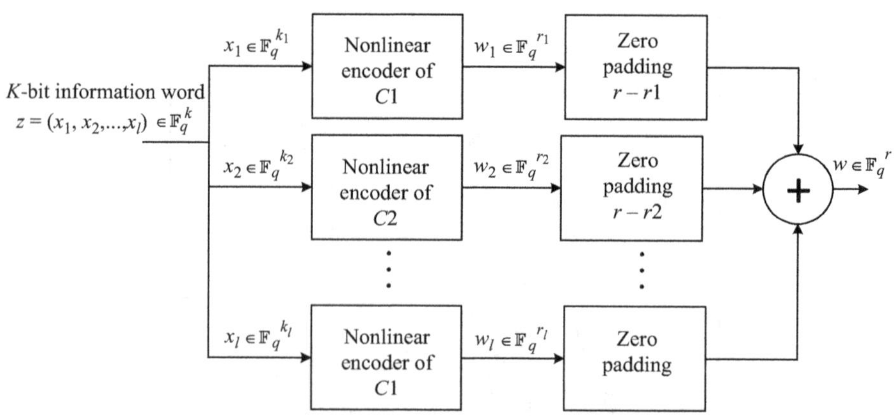

Figure 12.4 Compact protection code (CPC) structure

simplify the writing. For example, w_1 is computed in $\mathbb{F}_{2^3}$ but the vector $(w_1, 0)$ is treated as an element in $\mathbb{F}_2^4$. Another example, $(x_2)^3$, is computed in $\mathbb{F}_{2^5}$ but when multiplied by P it is treated as an element in $\mathbb{F}_2^5$. The second CPC consists of code words that satisfy

$$\mathscr{C} = \left\{ \begin{array}{l} (x, w) : x = (x_1 \in \mathbb{F}_{2^5}, x_2, x_3 \in \mathbb{F}_{2^6}) \in \mathbb{F}_2^{17}, \\ \qquad\qquad w = (x_1)^3 P_{5\times 4} \oplus (x_2)^3 P'_{6\times 4} \oplus (x_3)^3 P'_{6\times 4} \in \mathbb{F}_2^4 \end{array} \right\}$$

where P and P' are binary matrices of rank 4. The EMP of this code also is $\bar{Q} = 2^{-3}$.

In the next section, we examine how many information bits leak from a checker that utilizes robust codes and in particular CPC codes.

12.5 Information leakage from robust code-based checkers

We assume that the attacker knows the code and hence can know which errors will always be detected ($Q(e) = 0$) and which have a chance of passing unnoticed ($Q(e) > 0$). Therefore, from here on, we assume that an error e of the latter type is injected. We measure the information that leaks from the checker in terms of the MI between two random variables: the key Y and the checker output V.

Formally [42], let Z be a discrete random variable defined with a support set $\mathscr{Z}$ and denote by $p(z) = P_Z(z)$ the probability that Z will take the value z. Similarly, for two random variables $Z \in \mathscr{Z}, W \in \mathscr{W}$ denote by $p(z|w) = P_{Z|W}(z|w)$ the conditional probability that $Z = z$ given $W = w$. The entropy of a random variable Z is defined as

$$H(Z) \triangleq -\sum_{z\in\mathscr{Z}} p(z)\log_2(p(z)),$$

and the conditional entropy of Z given W is

$$H(Z|W) \triangleq -\sum_{z\in\mathscr{Z}, w\in\mathscr{W}} p(z, w)\log_2(p(z|w)).$$

The MI between two random variables Z and W is

$$I(Z; W) \triangleq \sum_{z\in\mathscr{Z}, w\in\mathscr{W}} p(z, w)\log_2\left(\frac{p(z, w)}{p(z)p(w)}\right) = H(Z) - H(Z|W).$$

Using these notations, the conditional MI between Y and V for a given error e and a chosen plaintext s is

$$I(Y; V|s, e) = \sum_{y,v\in\mathbb{F}_2^{k_b}} p(y, v|s, e)\log_2\left(\frac{p(y, v|s, e)}{p(y, v|s, e)p(y, v|s, e)}\right).$$

Similarly, if the plaintext is not chosen by the attacker, the MI is

$$I(Y; V|S, e) = \sum_{s\in\mathbb{F}_2^{k_b}} p(s)I(Y; V|s, e).$$

In this section, we consider two attack scenarios: an attack on the first round and an attack on the ith round for $i > 1$. In general, fault analysis attacks are performed on the middle/last rounds because of the trade-off between the number of required correct-fault pairs and the time-memory complexity [43,44]. Nevertheless, when error-detecting codes are embedded in hardware, an attacker who knows the coding scheme can derive information about the key even when the attack is performed on the very first cycle. The following example illustrates how the checker's output may be used to reduce the search space of the key.

Example 12.2. *For simplicity, consider a hypothetical 12-bit cipher. The inputs to the circuit implementing that cipher are a 12-bit plaintext s and a 12-bit key y. Assume that in the first cycle, this cipher computes*

$$x = (x_1, x_2, \ldots x_{12}) = f(s+y) = \begin{cases} (s+y)^{-1} & s+y \neq 0 \\ 0 & s+y = 0 \end{cases}$$

where the computation is performed in $\mathbb{F}_{2^{12}}$ with the primitive polynomial $D^{12} + D^6 + D^4 + D + 1$. Consider a hardware implementation protected by the QS code. The block we are referring to as the "original component" calculates $x(s,y)$, and in parallel, the predictor generates 2 redundant bits $w_1(s,y)$ and $w_2(s,y)$. When the two binary vectors x and w are treated as vectors over $\mathbb{F}_2^2$, i.e., $m = 2$, they form a code word $c = (\xi_1, \xi_2, \ldots \xi_6, \eta)$ with $k = k_b/m = 6$ information symbols and a single ($r = r_b/2$) redundant symbol. In this example, we use the following (intuitive) mapping $\xi_i = (x_{2i-1}, x_{2i})$ and $\eta = (w_1, w_2)$. Since we are using the QS code, every code word fulfills the following property:

$$\eta = \xi_1\xi_2 + \xi_3\xi_4 + \xi_5\xi_6. \tag{12.1}$$

The multiplication and additions in (12.1) *are over $\mathbb{F}_4$ with the primitive polynomial $D^2 + D + 1$ ($\mathbb{F}_4$ is isomorphic to $\mathbb{F}_2^2$). Note that the predictor generates the 2 redundant bits directly from s and y without explicitly computing the variable x, even though η and ξ fulfill* (12.1). *The code word stored in the register is then $c = (x, w)$. Our attack model assumes that the adversary can flip an arbitrary number of bits in arbitrary locations of both x and w.*

Assume that $s =$ (00 01 11 11 01 11), $y =$ (00 00 11 10 10 10) and that the attacker injects the error $e =$ (01 00 00 00 00 10, 11). Then, the code word written into the register is

$$c = (\underbrace{11\ 00\ 11\ 01\ 10\ 00}_{x}, \underbrace{11}_{w}) = (\underbrace{3\ 0\ 3\ 1\ 2\ 0}_{\xi}, \underbrace{3}_{\eta})$$

and the distorted word $\hat{c} = c + e = (\hat{x}, \hat{w}) = (\hat{\xi}, \hat{\eta})$ read from the register is

$$\hat{c} = c + e = (1\mathbf{0}\ 00\ 11\ 01\ 10\ 1\mathbf{0}, \mathbf{00}) = (\underbrace{\mathbf{2}\ 0\ 3\ 1\ 2\ \mathbf{2}}_{\hat{\xi}}, \underbrace{\mathbf{0}}_{\hat{\eta}})$$

(the erroneous bits and symbols appear in bold). The checker computes $\hat{\xi}_1\hat{\xi}_2 + \hat{\xi}_3\hat{\xi}_4 + \hat{\xi}_5\hat{\xi}_6 = 0$ and since it is equal to $\hat{\eta}$, it does not report a decoding error. The attacker

who knows the code sees that the attack has been masked (undetected). Therefore, the attacker infers that the following error-masking equation must be fulfilled:

$$(\xi_1 + 1)\xi_2 + \xi_3\xi_4 + \xi_5(\xi_6 + 2) = \eta + 3.$$

There are exactly 2^{10} *values of* x *that satisfy this equation; they are of the form*

$$\xi = (\xi_1,\ \xi_2 = 3 - 2\xi_5,\ \xi_3,\ \xi_4,\ \xi_5,\ \xi_6).$$

Since $y = (x^{-1} - s)$, y *can take* 2^{10} *values. Therefore, the search space for the key* y *has been narrowed from 12 to 10 bits, or by 2 bits.*

Note that if the value of the key had been such that the error would have been detected, the attacker could observe this outcome and narrow the search space for y *from* 2^{12} *down to* $(2^{12} - 2^{10})$ *possible values. However, the system would then receive an alarm and prevent the attacker from continuing to reduce the search space by further fault injections. For example, the system could initiate rekeying, in which case all the information that the attacker had acquired would be useless. Note that the error is not detected only for* $Q(e) = 1/4$ *of the keys for this error vector due to the properties of the QS code used. It is detected (leading to an alarm) for the remaining* $3/4$ *of the keys.* □

In what follows we prove that on average the information leakage is small. An attacker can observe the ciphertext, and hence he can determine whether the checker has output a different v. Moreover, the cipher output is a deterministic function of v. Therefore, we assume that v is observable by the attacker.

12.5.1 Fault attack on the first round

We distinguish between standard checkers that output a $\perp$ symbol when the word on their input is illegal, and infective checkers that output a random code word upon detecting such a problem. The following theorem states that the information that can be extracted from the checker's response in the presence of an error decreases exponentially with the number of redundant bits.

Theorem 12.1. *Let* $\mathcal{C}$ *be a systematic robust code of cardinality* 2^{k_b}, *of length* $n = k + r$, *and an EMP* $Q(e)$. *The MI (in bits) between the secret key* y *and a standard checker output* $v \in \mathbb{F}_2^{n_b}$ *for a given plaintext* $s \in \mathbb{F}_2^{k_b}$ *and an injected error* $e \in \mathbb{F}_2^{n_b}$ *is*

$$\begin{aligned} I_{\text{standard}}(Y; V|s, e) &= k_b Q(e) \\ &\quad -(1 - Q(e)) \cdot \log_2(1 - Q(e)). \end{aligned} \tag{12.2}$$

The MI between y *and an infective checker output* v *for a given plaintext* s *and an injected error* e *is*

$$\begin{aligned} I_{\text{infective}}(Y; V|s, e) &= k_b Q(e) \\ &\quad -Q(e)(2 - Q(e)) \log_2(2 - Q(e)) \\ &\quad -(1 - Q(e))^2 \log_2(1 - Q(e)). \end{aligned} \tag{12.3}$$

Proof. **Information leakage from standard checkers:**
The secret key $y \in \mathbb{F}_2^{k_b}$ is a uniformly distributed random variable; hence, so is $x = f(s, y)$, where f is a cryptographic function and thus bijective. Denote by $\mathcal{M}_e$ the set of x's associated with the code words that mask the error e; the size of this set is $|\mathcal{M}_e| = 2^{k_b}Q(e)$. The conditional probability distribution of the checker output is

$$P_V(v|e, x \in \mathcal{M}_e) = \begin{cases} 1 & v \in \mathcal{C}, v_x = x + e_x \\ 0 & \text{otherwise} \end{cases}$$

$$P_V(v|e, x \notin \mathcal{M}_e) = \begin{cases} 1 & v = \perp \\ 0 & \text{otherwise} \end{cases}$$

Thus,

$$P_V(v|e) = \begin{cases} 2^{-k_b} & v \in \mathcal{C}, v_x \in e_x + \mathcal{M}_e \\ 1 - Q(e) & v = \perp \\ 0 & \text{otherwise} \end{cases}$$

Under the assumption that the attacker knows the codebook and can choose s, it is clear that when the checker raises a $\perp$ flag it provides information about the key y. The MI between y and the checker output v for a given plaintext s and an injected error e is

$$\begin{aligned} I_{\text{standard}}(Y; V|s, e) &=_{(a)} I_{\text{standard}}(X; V|e) \\ &= H(V|e) - H(V|X, e) \\ &=_{(b)} H(V|e). \end{aligned}$$

The correctness of this equality follows from (a) v and s are statistically independent random variables, and (b) $H(V|X, e) = 0$ since the code is deterministic. The probability of $\perp$ symbol for a given error vector is $1 - Q(e)$; thus, for robust codes we have,

$$H(V|e) = Q(e) \cdot \log_2 (2^{k_b}) - (1 - Q(e)) \cdot \log_2 (1 - Q(e)).$$

Information leakage from infective checkers:
Infective checkers output a random symbol upon detecting an error. Therefore, we have:

$$P_{V,X}(v, x|e) = \begin{cases} 2^{-k_b} & v_x = e_x + x, x \in \mathcal{M}_e \\ 2^{-2k_b} & x \notin \mathcal{M}_e \\ 0 & \text{otherwise} \end{cases},$$

and since $2^{-k_b}|\mathcal{M}_e| = Q(e)$ we have:

$$P_V(v|e) = \begin{cases} 2^{-k_b}(2 - Q(e)) & v_x \in e_x + \mathcal{M}_e \\ 2^{-k_b}(1 - Q(e)) & v_x \notin e_x + \mathcal{M}_e \\ 0 & v = \perp \end{cases}.$$

The conditional entropy of V is

$$H(V|e) = k_b - Q(e)(2 - Q(e))\log_2(2 - Q(e)) \\ -(1 - Q(e))^2 \log_2(1 - Q(e))$$

and

$$H(V|X,e) = k_b(1 - Q(e)) + \underbrace{q^{-k}|\mathcal{M}_e| \cdot 1\log_2(1)}_{=0}.$$

The MI between y and the checker output for a given error for an infective checker is then

$$I_{\text{infective}}(Y; V|s,e) = I_{\text{infective}}(X; V|e) = H(V|e) - H(V|X,e).$$

□

Note that optimum (or close to optimum) robust codes have $Q(e) \leq 2^{-r_b+1}$ [27]; therefore, both $I_{\text{standard}}(Y; V|s,e)$ and $I_{\text{infective}}(Y; V|s,e)$ are smaller than $k_b 2^{-r_b+1}$. Hence, for a given key size and a bound for acceptable information leakage $I_{\max}$ (per injection), a code designer has to use at least $\lceil \log_2(k_b/I_{\max}) \rceil + 1$ redundant bits to stay below this bound.

Denote by $\bar{I}_{\text{standard}}$ and $\bar{I}_{\text{infective}}$ the maximal MI $I(Y; V|s,e)$ over all the nonzero errors e and plaintexts s. Tables 12.3 and 12.4 show the maximal EMP and the information leakage for different CPCs suitable for protecting 64-bit and 128-bit ciphers, respectively. The columns of the table are as follows. The first column is the number of redundant bits r_b. The second column shows, for each number of redundant bits, one possible construction of a CPC (explained further next). The EMP of each suggested CPC, $\bar{Q}$, computed as explained in [23] appears in column 3. The maximal number of bits that can be learned by observing the output of a standard and infective checker appears in columns 4 and 5, respectively. These values were calculated using (12.2) and (12.3).

In general, a CPC is constructed from $l \geq 1$ ground codes by concatenating their information portions and XORing their redundant portions. The ground codes $\mathscr{C}(\hat{n}_j, \hat{k}_j, \hat{m}_j)$, $j = 1, \ldots l$ are systematic robust codes of length $\hat{n}_j$ bits and size $2^{\hat{k}_j}$ defined over the finite field $\mathbb{F}_{2^{\hat{m}_j}}$. Here, we ignore the implementation cost of the codes and use (simple to describe) constructions that provide minimal EMP. For *optimal* constructions with *minimal implementation cost* refer to [23]. The acronyms used in Tables 12.3 and 12.4 are the following: QS stands for the QS code [27], PC stands for PC [29], and sQS and TQS for shortened QS and triple QS codes [23]. The following example explains one specific CPC from Table 12.3 and how the checker operates for that code.

Example 12.3. Consider a 64-bit key, protected by a CPC with 5 redundant bits ($r_b = 5$). There are several ways to construct such a code, the simplest one is shown in Table 12.3. The code is a composition of two ground codes: a QS(55,50,5) code and a PC(19,14,14) code. This means that the checker treats $\hat{x}$, the 64-bit vector at its input, as a mixed-alphabet vector composed of ten 5-bit symbols $\hat{\xi}_1, \ldots, \hat{\xi}_{10}$ from the finite field $\mathbb{F}_{2^5}$ and a single 14-bit symbol $\hat{\xi}_{11}$ from $\mathbb{F}_{2^{14}}$

$$\hat{x} = (\underbrace{\hat{x}_1, \ldots, \hat{x}_5}_{\hat{\xi}_1}, \underbrace{\hat{x}_6, \ldots \hat{x}_{10}}_{\hat{\xi}_2}, \ldots, \underbrace{\hat{x}_{46}, \ldots, \hat{x}_{50}}_{\hat{\xi}_{10}}, \underbrace{\hat{x}_{51}, \ldots \hat{x}_{64}}_{\hat{\xi}_{11}}).$$

Note that the error masking of a CPC is determined by the weakest ground code [23]. In our case $\bar{Q}_{\text{QS}} = 2^{-5}$ and $\bar{Q}_{\text{PC}} = 2^{-4}$, thus the EMP of the composed code is $2^{-4} = 0.0625$ (refer to Table 12.3).

Recall that x is the correct vector that was written into the register and $\hat{x}$ is the vector read from the register. In the case of a fault, $\hat{x}$ may be different from x. The checker computes

$$\begin{aligned} \psi_1 &= \hat{\xi}_1 \cdot \hat{\xi}_2 + \hat{\xi}_3 \cdot \hat{\xi}_4 + \cdots \hat{\xi}_9 \cdot \hat{\xi}_{10}, && \text{for the QS(55,50,5) code} \\ \psi_2 &= (\hat{\xi}_{11})^3 \cdot P_{14\times 5}, && \text{for the PC(19,14,14) code} \end{aligned}$$

(In this example, $P_{14\times 5}$ is a binary matrix of rank 5 that was chosen arbitrarily.) Then, the checker computes $\eta = \psi_1 + \psi_2$ and compares it to the predicted value $\hat{w}$ read from the 5-bit register. In the case of a fault, $\hat{w}$ may differ from w. If the computed value differs from $\hat{w}$, the checker outputs the symbol $\perp$, which will trigger an alarm.

Note that this construction is not the cost-optimal CPC because the 5 redundant bits of the QS code are computed from $k_b = 50$ information bits by multiplications over a small finite field ($\mathbb{F}_{2^5}$), whereas the 5 redundant bits of the PC code (to be XORed with the five bits of the QS code) are computed from the remaining 14 (= 64−50) information bits by multiplications over a large field ($\mathbb{F}_{2^{14}}$). Optimal and close to optimal constructions are given in [23]. □

It is interesting to compare the leakages from standard and infective checkers in Tables 12.3 and 12.4. Intuitively, one would expect that the infective checker would leak considerably less information than its counterpart, because it never generates the special symbol $\perp$ and is designed to prevent the adversary from receiving ciphertexts that are suitable for extracting information. The results suggest that there is a difference between the information leakages from the standard and the infective checkers for all considered CPCs, but that this difference is not that large. This can be explained by the low EMP of CPCs, which are optimal or close to optimal by construction. Therefore, the term $k_b Q(e) = k_b \cdot 2^{-r_b}$, which decreases exponentially in r_b, dominates the expressions in (12.2) and (12.3).

12.5.2 Fault attack on round $i > 1$

When attacking one of the internal rounds of a cipher, the attacker does not know the round state s_i or the round key y_i, but does know the functionality and can choose the error vector. Therefore, an indication that the error has been detected provides information about the subset (or subspace) that current $s_i + y_i$ belongs to. Formally, assume that no information is gained on the key prior to the fault injection on round i. This implies that for uniformly distributed keys, the round state s_i is also uniformly distributed. Hence, the information that can be learned is $I(Y_i; V_i|S, e) = I(Y_i; V_i|s_i, e)$.

Table 12.3 Error-masking probability and information leakage from the checker (in bits) for a 64-bit key

Number of redundant bits (r_b)	**CPC**	**EMP $\bar{Q}$**	**Standard checker $\bar{I}_{standard}$**	**Infective checker $\bar{I}_{infective}$**
2	QS(66,64,2)	2.50E−01	16.31128	15.88024
3	QS(63,60,3) + PC(7,4,4)	2.50E−01	16.31128	15.88024
4	QS(68,64,4)	6.25E−02	4.08729	3.96629
5	QS(55,50,5) + PC(19,14,14)	6.25E−02	4.08729	3.96629
6	TQS(60,54,6) + PC(16,10,10)	3.13E−02	2.04437	1.98286
7	QS(63,56,7) + PC(15,8,8)	1.56E−02	1.02237	0.99136
8	QS(72,64,8)	3.91E−03	0.25562	0.24783
9	QS(63,54,9) + PC(19,10,10)	3.91E−03	0.25562	0.24783
10	TQS(60,50,10) + PC(24,14,14)	1.95E−03	0.12782	0.12391
11	sQS(75,64,11)	9.77E−04	0.06391	0.06196
12	QS(60,48,12) + PC(28,16,16)	4.88E−04	0.03195	0.03098
13	TQS(52,39,13) + sQS(38,25,13)	2.44E−04	0.01598	0.01549
14	4 × PC(30,16,16)	1.22E−04	0.00799	0.00774
15	4 × PC(31,16,16)	6.10E−05	0.00399	0.00387
16	QS(80,64,16)	1.53E−05	0.00100	0.00097

Table 12.4 Error-masking probability and information leakage from the checker (in bits) for a 128-bit key

Number of redundant bits (r_b)	**CPC**	**EMP $\bar{Q}$**	**Standard checker $\bar{I}_{standard}$**	**Infective checker $\bar{I}_{infective}$**
2	QS(130,128,2)	2.50E−01	32.31128	31.88024
3	QS(126,123,3) + PC(8,5,5)	2.50E−01	32.31128	31.88024
4	QS(132,128,4)	6.25E−02	8.08729	7.96629
5	QS(125,120,5) + PC(13,8,8)	6.25E−02	8.08729	7.96629
6	QS(126,120,6) + PC(14,8,8)	3.13E−02	4.04437	3.98286
7	TQS(126,119,7) + PC(16,9,9)	1.56E−02	2.02237	1.99136
8	QS(136,128,8)	3.91E−03	0.50562	0.49783
9	QS(117,108,9) + 2 × PC(19,10,10)	3.91E−03	0.50562	0.49783
10	TQS(110,100,10) + 2 × PC(24,14,14)	1.95E−03	0.25282	0.24891
11	QS(99,88,11) + QS(50,40,10)	9.77E−04	0.12641	0.12446
12	QS(108,96,12) + 2 × PC(28,16,16)	4.88E−04	0.06320	0.06223
13	QS(117,104,13) + QS(36,24,12)	2.44E−04	0.03160	0.03111
14	QS(126,112,14) + PC(30,16,16)	1.22E−04	0.01580	0.01556
15	QS(126,112,14) + PC(31,16,16)	6.10E−05	0.00790	0.00778
16	QS(144,128,16)	1.53E−05	0.00198	0.00194

In fact, the attacker needs to have hypothesized on the previous $i-1$ round keys in order to make the use of this information on y_i. In other words, it is harder to extract information from an attack on the ith round than from attack on the first round.

Now consider a different scenario in which the attacker has narrowed the key's search space by ν bits prior to the attack. According to our assumption, the attacker knows the code and thus can *choose* an error e of minimal error-detecting probability. (Note that we ignore the fact that this cannot be done in polynomial time). In this case, the probability $Q^{(\nu)}(e)$ that a code will mask (i.e., not detect) e given that only $2^{k_b-\nu}$ out of its 2^{k_b} code words are equally likely to appear is upper bounded by

$$Q^{(\nu)}(e) \leq \frac{2^{k_b}\bar{Q}}{2^{k_b-\nu}} = 2^{\nu}\bar{Q}. \tag{12.4}$$

This implies that if the number of redundant bits is smaller than ν, an attacker whose computational resources are (almost) unlimited may be able to inject a fault without being noticed. The following example illustrates the computational effort involved in using the ν bits discovered so far to extract significant information about the remaining $(k_b-\nu)$ secret bits.

Example 12.4. Consider the CPC in Example 12.3 which protects the $k_b = 64$ state bits with 5 redundant bits. Assume that the attacker knows the first $\nu = 7$ bits of the key. Since $7 > 5$, it follows from (12.4) that it may be possible to find an error e that will be masked by the code word stored in the register. By injecting this error (if it exists), the attacker will be able to conduct a fault analysis attack without being noticed.

Let $\mathscr{A}$ denote a subset of size $2^{57} = 2^{64-\nu}$ that contains all the possible values that the x can take at the i'th round. Assume that the attacker has the resources (processing power and memory space) to construct $\mathscr{A}$. Given $\mathscr{A}$, the attacker has to choose an error vector

$$e = (e_1, \ldots, e_{64}, e_{65}, \ldots e_{69}) = (\varepsilon_1, \ldots, \varepsilon_{10}, \varepsilon_{11}, \varepsilon_{12})$$

of maximal $Q^{(\nu)}(e)$. To do so, for each e (which has not been injected prior to the current injection) the attacker has to compute the size of $\mathscr{A} \cap \mathscr{B}$ where $\mathscr{B}$ is the subset of code words (ξ, η) that satisfy the following equation:

$$\begin{aligned}\eta + \varepsilon_{12} = {} & (\xi_1 + \varepsilon_1)\cdot(\xi_2 + \varepsilon_2) + \cdots + (\xi_9 + \varepsilon_9)\cdot(\xi_{10} + \varepsilon_{10}) \\ & + (\xi_{11} + \varepsilon_{11})^3 \cdot P_{14\times 5}\end{aligned}$$

where $\varepsilon_i \in \mathbb{F}_2^5$ for $i \neq 11$ and $\varepsilon_{11} \in \mathbb{F}_2^{14}$. □

Clearly the complexity of calculating the size of $\mathscr{A} \cap \mathscr{B}$, even for a single e, is as large as the complexity of a brute force search of the unknown $(64-\nu)$ key bits. In fact, there is no viable (i.e., low-complexity) way in which the attacker can make the use of the known ν key bits to increase its chances of not being detected by a CPC checker. Therefore, there seems to be no practical way for the attacker to use the additional knowledge of key bits for injection of faults that will lead to a larger information leakage than in the first round.

To conclude, one of the most effective ways to mitigate fault injection attacks is by embedding security-oriented codes in hardware. This chapter examined whether the use of robust codes with a deterministic encoder can degrade security. By using information theoretic tools, proofs were provided that the probability that an attack will go undetected is exponentially smaller than the entropy loss due to information leak from the checker. Given a bound for acceptable information leakage, analysis thus enables the designer to easily choose the number of redundant bits required to stay below that bound.

Acknowledgment

This research was supported by the ISRAEL SCIENCE FOUNDATION (grant No. 923/16) and by the DFG (German Research Foundation) project Po 1220/7-2 "Algebraic Fault Attacks".

References

[1] M. Tunstall, D. Mukhopadhyay, and S. Ali, "Differential fault analysis of the advanced encryption standard using a single fault," in *WISTP*, vol. 6633 of *Lecture Notes in Computer Science*, pp. 224–233, Berlin, Heidelberg: Springer, 2011.

[2] F. D. Santis, O. M. Guillen, E. Sakic, and G. Sigl, "Ciphertext-only fault attacks on PRESENT," in *LightSec*, vol. 8898 of *Lecture Notes in Computer Science*, pp. 85–108, Cham: Springer, 2014.

[3] P. Jovanovic, M. Kreuzer, and I. Polian, "A fault attack on the LED block cipher," in *COSADE*, vol. 7275 of *Lecture Notes in Computer Science*, pp. 120–134, Berlin, Heidelberg: Springer, 2012.

[4] A. Alkhoraidly, A. Dominguez-Oviedo, and M. A. Hasan, "Fault attacks on elliptic curve cryptosystems," in *Fault Analysis in Cryptography*, Information Security and Cryptography, pp. 137–155, Berlin: Springer, 2012.

[5] J. Krämer and M. Loiero, "Fault attacks on UOV and rainbow," in *COSADE*, vol. 11421 of *Lecture Notes in Computer Science*, pp. 193–214, Cham: Springer, 2019.

[6] F. Valencia, I. Polian, and F. Regazzoni, "Fault sensitivity analysis of lattice-based post-quantum cryptographic components," in *Embedded Computer Systems: Architectures, Modeling, and Simulation*, pp. 107–123, Cham: Springer, 2019.

[7] T. Fuhr, E. Jaulmes, V. Lomné, and A. Thillard, "Fault attacks on AES with faulty ciphertexts only," in *2013 Workshop on Fault Diagnosis and Tolerance in Cryptography*, pp. 108–118, Los Alamitos, CA: IEEE Computer Society, 2013.

[8] B. Karp, M. Gay, O. Keren, and I. Polian, "Security-oriented code-based architectures for mitigating fault attacks," in *2018 Conference on Design of Circuits and Integrated Systems (DCIS)*, pp. 1–6, 2018.

[9] R. Cramer, Y. Dodis, S. Fehr, C. Padró, and D. Wichs, "Detection of algebraic manipulation with applications to robust secret sharing and fuzzy extractors," in *Advances in Cryptology – EUROCRYPT 2008*, pp. 471–488, Berlin, Heidelberg: Springer, 2008.

[10] Z. Wang and M. Karpovsky, "Algebraic manipulation detection codes and their applications for design of secure cryptographic devices," in *On-Line Testing Symposium (IOLTS), 2011 IEEE 17th International*, pp. 234–239, Los Alamitos, CA: IEEE Computer Society, 2011.

[11] X. T. Ngo, S. Bhasin, J. Danger, S. Guilley, and Z. Najm, "Linear complementary dual code improvement to strengthen encoded circuit against hardware Trojan horses," in *IEEE International Symposium on Hardware Oriented Security and Trust, HOST 2015*, Los Alamitos, CA: IEEE Computer Society, pp. 82–87, 5–7 May, 2015.

[12] S. Dziembowski, K. Pietrzak, and D. Wichs, "Non-malleable codes." Cryptology ePrint Archive, Report 2009/608, 2009. http://eprint.iacr.org/2009/608.

[13] O. Keren and M. Karpovsky, "Relations between the entropy of a source and the error masking probability for security-oriented codes," *IEEE Transactions on Communications*, vol. 63, no. 1, pp. 206–214, 2015.

[14] S. Saha, D. Jap, J. Breier, S. Bhasin, D. Mukhopadhyay, and P. Dasgupta, "Breaking redundancy-based countermeasures with random faults and power side channel," in *2018 Workshop on Fault Diagnosis and Tolerance in Cryptography (FDTC)*, Los Alamitos, CA: IEEE Computer Society, pp. 15–22, 2018.

[15] J. Riha, V. Miskovský, H. Kubatova, and M. Novotný, "Influence of fault-tolerance techniques on power-analysis resistance of cryptographic design," in *2017 Euromicro Conference on Digital System Design (DSD)*, Los Alamitos, CA: IEEE Computer Society, pp. 260–267, 2017.

[16] D. Boneh, R. A. DeMillo, and R. J. Lipton, "On the importance of eliminating errors in cryptographic computations," *Journal of Cryptology*, vol. 14, no. 2, pp. 101–119, 2001.

[17] A. Barenghi, L. Breveglieri, I. Koren, and D. Naccache, "Fault injection attacks on cryptographic devices: Theory, practice, and countermeasures," *Proceedings of the IEEE*, vol. 100, no. 11, pp. 3056–3076, 2012.

[18] I. Polian and F. Regazzoni, "Counteracting malicious faults in cryptographic circuits," in *ETS*, pp. 1–10, Los Alamitos, CA: IEEE Computer Society, 2017.

[19] S. Tajik, H. Lohrke, F. Ganji, J. Seifert, and C. Boit, "Laser fault attack on physically unclonable functions," in *2015 Workshop on Fault Diagnosis and Tolerance in Cryptography (FDTC)*, Los Alamitos, CA: IEEE Computer Society, pp. 85–96, 2015.

[20] Y. Li, K. Sakiyama, S. Gomisawa, T. Fukunaga, J. Takahashi, and K. Ohta, "Fault sensitivity analysis," in *CHES*, vol. 6225 of *Lecture Notes in Computer Science*, pp. 320–334, Berlin, Heidelberg: Springer, 2010.

[21] N. F. Ghalaty, B. Yuce, M. M. I. Taha, and P. Schaumont, "Differential fault intensity analysis," in *FDTC*, pp. 49–58, Los Alamitos, CA: IEEE Computer Society, 2014.

[22] F. Regazzoni, L. Breveglieri, P. Ienne, and I. Koren, "Interaction between fault attack countermeasures and the resistance against power analysis attacks," in *Fault Analysis in Cryptography*, Berlin, Heidelberg: Springer, pp. 257–272, 2012.

[23] H. Rabii, Y. Neumeier, and O. Keren, "High rate robust codes with low implementation complexity," *IEEE Transactions on Dependable and Secure Computing*, DOI: 10.1109/TDSC.2018.2816638, 2018.

[24] B. Karp, M. Gay, O. Keren, and I. Polian, "Detection and correction of malicious and natural faults in cryptographic modules," in *PROOFS 2018, 7th International Workshop on Security Proofs for Embedded Systems*, pp. 68–82, EasyChair, 2018.

[25] F. Busaba, P. K. Lala, and A. Walker, "On self-checking design of CMOS circuits for multiple faults," *VLSI Design*, vol. 1998, no. 2, pp. 151–161, 1998.

[26] A. Burg and O. Keren, "On the efficiency of berger codes against error injection attacks on parallel asynchronous communication channels," *Information Security Journal: A Global Perspective*, vol. 22, no. 5–6, pp. 208–215, 2013.

[27] M. G. Karpovsky, K. J. Kulikowski, and Z. Wang, "Robust error detection in communication and computational channels," in *Spectral Methods and Multirate Signal Processing. SMMSP'2007. 2007 International Workshop on*, Boston University, CiteSeer, 2007.

[28] N. Admaty, S. Litsyn, and O. Keren, "Puncturing, expurgating and expanding the q-ary BCH based robust codes," in *Electrical Electronics Engineers in Israel (IEEEI), 2012 IEEE 27th Convention of*, Los Alamitos, CA: IEEE Computer Society, pp. 1–5, Nov 2012.

[29] Y. Neumeier and O. Keren, "Robust generalized punctured cubic codes," *IEEE Transactions on Information Theory*, vol. 60, no. 5, pp. 2813–2822, 2014.

[30] Z. Wang, M. Karpovsky, and K. J. Kulikowski, "Design of memories with concurrent error detection and correction by nonlinear SEC-DED codes," *Journal of Electronic Testing*, vol. 26, no. 5, pp. 559–580, 2010.

[31] Z. Wang, M. Karpovsky, and A. Joshi, "Reliable MLC NAND flash memories based on nonlinear t-error-correcting codes," in *Dependable Systems and Networks (DSN), 2010 IEEE/IFIP International Conference on*, pp. 41–50, Los Alamitos, CA: IEEE Computer Society, 2010.

[32] H. Rabii and O. Keren, "A new class of security oriented error correcting robust codes," *Cryptography and Communications*, vol. 11, pp. 965–978, 2019.

[33] Y. L. Vasil'ev, "On nongroup close-packed codes," *Probleme der Kybernetik*, vol. 8, pp. 375–378, 1962.

[34] Z. Wang, M. G. Karpovsky, and K. J. Kulikowski, "Replacing linear Hamming codes by robust nonlinear codes results in a reliability improvement of memories," in *Dependable Systems & Networks, 2009. DSN'09. IEEE/IFIP International Conference on*, pp. 514–523, Los Alamitos, CA: IEEE Computer Society, 2009.

[35] K. Phelps, "A combinatorial construction of perfect codes," *SIAM Journal on Algebraic Discrete Methods*, vol. 4, no. 3, pp. 398–403, 1983.

[36] T. Etzion and A. Vardy, "Perfect binary codes: Constructions, properties, and enumeration," *IEEE Transactions on Information Theory*, vol. 40, no. 3, pp. 754–763, 1994.

[37] K. T. Phelps and M. Levan, "Kernels of nonlinear hamming codes," *Designs, Codes and Cryptography*, vol. 6, no. 3, pp. 247–257, 1995.

[38] M. Karpovsky and A. Taubin, "New class of nonlinear systematic error detecting codes," *IEEE Transactions on Information Theory*, vol. 50, no. 8, pp. 1818–1819, 2004.

[39] S. Engelberg and O. Keren, "A comment on the Karpovsky–Taubin code," *IEEE Transactions on Information Theory*, vol. 57, no. 12, pp. 8007–8010, 2011.

[40] K. J. Kulikowski, M. G. Karpovsky, and A. Taubin, "Fault attack resistant cryptographic hardware with uniform error detection," in *Fault Diagnosis and Tolerance in Cryptography*, pp. 185–195, Berlin, Heidelberg: Springer, 2006.

[41] K. J. Kulikowski, Z. Wang, and M. G. Karpovsky, "Comparative analysis of robust fault attack resistant architectures for public and private cryptosystems," in *Fault Diagnosis and Tolerance in Cryptography, 2008. FDTC'08. 5th Workshop on*, pp. 41–50, Los Alamitos, CA: IEEE Computer Society, 2008.

[42] R. Gallager and L. Laboratory, *Sequential Decoding for Binary Channels with Noise and Synchronization Errors*. Group Report, Massachusetts Institute of Technology, Lincoln Laboratory, 1961.

[43] M. Rivain, "Differential fault analysis on des middle rounds," in *CHES*, pp. 457–469, Berlin, Heidelberg: Springer, 2009.

[44] P. Derbez, P.-A. Fouque, and D. Leresteux, "Meet-in-the-middle and impossible differential fault analysis on AES," in *International Workshop on Cryptographic Hardware and Embedded Systems*, pp. 274–291, Berlin, Heidelberg: Springer, 2011.

Part III

Physical-layer security

Chapter 13

Confidential and energy-efficient cognitive communications by physical-layer security

Pin-Hsun Lin[1] *and Eduard A. Jorswieck*[1]

13.1 Introduction

Traditionally, the security of data transmission has been ensured by key-based enciphering. However, this application-layer approach to cryptography requires additional complexity and resources to perform the encryption/decryption tasks, which increases the communication delay. Moreover, a centralized network infrastructure is required to either distribute the cryptographic keys (for private-key approaches) or verify public keys and digital signatures (for public-key approaches). Finally, the need for key distribution/verification leads to undesirable communication and feedback overheads. These drawbacks make traditional cryptography not suitable to future IoT and cellular networks [1,2] that will have to serve more than 50 billion nodes by 2020, and where self-organization, high data rates, low latency, and low energy consumptions will be critical design goals.

A promising approach to overcome these issues and secure future networks is to place the security module directly at the physical layer. This approach, known as physical-layer security [3,4], has the potential to replace and complement traditional application layer cryptography, without requiring the use of any cryptographic key, and dispensing with central authorities. Instead of protecting the communication by encrypting the communication by cryptographic keys, physical-layer security exploits the unpredictability and randomness of the wireless communication channel to prevent eavesdropping. The fundamental building block of the theory of physical-layer security is represented by the wire-tap channel, defined as a three-node system in which a legitimate transmitter, conventionally labeled Alice, wishes to communicate to a legitimate receiver, labeled Bob, while a malicious node, labeled Eve, eavesdrops the communication. In [3] the discrete, memoryless, degraded wiretap channel is characterized, and the largest rate at which Alice and Bob can communicate is determined with Eve eavesdropping *no information*. This fundamental performance limit is defined as the channel *secrecy capacity* and defines the maximum amount of information that can be reliably and confidentially transmitted between two nodes in the

[1]Department of Information Theory and Communications Systems, Institute for Communications Technology, Technische Universität Braunschweig, Braunschweig, Germany

presence of an eavesdropper. In [4], the results of [3] are extended to non-degraded wiretap channels. More recently, with the advent of multiple antenna techniques, physical-layer security has started being investigated also with reference to multiple-antenna communication systems. In [5], the secrecy capacity of a Gaussian channel with a 2×2 legitimate channel and one single-antenna-eavesdropper is characterized. Following works extended this result to the case in which the legitimate nodes, as well as the eavesdropper, have an arbitrary number of antennas [6–10]. Such a scenario is commonly referred to as a Gaussian multiple-input multiple-output multiple-antenna-eavesdropper channel. A common assumption made by the previous works, which is critical to obtain the derived capacity results is the knowledge of perfect CSI at the legitimate transmitter (CSIT) of both Bob's and Eve's channels. This assumption might be not fulfilled in real-world systems, especially as far as the channel to the eavesdropper is concerned. Unavoidable channel estimation errors and the limited capacity of feedback channels can cause the imperfect knowledge of the legitimate channel, and even more, the eavesdropper can be a hidden node unwilling to send any feedback signal to the legitimate transmitter. Moreover, in fast-fading scenarios, tracking the instantaneous channel realizations is not feasible. Therefore, the results from previously cited papers, which require perfect CSIT, represent a performance upper bound for practical system designs.

Recently, there has been substantial interest in the secrecy of multiuser systems [11], with a particular emphasis on potential cooperation between users to enhance the secrecy of communications. Cooperation in communication networks is an emerging technique to improve the reliability of wireless communication systems, and it involves multiple parties assisting each other in the transmission of messages, see, e.g., [12]. Assuming that the cooperative node(s) can be trusted and that they aim at increasing the secrecy of the original transmission in the presence of a possible external eavesdropper, several cooperative strategies have been proposed [13–16]. Many works have considered the impact of different variants of interference injection, such as noise-forwarding [13], cooperative jamming (CJ) [14], or interference-assisted secure communications [17]. While CJ with Gaussian noise has the advantage of simplicity, the non-decodability of the noisy signals is always hurting the legitimate receiver. Consequently, more elaborate CJ strategies have been recently proposed [18] to mitigate this negative effect. The second type corresponds to the classical sense of cooperation, where the cooperative nodes strengthen the main transmission by using common relaying techniques such as decode-and-forward, amplify-and-forward [15], or compress-and-forward [19]. A comprehensive review of the main results for multiuser networks with secrecy can be found in [11]. As one kind of cooperative communications schemes [20], cognitive radio (CR) technology has been proposed by Mitola in [21] as an efficient way to enhance the spectrum efficiency which has considerable development over the last few decades. The state-of-the-art information theoretical analysis of CR systems can be found in [22,23]. The concept of cooperation for secrecy and the corresponding cooperative techniques can naturally be applied to the CR network (CNN).

As briefly mentioned at the beginning of this section, one critical requirement of future cellular communications will be to ensure high data rates, while at the same time reducing the energy consumptions. More specifically, sustainable growth and ecological concerns are putting forth the issue of energy efficiency as one key-performance indicator for communication networks. It is foreseen that the number of connected nodes will reach 50 billion by 2020 and the corresponding energy demand will soon become unmanageable [24]. Moreover, the resulting greenhouse gas emissions and electromagnetic pollution will exceed safety thresholds. There is a general consensus that future 5G networks will be required to provide a 1,000× data-rate increase, but at half of today's energy consumption. This requires a 2,000× increase of the bit/J energy efficiency, defined as the number of bits which can be reliably transmitted per Joule of consumed energy [25,26]. This performance metric has received much attention for communication networks without confidentiality constraints [27], but its study in networks with secure communications is still open, and very little prior works have appeared which jointly consider energy-efficient (EE) and secrecy issues. In [28], the secrecy-energy trade-off in single-antenna, single-carrier Gaussian wiretap channels is studied in terms of an actual communication rate and consumed energy, while [29] provides an information-theoretic analysis of secret communications per transmission cost. In [30], the ratio of the system outage capacity and the consumed energy is optimized for a downlink OFDMA network, while [31,32] consider the ratio between the secrecy outage capacity and the consumed power is analyzed for multiple-antenna systems.

When confidentiality is required, a natural extension of the EE concept is to consider the number of bits that can be reliably *and confidentially* transmitted per Joule of consumed energy, which leads to defining the *secure bit/J* energy efficiency secrecy EE (SEE) [33], as the ratio between the system secrecy capacity (or achievable rate) and the consumed power. Given their fractional nature, EE metrics are naturally optimized by means of fractional programming theory, the branch of optimization theory concerned with the properties, and optimization of ratios [27]. The SEE is no exception, and indeed this chapter will show how fractional programming can be used to perform radio resource allocation for SEE maximization in a CR channel (CRC). More specifically, we consider an EE problem of a four-node overlay CNN based on [34], where the secondary receiver is treated as a potential eavesdropper with respect to the primary transmission. In exchange for cooperation from the secondary user to improve his own secrecy rate, the primary user allows the secondary user to share part of the spectrum. Some important and related works optimizing the secrecy rate for CRCs are compared in the following. In [35], the secondary user wants to keep his message confidential to the primary network. That is, the primary receiver is viewed as an eavesdropper from the secondary network perspective. Hence, the CR transmitter should make sure that the message is not leaked to the primary receiver. In [36], the authors partially generalize the model of [35] by additionally considering the secrecy of the primary message and derive the rate equivocation region of messages from both users. In [37], the authors improve [36] by rate splitting: part of the message is transmitted by the primary transmitter with a deterministic encoder, and the other part is transmitted by the cognitive transmitter with a stochastic encoder. Considering the primary receiver as an eavesdropper, [38] minimized the transmission power at the secondary transmitter under the eavesdropping rate constraint. In [39], the authors studied a scenario where the secondary transmitter relays the primary packet and meanwhile assigns a portion of its power to transmit a jamming signal to boost the secrecy rate of PUs. When the SEE is the objective to be optimized, regarding the secondary receiver as a potential eavesdropper, [40] studied the optimal power allocation at the secondary transmitter for an underlay CRNs to maximize the secondary energy efficiency under the primary secrecy constraint. In [41], the authors investigate the trade-off between the secrecy throughput and the energy efficiency of secure underlay CRN.

The main difference of our work to the aforementioned papers is that we additionally consider *secure coexistence conditions* that can simplify the design of the legacy system and also broaden the applicable usage scenarios, if the primary systems employ wiretap coding, e.g., in some scenarios of the 5G networks. More specifically, the secure coexistence conditions are to guarantee that the primary user's secure transmission scheme, including the way to encode, decode, and the code rates, are all kept intact independently of the CR being active or not. Without this constraint, the legacy system must be able to adapt themselves to scenarios where the secondary systems enter or exit the spectrum. However, when considering these additional conditions, a deterministic encoder at the primary system [35,37] with rate splitting is incapable of guaranteeing the secrecy of the primary message. Furthermore, we investigate the interrelation between channels observed by the primary transmitter when the cognitive transmitter is active or not to guarantee the secrecy constraint, which is a novel contribution compared to the cited papers.

The rest of this chapter is organized as follows. The considered system model and an overview of the fractional programming, which is used to deal with the energy efficiency optimization problem, are given in Section 13.2. In Section 13.3, we first overview the achievable rate result of an overlay CRC with an additional coexistence constraint followed by the problem formulation with respect to energy efficiency. The optimization tools are described in the same section. Finally, a numerical performance assessment is carried out in Section 13.4. Concluding remarks are provided in Section 13.5.

13.2 Preliminaries

In this section, we provide an overview of physical-layer security followed by a brief introduction to the fractional programming theory.

13.2.1 System model

We investigate the CRC with a confidential message (CM) at the primary user described in Figure 13.1, which consists of the following single-antenna half-duplex nodes: the primary/secondary (cognitive) transmitters T_1/T_2* with finite and discrete

*In this chapter, T_1 and U_1 are used interchangeably with the primary transmitter and primary receiver, respectively. Also, T_2 and U_2 are used interchangeably with the secondary (CR) transmitter and secondary (CR) receiver, respectively.

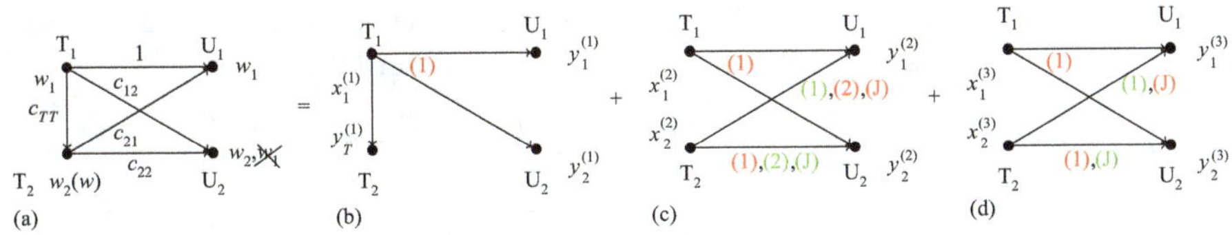

Figure 13.1 Multiphase transmission scheme: (a) complete model, (b) the first phase, (c) the second phase, (d) the third phase. (J) refers to jamming, (1) to w_1, and (2) to w_2. Green indicates a positive effect (e.g., jamming and relaying), while red indicates a negative effect (e.g., interference and eavesdropping).

channel input alphabets $\mathscr{X}_1$ and $\mathscr{X}_2$, respectively, and the primary/secondary receivers U_1/U_2 with finite channel output alphabets $\mathscr{Y}_1$ and $\mathscr{Y}_2$, respectively. T_1 wishes to transmit the secret message w_1 to U_1, which should be kept secret from U_2. Meanwhile, T_2 wants to transmit the message w_2 (without secrecy constraints) to U_2. In this work, we consider the following two requirements coined as the *secure coexistence conditions* inherited from [42], which considers the coexistence conditions without the secrecy constraint.

Definition 13.1. *The secure coexistence conditions require the following:*

1. *the transmission of T_2 does not degrade the primary user's secrecy rate R_{S1} and*
2. *the encoder and decoder at T_1 and U_1, respectively, are left intact whether T_2 transmits or not.*

The reasons to consider the secure coexistence conditions are 2-fold. First, to utilize the time-frequency slot in the overlaying sense, CR systems are obligated not to interfere with the primary systems, which is common in CR systems design. Second, with the condition (2), CRs are backward compatible with the legacy systems, which cannot sense and adapt to the environment agilely. This condition makes the CR capable of operating in broader scenarios. Note that the abovementioned conditions are the main differences between our model and those in [35–37]. Furthermore, to attain the result, the derivations to accommodate the secure coexistence conditions require a different analysis.

One of the possible practical scenarios of the considered model is that the primary users belong to a licensed system who sells rights of the spectrum usage to a femtocell system. Here, we can let the secondary transmitter and receiver be the femtocell base station and users, respectively. However, the femtocell operator may not be able to guarantee that the femtocell users are malicious or not. Thus, to provide a secrecy transmission to the primary users, not only the primary base station needs to use the wiretap coding, but also the femtocell base station needs to help to maintain that secrecy transmission for the primary system. These considerations are included in the secure coexistence conditions and will be discussed in detail later.

Denote Y_1'/Y_2' and Y_1/Y_2 as the received signals at U_1/U_2 when T_2 transmits or not, respectively. A discrete memoryless CRC-CM is described as

$$P(\boldsymbol{y}_1', \boldsymbol{y}_2' | \boldsymbol{x}_1, \boldsymbol{x}_2) = \prod_{k=1}^{n} P(y_{1,k}', y_{2,k}' | x_{1,k}, x_{2,k}), \tag{13.1}$$

where the subscript k denotes the kth symbol, $\boldsymbol{x}_1 \in \mathscr{X}_1^n$ and $\boldsymbol{x}_2 \in \mathscr{X}_2^n$ are channel inputs from T_1 and T_2, respectively; $\boldsymbol{y}_1' \in \mathscr{Y}_1^n$ and $\boldsymbol{y}_2' \in \mathscr{Y}_2^n$ are channel outputs at U_1 and U_2, respectively, when T_2 transmits. The primary transmitter T_1 intends to send a message $W_1 \in \mathscr{W}_1 = \{1, \ldots, 2^{nR_{S1}}\}$ to U_1 in n channel uses. A $(2^{nR_{S1}}, n)$ *primary user's code* is a choice of encoding and decoding rules

$$E_1^n : \{1, \ldots, 2^{nR_{S1}}\} \rightarrow \mathscr{X}_1^n \text{ and } D_1^n : \mathscr{Y}_1^n \rightarrow \{1, \ldots, 2^{nR_{S1}}\}.$$

The weak secrecy rate R_{S1} is achievable for the primary user for the wiretap channel if there exists a sequence of $(2^{nR_{S1}}, n)$ codes such that the following two constraints are satisfied

$$\lim_{n\to\infty} \frac{1}{2^{nR_{S1}}} \sum_{w_1=1}^{2^{nR_{S1}}} Pr[D_1^n(Y_1') \neq w_1 | w_1 \text{ is sent}] \rightarrow 0, \tag{13.2}$$

$$\lim_{n\to\infty} \left(R_{S1} - \frac{1}{n} H(W_1 | \boldsymbol{Y}_2') \right) \rightarrow 0^+, \tag{13.3}$$

where (13.2) means a vanishing average error probability with block length n, and (13.3) means a weakly vanishing information leakage to the eavesdropper.

In the considered CR system, the primary user should not be required to change his encoder and decoder due to the presence of secondary users. Therefore, T_1 and U_1 are restricted to use the usual wiretap encoder and decoder, which are designed for a three-node wiretap channel: one legitimate transmitter, one legitimate receiver, and one eavesdropper. The secrecy rate R_{S1} that can be achieved under this condition is defined as a *single-user achievable secrecy rate*. On the other hand, T_2 intends to send an independent message $W_2 \in \mathscr{W}_2 = \{1, \ldots, 2^{nR_2}\}$ to the secondary receiver U_2 in n channel uses with the following encoding and decoding rules:

$$E_2^n : E_1^n \times \mathscr{W}_1 \times \mathscr{W}_2 \rightarrow \mathscr{X}_2^n \text{ and } D_2^n : \mathscr{Y}_2'^n \rightarrow \mathscr{W}_2.$$

A rate R_2 is achievable for the secondary user if there exists a sequence of $(2^{nR_2}, n)$ codes such that

$$\lim_{n\to\infty} \frac{1}{2^{nR_2}} \sum_{w_2=1}^{2^{nR_2}} Pr[D_2^n(Y_2') \neq w_2 | w_2 \text{ is sent}] \rightarrow 0 \tag{13.4}$$

while satisfying the secure coexistence conditions, where (13.4) means a vanishing error probability at the CR receiver with block length n.

Based on the previous constraints, we will derive an achievable rate R_2 in the next section with a multiphase transmission scheme, which is an optimization problem over several system parameters, including the duration of each phase, the ratios of power for relaying, transmitting CR's own signal, and also for jamming signal.

13.2.2 Fractional programming theory

The basics of fractional programming theory are reviewed here. For a more detailed review with applications to wireless communications, we refer to [27].

Definition 13.2 ((Strict) pseudo-concavity). *Let $\mathscr{C} \subseteq \mathbb{R}^n$ be a convex set. Then $r : \mathscr{C} \to \mathbb{R}$ is pseudo-concave if and only if, for all $\mathbf{x}_1, \mathbf{x}_2 \in \mathscr{C}$, it is differentiable and*

$$r(\mathbf{x}_2) < r(\mathbf{x}_1) \Rightarrow \nabla(r(\mathbf{x}_2))^T(\mathbf{x}_1 - \mathbf{x}_2) > 0 \, . \tag{13.5}$$

Strict pseudo-concavity holds if, for all $\mathbf{x}_1 \neq \mathbf{x}_2 \in \mathscr{C}$,

$$r(\mathbf{x}_2) \leq r(\mathbf{x}_1) \Rightarrow \nabla(r(\mathbf{x}_2))^T(\mathbf{x}_1 - \mathbf{x}_2) > 0 \, . \tag{13.6}$$

The interest for pseudo-concave functions stems from the following result.

Proposition 13.1. *Let $r : \mathscr{C} \to \mathbb{R}$ be a pseudo-concave function.*

1. *If $\mathbf{x}^*$ is a stationary point for r, then it is a global maximizer for r.*
2. *The Karush–Kuhn–Tucker (KKT) conditions for the problem of maximizing r subject to convex constraints are necessary and sufficient conditions for optimality.*
3. *If r is strictly pseudo-concave, then a unique maximizer $\mathbf{x}^*$ exists.*

Pseudo-concavity plays a key role in the optimization of fractional functions, due to the following result.

Proposition 13.2. *Let $r(x) = (f(x)/g(x))$, with $f : \mathscr{C} \subseteq \mathbb{R}^n \to \mathbb{R}$ and $g : \mathscr{C} \subseteq \mathbb{R}^n \to \mathbb{R}_+$. If f is nonnegative, differentiable, and concave, while g is differentiable and convex, then r is pseudo-concave. If g is affine, the nonnegativity of f can be relaxed. Strict pseudo-concavity holds if either f is strictly concave, or g is strictly convex.*

Finally, let us introduce the definition of a fractional program.

Definition 13.3 (Fractional program). *Let $\mathscr{X} \subseteq \mathbb{R}^n$, and consider the functions $f : \mathscr{X} \to \mathbb{R}_0^+$ and $g : \mathscr{X} \to \mathbb{R}^+$. A fractional program is the optimization problem*

$$\max_{x \in \mathscr{X}} \frac{f(x)}{g(x)} \, . \tag{13.7}$$

The following result relates the solution of (13.7) to the auxiliary function $F(\beta) = \max_{x \in \mathscr{X}} \{f(x) - \beta g(x)\}$.

Theorem 13.1 ([43]). *An $x^* \in \mathscr{X}$ solves (13.7) if and only if $x^* = \arg\max_{x \in \mathscr{X}} \{f(x) - \beta^* g(x)\}$, with β^* being the unique zero of $F(\beta)$. Moreover, β^* coincides with the global maximum of (13.7), while $F(\beta) > 0$ for any $\beta < \beta^*$ and $F(\beta) < 0$ for all $\beta > \beta^*$.*

This result allows us to solve (13.7) by finding the zero of $F(\beta)$. An efficient algorithm to do so is Dinkelbach's algorithm, which is reported here in Algorithm 13.1.

If $f(x)$ and $g(x)$ are concave and convex, respectively, and if $\mathscr{X}$ is a convex set, then Dinkelbach's algorithm requires to solve one convex problem in each iteration.

Algorithm 13.1: Dinkelbach's algorithm

Set $\varepsilon > 0$; $\beta = 0$; $F > \varepsilon$;
repeat
  $\boldsymbol{x}^* = \arg\max_{\boldsymbol{x}\in\mathscr{X}} \{f(\boldsymbol{x}) - \beta g(\boldsymbol{x})\}$
  $F = f(\boldsymbol{x}^*) - \beta g(\boldsymbol{x}^*)$;
  $\beta = f(\boldsymbol{x}^*)/g(\boldsymbol{x}^*)$;
until $F \geq \varepsilon$

If one of these assumptions is not fulfilled, the auxiliary problem to be solved in each iteration is non-convex, which in general requires an exponential complexity to be solved. It should also be stressed that in this case Dinkelbach's algorithm is still guaranteed to converge to the global solution, assuming however that the auxiliary problem is *globally* solved in each iteration. More in general, in this scenario no computationally efficient algorithm is known to find the global solution of (13.7).

As for the number I of iterations required for Algorithm 13.1 to converge, Dinkelbach's algorithm enjoys a superlinear convergence rate [44]. However, a closed-form expression for the number of iterations required to obtain a given tolerance ε is not available. Instead, this can be obtained by a simple modification of Algorithm 13.1, which makes use of the bisection method to determine the zero of the auxiliary function. Denoting by U and L the initial positive and negative values for $F(\beta)$, a very similar algorithm as Algorithm 13.1 can be considered, with the only difference that β is updated in each iteration according to the bisection rule. Then, the number of required iterations to obtain the tolerance ε can be expressed as

$$I = \left\lceil \log_2 \frac{U - L}{\varepsilon} \right\rceil . \tag{13.8}$$

Combining the knowledge introduced in Sections 13.2.1 and 13.2.2, we can formulate our problem abstractly as follows:

$$\begin{aligned} \max \quad & R_2 \\ \text{s.t.} \quad & \text{secure coexistence conditions at primary user,} \\ & \text{power constraint at CR system.} \end{aligned} \tag{13.9}$$

13.3 Radio resource allocation for EE maximization

In this section, we first review the derived achievable rate of the overlay CR model from [34]. After that, we formulate the secrecy energy efficiency resource allocation problem based on the derived achievable rate and approximate it. Then we are able to solve the approximated problem by the sequential optimization with Dinkelbach's algorithm. An alternating algorithm is proposed.

13.3.1 The achievable rate and EE formulation

The motivations of the three-phase transmission scheme are 2-fold: (1) to provide an additional degree of freedom by different transmission schemes to improve the secondary user's rate depending on the location of nodes and their channels; (2) to take into account the latency of successful decoding of w_1 at T_2 to accommodate the practical scenarios where w_1 cannot be known non-causally at T_2.

13.3.1.1 Discrete memoryless channels

In the following, we show the result of a three-phase transmission [34], where the distribution of each phase can be respectively factorized from (13.1) as follows:

$$\begin{aligned} P_{Y_1'^{(1)},Y_2'^{(1)},V_1} &= P_{Y_1'^{(1)},Y_2'^{(1)}|V_1}P_{V_1}, \\ P_{Y_1'^{(2)},Y_2'^{(2)},X_2^{(2)},V_1,V_2} &= P_{Y_1'^{(2)},Y_2'^{(2)}|X_2^{(2)},V_1}P_{X_2^{(2)}|V_1,V_2}P_{V_2|V_1}P_{V_1}, \\ P_{Y_1'^{(3)},Y_2'^{(3)},X_2^{(3)},V_1} &= P_{Y_1'^{(3)},Y_2'^{(3)}|X_2^{(3)},V_1}P_{X_2^{(3)}|V_1}P_{V_1}, \end{aligned}$$

where the notation $A^{(k)}$ denotes random variable A in the kth phase. Note that distributions of V_1 and V_2 are fixed over the three phases; thus, we omit the specification of the phase to simplify the notation.

Assume the time index $t \in \mathbb{N}$. We define the sets of the three phases as $\mathsf{T}_1 = \{t : 1 \le t \le \lfloor \eta_1 n \rfloor\}$, $\mathsf{T}_2 = \{t : \lfloor \eta_1 n \rfloor + 1 \le t \le \lfloor (\eta_1 + \eta_2) n \rfloor\}$, $\mathsf{T}_3 = \{t : \lfloor (\eta_1 + \eta_2) n \rfloor + 1 \le t \le n\}$, respectively, with $\eta_1 + \eta_2 + \eta_3 = 1$, $0 < \eta_1 < 1$, $0 \le \eta_2$, and $0 \le \eta_3$. Note that $\lfloor \eta_1 n \rfloor$ must be no less than the time that T_2 needs to successfully decode w_1 [45] and η_1 must be less than 1 for the possibility of T_2's own transmission, i.e., a nonzero duration of the second phase. The former condition can be modeled by letting T_1, T_2, U_1, and U_2 form a degraded wiretap channel, where T_2 and U_1 are legitimate receivers and U_2 is the eavesdropper.† Besides, these ratios are fixed before each transmission according to the optimization results which will be discussed at the end of this section. We want to design T_2's transmit signal $\boldsymbol{x}_2$ where the secure coexistence conditions are satisfied and we consider the weak secrecy constraint.

Proposition 13.3. [34] *Assume that the primary user's coding scheme is fixed with code rate $R_{S1} \triangleq I(V_1; Y_1) - I(V_1; Y_2)$, the rate per bin is $R'_{S1} \triangleq I(V_1; Y_2) - 2\varepsilon$, and T_1, T_2, and U_1 form a degraded wiretap channel. Then rate R_2 is achievable such that the reliable decoding and the weak secrecy can be guaranteed where*

$$R_2 = \max_{\substack{p_{V_2|V_1} \\ p_{X_2|V_2,V_1}}} \eta_2 \left(I(V_2; Y_2'^{(2)}) - I(V_2; V_1) \right),$$

$$\text{s.t.} \quad R_{S1} + R'_{S1} \le \sum_{k=1}^{3} \eta_k I\left(V_1; Y_1'^{(k)}\right), \tag{13.10}$$

$$R'_{S1} = \eta_1 I(V_1; Y_2'^{(1)}) + \eta_2 I\left(V_1; V_2, Y_2'^{(2)}\right) + \eta_3 I\left(V_1; Y_2'^{(3)}\right). \tag{13.11}$$

†Here we consider the stochastic degradedness [19], i.e., there exists a random variable $\tilde{Y}_T$ denoting the received signal at T_2 such that (1) $\tilde{Y}_T | \{X = x\} \sim p_{\tilde{Y}_T|X}(\tilde{y}_T|x)$ and (2) $X \to \tilde{Y}_T \to Y_1 \to Y_2$ form a Markov chain.

The result in Proposition 13.3 can be derived by applying the concept of parallel channels [46] such that the achievable secrecy rate takes the average over the considered three transmission phases in addition to the fact that V_2 only exists in the second phase. See [46] for a proof.

13.3.1.2 Additive white Gaussian noise channels

The three-phase transmission for AWGN channels is explained as follows:

Phase 1 ($t \in \mathsf{T}_1$): The following decodability constraint is required [45]

$$|c_{TT}| > |c_{11}|, \tag{13.12}$$

which can be easily seen from the degradedness assumption of Y_1 and Y_T. In the following, we normalize $|c_{11}|$ to 1.

Phase 2 ($t \in \mathsf{T}_2$): The transmit power of T_2 in this phase $P_2^{(2)}$ is divided into three parts:

1. **Jamming:** The jamming signal $a_2(t)$ has power $P_{2a}^{(2)} = \rho_2 P_2^{(2)}$ to confuse the eavesdropping secondary user U_2. The parameter $\rho_2 \in [0, 1)$ denotes the fraction of the power used for jamming.
2. **Relaying of the primary message:** If (13.12) is valid, T_2 helps to relay $\{x_1(t)\}_{t\in\mathsf{T}_2}$ in Phase 2 while simultaneously transmitting its own message w_2. The relay power is $P_{2,1}^{(2)} = \gamma(1 - \rho_2)P_2^{(2)}$, where γ is the ratio of the remaining power for relaying.
3. **Transmission of the secondary message:** w_2 is encoded into $\boldsymbol{v}_2^{(2)}$ with power $(1 - \gamma)(1 - \rho_2)P_2^{(2)}$ to be decoded by U_2 only.

Phase 3 ($t \in \mathsf{T}_3$): T_2 relays $\{x_1(t)\}_{t\in\mathsf{T}_3}$ with power $(1 - \rho_3)P_2^{(3)}$ and transmits the jamming signal with power $\rho_3 P_2^{(3)}$, but without superimposing its own signal $v_2(t)$.

The average transmit power constraints for T_k are

$$\frac{1}{n}\sum_{i=1}^{n} |x_k(i)|^2 \le P_k \quad \text{for} \quad k \in \{1, 2\}. \tag{13.13}$$

More specifically, the transmit power constraint at T_2 is

$$\eta_2 P_2^{(2)} + \eta_3 P_2^{(3)} \le P_2, \tag{13.14}$$

where $P_2^{(k)} \triangleq (1/|\mathsf{T}_k|)\sum_{t\in\mathsf{T}_k} |x_2(t)|^2$, $k = 2$ and 3.

Assume that the noises at all nodes are independent and identically distributed circularly symmetric complex AWGN with zero mean, unit variance and are mutually independent for all t. We also assume that T_1 perfectly knows the channel states from T_1 to U_1 and from T_1 to U_2, while T_2 knows all channel states. Note that when $|c_{TT}| \le 1$, T_2 cannot know what is transmitted due to the wiretap code used at T_1. On the contrary, if $|c_{TT}| > 1$, T_2 can accomplish the successful decoding in the same way as at the legitimate receiver U_1, since $|c_{TT}|$ is large enough for T_2 to decode all the code words in the binning codebook successfully.[‡]

When T_2 does not transmit, the secrecy capacity of the primary user is $(\mathscr{C}(P_1) - \mathscr{C}(|c_{12}|^2 P_1))^+$. Define equivalent channel gains as follows:

$$c_{11}^{(2)} \triangleq 1 + c_{21} e^{-j\phi_{21}} \sqrt{(1-\rho_2)\gamma P_2^{(2)}/P_1}, \tag{13.15a}$$

$$c_{12}^{(2)} \triangleq c_{12} + c_{22} e^{-j\phi_{21}} \sqrt{(1-\rho_2)\gamma P_2^{(2)}/P_1}, \tag{13.15b}$$

$$c_{11}^{(3)} \triangleq 1 + c_{21} e^{-j\phi_{21}} \sqrt{(1-\rho_3) P_2^{(3)}/P_1}, \tag{13.15c}$$

$$c_{12}^{(3)} \triangleq c_{12} + c_{22} e^{-j\phi_{21}} \sqrt{(1-\rho_3) P_2^{(3)}/P_1}, \tag{13.15d}$$

where ϕ_{21} is the phase of c_{21}. The secondary user's rate with dirty paper coding (DPC) is given in the following proposition.

Proposition 13.4. [34] *Given primary user's rate as* $R_{S1} = (\mathscr{C}(P_1) - \mathscr{C}(|c_{12}|^2 P_1))^+$, *whose codebook is fixed. When DPC is used at* T_2, *the rate* R_2 *is achievable such that the secure coexistence conditions can be guaranteed where*

$$R_2 = \eta_2 \mathscr{C}\left(\frac{|c_{22}|^2 P_{U_2}}{1 + |c_{22}|^2 \rho_2 P_2^{(2)}}\right), \tag{13.16}$$

if (13.12) *and the following constraints*

$$\eta_2 \mathscr{C}\left(\frac{|c_{11}^{(2)}|^2 P_1}{1 + |c_{21}|^2 (\rho_2 P_2^{(2)} + P_{U_2})}\right) + \eta_3 \mathscr{C}\left(\frac{|c_{11}^{(3)}|^2 P_1}{1 + |c_{21}|^2 \rho_3 P_2^{(3)}}\right) \geq (1-\eta_1)\mathscr{C}(P_1), \tag{13.17}$$

$$\eta_2 \mathscr{C}\left(\frac{|c_{12}^{(2)}|^2 P_1}{1 + |c_{22}|^2 (\rho_2 P_2^{(2)} + P_{U_2})}\right) + \eta_3 \mathscr{C}\left(\frac{|c_{12}^{(3)}|^2 P_1}{1 + |c_{22}|^2 \rho_3 P_2^{(3)}}\right) = (1-\eta_1)\mathscr{C}(|c_{12}|^2 P_1), \tag{13.18}$$

are fulfilled, where $P_{U_2} \triangleq (1-\rho_2)(1-\gamma)P_2^{(2)}$.

Based on Proposition 13.3 and the fact that a smaller η_1 results in a longer duration for transmitting the secondary user's signal and clean relaying, we set

[‡]The celebrated Shannon's random coding is adopted here, which guarantees the existence of codebook with decodable w_1 in Phase 1 as in the seminal papers [47,48].

$\eta_1 = \mathscr{C}(P_1)/\mathscr{C}(|c_{TT}|^2 P_1)$, where $\lfloor \eta_1 n \rfloor$ is the smallest duration for T_2 to successfully decode T_1's message. We then define the optimization problem for this case as

Problem 1:

$$\max_{\substack{\eta_2,\rho_2,\rho_3\\ \gamma,P_2^{(2)},P_2^{(3)}}} \quad R_2 \tag{13.19}$$

$$\begin{aligned} \text{s.t.} \quad & (13.12),\ (13.17),\ (13.18),\ \text{and} \\ & \eta_2 P_2^{(2)} + (1-\eta_1-\eta_2)P_2^{(3)} \le P_2,\ P_2^{(2)} \ge 0,\ P_2^{(3)} \ge 0, && (13.20)\\ & \rho_2 \ge 0,\ \rho_3 \ge 0,\ \gamma \ge 0, && (13.21)\\ & \eta_2 + \eta_3 = 1-\eta_1,\ \eta_2 \ge 0,\ \eta_3 \ge 0, && (13.22) \end{aligned}$$

where the first inequality in (13.20) is the average power constraint. Instead of focusing on solving the non-convex Problem 1, we will thoroughly assess it by numerical illustrations in Section 13.4 and investigate the impact of the optimization variables on the rates for relevant scenarios.

Remark 13.1. *Due to the secure coexistence condition (1) in Definition 13.1, pure CJ cannot be used here. This is because there are $2^{n(I(V_1;Y_1)-\varepsilon)}$ code words in the primary user's codebook that is fixed, and U_1 must be able to successfully decode all the code words. When pure CJ is used, the capacity of the main channel will be smaller than the case in which CJ is not used. Then the reliability constraint of the transmission to the legitimate receiver T_1 will be violated.*

Finally, given R_2 and P_2, the system EE is written as

$$\text{EE} = \frac{R_2}{P_{\max}}\ \text{(bit/J)}\,, \tag{13.23}$$

with $P_{\max}$ being the total consumed power. It is seen that (13.23) is measured in bit/J, thus representing a measure of the amount of information reliably and confidentially transmitted to the receiver, per Joule of consumed energy. Note that the maximization of the EE is quite different from the traditional maximization of the achievable rate. The main difference is that, unlike the achievable rate, the EE is not monotonically increasing with the transmit power but actually tends to zero for increasing transmit powers. Indeed, while the numerator grows at most logarithmically with the transmit power, the denominator grows linearly. This implies that while the achievable rate maximization typically leads to using all the available transmit power, the EE function is maximized by a finite power level [49]. Once the maximum feasible power is larger than this finite power level, then using full power is a suboptimal choice as far as EE maximization is concerned. On the other hand, for lower maximum feasible powers, it is possible that using full power is optimal also for EE maximization, and in this case, optimizing the EE is equivalent to optimizing the rate at the numerator.

The radio resource allocation problem of the EE CR is thus formulated as follows:

$$\textbf{Problem 2:} \quad \max_{\substack{\eta_2, \rho_2, \rho_3 \\ \gamma, P_2^{(2)}, P_2^{(3)}}} \quad \frac{R_2}{\mu\left(\eta_2 P_2^{(2)} + (1 - \eta_1 - \eta_2) P_2^{(3)}\right) + P_c} \tag{13.24a}$$

$$\text{s.t.} \quad (13.12), (13.17), (13.18), (13.20), (13.21), (13.22). \tag{13.24b}$$

The rest of this chapter will describe how Problem 2 can be solved.

13.3.2 Radio resource allocation for EE CR systems

We first introduce the sequential optimization followed by an approximation of Problem 2 to make it comparable to the sequential optimization.

13.3.2.1 Sequential optimization

The main challenge posed by Problem 2 is that it is a fractional program with non-concave numerator and denominator. We combine two schemes to approach the solution: (1) alternating optimization: to make the denominator a linear function of the variables and meanwhile to make the numerator into a form that can be dealt by the sequential optimization in the second scheme; (2) the fractional programming: the framework described in Section 13.2 cannot be directly employed with affordable complexity. To circumvent this problem, fractional programming will be complemented with the framework of sequential optimization [50–52], to develop an optimization approach able to trade-off optimality and complexity, being able to yield a resource allocation enjoying strong optimality properties while at the same time requiring affordable complexity. Sequential optimization has been used for resource allocation in communication networks in several works, for both rate and energy efficiency optimization [53–56].

The idea behind the sequential optimization tool is to tackle a difficult maximization problem by solving a sequence of easier maximization problems. More formally, consider a maximization problem $\mathscr{P}$ with differentiable objective f_0 and constraint functions $\{f_i\}_{i=1}^{I}$. Then, let us consider a sequence of approximate problems $\{\mathscr{P}_\ell\}_\ell$ with differentiable objectives $\{f_{0,\ell}\}_\ell$ and constraint functions $\{f_{\ell,i}\}_\ell$, $i = 1, \ldots, I$, such that the following three properties are fulfilled, for all ℓ:

(P1) $f_{i,\ell}(\mathbf{x}) \leq f_i(\mathbf{x})$, for all $\mathbf{x}$, $i = 0, 1, \ldots, I$;
(P2) $f_{i,\ell}(\mathbf{x}^{(\ell-1)}) = f_i(\mathbf{x}^{(\ell-1)})$, with $\mathbf{x}^{(\ell-1)}$ the maximizer of $f_{\ell-1}$, $i = 0, 1, \ldots, I$;
(P3) $\nabla f_{i,\ell}(\mathbf{x}^{(\ell-1)}) = \nabla f_i(\mathbf{x}^{(\ell-1)})$, $i = 0, 1, \ldots, I$.

Now, if properties **(P1)**, **(P2)**, and **(P3)** are fulfilled, the sequence $\{f(\mathbf{x}^{(\ell)})\}_\ell$ is monotonically increasing and convergent. Moreover, upon convergence of the sequence $\{f(\mathbf{x}^{(\ell)})\}_\ell$, the resulting point $\mathbf{x}^*$ fulfills the KKT conditions of the original Problem $\mathscr{P}$ [50].

Some recent refinements of the previous results are also available in the literature. If the feasible set of the original problem is not modified, then every limit point of the sequence $\{\mathbf{x}^{(\ell)}\}_\ell$ is a KKT point of the original Problem $\mathscr{P}$ [52]. Another result from [51] states that if the original objective is strictly concave,[§] then again every limit point of the sequence $\{\mathbf{x}^{(\ell)}\}_\ell$ is a KKT point of the original Problem $\mathscr{P}$, also when the original feasible set is approximated in each iteration.

This approach appears to be very powerful, but the critical point is to find suitable functions $\{f_\ell\}_\ell$ fulfilling properties **(P1)**, **(P2)**, and **(P3)**, and that can also be maximized with limited complexity. This latter constraint is not directly related to the theoretical properties of the method and indeed was not mentioned in the previous description, but clearly it is a practical requirement that must be fulfilled in order for the method to be of practical relevance. For the case of Problem 2, all these requirements can be fulfilled as described next.

13.3.2.2 Approximation of Problem 2

We can easily see that the objective function and also the constraints (13.17) and (13.18) of Problem 2 are not convex, which hinders us to directly invoke Dinkelbach's algorithm to solve it. To proceed, let us reexpress R_2 as

$$R_2 = \eta_2 \left(\underbrace{\log_2\left(1 + |c_{22}|^2 (P_{U_2} + \rho_2 P_2^{(2)})\right)}_{R^+} - \underbrace{\log_2\left(1 + |c_{22}|^2 \rho_2 P_2^{(2)}\right)}_{R^-} \right), \quad (13.25)$$

which is a product of η_2 and the difference of the two concave functions $R^+(\rho_2, \gamma, P_2^{(2)})$ and $R^-(\rho_2, P_2^{(2)})$. Recall that $P_{U_2} = (1-\rho_2)(1-\gamma)P_2^{(2)}$. Thus, for any given point $(\rho_{2,0}, \gamma_0, P_{2,0}^{(2)})$, the numerator of (13.24a) can be lower bounded by replacing $R^-(\rho_2, P_2^{(2)})$ by its first-order Taylor series expansion around $(\rho_{2,0}, \gamma_0, P_{2,0}^{(2)})$, which yields the following:

$$\begin{aligned}
R_2 &= \eta_2 \left(R^+(\rho_2, \gamma, P_2^{(2)}) - R^-(\rho_2, P_2^{(2)}) \right) \\
&\geq \eta_2 \left(R^+(\rho_2, \gamma, P_2^{(2)}) - R^-(\rho_{2,0}, P_{2,0}^{(2)}) - \left.\frac{\partial R^-}{\partial \rho_2}\right|_{\substack{\rho_2=\rho_{2,0} \\ P_2^{(2)}=P_{2,0}^{(2)}}} \cdot (\rho_2 - \rho_{2,0}) \right. \\
&\qquad \left. - \left.\frac{\partial R^-}{\partial P_2^{(2)}}\right|_{\substack{\rho_2=\rho_{2,0} \\ P_2^{(2)}=P_{2,0}^{(2)}}} \cdot (P_2^{(2)} - P_{2,0}^{(2)}) \right) \\
&= \eta_2 \left(R^+(\rho_2, \gamma, P_2^{(2)}) - \log\left(1 + |c_{22}|^2 \rho_{2,0} P_{2,0}^{(2)}\right) - \right. \\
&\qquad \left. \frac{|c_{22}|^2}{1 + |c_{22}|^2 \rho_{2,0} P_{2,0}^{(2)}} \left(P_{2,0}^{(2)}(\rho_2 - \rho_{2,0}) + \rho_{2,0}(P_2^{(2)} - P_{2,0}^{(2)}) \right) \right) \\
&\triangleq \tilde{R}_2.
\end{aligned}$$

[§]Reference [51] assumes strict convexity, as minimization problems are discussed instead of maximization problems.

It can be seen that $\tilde{R}_2$ is a concave lower bound of R_2. Now we consider the two constraints (13.17) and (13.18) and denote the left-hand sides of them by Δ_1 and Δ_2, respectively. We first expand Δ_1 and Δ_2 as follows:

$$\begin{aligned}\Delta_1 =& \underbrace{\eta_2 \log\left(1 + |c_{11}^{(2)}|^2 P_1 + |c_{21}|^2\left(\rho_2 P_2^{(2)} + P_{U_2}\right)\right)}_{\Delta_{11}} \\ &+ \underbrace{\eta_3 \log\left(1 + |c_{11}^{(3)}|^2 P_1 + |c_{21}|^2 \rho_3 P_2^{(3)}\right)}_{\Delta_{13}} \\ &- \underbrace{\eta_2 \log\left(1 + |c_{21}|^2\left(\rho_2 P_2^{(2)} + P_{U_2}\right)\right)}_{\Delta_{12}} - \underbrace{\eta_3 \log\left(1 + |c_{21}|^2 \rho_3 P_2^{(3)}\right)}_{\Delta_{14}}, \\ \Delta_2 =& \underbrace{\eta_2 \log\left(1 + |c_{12}^{(2)}|^2 P_1 + |c_{22}|^2\left(\rho_2 P_2^{(2)} + P_{U_2}\right)\right)}_{\Delta_{21}} \\ &+ \underbrace{\eta_3 \log\left(1 + |c_{12}^{(3)}|^2 P_1 + |c_{22}|^2 \rho_3 P_2^{(3)}\right)}_{\Delta_{23}} \\ &- \underbrace{\eta_2 \log\left(1 + |c_{22}|^2\left(\rho_2 P_2^{(2)} + P_{U_2}\right)\right)}_{\Delta_{22}} - \underbrace{\eta_3 \log\left(1 + |c_{22}|^2 \rho_3 P_2^{(3)}\right)}_{\Delta_{24}}.\end{aligned}$$

Similar to the steps to derive $\tilde{R}_2$, we can linearize Δ_{12} and Δ_{14} as Δ'_{12} and Δ'_{14} through the first-order Taylor series expansion at $\left(\rho_{2,0}, P_{2,0}^{(2)}, \gamma_0\right)$ and $\left(\rho_{3,0}, P_{2,0}^{(3)}\right)$, respectively:

$$\begin{aligned}\Delta'_{12} =& \Delta_{12}\left(\rho_{2,0}, P_{2,0}^{(2)}, \gamma_0\right) + \frac{\eta_2 |c_{21}|^2}{1 + |c_{21}|^2\left(\rho_{2,0} P_{2,0}^{(2)} + P_{U_2}\left(\rho_{2,0}, \gamma_0, P_{2,0}^{(2)}\right)\right)} \cdot \\ &\left\{\left(P_{2,0}^{(2)} - (1-\gamma_0) P_{2,0}^{(2)}\right)(\rho_2 - \rho_{2,0}) + (\rho_{2,0} + (1-\gamma_0)(1-\rho_{2,0}))\left(P_2^{(2)} - P_{2,0}^{(2)}\right) + \right. \\ &\left.\left(-(1-\rho_{2,0}) P_{2,0}^{(2)}\right)(\gamma - \gamma_0)\right\}, \\ \Delta'_{14} =& \Delta_{14}\left(\rho_{3,0}, P_{2,0}^{(3)}\right) + \frac{\eta_3 |c_{21}|^2}{1 + |c_{21}|^2 \rho_{3,0} P_{2,0}^{(3)}} \left(P_{2,0}^{(3)}(\rho_3 - \rho_{3,0}) + \rho_{3,0}\left(P_2^{(3)} - P_{2,0}^{(3)}\right)\right).\end{aligned}$$

After the linearization, we define

$$\Delta'_1 = \Delta_{11} + \Delta_{13} - \Delta'_{12} - \Delta'_{14}. \tag{13.26}$$

Due to the first-order Taylor expansion, it is clear that $\Delta'_{12} \geq \Delta_{12}$ and $\Delta'_{14} \geq \Delta_{14}$. Therefore, we have

$$\Delta_1 \geq \Delta'_1. \tag{13.27}$$

Similarly, we can linearize the negative terms in Δ_2 to get Δ_2' expressed as follows:

$$\Delta_2' \triangleq \Delta_{21} + \Delta_{23} - \Delta_{22}' - \Delta_{24}', \tag{13.28}$$

where

$$\begin{aligned}
\Delta_{22}' &= \Delta_{22}\left(\rho_{3,0}, P_{2,0}^{(3)}, \gamma_0\right) + \frac{\eta_2 |c_{22}|^2}{1 + |c_{22}|^2\left(\rho_{2,0} P_{2,0}^{(2)} + P_{U_2}\left(\rho_{2,0},\ \gamma_0,\ P_{2,0}^{(2)}\right)\right)} \cdot \\
&\quad \left\{\left(P_{2,0}^{(2)} - (1-\gamma_0)P_{2,0}^{(2)}\right)(\rho_2 - \rho_{2,0}) + \rho_{2,0} + (1-\gamma_0)(1-\rho_{2,0})\left(P_2^{(2)} - P_{2,0}^{(2)}\right)\right. \\
&\quad \left. +\left(-(1-\rho_{2,0})P_{2,0}^{(2)}\right)(\gamma - \gamma_0)\right\}, \\
\Delta_{24}' &= \Delta_{24}\left(\rho_{3,0}, P_{2,0}^{(3)}\right) + \frac{\eta_3 |c_{22}|^2}{1 + |c_{22}|^2 \rho_{3,0} P_{2,0}^{(3)}}\left(P_{2,0}^{(3)}(\rho_3 - \rho_{3,0}) + \rho_{3,0}\left(P_2^{(3)} - P_{2,0}^{(3)}\right)\right), \\
\Delta_{22} &= \eta_2 \log\left(1 + |c_{22}|^2\left(\rho_2 P_2^{(2)} + P_{U_2}\right)\right), \\
\Delta_{24} &= \eta_3 \log\left(1 + |c_{22}|^2 \rho_3 P_2^{(3)}\right).
\end{aligned}$$

Then we have $\Delta_2 \geq \Delta_2'$. Combining the previous, we can formulate a new optimization problem as follows:

Problem 3: (13.29a)

$$\max_{\substack{\eta_2,\rho_2,\rho_3\\ \gamma, P_2^{(2)}, P_2^{(3)}}} \frac{\eta_2 \tilde{R}_2(\rho_2,\ \gamma,\ P_2^{(2)})}{\mu\left(\eta_2 P_2^{(2)} + (1-\eta_1-\eta_2)P_2^{(3)}\right) + P_c} \tag{13.29b}$$

$$\text{s.t.}\quad \Delta_1' \geq (1-\eta_1)\mathscr{C}(P_1), \tag{13.29c}$$

$$\Delta_2' = (1-\eta_1)\mathscr{C}(|c_{12}|^2 P_1), \tag{13.29d}$$

$$|c_{TT}| > |c_{11}|, \tag{13.29e}$$

$$\eta_2 P_2^{(2)} + (1-\eta_1-\eta_2)P_2^{(3)} \leq P_2,\ P_2^{(2)} \geq 0,\ P_2^{(3)} \geq 0, \tag{13.29f}$$

$$\rho_2 \geq 0,\ \rho_3 \geq 0,\ \gamma \geq 0, \tag{13.29g}$$

$$\eta_2 + \eta_3 = 1 - \eta_1,\ \eta_2 \geq 0,\ \eta_3 \geq 0. \tag{13.29h}$$

We can easily prove that the solution of Problem 3 leads to an EE which is a lower bound of that of Problem 2.

Proposition 13.5. *The EE of Problem 3 is smaller than the EE of Problem 2.*

As we can easily see that there are terms in the form of product of several variables, which makes the objective function and constraints non-convex. To transform it into a convex one, we consider the alternating algorithm by considering four groups of variables: (1) η_2; (2)ρ_2 and ρ_3; (3) γ; and (4) $P_2^{(2)}$ and $P_2^{(3)}$. The proposed algorithm is

---

Algorithm 13.2: EE maximization for a cognitive radio channel with a secure coexistence constraint

---

$\ell = 0$; select a feasible point $(\eta_2^{(0)}, \rho_2^{(0)}, \rho_3^{(0)}, \gamma_0^{(0)}, P_2^{(2),(0)}, P_2^{(3),(0)})$ as initial values;
repeat
1) Given $(P_2^{(2),(l-1)}, P_2^{(3),(l-1)}, \gamma^{(l-1)}, \rho_2^{(l-1)}, \rho_3^{(l-1)})$, solve Problem 3 in (13.29) w.r.t. η_2 by Algorithm 13.1.
2) Given $(\eta_2^{(l)}, \rho_2^{(l-1)}, \rho_3^{(l-1)}, \gamma^{(l-1)})$, solve Problem 3 in (13.29) w.r.t. $(P_2^{(2)}, P_2^{(3)})$ by Algorithm 13.1.
3) Given $(P_2^{(2),(l)}, P_2^{(3),(l)}, \gamma^{(l-1)})$, solve Problem 3 in (13.29) w.r.t. (ρ_2, ρ_3) by Algorithm 13.1.
4) Given $(P_2^{(2),(l)}, P_2^{(3),(l)}, \eta_2^{(l)}, \rho_2^{(l)}, \rho_3^{(l)})$, solve Problem 3 in (13.29) w.r.t. γ by Algorithm 13.1.
until Convergence

---

summarized in Algorithm 13.2, where for the lth round of iteration, the variables are expressed by $\eta_2^{(l)}, \rho_2^{(l)}, \rho_3^{(l)}, \gamma_0^{(l)}, P_2^{(2),(l)}$, and $P_2^{(3),(l)}$. From [57, Proposition 2.3.1], we know that if there is a unique solution to each of the coordinate-wise minimization problems, then any limit point of the sequence of solutions is a stationary point (which could be a minimum or a saddle point).

By applying the alternating algorithm, even though Problem 3 is still a fractional program, it has a concave numerator in the objective function, in contrast to (13.24a) and (13.24b). This property enables us to solve the four subproblems in Algorithm 13.2 with polynomial complexity by means of standard fractional programming approaches, as described in Section 13.2.2.

13.4 Numerical experiments and assessments

In this section, we investigate thoroughly the optimization of Problem 3. We first describe our geometrical setup in Section 13.4.1. We then show the performance of our three-phase clean relaying scheme through numerical simulation, with DPC encoding at T_2 in Section 13.4.2.

13.4.1 Setup

For the numerical illustrations of the theoretical results, we consider the following four scenarios:

1. EE of the CR systems versus different locations of CR receiver, while the locations of the primary transmitter and receiver and also the CR transmitter are fixed. In particular, we fix the locations of the primary transmitter T_1 and receiver U_1 at the coordinates $(0, 0)$ and $(1, 0)$, respectively. The secondary transmitter is also fixed, which is at $(0, 0.16)$.

2. EE of the CR systems versus different locations of the CR transmitter, while the locations of the primary transmitter and receiver and also the CR receiver are fixed. In particular, we fix the locations of the primary transmitter T_1 and receiver U_1 at the coordinates $(0, 0)$ and $(1, 0)$, respectively. The secondary receiver is at $(0, 1)$.
3. EE of the CR systems versus transmit power for selected locations of the CR receiver where we consider two different objective functions: maximizing the EE and the achievable rate. In particular, we select the following four locations for U_2: (0.3, 1), (0.5, 1), (0.7, 1), and (1, 1).
4. Achievable rate of the CR systems versus signal-to-noise ratio (SNR) where again we consider two different objective functions: maximizing the EE and the achievable rate. In particular, we select the following four locations for U_2: (0.3, 1), (0.5, 1), (0.7, 1), and (1, 1).

We assume a path-loss model with path-loss exponent $\alpha = 3$. Note that P_1 in the expression of $\mathscr{C}(P_1)$ in Proposition 13.4 is, in fact, the SNR, because we normalize the noise power to one with the same unit of the transmit power, while P_1 which is not inside the function $\mathscr{C}(.)$ is the transmit power. Here we select the transmit power constraints at T_1 as $P_1 = 10$ mW and the noise power as 1 mW such that the SNR is 10 dB. Note that we also include power control as a possible design parameter for T_2, i.e., the transmit power utilized is not necessarily fixed to its maximum P_2. The power consumption parameters are $P_c = 10$ mW and $\mu = 2.5$. The unit of rate results is bit per channel use. Note that we use the solver CVX from [58,59] to solve the optimization problem numerically. It cannot guarantee a globally optimal solution for the considered problems Problem 2 and Problem 3 that are non-convex. In contrast, Problem 3 solved by Algorithm 13.2 is convex, and from the sequential optimization we can get the global solution for it. However, this solution is local for Problem 2.

13.4.2 Numerical results

Recall that we assume that all nodes are half duplex. Note that we set a forbidden zone around the primary transmitter such that the CR receiver cannot enter. Therefore, the EE of the smallest square enclosing T_1 is zero. The color bar is with a unit in bits/channel use/mW.

The results of the first scenario are shown in Figure 13.2. Without the relay function at T_1 and also the half-duplex assumption, U_2 inside the circle (shown in orange in Figure 13.2) where T_1 is the origin with radius 1 will cause a degraded wiretap channel where U_2 will have a stronger channel than that between T_1 and U_1. As a result, the secrecy rate of the primary user will be zero and then the EE will also be zero. In contrast, when T_2 is equipped with the relay function, U_2 is allowed to operate inside the aforementioned circle, while the primary user's secure coexistence constraint can be guaranteed. It can be observed that when the CR receiver is closer to the CR transmitter T_2, the higher EE we can attain, which is consistent with the intuition.

The results of the second scenario are shown in Figure 13.3. It can be observed that better EE for the CR system happens if the CR transmitter is close to T_1. This

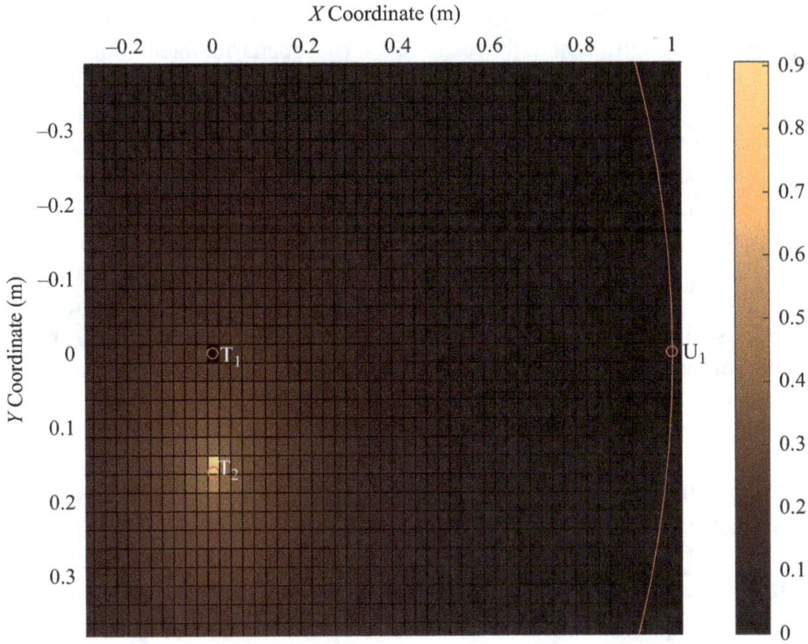

Figure 13.2 Achievable EE versus geo-location. T_1 is at (0,0), U_1 is at (1,0), T_2 is at (0.16,0), and U_2 is at any location except (0,0) and (0.16,0). $P_1 = 10\,dBm$.

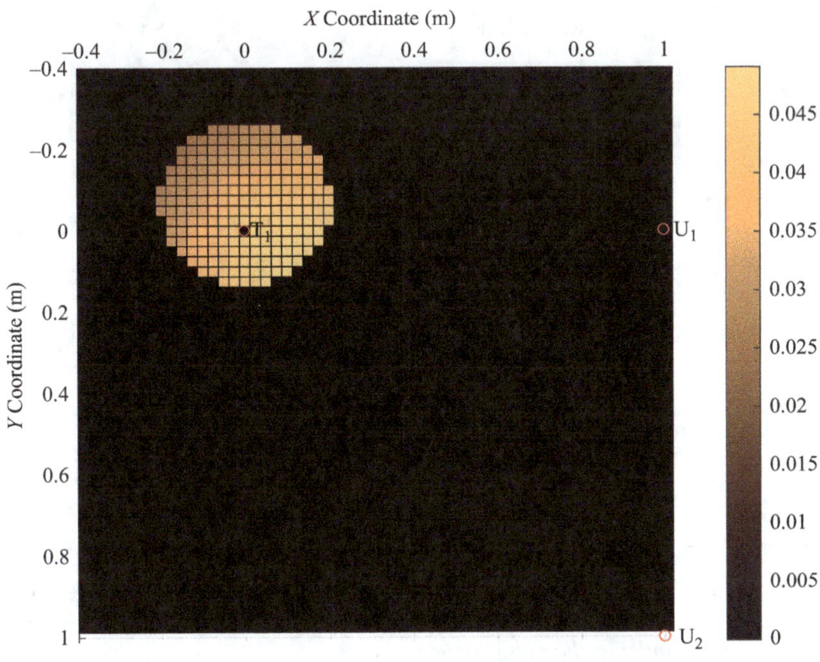

Figure 13.3 Achievable EE versus geo-location. T_1 is at (0,0), U_1 is at (1,0), U_2 is at (1,1), and T_2 is at any location except (0,0) and (1,1). $P_1 = 10\,dBm$.

behavior is similar to figure 2 in [34], which does not consider the secrecy constraint at the primary user. Different to the result of the first scenario, interestingly the higher EEs are not obtained for T_2 located close to U_2 which can be explained in two ways. First, due to the decodability condition, the location of T_2 is bounded inside the circle around T_1 with a radius as the distance between T_1 and U_1. Second, as T_2 must maintain primary user's secrecy rate unchanged, T_2 must help the primary transmission in the classical relaying sense, for which the optimal location of T_2 is between T_1 and U_1.

The results of the third scenario are shown in Figure 13.4, which illustrates the EE curves versus SNR under different locations of the secondary receiver. We can observe that all gaps among the curves corresponding to EE maximizations remain constant for sufficiently high $P_{\max}$. The reason for this behavior is that for high $P_{\max}$ the EE-maximizing resource allocation strategy is not to transmit at maximum power. As a consequence, for large $P_{\max}$, the actual transmit power does not grow large but remains constantly equal to the finite maximizer of the EE. For the simulated scenario, this happens at $P_{\max} = 2, 2, 4, 6$ dBm, respectively, for CR receiver at (0.3, 0.2), (0.5, 0.2), (0.7, 0.2), and (1, 0.2), respectively. If $P_{\max}$ is further increased, the excess power is not used as it would lead to a decrease in the EE. Instead, it can be seen that the EE achieved by rate maximization is increasing and equal to that obtained by EE maximization up to $P_{\max} = 2, 2, 4, 6$ dBm, respectively, and then starts decreasing.

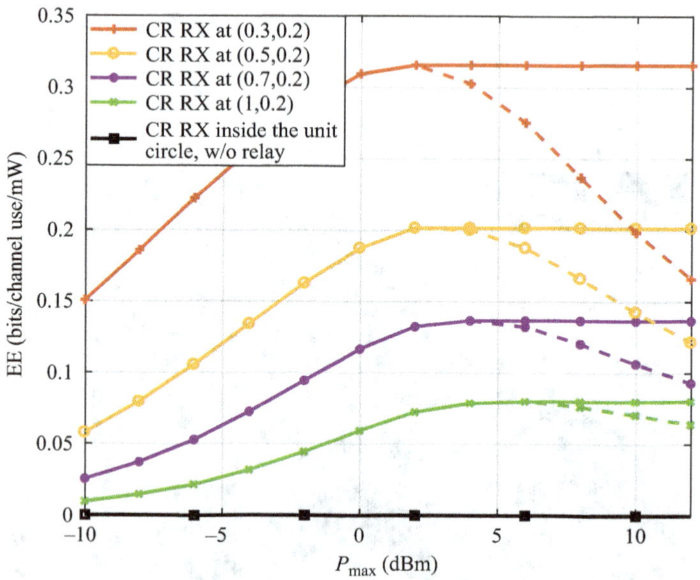

Figure 13.4 Achievable EE versus $P_{\max}$ for four different locations of U_2: (0.3, 0.2), (0.5, 0.2), (0.7, 0.2), (1, 0.2), while T_1, U_1, and T_2 are fixed at (0,0), (1,0), and (−0.16,0), respectively. $P_1 = 10$ dBm. Solid lines are solved from maximizing EEs, while dashed lines are solved from maximizing the achievable rates.

This is because, unlike the EE, for rate maximization, we need to use the excess power. However, this will degrade the EE value. In contrast, when P_{max} is low, full power transmission is the optimal strategy for both EE and secrecy rate maximization. We also show the case in which the CR transmitter has no function of relay and the CR receiver is inside the unit circle. As mentioned that in this case the channel between the primary transmitter and the CR receiver is stronger channel than that between the primary transmitter and the primary receiver, without the relaying the coexistence conditions cannot be fulfilled. This leads to zero EEs for all P_{max}.

The result of the fourth scenario is shown in Figure 13.5. The achievable rate obtained by EE maximization eventually saturates with P_{max}. Instead, the achievable rate obtained by rate maximization increases with P_{max}, showing that the achievable rate is monotonically increasing with P_{max}. More specifically, after expansion of the right-hand side of (13.16), we can see that R_2 is convex of $P_2^{(2)}$. In addition, if we fix η_2, ρ_2, ρ_3, and γ, increasing P_{max} will result in a larger feasible range of $P_2^{(2)}$, which will lead to this phenomenon. We also show the case in which the CR transmitter has no function of relay, and the CR receiver is inside the unit circle. Similar to Case 3, the coexistence conditions cannot be fulfilled without the relaying, which leads to zero rates for all P_{max}.

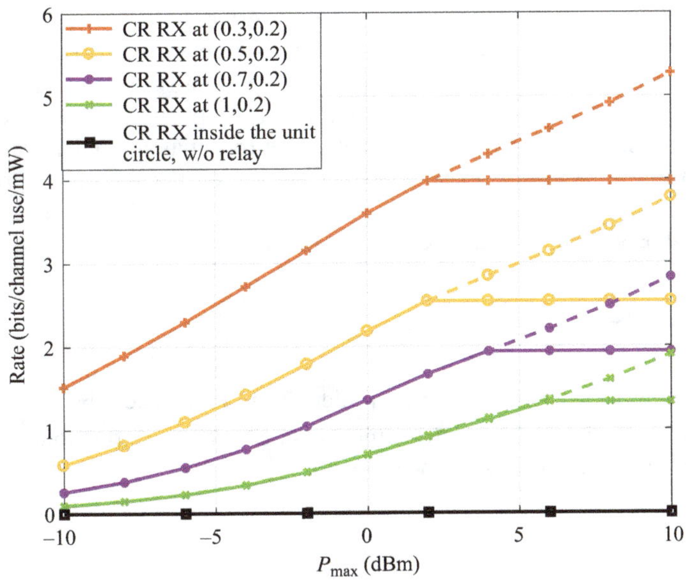

Figure 13.5 Achievable rate versus P_{max} for four different locations of T_2: (0.3, 0.2), (0.5, 0.2), (0.7, 0.2), (1, 0.2), while T_1, U_1, and T_2 are fixed at (0,0), (1,0), and (−0.16,0), respectively. P_1 = 10 dBm. Solid lines are solved from maximizing EEs, while dashed lines are solved from maximizing the achievable rates.

13.5 Conclusions

This chapter focused on the issue of radio resource allocation in overlay CR systems, where the primary system employs physical-layer security techniques. Unlike most previous contributions that focus on traditional and non-EE performance metrics such as achievable rates, this chapter has described a radio resource allocation framework to optimize both the energy efficiency of a cognitive communication system while keeping the confidentiality of the primary system. To this end, a key step has been the introduction of the performance metric of EE, defined as the ratio between the system capacity (or achievable rate) and the consumed power, including both radiated and hardware power. This metric measures the amount of information that can be reliably transmitted per Joule of consumed energy, thus being a natural measure of the energy efficiency of the communication. Both the ability of cognition and EE are anticipated to be particularly relevant in future communication networks, which will be composed of an unprecedented amount of devices.

The resulting fractional and NP-hard optimization problems have been tackled by a combination of fractional programming and sequential optimization theories. The described alternating algorithm, although not being theoretically guaranteed to be globally optimal, is able to monotonically increase the EE value after each iteration, also enjoying first-order KKT optimality properties while at the same time enjoying limited complexity.

The application of the considered model, performance metric, and the proposed scheme can be applied to a legacy system that uses wiretap coding scheme to ensure the secrecy, e.g., in some of the scenarios in 5G systems, to provide an EE design. In particular, such legacy system is not required to cooperate with the CRN and the CRN can be plugged and play independently. The formulated problem can be easily specialized to the case without secrecy constraint, i.e., the secrecy is provided by cryptography and the objective can also be easily specialized to the rate instead of EE, which shows the diverse applications of this work.

Appendix I Proof of Proposition 13.5

Proof. Since $\Delta_1 \geq \Delta_1'$ and $\Delta_2 \geq \Delta_2'$, Problem 2 has a larger feasible range than that of Problem 3. In addition, under the same ρ_2, γ, $P_2^{(2)}$, $R_2 \geq \tilde{R}_2$ due to the first-order Taylor expansion, which finishes the proof. □

References

[1] Liang Y and Poor HV. Multiple Access Channels With Confidential Messages. IEEE Transactions on Information Theory. 2008;54:976–1002.

[2] Zhou X, Ganti RK, and Andews JG. Secure Wireless Network Connectivity with Multi-Antenna Transmission. IEEE Transactions on Wireless Communications. 2011;10(2):425–430.

[3] Wyner AD. The Wiretap Channel. Bell Labs Technical Journal. 1975;54: 1355–1387.

[4] Csiszár I and Korner J. Broadcast Channels With Confidential Messages. IEEE Transactions on Information Theory. 1978;24(3):339–348.

[5] Shafiee S and Ulukus S. Towards the Secrecy Capacity of the Gaussian MIMO Wire-Tap Channel: The 2-2-1 Channel. IEEE Transactions on Information Theory. 2009;55(9):4033–4039.

[6] Khisti A and Wornell GW. Secure Transmission With Multiple Antennas-II: The MIMOME Wiretap Channel. IEEE Transactions on Information Theory. 2010;56(11):5515–5532.

[7] Oggier F and Hassibi B. The Secrecy Capacity of the MIMO Wiretap Channel. IEEE Transactions on Information Theory. 2011;57(8):4961–4972.

[8] Liu T and Shamai (Shitz) S. A Note on the Secrecy Capacity of the Multi-Antenna Wiretap Channel. IEEE Transactions on Information Theory. 2009;55(6):2547–2553.

[9] Liang Y, Poor V, and Shamai (Shitz) S. Secure Communication Over Fading Channels. IEEE Transactions on Information Theory. 2008;54(6):2470–2492.

[10] Bashar S, Ding Z, and Xiao C. On Secrecy Rate Analysis of MIMO Wiretap Channels Driven by Finite-Alphabet Input. IEEE Transactions on Communications. 2012;60(12):3816–3825.

[11] Liang Y, Poor HV, and Shamai (Shitz) S. Information Theoretic Security. Foundations and Trends in Communications and Information Theory. 2009;5: 355–580.

[12] Laneman JN, Tse D, and Wornell GW. Cooperative Diversity in Wireless Networks: Efficient Protocols and Outage Behavior. IEEE Transactions on Information Theory. 2004;50(12):3062–3080.

[13] Lai L and El Gamal H. The Relay—Eavesdropper Channel: Cooperation for Secrecy. IEEE Transactions on Information Theory. 2008;54(9):4005–4019.

[14] Tekin E and Yener A. The General Gaussian Multiple-Access and Two-Way Wiretap Channels: Achievable Rates and Cooperative Jamming. IEEE Transactions on Information Theory. 2008;54(6):2735–2751.

[15] Dong L, Han Z, Petropulu A, *et al.* Improving Wireless Physical Layer Security via Cooperating Relays. IEEE Transactions on Signal Processing. 2010;58(3):1875–1888.

[16] Gabry F, Li N, Girnyk M, *et al.* On the Optimization of the Secondary Transmitter's Strategy in Cognitive Radio Channels with Secrecy. IEEE Journal on Selected Areas in Communications. 2014;32(3):451–463.

[17] Tang X, Liu R, Spasojevic P, *et al.* Interference Assisted Secret Communication. IEEE Transactions on Information Theory. 2011;57(5):3153–3167.

[18] Bassily R and Ulukus S. Deaf Cooperation and Relay Selection Strategies for Secure Communication in Multiple Relay Networks. IEEE Transactions on Signal Processing. 2013;61(6):1544–1554.

[19] Kim T and Poor V. On the Secure Degrees of Freedom of Relaying with Half-Duplex Feedback. IEEE Transactions on Information Theory. 2011;57(1): 291–302.

[20] Gamal AE, and Kim YH. Network Information Theory. New York, NY: Cambridge University Press, 2012.

[21] Mitola J. Cognitive Radio: An Integrated Agent Architecture for Software Defined Radio. KTH; 2000.

[22] Rini S, Tuninetti D, and Devroye N. New Inner and Outer Bounds for the Discrete Memoryless Cognitive Interference Channel and Some Capacity Results. IEEE Transactions on Information Theory. 2011;57(7):4087–4109.

[23] Rini S, Tuninetti D, and Devroye N. Inner and Outer Bounds for the Gaussian Cognitive Interference Channel and New Capacity Results. IEEE Transactions on Information Theory. 2012;58(2):820–848.

[24] Auer G, Giannini V, Desset C, *et al.* How Much Energy Is Needed to Run a Wireless Network? IEEE Wireless Communications. 2011;18(5):40–49.

[25] NGMN 5G White Paper; 2015. Available from: https://www.ngmn.org/wp-content/uploads/NGMN_5G_White_Paper_V1_0.pdf.

[26] Buzzi S, Chih-Lin I, Klein TE, *et al.* A Survey of Energy-Efficient Techniques for 5G Networks and Challenges Ahead. IEEE Journal on Selected Areas in Communications. 2016;34(4):697–709.

[27] Zappone A and Jorswieck E. Energy Efficiency in Wireless Networks via Fractional Programming Theory. Foundations and Trends in Communications and Information Theory. 2015;11(3-4):185–396.

[28] Comaniciu C and Poor HV. On Energy-Secrecy Trade-Offs for Gaussian Wiretap Channels. IEEE Transactions on Information and Forensics and Security. 2013;8(2):314–323.

[29] El-Halabi M, Liu T, and Georghiades CN. Secrecy Capacity per Unit Cost. IEEE Journal on Selected Areas in Communications. 2013;31(9):1909–1920.

[30] Ng DWK, Lo ES, and Schober R. Energy-Efficient Resource Allocation for Secure OFDMA Systems. IEEE Transactions on Vehicular Technology. 2012;61(6):2572–2585.

[31] Chen X and Lei L. Energy-Efficient Optimization for Physical Layer Security in Multi-Antenna Downlink Networks with QoS Guarantee. IEEE Communications Letters. 2013;17(4):637–640.

[32] Chen X, Zhonga C, Yuen C, *et al.* Multi-Antenna Relay Aided Wireless Physical Layer Security. IEEE Communications Magazine. 2015;53(12):40–46.

[33] Zappone A, Lin PH, and Jorswieck EA. Energy Efficiency of Confidential Multi-Antenna Systems With Artificial Noise and Statistical CSI. IEEE Journal of Selected Topics in Signal Processing. 2016;10(8):1462–1477.

[34] Lin P, Gabry F, Thobaben R, *et al.* Multi-Phase Smart Relaying and Cooperative Jamming in Secure Cognitive Radio Networks. IEEE Transactions on Cognitive Communications and Networking. 2016;2(1):38–52.

[35] Liang Y, Somekh-Baruch A, Poor HV, *et al.* Capacity of Cognitive Interference Channels With and Without Secrecy. 2009;55(2):604–619.

[36] Bafghi HG, Salimi S, Seyfe B, *et al.* Cognitive Interference Channel with Two Confidential Messages. In: Proc. IEEE International Symposium on Information Theory and Its Applications (ISITA), Taichung, Taiwan; 2010.

[37] Farsani RK and Ebrahimpour R. Capacity Theorems for the Cognitive Radio Channel With Confidential Messages. In: Proc IEEE International Symposium on Information Theory (ISIT). Honolulu, HI, USA; 2014.

[38] Wu Y and Chen X. Robust Beamforming and Power Splitting for Secrecy Wireless Information and Power Transfer in Cognitive Relay Networks. IEEE Communications Letters. 2016;20(6):1152–1155.

[39] Al-Talabani A, Deng Y, Nallanathan A, *et al.* Enhancing Secrecy Rate in Cognitive Radio Networks via Stackelberg Game. IEEE Transactions on Communications. 2016;64(11):4764–4775.

[40] Gabry F, Zappone A, Thobaben R, *et al.* Energy Efficiency Analysis of Cooperative Jamming in Cognitive Radio Networks With Secrecy Constraints. IEEE Wireless Communications Letters. 2015;4(4):437–440.

[41] Liu X, Zheng K, Fu L, *et al.* Energy Efficiency of Secure Cognitive Radio Networks with Cooperative Spectrum Sharing. IEEE Transactions on Mobile Computing. 2019;18(2):305–318.

[42] Jovicic A and Viswanath P. Cognitive Radio: An Information-Theoretic Perspective. IEEE Transactions on Information Theory. 2009;55(9):3945–3958.

[43] Dinkelbach W. On Nonlinear Fractional Programming. Management Science. 1967;13(7):492–498.

[44] Crouzeix JP, Ferland JA, and Schaible S. An Algorithm for Generalized Fractional Programs. Journal of Optimization Theory and Applications. 1985;47(1):35–49.

[45] Azarian K, El Gamal H, and Schniter P. On the Achievable Diversity-Multiplexing Tradeoff in Half-Duplex Cooperative Channels. IEEE Transactions on Information Theory. 2005;51(12):4152–4172.

[46] Liang YB, Poor HV, and Shamai S. Secure Communication Over Fading Channels. IEEE Transactions on Information Theory. 2008;54(6):2470–2492.

[47] Mitran IP, Ochiai H, and Tarokh V. Space-Time Diversity Enhancements Using Collaborative Communications. IEEE Transactions on Information Theory. 2005;51(6):2041–2057.

[48] Azarian K, El-Gamal H, and Schniter P. On the Achievable Diversity Multiplexing Tradeoff in Half-Duplex Cooperative Channels. IEEE Transactions on Information Theory. 2005;51(12):4152–4172.

[49] Zappone A and Jorswieck EA. Energy Efficiency in Wireless Networks via Fractional Programming Theory – Foundations and Trends in Communications and Information Theory. vol. 11. 2nd ed. Netherlands: Now; 2015.

[50] Marks BR and Wright GP. A General Inner Approximation Algorithm for Non-Convex Mathematical Programs. Operations Research. 1978;26(4):681–683.

[51] Beck A, Ben-Tal A, and Tetruashvili L. A Sequential Parametric Convex Approximation Method With Applications to Non-Convex Truss Topology Design Problems. Journal of Global Optimization. 2010;47(1).

[52] Razaviyayn M, Hong M, and Luo ZQ. A Unified Convergence Analysis of Block Successive Minimization Methods for Nonsmooth Optimization. SIAM Journal on Optimization. 2013;23(2).

[53] Chiang M, Wei C, Palomar DP, *et al.* Power Control by Geometric Programming. IEEE Transactions on Wireless Communications. 2007;6(7):2640–2651.
[54] Venturino L, Prasad N, and Wang X. Coordinated Scheduling and Power Allocation in Downlink Multicell OFDMA Networks. IEEE Transactions on Vehicular Technology. 2009;58(6):2835–2848.
[55] Venturino L, Zappone A, Risi C, *et al.* Energy-Efficient Scheduling and Power Allocation in Downlink OFDMA Networks With Base Station Coordination. IEEE Transactions on Wireless Communications. 2015;14(1):1–14.
[56] Zappone A, Jorswieck EA, and Buzzi S. Energy Efficiency and Interference Neutralization in Two-Hop MIMO Interference Channels. IEEE Transactions on Signal Processing. 2014;62(24):6481–6495.
[57] Bertsekas DP. Nonlinear Programming. 3rd ed. Massachusetts: Wiley; 2016.
[58] Grant M and Boyd S. CVX: MATLAB® Software for Disciplined Convex Programming, version 2.1; 2014. http://cvxr.com/cvx.
[59] Grant M and Boyd S. Graph Implementations for Nonsmooth Convex Programs. In: Blondel V, Boyd S, Kimura H, editors. Recent Advances in Learning and Control. Lecture Notes in Control and Information Sciences. Springer-Verlag Limited; 2008. p. 95–110. http://stanford.edu/~boyd/graph_dcp.html.

Chapter 14

Physical-layer security for mmWave massive MIMO communications in 5G networks

Ning Wang[1], *Long Jiao*[1], *Jie Tang*[1], *and Kai Zeng*[1]

Wireless communications play an important role in meeting the continuously increasing demands for transmission of important/private information, spanning credit card information, e-health data, control messages, etc. However, on the other hand, due to the broadcast nature of the wireless medium, wireless networks are subject to physical-layer threats such as eavesdropping, spoofing, and jamming attacks. In the fifth-generation (5G) wireless communication networks, information security is a critical issue [1]. Basically, security strategies today rely on cryptographic techniques and associated protocols. Nevertheless, these solutions have drawbacks, for example, standardized protections within public wireless networks are not secure enough, and many of their weaknesses are well known; even if enhanced protocols exist, they incur strong constraints and high additional costs for the users of public networks [2]. Therefore, physical-layer security (PLS), a new security method, which safeguards wireless communications based on the information theoretic approach has received significant research interest recently, especially in the 5G communication techniques.

In the list of the wireless 5G enablers, massive multiple-input–multiple-output (massive MIMO) and millimeter wave (mmWave) communications are two dominant technologies [3]. With a large number of antennas, the ability of massive MIMO to create extremely narrow beams for many users is gaining the attention of industry and academia. Massive MIMO has the characteristic of beam spatial multiplexing, which enables simultaneous communication with different terminals at different geographical locations. This method can greatly improve spectrum efficiency [4]. With the carrier frequencies of 30–300 GHz, the mmWave technique would allow service providers to significantly expand the channel bandwidths. Thus, the data capacity of mmWave is significantly increased and the latency for data traffic is greatly decreased, which can dramatically facilitate internet-based access and applications that require minimal latency [5].

Furthermore, massive MIMO and mmWave can complement each other, as shown in Figure 14.1. The tiny wavelength in mmWave allows for hundreds of antenna elements to be placed in an array on a relatively small physical platform [6].

[1]Department of Electrical and Computer Engineering, George Mason University, Fairfax, VA, USA

Meanwhile, high frequencies in mmWave communications require the beamforming technique from massive MIMO to overcome the path loss [3].

Considering the unique characteristics of mmWave massive MIMO 5G communications, PLS approaches are facing both challenges and opportunities. On the one hand, massive MIMO and mmWave communications bring new and stubborn physical-layer threats to 5G applications. For example, massive MIMO needs to collect accurate channel state information (CSI) for selecting optimal beamforming. In the channel training phase, the channel estimation could be compromised by a pilot-contamination attacker, which can imitate and send the same pilot signals as that of legitimate users (LUs). As a result, the attacker could acquire an illegitimate advantage in the next communication phase [7]. On the other hand, the special characteristics in massive MIMO and mmWave provide PLS new opportunities to counter these physical-layer threats. In mmWave communications, high propagation loss and directionality can be used to counter spoofing attacks and eavesdropping attacks [8], and the unique synthesis properties of mmWave massive MIMO systems can enable an efficient pilot-contamination attack detection [9].

In this chapter, we provide a comprehensive summarization of the recent PLS research about massive MIMO and mmWave 5G communications. In particular, we review the typical physical-layer threats in 5G wireless networks, including eavesdropping, contamination, spoofing, and jamming attacks, and survey state-of-the-art works about PLS techniques to alleviate the physical-layer threats in massive MIMO and mmWave 5G networks. Furthermore, PLS schemes based on mmWave and massive MIMO in other 5G communication scenarios are discussed, involving unmanned aerial vehicle (UAV) communications, non-orthogonal multiple access (NOMA) communications, full-duplex communications, and energy harvesting (EH) communications.

The rest of this chapter is organized as follows. Section 14.1 introduces the possible physical-layer threats in mmWave massive MIMO communications. Section 14.2 describes the latest PLS schemes based on mmWave techniques. In

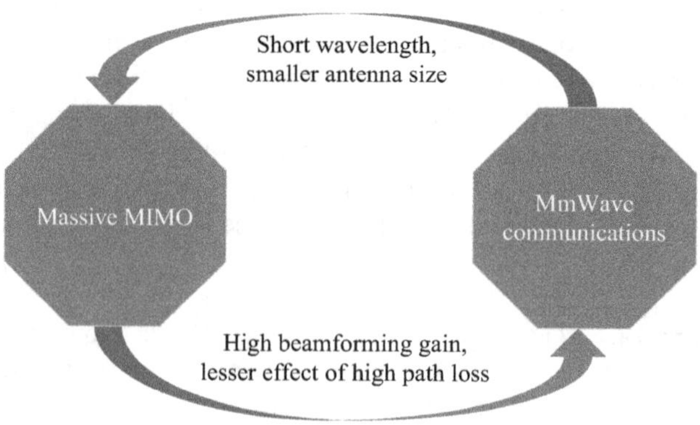

Figure 14.1 Massive MIMO and mmWave communications

Section 14.3, the investigation of PLS schemes based on massive MIMO is provided. PLS schemes in other 5G communication scenarios considering mmWave and massive MIMO technologies are introduced in Section 14.4.

14.1 Physical-layer threats in mmWave massive MIMO

In this section, we offer a window into the typical PLS threats in wireless networks, where these attacks may compromise mmWave massive MIMO communication networks, including eavesdropping, contaminating, spoofing, and jamming, as shown in Figure 14.2. Furthermore, we provide an overview of the corresponding PLS countermeasures and discuss the impact of mmWave massive MIMO on these physical-layer threats.

14.1.1 Eavesdropping

Eavesdropping attackers aim for intercepting some confidential information from legitimate communications. It is hard for the legitimate transmitters or receivers to detect or locate the eavesdroppers, since this type of attacker does not transmit any signal. This physical-layer threat can be divided into two types based on the main manner of the attacker: interception and traffic analysis [10,11].

- **Interception:** Monitoring and eavesdropping are the most common attacks on privacy for wireless communications. The attacker could easily discover legitimate communication by snooping the nearby wireless environment. When the traffic conveys control information about the wireless network configuration, which contains potentially more detailed information than the information accessible through the location server, eavesdropping attackers can act effectively against privacy protection [10].

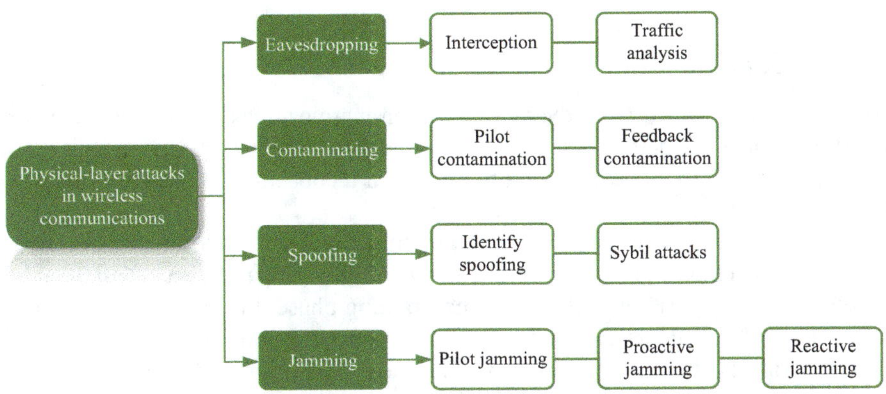

Figure 14.2 Physical-layer security threats in wireless communications

- **Traffic analysis:** Critical information in legitimate communication may be encrypted by cryptographic algorithms. In this case, while the attacker is able to intercept the transmitted signal, the important contents can be protected by the encryption algorithm. However, wireless devices' activities may potentially reveal enough information to enable an attacker to harm the radio networks. Traffic analysis may be useful for tracking the communication patterns to launch other forms of attacks [11].

PLS strategies to counter eavesdropping—To counter eavesdropping attacks, the basic idea in PLS schemes is to create a channel security capacity between LUs, where the transmitter delivers a positive information rate (channel security capacity) to the legitimate receivers and ensures that the eavesdropper cannot obtain any information. Generally, there are two methods to achieve this channel security capacity, involving PLS based on artificial noise (AN) and PLS channel coding. The PLS using AN provides artificial interference to the eavesdroppers to decrease the channel capacity of eavesdropping channels and improve the channel security capacity [12]. The PLS channel coding can achieve this purpose by using the function of error correction in the channel coding, where the probability of the decoding error under the channel between the eavesdropper and LUs will become higher than that of the channels between LUs. The existing PLS channel coding includes LDPC codes [13], Polar codes [14], and Lattice codes [15].

Under mmWave massive MIMO—For mmWave massive MIMO 5G networks, the potential of a PLS solution to this passive eavesdropping attack will be increased dramatically. Per the investigation in [7], resorting to massive MIMO technology, the received signal power at LUs could be several orders of magnitude larger than the received signal power at the eavesdropper. As a result, the channel security capacity offered by PLS schemes is close to the full transmission capacity of the LUs. Thus, if the passive eavesdropping attacker is not very close to the LUs, it is hard to obtain useful information from the legitimate communication under the massive MIMO scenario. Therefore, this physical-layer threat is curbed to a certain degree under 5G networks with massive MIMO.

14.1.2 Contaminating

In order to obtain illegitimate advantages, contaminating attackers attempt to contaminate the channel estimation using faked information in channel estimation phase [16]. This type of attacks can be divided into pilot- and feedback-contamination attacks.

- **Pilot contamination:** In pilot-contaminating attacks, it is assumed that a smart active attacker has precise prior knowledge of pilot information used in the wireless networks [17]. During the channel training phase, the attacker can send the same pilot signals as that of LUs to confuse access point (AP) or base station (BS). In this case, the AP or BS would make a wrong CSI estimation between itself and LUs. This wrong CSI information will in turn yield wrong precoding, beamforming, or successive interference cancellation (SIC) at the AP or BS.

- **Feedback contamination:** Existing mmWave massive MIMO beam-training protocols, like IEEE 802.11ad [18], are to find the optimal antenna steering. The best received probing frame is reported back so that the transmitter can select the corresponding beam for transmissions. To steer these beams to attackers other than the intended users, attackers can inject a forged feedback to induce the transmitter. Even if the transmitter steers the beams in an unintended direction, it could cause a denial-of-service and prevent an association process in beam-training protocols [16].

PLS strategies to counter contaminating—In general, physical-layer authentication (PLA) can be used to counter pilot-contamination attacks, for example, pilot-contamination attack detection and RF/hardware-based PLA. Existing pilot-contamination attack detections emphasize the detection by using special signals, such as pilots design [19] or beamformer design [20]. RF/hardware-based PLA is based on the fact that different wireless transceivers emit RF signals with unique features/patterns in analog and modulation domains, and these features can be explored against the contaminating attacks [21]. For feedback contamination attacks, a countermeasure based on a special sector sweep with authentication was proposed in [16].

Under mmWave massive MIMO—In massive MIMO communications, the beamforming designs at BS/AP are based on the CSI that is estimated through pilot sequences. If the pilot signals are contaminated by attackers, the latter will obtain more chances to eavesdrop on legitimate communication or significantly degrade the communication performance of LUs. Though several methods have been proposed to detect this type of attack [19,20], due to the complex pilot design and retransmission mechanisms, these methods are not desirable PLS solutions for the low-cost 5G IoT devices. Meanwhile, there has not been many works done on how to achieve an efficient RF/hardware-based PLA in 5G IoT networks. Hence, the contaminating attack is still a challenging physical-layer threat for mmWave massive MIMO 5G networks.

14.1.3 Spoofing

The aim of spoofing attackers is to join or destroy legitimate communications. Using a forged identity, spoofing attackers could transmit a deceiving signal with a higher power regardless of the activity of legitimate users. Furthermore, a smarter spoofing attacker can listen to the victim and find the best timing to send the deceiving signal. In general, there are two typical spoofing attacks: identity spoofing and Sybil.

- **Identity spoofing:** Identity spoofing attacks are easy to launch in radio networks. By using a fake identity such as the media access control (MAC) or internet protocol (IP) address of the LU, an identity spoofing attacker can claim to be another legitimate wireless device. The attacker may gain illegal access to the network and launch a more advanced attack, such as man-in-the-middle attacks and denial-of-service attacks [21].
- **Sybil attacks:** In Sybil attack, a malicious node can impersonate other nodes or claim false identities, and the attacker may generate an arbitrary number of

additional node identities, using only one physical device [23,24]. In the presence of Sybil attacks, the radio systems may generate wrong reports, the legitimate user may receive spam and their privacy could be compromised [25].

PLS strategies to counter spoofing—PLA still is the main technology to counter the spoofing attacks in PLS strategies [22]. Other than RF-based/hardware-based PLA, channel-based/location-based PLA techniques offer a way to detect identity spoofing or Sybil attacks, where it can work even with coarse information, such as received signal strength (RSS). If an attacker is at a different location as the legitimate device, channel-based/location-based PLA techniques can efficiently detect it by examining the RSS fingerprint of the signals. Even in Sybil attacks, where an attacker can pretend to be multiple devices by transmitting packets using multiple identities, channel-based/location-based PLA techniques can efficiently detect this type of attack by recognizing that the received signals corresponding to different identities are eccentrically transmitted from the same location.

Under mmWave massive MIMO—Resorting to mmWave massive MIMO communication techniques, LUs could counter this physical-layer threat. Based on spatial distinction, beamforming technologies in massive MIMO and directionality in mmWave could be explored to curb the damage of spoofing attacks. Exploiting the unique features of mmWave massive MIMO technologies to achieve an effective solution for spoofing attacks is an interesting topic.

14.1.4 Jamming

A jamming attacker is to block legitimate communications using noise signals [26]. For this purpose, the adversary could transfer radio signals continuously on a wireless channel to disrupt communication by decreasing the signal-to-noise ratio (SNR). This could lead to denial-of-service attacks at physical layer [27]. Jamming attacks can be classified into three types: pilot jamming, proactive jamming, and reactive jamming.

- **Pilot jamming:** The pilot jamming attack can be viewed as a special case of jamming attacks launched during channel training phase [28]. Its purpose is to corrupt the legitimate communication even without the exact pilot sequences. Only possessing the prior knowledge of the pilot length and pilot sequence codebook, an adversary can launch a pilot jamming attack. This jamming attack can be very energy efficient since the attacker only needs to corrupt the pilot signals, not the entire communication.
- **Proactive jamming:** Proactive jamming attackers would send jamming or interfering signals no matter the legitimate data communication is presenting on the channel or not [29,30]. In order to save energy and toggle between the sleep phase and jamming phase, attacker sporadically spread either arbitrary bits or normal packets into networks. This jamming type can be further classified into spot jamming, sweep jamming, barrage jamming, and deceptive jamming [29].
- **Reactive jamming:** The reactive jamming attackers can monitor the activity of the legitimate channel. If there is an activity, the adversary immediately sends out

a random signal to collide with the existing signal on the channel [31]. Reactive jamming is usually more energy efficient than proactive jamming.

PLS strategies to counter jamming—Generally, PLS strategies to address jamming attacks include frequency-hopping spread spectrum, direct sequence spread spectrum, and ultrawide-band technology [26]. Meanwhile, these countermeasures to jamming attacks can be categorized into jamming detection techniques [32], reactive countermeasures [33], and proactive countermeasures [34]. The detection techniques cannot cope with jamming alone but can instantly detect jamming attacks and provide valuable data to enhance jamming protection with other countermeasures. Reactive countermeasures enable reaction only upon the incident of a jamming attack sensed by the jamming detection. In contrast, the proactive countermeasures will perform the protection mechanism regardless of the activity of jamming attackers. However, this will cost more energy than the reactive countermeasure.

Under mmWave massive MIMO—With regard to mmWave massive MIMO 5G networks, new strategies to defend against jamming attacks are emerging. For example, a new countermeasure is proposed in [35], where the BS estimates not only the legitimate channel in the pilot phase but also the jamming channel by exploiting a purposely unused pilot sequence. The jamming channel estimation is used to construct linear receiver filters that reject the impact of the jamming signal.

14.2 Physical-layer security in mmWave

14.2.1 mmWave communications

Generally speaking, the mmWave frequency band has an abundant bandwidth of up to 5 GHz for per channel, where the high-frequency range of the spectrum is 30–300 GHz and the range of the corresponding wavelength is 1–10 mm [3]. These mmWave bands will open up 100 GHz new spectrum for mobile broadband applications [36]. In the 30–100-GHz bands, about 23-GHz bandwidth is being identified for mmWave cellular, including the unlicensed 60-GHz band (i.e., 57–64-GHz oxygen absorption band), which is best suited for indoor wireless communications [37]. The mmWave communication technique can significantly enhance wireless transmission in many 5G application scenarios. Figure 14.3 shows a typical mmWave application scenario, where a router (Talon AD7200 with 60 GHz) with the mmWave technique is used to connect with other wireless devices in a smart home.

14.2.2 PLS schemes based on mmWave communication

As an important 5G wireless communication technology, mmWave can improve the transmission in 5G networks [6,38,39] and has been considered in many PLS works [9]. Compared to micro-Wave, mmWave has various new characteristics such as blocking effect and highly directional transmissions [8]. Under these new characteristics of mmWave channels, the efficiency of traditional PLS techniques could be

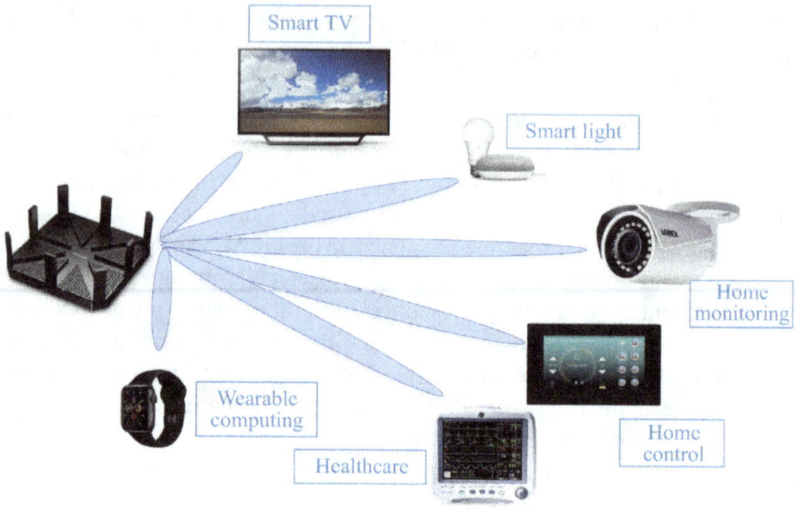

Figure 14.3 mmWave communications in a smart home

improved [8,9]. Here, we introduce some recent typical PLS schemes based on mmWave communications, involving key-generation, AN, hybrid analog-digital designs, countermeasures to multiple eavesdroppers, satellite communications, and directional beamforming.

14.2.2.1 Key generation

MmWave can be exploited to achieve key generation, where key sequence can be generated based on the wireless physical-layer features in mmWave channels. The authors in [40] proposed to use channel characteristics of mmWave, i.e., virtual AoA (angle of arrival) and AoD (angle of departure), to generate a shared secret key between two devices. Other than reciprocity, AoA and AoD of the virtual channel present sparsity and are robust against noise. The proposed key generation scheme can achieve high key generation rates and low bit disagreement ratios even under low SNR (e.g., –10 dB) [40]. The authors claimed that the bit agreement ratio becomes higher when the number of antennas increases, along with higher key generation rates achieved per channel sounding.

14.2.2.2 AN-based mmWave communication

AN is a popular approach in PLS schemes, in which a deliberate noise generated by LUs is exploited to improve the security capacity in wireless communications. Wang *et al.* investigated the secrecy performance of a mmWave cellular network [8]. They proposed a systematic secrecy performance analysis approach for the mmWave cellular communication based on stochastic geometry framework and blockage model, where a secrecy performance analysis of both noise-limited and AN-assisted mmWave communication is conducted. Ju *et al.* explained how to apply the AN-based scheme to mmWave systems [41,42]. In this scheme, when partial eavesdropper's CSI is known at the transmitter, an AN-based transmission strategy is proposed, where this strategy depends on the directions of the destination's and the eavesdropper's propagation paths. The authors also analyze the secrecy outage probability (SOP) through an on–off transmission scheme. A closed-form optimal power allocation between the information bearing signal and AN is derived considering the minimization of SOP subject to a secrecy rate constraint. In [43], three secure transmission schemes were comprehensively investigated over the mmWave channel by evaluating both the SOP and the secrecy throughput. In this PLS scheme, the authors analyzed the new secrecy properties of the maximum ratio transmission (MRT) for the mmWave system. Moreover, AN-based beamforming and partial MRT-based beamforming were proposed.

14.2.2.3 Hybrid analog-digital designs

Hybrid analog-digital precoders can be explored to reduce hardware complexity and power consumption. A hybrid precoders for PLS under two types of channel knowledge was investigated in [44]. In this PLS scheme, the hybrid precoder with full channel knowledge was designed to maximize the secrecy rate. By maximizing the average projection between the fully digital precoder and the hybrid precoder, the authors proposed a low-complexity closed-form hybrid precoder. Furthermore, if there is only partial channel knowledge at the transmitter, the authors provided an SOP upper-bound, where the secrecy throughput maximization is converted into a sequence of SOP minimization problems. Thus, the hybrid precoder is designed to minimize the SOP by an iterative hybrid precoder design. Tian *et al.* considered the problem of hybrid precoder and combiner design for secure transmission in mmWave MIMO systems in order to protect the legitimate transmission from eavesdropping [45]. If eavesdropper's CSI is known, the authors proposed a joint analog precoder and combiner design algorithm. This method can prevent the information leakage to the eavesdropper. In this scheme, in order to further maximize the secrecy rate, the digital precoder and combiner are computed according to the obtained effective baseband channel. Moreover, the authors presented an AN-based hybrid beamforming approach when the prior knowledge of the eavesdropper's CSI is unavailable. Thus, the proposed approach can counter eavesdropper and maintain the quality-of-service of the intended receiver.

14.2.2.4 Countermeasure to multiple eavesdroppers

Multiple eavesdroppers are considered in recent PLS scheme based on mmWave communications. The authors in [46] proposed a network model to characterize the conditional connection outage probability and SOP of hybrid networks under multiple eavesdroppers. In this PLS scheme, the secrecy outage of mmWave networks under the impact of blockages is studied. In [47], the authors investigated PLS in an mmWave MISO scenario, where multiple eavesdroppers with single-antenna are randomly situated in the network. In this work, both MRT beamforming and AN-based beamforming are jointly considered, in which the achievable secrecy level is closely affected by eavesdroppers' density in the network as well as the spatially resolvable paths of both LUs' and eavesdroppers' channels in the mmWave system.

14.2.2.5 Satellite communications

The mmWave communication also can be used in satellite networks. Xiao *et al.* [48] provided a PLS solution of mmWave satellite communications, in which a ray-tracing-based mmWave satellite communication channel model is presented. The authors claimed that the secrecy capacity in the mmWave band widely depends on the richness of the RF environment. Considering a satellite network, Lin *et al.* [49] investigated how to achieve secure beamforming for 5G cellular system operating at mmWave. In this work, it is assumed that the CSI of multiple eavesdroppers based on imperfect AoA is known at the BS, and to maximize the secrecy rate of the worst case for the cellular user, a constrained optimization problem considering the interference and transmit power of BS is formulated. Furthermore, in the case of uncoordinated and coordinated eavesdroppers, two beamforming approaches are provided to solve the optimization problem.

14.2.2.6 Directional beamforming

In recent works, directional beamforming is utilized in mmWave PLS schemes. In [50], the authors investigated the PLS scheme for large-scale mmWave ad hoc networks that can be modeled based on stochastic geometry. The directional beamforming is used between the transmitters and the desired receivers, and the corresponding average secrecy rate is provided. The authors also claimed that using more antennas at the transmitters is beneficial for suppressing the array gains at the eavesdroppers' side under the case of the uniform linear array. Moreover, the power allocation of the transmitted signal and AN is discussed in this work. The authors in [8] considered the downlink secure communication for mmWave cellular networks, where the BS uses the directional beamforming with the intended users' perfect CSI. Here, LU and eavesdroppers in the networks are assumed to be equipped with a single omnidirectional antenna. This work shows that to improve the secrecy performance of mmWave, narrowing the beam width of the directional beamforming antenna with more focused array gain is helpful.

14.3 Physical-layer security in massive MIMO

14.3.1 Massive MIMO communications

Massive MIMO wireless communications refer to the idea of using a very large number of antennas. One of the most important functions of massive MIMO is the ability to form beamforming, which has been shown to potentially allow for orders of magnitude improvement in spectrum. Based on a large number of antennas, the beamforming technique adjusts the parameters of the antenna element in the phase array so that signals of certain angles obtain constructive interference, while signals of other angles acquire destructive interference. Using this feature, massive MIMO can help one to improve the performance of many 5G network applications. For example, Figure 14.4 shows a typical massive MIMO application scenario, where massive MIMO techniques are used to facilitate Internet of Vehicles. In this case, based on beamforming, a BS with multiple antennas simultaneously serves a set of users, and the multiplexing gain can be shared by all users.

14.3.2 PLS schemes based on massive MIMO

As one of the core 5G technologies, massive MIMO has received huge attention on PLS research [4,51]. The investigation of MIMO techniques for PLS is discussed in [52,53]. The authors in [7] investigated the passive eavesdropping and active attacks in massive MIMO. They presented that the potential of PLS against passive eavesdropping attacks in massive MIMO technique can be dramatically increased while facing serious threats under pilot-contamination attacks [17]. In pilot-contamination attacks, an active attacker can send the same pilot sequence as LUs' and cover the legitimate pilot signals during the channel training phase. Thus, the active attacker has a greater chance to disturb the beamforming and degrade the channel gain at LUs while even improving the attacker's channel gain. Therefore, the pilot-contamination attack is a fatal physical-layer threat for massive MIMO communication. Here, we introduce some typical PLS works based on massive MIMO, including pilot-contamination attack detection, countermeasures to jamming attacks, AN, relay-aided MIMO, secure transmission under finite alphabet and hardware impairments, and directional modulation.

14.3.2.1 Pilot-contamination attack detection

In existing researches of PLS, several schemes have been proposed to counter the pilot-contamination attacks in massive MIMO communications. One is a random pilot scheme that introduces controllable randomness in pilot signals to detect this active attacker. There are mainly two variations: random phase-shift keying (PSK) pilot and random frequency shifts (RFSs) pilot. Random PSK pilot is first proposed in [19]. An improved work based on PSK considering three or more observations of channel

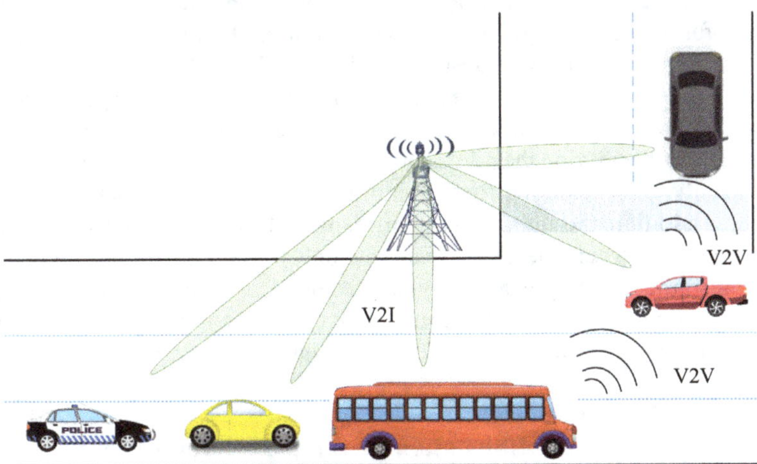

Figure 14.4 Massive MIMO in an Internet of Vehicle

measurement is provided in [54]. First, an LU transmits a sequence of random PSK pilot symbols that are chosen independently from an N-PSK constellation. Second, after obtaining these PSK symbols, AP detects the phase change between two symbols and estimates whether an active attacker exists over this period. If the received signals have valid PSK symbols, there is no active attacker presenting, otherwise, an active attacker is detected. Another random pilot scheme is based on RFSs [20]. An LU deliberately introduces multiple RFSs when transmitting the pilot sequence. Then AP can detect whether there is an active attacker by calculating the autocorrelation matrix for the training phase. In addition, a special beamformer has been designed to detect the activity of pilot-contamination attackers [54]. LU transmits a training signal to AP who uses this training signal to estimate the channel. In the next phase, AP establishes a specialized beamformer that ensures an agreed value of the received signal at LU approaching to 1 and transmits the signals to LU by using the same beamformer. If the attacker is active, the channel estimation in the second phase will inevitably be affected, i.e., the agreed value at LU will sharply decline under the case of active attackers.

In addition, to counter pilot-contamination attacks in massive MIMO, the authors in [55] presented a secret key agreement protocol for a multiuser time-division duplex system, where a BS with a large antenna array aim to generate shared secret keys with users in the presence of non-colluding eavesdroppers. The BS transmits random sequences to LUs to improve the randomness of secret keys. To gain an eavesdropping advantage, each eavesdropper transmits its target user's training sequence. In this work, the authors provided an estimator for the channel gain from the BS to the eavesdropper and proposed a rate-adaptation scheme for adjusting the length of the secret key under the pilot-contamination attack.

14.3.2.2 Countermeasures to jamming attacks

To counter jamming attacks, a pilot retransmission scheme in the channel training phase was proposed [56]. In this work, AP first obtains an achievable massive MIMO uplink rate for a single user with the presence of a pilot jamming attacker. AP then decides whether LU needs to retransmit its pilot signal based on the estimation of the jamming pilot. It is assumed that the jamming attacker may have the prior knowledge of the pilot length and pilot sequence codebook but lacks the knowledge of exact pilot sequence. By retransmitting the pilot sequence, AP could find a proper pilot sequence to minimize the pilot jamming to a certain degree; thus, the impact of jamming is mitigated. In addition, the authors in [35] investigated jamming attacks using massive MIMO technology, where the BS estimates not only the legitimate channel in the pilot phase but also the jamming channel by using a purposely unused pilot sequence. In order to decrease the impact of the jamming signal, the jamming channel estimation is used to construct a linear receiver filter to counter the jamming attacks.

14.3.2.3 AN-based massive MIMO

PLS schemes for massive MIMO systems with passive eavesdroppers have been recently investigated. Zhu *et al.* studied secure massive MIMO transmissions for multicell multiuser systems over i.i.d. Rayleigh fading channel [53], where a passive eavesdropper tries to decode the information sent to one of the users. The authors considered secure downlink transmission in a multicell massive MIMO system with matched filter precoding in the presence of a passive multi-antenna eavesdropper. Under the cases of the perfect training and pilot contamination, the resulting achievable secrecy rate and the SOP are studied. Wang *et al.* [57] proposed a PLS scheme based on AN to protect massive MIMO transmission over i.i.d. Rician fading channel, where the SOP is derived and a jamming-beneficial range is determined. In this work, the authors proposed a directional jamming method that generates jamming signals at selected beams only. In addition, the optimal power allocation between messages and AN is discussed for two different scenarios, including the cases with and without the information of the possible locations of eavesdroppers.

14.3.2.4 Relay-aided MIMO

Chen *et al.* [58] considered a massive MIMO relaying system, where an information source sends the message to its intended destination aided by a massive MIMO relay. It is assumed that a passive eavesdropper tries to intercept the information forwarded by the relay. This work provided a thorough secrecy performance analysis and comparison of two classic relaying techniques, amplify-and-forward (AF) and decode-and-forward (DF). The authors claimed that under large transmit powers, AF is a better choice than DF from the perspectives of both secrecy performance and implementation complexity, and there is an optimal transmit power at the medium regime that maximizes the secrecy outage capacity. The authors in [59] investigated the problem of joint power and time allocation for secure communications in a DF massive MIMO relaying system in the presence of a passive eavesdropper. This work gave a performance analysis of secrecy outage capacity and showed the minimum required number of relay antennas to achieve a positive secrecy outage capacity. In addition, considering the goal of maximizing the secrecy outage capacity, the authors proposed an optimization framework to jointly optimize source transmit power, relay transmit power, and transmission time.

14.3.2.5 Finite alphabet and hardware impairments

The work in [60] investigated secure transmission over massive MIMO wiretap channel with finite alphabet inputs. Considering that finite alphabet inputs may exhibit a severe performance loss in a high SNR, the authors proposed a generalized singular value decomposition (GSVD)-based design as the CSI of the eavesdropper is known at the transmitter. In addition, a per-Group-GSVD (PG-GSVD) design was presented to compensate for the performance loss caused by the GSVD design. Furthermore, this PG-GSVD design is extended to the case where only statistical CSI of the eavesdropper is available at the transmitter. Zhu *et al.* [61] studied the security problem for the case that the number of RF chains at the BSs may be much smaller than the number of BS antennas. To improve the security performance of massive MIMO, a design framework for joint data and AN-based precoding in a multiuser massive MIMO system with the limited number of RF chains was proposed. With imperfect CSI, the authors provided an achievable lower bound on the secrecy rate of any mobile terminal for both analog and hybrid precoding schemes, and the closed-form lower bound was derived to determine optimal power allocation between messages and AN that maximizes the secrecy rate. The authors in [61] investigated the impact of hardware impairments on the secrecy performance of downlink massive MIMO systems in the presence of a passive eavesdropper with multiple antennas. In this work, the joint effects of multiplicative phase noise, additive distortion noise, and amplified receiver noise are taken into account. It is assumed that the eavesdropper employs ideal hardware. In addition, a lower bound for the secrecy rate is provided, where matched filter data precoding and AN transmission are exploited at the BS. Furthermore, this work investigated the impact of the various system parameters on the secrecy rate and optimized both the pilot sets used for uplink training and the AN precoding.

14.3.2.6 Directional modulation

Valliappan *et al.* [62] proposed a low-complexity directional modulation technique, called antenna subset modulation (ASM), for point-to-point mmWave massive MIMO communications. In this work, two methods for implementing ASM were proposed. In the first method, an antenna subset is randomly selected for every symbol, where an expression for the average uncoded symbol error rate is derived as a function of the observation angle. In the second method, an optimized antenna subset selection procedure based on simulated annealing is provided to achieve superior performance compared with random selection.

14.4 PLS schemes integrating mmWave massive MIMO with other 5G scenarios and techniques

As the development of 5G communication, different advanced communication technologies are expected to integrate into the 5G framework. Therefore, considering the combination of mmWave massive MIMO and other 5G technologies will be an important research direction in the development of PLS schemes.

In this section, we will discuss the PLS approaches integrating mmWave massive MIMO with other 5G application scenarios and techniques, involving UAV communications, NOMA communications, full-duplex communications, and EH communications.

14.4.1 UAV communications

UAVs have enormous potential in the public and civil domains, and UAV-aided communications. It can be used for information dissemination, relaying and collection to/from the radio devices because the low-altitude UAVs can be rapidly deployed as UAV-enabled BSs in 5G system with high mobility and fast deployment. Compared favorably with terrestrial systems, UAVs can overcome the propagation and blockage constraints due to terrain characteristics, especially for mmWave, and augment the coverage area by optimizing the UAVs' location (e.g., altitude) in the 3D model [63].

The mmWave massive MIMO can help UAVs to achieve PLS schemes. Khosravi *et al.* [64] analyzed the secrecy performance of 3D UAV-enabled mmWave networks considering air-to-ground channel features (e.g., 3D antenna gain), improved by using part of UAVs to send jamming signals, and used the Matern hardcore point process to guarantee the safety distance among the randomly deployed UAV BSs. Another work investigated the secrecy rates of UAV-based mmWave communication network considering NOMA together with highly directional multi-antenna and optimized a protected zone to enhance the secrecy [65]. The mmWave vehicular communication systems have the potential to improve traffic efficiency and safety [66]. To secure communication links, a physical-layer precoding technique for mmWave vehicular communication systems was proposed in this work. The proposed technique exploited the large dimensional antenna arrays available at mmWave systems to produce direction-dependent transmission. This results in coherent transmission to the legitimate receiver and jams eavesdroppers with sensitive receivers.

14.4.2 NOMA communications

Via non-orthogonal resource allocation, NOMA can improve spectral efficiency and achieve massive connectivity with low transmission latency and signaling cost [67]. It can be used in many applications, from smart manufacturing to smart farming, where there are hordes of sensors and actors [68]. More specifically, user pairing, i.e., assigning two users (near user and far user) to a single orthogonal resource block [69], is developed to trade off the complexity and efficiency. Superposed coding technology is applied to achieve the capability of superposing multiple signals into one orthogonal resource. Accordingly, to extract the corresponding information for separate LUs, SIC is used to eliminate the redundant signals based on corresponding CSI. However, if the pilot sequence is falsified or distorted by pilot-contamination attack, it is more likely to cause the failure of SIC or lead to confidential information leakage.

The authors in [9] presented a new pilot-contamination detection scheme that exploited the characteristics of mmWave massive MIMO communications. This scheme relies only on the signal processing at AP without requiring any additional signal design or change at LU (i.e., IoT device). By harnessing the sparseness and statistics of mmWave and massive MIMO virtual channel, two effective pilot-contamination attack detection schemes are proposed for tackling static and dynamic environments, respectively. For the static environment, the problem of pilot-contamination attack detection is formulated as a binary hypothesis test of virtual channel sparseness. For the dynamic environment, the statistic of the peak in the virtual channel is leveraged to distinguish the contamination state from the normal state. A peak estimation algorithm and a machine-learning-based detection scheme are proposed to achieve high detection performance.

14.4.3 Full-duplex communications

Full-duplex radio, i.e., in-band full duplex [70], can simultaneously transmit and receive data signals within the same frequency band. Thus, this technique can bring radio applications double ergodic capacity and reduce feedback delay, which can enhance virtual reality and augmented reality [70–72].

For PLS solutions, integrating mmWave massive MIMO with full duplex can be used to counter eavesdropping attacks. For example, the authors in [73] considered a multiple-antenna full-duplex BS that communicates with a single-antenna transmitter and a single-antenna receiver simultaneously facing the single-antenna eavesdroppers. This work investigated the joint precoding and AN generation design at the BS with global perfect CSI and perfect SIC to guarantee both the uplink and downlink transmission security. Sun *et al.* [74] studied a more general system where a multiple-antenna full-duplex BS receives information from multiple single-antenna uplink users and transmit information to multiple single antenna downlink users simultaneously in the presence of multiple potential eavesdroppers.

14.4.4 EH communications

In the era of 5G and beyond, EH technique, i.e., simultaneous wireless information and power transfer (SWIPT) technology, is an important and subversive technology, especially in 5G IoT systems where there are lots of power-limited devices. The concept of SWIPT is that the RF signals can be simultaneously utilized to power multiple users for EH and information decoding purpose in a relatively long distance [75].

PLS scheme can be achieved under massive MIMO and EH. For instance, Salem *et al.* [76] analyzed the performance of a multi-antenna energy-constrained AF relay network in the context of PLS. This work derived exact analytical expressions for the ergodic secrecy capacity for various well-known EH relaying protocols, namely, time-switching relaying, power-splitting relaying, and ideal relaying receiver. The benefits of a massive MIMO relay were exploited in [77] for providing communication security in SWIPT networks. It was shown in [77] that the power transfer efficiency is improved significantly by exploiting a large array gain of massive MIMO. Also, information leakage is decreased sharply due to the use of high-resolution spatial beamformer. Furthermore, an energy-efficient power allocation scheme was proposed for such wireless powered massive MIMO secure relaying system in [78], which partially solved the problem of energy efficiency for SWIPT [52].

Acknowledgment

This work was supported in part by the Commonwealth Cyber Initiative (CCI) and its Northern Virginia (NOVA) Node, an investment in the advancement of cyber R&D, innovation and workforce development (for more information about CCI, visit cyberinitiative.org).

References

[1] Ahmad I, Kumar T, Liyanage M, *et al.* Overview of 5G security challenges and solutions. IEEE Communications Standards Magazine. 2018;2(1):36–43.

[2] Wu Y, Khisti A, Xiao C, *et al.* A survey of physical layer security techniques for 5G wireless networks and challenges ahead. IEEE Journal on Selected Areas in Communications. 2018.

[3] Busari SA, Huq KMS, Mumtaz S, *et al.* Millimeter-wave massive MIMO communication for future wireless systems: A survey. IEEE Communications Surveys & Tutorials. 2017;20(2):836–869.

[4] Araujo DC, Maksymyuk T, de Almeida AL, *et al.* Massive MIMO: Survey and future research topics. IET Communications. 2016;10(15):1938–1946.

[5] Akdeniz MR, Liu Y, Samimi MK, *et al.* Millimeter wave channel modeling and cellular capacity evaluation. IEEE Journal on Selected Areas in Communications. 2014;32(6):1164–1179.

[6] Rappaport TS, Sun S, Mayzus R, *et al.* Millimeter wave mobile communications for 5G cellular: It will work!. IEEE Access. 2013;1:335–349.

[7] Kapetanovic D, Zheng G, and Rusek F. Physical layer security for massive MIMO: An overview on passive eavesdropping and active attacks. IEEE Communications Magazine. 2015;53(6):21–27. Available from: ⟨GotoISI⟩://WOS:000356157000004.

[8] Wang C and Wang HM. Physical layer security in millimeter wave cellular networks. IEEE Transactions on Wireless Communications. 2016;15(8):5569–5585.

[9] Wang N, Jiao L, and Zeng K. Pilot contamination attack detection for NOMA in Mm-Wave and massive MIMO 5G communication. In: 2018 IEEE Conference on Communications and Network Security (CNS). Beijing: IEEE; 2018. p. 1–9.

[10] Padmavathi G and Shanmugapriya D. A survey of attacks, security mechanisms and challenges in wireless sensor networks. arXiv preprint arXiv: 0909.0576. 2009.

[11] Shiu YS, Chang SY, Wu HC, *et al.* Physical layer security in wireless networks: A tutorial. IEEE Wireless Communications. 2011;18(2).

[12] Romero-Zurita N, Ghogho M, and McLernon D. Outage probability based power distribution between data and artificial noise for physical layer security. IEEE Signal Processing Letters. 2012;19(2):71–74.

[13] Thangaraj A, Dihidar S, Calderbank AR, *et al.* Applications of LDPC codes to the wiretap channel. IEEE Transactions on Information Theory. 2007;53(8):2933–2945.

[14] Liang X, Zhang M, and Han D. Security performance of polar codes in UV wireless communications. In: 2018 Asia Communications and Photonics Conference (ACP). Hangzhou: IEEE; 2018. p. 1–3.

[15] Oggier F, Solé P, and Belfiore JC. Lattice codes for the wiretap Gaussian channel: Construction and analysis. IEEE Transactions on Information Theory. 2016;62(10):5690–5708.

[16] Steinmetzer D, Ahmad S, Anagnostopoulos N, *et al.* Authenticating the sector sweep to protect against beam-stealing attacks in IEEE 802.11 ad networks. In: Proceedings of the 2nd ACM Workshop on Millimeter Wave Networks and Sensing Systems. New Delhi: ACM; 2018. p. 3–8.

[17] Zhou X, Maham B, and Hjorungnes A. Pilot contamination for active eavesdropping. IEEE Transactions on Wireless Communications. 2012;11(3):903–907.

[18] Nitsche T, Cordeiro C, Flores AB, *et al.* IEEE 802.11 ad: Directional 60 GHz communication for multi-Gigabit-per-second Wi-Fi. IEEE Communications Magazine. 2014;52(12):132–141.

[19] Kapetanovic D, Zheng G, Wong KK, *et al.* Detection of pilot contamination attack using random training and massive MIMO. In: 2013 IEEE 24th International Symposium on Personal, Indoor, and Mobile Radio Communications (PIMRC); 2013. p. 13–18. Available from: ⟨GotoISI⟩: //WOS:000346481200003.

[20] Zhang W, Lin H, and Zhang R. Detection of pilot contamination attack based on uncoordinated frequency shifts. IEEE Transactions on Communications. 2018.

[21] Zeng K, Govindan K, and Mohapatra P. Non-cryptographic authentication and identification in wireless networks [security and privacy in emerging wireless networks]. IEEE Wireless Communications. 2010;17(5).

[22] Yılmaz MH and Arslan H. A survey: Spoofing attacks in physical layer security. In: Local Computer Networks Conference Workshops (LCN Workshops), 2015 IEEE 40th. Clearwater Beach, FL: IEEE; 2015. p. 812–817.

[23] Newsome J, Shi E, Song D, *et al.* The Sybil attack in sensor networks: Analysis & defenses. In: Proceedings of the 3rd International Symposium on Information Processing in Sensor Networks. Berkeley, CA: ACM; 2004. p. 259–268.

[24] Xiao L, Greenstein LJ, Mandayam NB, *et al.* Channel-based detection of Sybil attacks in wireless networks. IEEE Transactions on Information Forensics and Security. 2009;4(3):492–503.

[25] Zhang K, Liang X, Lu R, *et al.* Sybil attacks and their defenses in the Internet of Things. IEEE Internet of Things Journal. 2014;1(5):372–383.

[26] Mpitziopoulos A, Gavalas D, Konstantopoulos C, *et al.* A survey on jamming attacks and countermeasures in WSNs. IEEE Communications Surveys & Tutorials. 2009;11(4).

[27] Dener M. Security analysis in wireless sensor networks. International Journal of Distributed Sensor Networks. 2014;10(10):303501.

[28] Clancy TC. Efficient OFDM denial: Pilot jamming and pilot nulling. In: Communications (ICC), 2011 IEEE International Conference on. Kyoto: IEEE; 2011. p. 1–5.

[29] Mpitziopoulos A, Gavalas D, Konstantopoulos C, *et al.* A survey on jamming attacks and countermeasures in WSNs. IEEE Communications Surveys and Tutorials. 2009;11(4).

[30] Yang H, Shi M, Xia Y, *et al.* Security research on wireless networked control systems subject to jamming attacks. IEEE Transactions on Cybernetics. 2018.

[31] Wood AD, Stankovic JA, and Zhou G. DEEJAM: Defeating energy-efficient jamming in IEEE 802.15. 4-based wireless networks. In: Sensor, Mesh and Ad Hoc Communications and Networks, 2007. SECON'07. 4th Annual IEEE Communications Society Conference on. San Diego, CA: IEEE; 2007. p. 60–69.

[32] Li M, Koutsopoulos I, and Poovendran R. Optimal jamming attacks and network defense policies in wireless sensor networks. In: IEEE INFOCOM 2007—26th IEEE International Conference on Computer Communications. Anchorage, AK: IEEE; 2007. p. 1307–1315.

[33] Xu W, Trappe W, and Zhang Y. Channel surfing: Defending wireless sensor networks from interference. In: Proceedings of the 6th International Conference on Information Processing in Sensor Networks. Cambridge, MA: ACM. p. 499–508.

[34] Law YW, Palaniswami M, Hoesel LV, *et al.* Energy-efficient link-layer jamming attacks against wireless sensor network MAC protocols. ACM Transactions on Sensor Networks (TOSN). 2009;5(1):6.

[35] Do TT, Björnson E, Larsson EG, *et al.* Jamming-resistant receivers for the massive MIMO uplink. IEEE Transactions on Information Forensics and Security. 2017;13(1):210–223.

[36] Shi Z, Lu R, Chen J, *et al.* Three-dimensional spatial multiplexing for directional millimeter-wave communications in multi-cubicle office environments. In: 2013 IEEE Global Communications Conference (GLOBECOM). Atlanta, GA: IEEE; 2013. p. 4384–4389.

[37] Khan F and Pi Z. mmWave mobile broadband (MMB): Unleashing the 3–300GHz spectrum. In 34th IEEE Sarnoff Symposium. Princeton, NJ: IEEE; 2011. p. 1–6.

[38] Niu Y, Li Y, Jin D, *et al.* A survey of milimeter wave (mmwave) communications for 5G: Opportunities and challenges. Computer Science-Networking and Internet Architecture. 2015.

[39] Heath RW, Gonzalez-Prelcic N, Rangan S, *et al.* An overview of signal processing techniques for millimeter wave MIMO systems. IEEE Journal of Selected Topics in Signal Processing. 2016;10(3):436–453.

[40] Jiao L, Tang J, and Zeng K. Physical layer key generation using virtual AoA and AoD of mmWave massive MIMO channel. In: 2018 IEEE Conference on Communications and Network Security (CNS). Beijing: IEEE; 2018. p. 1–9.

[41] Ju Y, Wang HM, Zheng TX, *et al.* Secrecy throughput maximization for millimeter wave systems with artificial noise. In: 2016 IEEE 27th Annual International Symposium on Personal, Indoor, and Mobile Radio Communications (PIMRC). Valencia: IEEE; 2016. p. 1–6.

[42] Ju Y, Wang HM, Zheng TX, *et al.* Secure transmission with artificial noise in millimeter wave systems. In: 2016 IEEE Wireless Communications and Networking Conference. Philadelphia, PA: IEEE; 2016. p. 1–6.

[43] Ju Y, Wang HM, Zheng TX, *et al.* Secure transmissions in millimeter wave systems. IEEE Transactions on Communications. 2017;65(5):2114–2127.

[44] Ramadan YR, Minn H, and Ibrahim AS. Hybrid analog-digital precoding design for secrecy mmWave MISO-OFDM systems. IEEE Transactions on Communications. 2017;65(11):5009–5026.

[45] Tian X, Li M, Wang Z, *et al.* Hybrid precoder and combiner design for secure transmission in mmWave MIMO systems. In: GLOBECOM 2017—2017 IEEE Global Communications Conference. Singapore: IEEE; 2017. p. 1–6.

[46] Vuppala S, Biswas S, and Ratnarajah T. An analysis on secure communication in millimeter/micro-wave hybrid networks. IEEE Transactions on Communications. 2016;64(8):3507–3519.

[47] Ju Y, Wang HM, Zheng TX, *et al.* Safeguarding millimeter wave communications against randomly located eavesdroppers. IEEE Transactions on Wireless Communications. 2018;17(4):2675–2689.

[48] Xiao K, Zhang S, Michel K, *et al.* Study of physical layer security in mmWave satellite networks. In: 2018 IEEE International Conference on Communications (ICC). Kansas City, MO: IEEE; 2018. p. 1–6.

[49] Lin Z, Lin M, Wang JB, *et al.* Robust secure beamforming for 5G cellular networks coexisting with satellite networks. IEEE Journal on Selected Areas in Communications. 2018;36(4):932–945.

[50] Zhu Y, Wang L, Wong KK, *et al.* Secure communications in millimeter wave ad hoc networks. IEEE Transactions on Wireless Communications. 2017;16(5):3205–3217.

[51] Lu L, Li GY, Swindlehurst AL, *et al.* An overview of massive MIMO: Benefits and challenges. IEEE Journal of Selected Topics in Signal Processing. 2014;8(5):742–758.

[52] Chen X, Ng DWK, Gerstacker WH, *et al.* A survey on multiple-antenna techniques for physical layer security. IEEE Communications Surveys & Tutorials. 2017;19(2):1027–1053.

[53] Zhu J, Schober R, and Bhargava VK. Secure transmission in multicell massive MIMO systems. IEEE Transactions on Wireless Communications. 2014;13(9):4766–4781.

[54] Kapetanovic D, Al-Nahari A, Stojanovic A, *et al.* Detection of active eavesdroppers in massive MIMO. In: 2014 IEEE 25th Annual International Symposium on Personal, Indoor, and Mobile Radio Communication (PIMRC); 2014. p. 585–589. Available from: ⟨GotoISI⟩://WOS:000392729300113.

[55] Im S, Jeon H, Choi J, *et al.* Secret key agreement with large antenna arrays under the pilot contamination attack. IEEE Transactions on Wireless Communications. 2015;14(12):6579–6594.

[56] Do TT, Ngo HQ, Duong TQ, *et al.* Massive MIMO pilot retransmission strategies for robustification against jamming. IEEE Wireless Communications Letters. 2017;6(1):58–61.

[57] Wang J, Lee J, Wang F, *et al.* Jamming-aided secure communication in massive MIMO Rician channels. IEEE Transactions on Wireless Communications. 2015;14(12):6854–6868.

[58] Chen X, Lei L, Zhang H, *et al.* Large-scale MIMO relaying techniques for physical layer security: AF or DF?. IEEE Transactions on Wireless Communications. 2015;14(9):5135–5146.

[59] Chen J, Chen X, Gerstacker WH, *et al.* Resource allocation for a massive MIMO relay aided secure communication. IEEE Transactions on Information Forensics and Security. 2016;11(8):1700–1711.

[60] Wu Y, Wang JB, Wang J, *et al.* Secure transmission with large numbers of antennas and finite alphabet inputs. IEEE Transactions on Communications. 2017;65(8):3614–3628.

[61] Zhu J, Xu W, and Wang N. Secure massive MIMO systems with limited RF chains. IEEE Transactions on Vehicular Technology. 2016;66(6):5455–5460.

[62] Valliappan N, Lozano A, and Heath RW. Antenna subset modulation for secure millimeter-wave wireless communication. IEEE Transactions on communications. 2013;61(8):3231–3245.

[63] Al-Hourani A, Kandeepan S, and Lardner S. Optimal LAP altitude for maximum coverage. IEEE Wireless Communications Letters. 2014;3(6): 569–572.

[64] Khosravi Z, Gerasimenko M, Andreev S, and Koucheryavy Y. Performance evaluation of UAV-assisted mmWave operation in mobility-enabled urban deployments. In 2018 41st International Conference on Telecommunications and Signal Processing (TSP). Athens: IEEE; 2018. p. 1–5.

[65] Rupasinghe N, Yapıcı Y, Güvenç I, Dai H, and Bhuyan A. Enhancing physical layer security for NOMA transmission in mmWave drone networks. In 2018 52nd Asilomar Conference on Signals, Systems, and Computers. Pacific Grove, CA: IEEE; 2018. p. 729–733.

[66] Eltayeb ME, Choi J, Al-Naffouri TY, *et al.* On the security of millimeter wave vehicular communication systems using random antenna subsets. In: 2016 IEEE 84th Vehicular Technology Conference (VTC-Fall). Nanjing: IEEE; 2016. p. 1–5.

[67] Ding ZG, Lei XF, Karagiannidis GK, *et al.* A survey on non-orthogonal multiple access for 5G networks: Research challenges and future trends. IEEE Journal on Selected Areas in Communications. 2017;35(10):2181–2195. Available from: ⟨GotoISI⟩://WOS:000410646800001.

[68] Ding Zg, Xu M, Chen Y, *et al.* Embracing non-orthogonalmultiple access in future wireless networks. Frontiers of Information Technology & Electronic Engineering. 2018;19(3):322–339.

[69] Ding Z, Fan P, and Poor HV. Impact of user pairing on 5G nonorthogonal multiple-access downlink transmissions. IEEE Transactions on Vehicular Technology. 2016;65(8):6010–6023.

[70] Kim D, Lee H, and Hong D. A survey of in-band full-duplex transmission: From the perspective of PHY and MAC layers. IEEE Communications Surveys & Tutorials. 2015;17(4):2017–2046.

[71] Abari O, Bharadia D, Duffield A, and Katabi D. Enabling high-quality untethered virtual reality. In 14th USENIX Symposium on Networked Systems Design and Implementation (NSDI 17). Boston, MA: USENIX Association; 2017. p. 531–544.

[72] Bharadia D and Katti S. FastForward: Fast and constructive full duplex relays. ACM SIGCOMM Computer Communication Review. 2015;44(4): 199–210.

[73] Zhu F, Gao F, Yao M, *et al.* Joint information-and jamming-beamforming for physical layer security with full duplex base station. IEEE Transactions on Signal Processing. 2014;62(24):6391–6401.

[74] Sun Y, Ng DWK, Zhu J, *et al.* Multi-objective optimization for robust power efficient and secure full-duplex wireless communication systems. IEEE Transactions on Wireless Communications. 2016;15(8):5511–5526.

[75] Lu X, Wang P, Niyato D, *et al.* Wireless networks with RF energy harvesting: A contemporary survey. IEEE Communications Surveys & Tutorials. 2015;17(2):757–789.

[76] Salem A, Hamdi KA, and Rabie KM. Physical layer security with RF energy harvesting in AF multi-antenna relaying networks. IEEE Transactions on Communications. 2016;64(7):3025–3038.

[77] Chen X, Chen J, and Liu T. Secure transmission in wireless powered massive MIMO relaying systems: Performance analysis and optimization. IEEE Transactions on Vehicular Technology. 2015;65(10):8025–8035.

[78] Du C, Chen X, and Lei L. Energy-efficient power allocation for secure communications in wireless powered massive MIMO relaying systems. In: 2015 International Conference on Wireless Communications & Signal Processing (WCSP). Nanjing: IEEE; 2015. p. 1–6.

Chapter 15
Security of in-vehicle controller area network: a review and future directions

Zhaojun Lu[1], *Qian Wang*[1], *Gang Qu*[1], *and Zhenglin Liu*[2]

Modern vehicles have about 70 Electronic Control Units (ECUs) on average that communicate through the Controller Area Network (CAN) bus, the de facto standard for In-Vehicle Networks (IVNs). Unfortunately, the lack of security services makes CAN bus vulnerable to a wide range of attacks such as the well-known remote exploitation of a 2014 Jeep Cherokee. Since the report of this sensational hacking, more and more practical attack models have been proposed and successfully launched on various brands of vehicles, including BMW and Tesla. Meanwhile, corresponding countermeasures have also been studied. This chapter is carefully positioned to avoid overlap with existing surveys by filling the gaps and reporting the latest advances on the security of CAN bus while keeping it self-explained. We start with a brief introduction of CAN protocol and then assess its vulnerabilities and potential attack interfaces. Next, we give an in-depth review of the attack models with focus on those proposed in the past 5 years and analyze two main types of defense methods. We argue that it is challenging for current IVN protocols to meet the performance and security requirements of the forthcoming Intelligent Transportation Systems (ITSs). As a conclusion, we point out future research directions that can fundamentally address the security issue of IVNs.

15.1 Introduction

The advances in automotive electronics lead to the rapid increase in complexity and diversity of the in-vehicle environment. A modern vehicle, even not fully featured, already has 70–100 ECUs with over 2,500 signals to transmit internally [1]. The IVNs composed of several kinds of bus protocols play a critical role in the ITSs [2], which aim to improve road safety and driving conditions. With numerous sensors, actors, and processors being connected on IVNs, a vehicle provides drivers and manufacturers with the data-processing services, including infotainment, vehicle diagnostics, Firmware Over-The-Air [3], and automatic driving [4].

[1]Department of Electrical and Computer Engineering and Institute for Systems Research, University of Maryland, College Park, MD, USA

[2]School of Optical and Electronic Information, Huazhong University of Science and Technology, Wuhan, China

Five bus protocols are widely used for IVN communications [5]. The Local Interconnection Network is used in low-speed applications that do not require stringent timing performance [6]. The Media-Oriented Systems Transport is primarily deployed in premium vehicles for the infotainment data [6]. The FlexRay is usually used in the advanced chassis control and communication backbones due to its high determinism and fault-tolerance [7]. The automotive Ethernet is a new protocol with high bandwidth and very limited latency and jitter [8]. The CAN is the de facto standard due to its dramatically decreased communication lines and higher data transmission reliability [9].

Figure 15.1 is the typical IVNs based on CAN bus. The high-speed CAN bus (on the right) has a data rate of about 500 kbit/s and is used for the time-critical ECUs, while the low-speed CAN bus (on the left) is used for other ECUs that have no real-time requirement. The original CAN protocol is designed under the assumption that all ECUs are legitimate, trustworthy, and operating according to their specifications [10]. Without considering security, this results in several intrinsic vulnerabilities for the CAN protocol. Moreover, external interfaces such as the Second On-Board Diagnostic (OBD-II), Bluetooth, Wi-Fi, and Global Positioning System (GPS) provide opportunities for adversaries to break into the open in-vehicle systems through the unprotected CAN bus [11]. In 2015, this threat became a reality when Miller and Valasek [12] remotely controlled the vulnerable 2014 Jeep Cherokee without any physical alteration, which attracted much attention from academic and industry on the security issue of IVNs. Since then, various attack models against CAN bus have been proposed and successfully launched on many brands of vehicles. Corresponding countermeasures have been proposed to defend these attacks.

Several excellent surveys about IVNs have been published in the past few years. Each has its own competitive advantages and shortcomings. The 2010 survey by

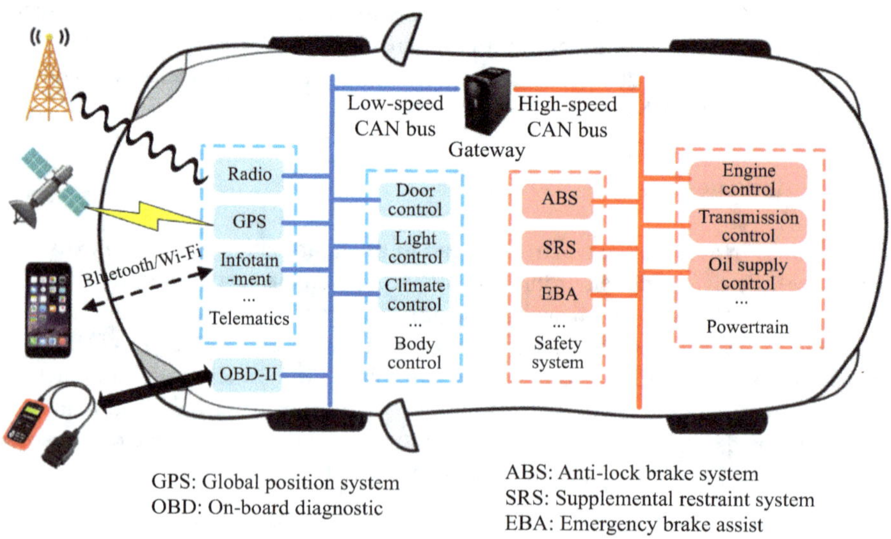

Figure 15.1 Architecture of the in-vehicle CAN bus

Koscher *et al.* [9] and the 2011 survey by Checkoway *et al.* [11] were the seminal paper in the field of automotive vulnerability assessment. They both experimentally demonstrated that it was possible to control the vehicle through accessing the available interfaces. The 2015 survey by Tuohy *et al.* [13] and the 2016 survey by Zeng and Khalid [5] focused on the different types of bus protocols in terms of system cost, data transmission capacity, and fault-tolerance capability, whereas they understated the security requirements. The 2017 survey by Liu *et al.* [14] explained the basic concepts about the vulnerabilities of the CAN bus, the attack methodologies, and the countermeasures. However, limited by the scope, it did not provide enough details in every aspect. The 2018 survey by Lu *et al.* [2] gave a comprehensive overview of the security, trust, and privacy issues related to vehicular network with emphasis on the vulnerabilities and attacks on IVNs and CAN.

This survey provides a detailed guidance to the security issues of IVNs. We mainly focus on the CAN protocol for two reasons. First, CAN is the most representative in-vehicle bus protocol as it has been required to be deployed for powertrain and safety system in all the vehicles of the United States since 2008 [9]. Second, most of the attack models exploit the vulnerabilities of the CAN bus to compromise the life-critical ECUs. In this survey, we give a brief description of the CAN protocol in Section 15.2 and then assess the intrinsic vulnerabilities and potential attack interfaces in Section 15.3. In Section 15.4, we elaborate on how to compromise an ECU through the external interfaces and attack the entire CAN bus using the compromised ECUs. We also report the technical details of several representative hacking cases in the past 5 years that involves some well-known automobile manufacturers that indicate the imperious demand to protect IVNs against the practical attack models. Next, we provide a comprehensive analysis of the countermeasures in Section 15.5, including the three main types of Intrusion Detection Systems (IDSs) and the encryption and authentication enhancements for the CAN frames. Unfortunately, no single solution is able to prevent all the existing attack models. Moreover, unknown attack modes keep emerging to bypass the existing countermeasures. Therefore, we present promising research directions to fundamentally secure IVNs in Section 15.6. On the one hand, we suggest developing new bus protocols with security constraints to replace the conventional CAN protocol. On the other hand, the next-generation gateway is worth studying since it will play an important role in the future architecture of IVN communications. We summarize this survey in Section 15.7.

15.2 Overview of CAN protocol

Before assessing the vulnerabilities, we provide an overview of the CAN protocol in this section, including the format of the CAN frame, the bus arbitration, and the error management.

15.2.1 Format of the CAN frame

A CAN frame is used to transmit information from a transmitter ECU to one or more receivers ECUs [15]. Figure 15.2 shows the format of a CAN frame, which consists

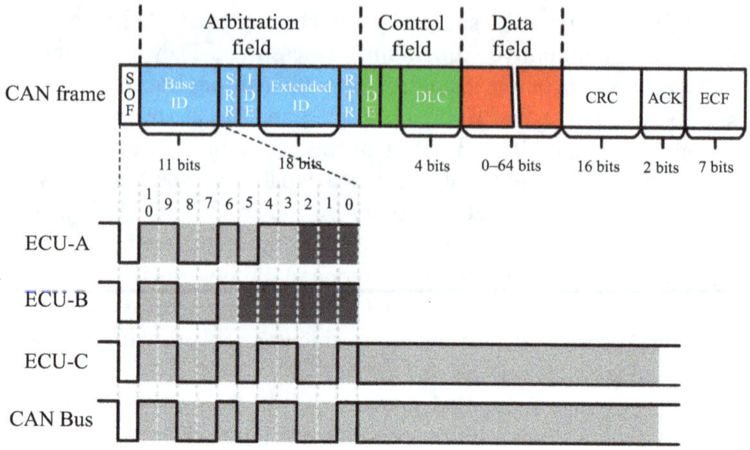

Figure 15.2 Format of the CAN frame and an example of CAN bus arbitration

of the Start-of-Frame (SOF), the arbitration field, the control field, the data field, the Cyclic Redundancy Check (CRC), the Acknowledge (ACK), and the End of Frame.

The CAN protocol uses the Identifier (ID) instead of an explicit address to identify the receivers. There are two different types of CAN frames: the standard and the extended [16]. A standard frame is defined with an 11-bit ID, while an extended frame is defined with a 29-bit ID. Both the standard and extended frames can be simultaneously transmitted on the CAN bus. The arbitration field starts with the 11-bit ID, followed by the 1-bit Remote Transmission Request (RTR) in a standard frame or by the 1-bit Substitute Remote Request (SRR) in an extended frame. RTR is used to distinguish the data frames (logic 0 as dominant) from the remote request frames (logic 1 as recessive). SRR keeps recessive to guarantee the deterministic resolution of the arbitration field between a standard frame and an extended frame.

There are 6 bits in the control field [17]. The first 2 bits contain a 1-bit ID Extension (IDE) and a reserved bit. The bits of IDE are dominant in a standard frame and recessive in an extended frame, which ensures the deterministic resolution of the contention when the first 11-bit IDs of two frames are the same. The following 4-bit Data Length Content (DLC) defines the length of the data field in bytes (0–64 bytes).

CRC consists of 16 bits, in which the first 15 bits are used for the CRC algorithm and the last 1 bit is used as delimiter [17]. ACK consists of 2 bits, in which the first bit is used to record acknowledgment from the receiver and the other bit is a delimiter. The receivers report that the received frame is valid by overwriting the ACK slot with a dominant bit.

15.2.2 Bus arbitration

The arbitration mechanism for the CAN bus is priority-based non-preemptive. When a frame is being transmitted, the transmission will not be preempted by other frames with higher priority. To meet the real-time constraint, the priority of each frame is defined by its ID [18]. Figure 15.2 depicts a situation that three ECUs compete to get access to the shared medium. The IDs of ECU-A, ECU-B, and ECU-C are 11001011111, 11001111111, and 11001011001 in binary, respectively. The contention is synchronized by the SOF bit and followed by the ID bits. As the first 5 bits are the same, all ECUs detect no difference and proceed. At the sixth bit, ECU-A and ECU-C transmit a bit at 0, while ECU-B transmits a bit at 1. The CAN bus stays dominant at 0 that stops the transmission of ECU-B and switches ECU-B to the listening mode. Similarly, at the ninth bit, ECU-C transmits a bit at 0, while ECU-A transmits a bit at 1. As a result, ECU-A withdraws from the competition and ECU-C is the winner. Then, ECU-C keeps transmitting the remaining bits of the frame.

15.2.3 Error management

The CAN protocol provides mechanisms for error detection, self-diagnostics, and signaling, which prevent faulty ECUs from impacting on the status of the CAN bus. The corrupted CAN frames are flagged by any ECU detecting an error and will be retransmitted automatically [16].

The CAN protocol defines five different error types:

- **Bit error.** A bit error is detected by the transmitter ECU when the bit value read from the CAN bus is different from the bit value it transmitted, except in the ID field or in the ACK slot.
- **Stuff error.** A stuff error is detected at the sixth consecutive occurrence of the same bit in a CAN frame that is subject to bit stuffing.
- **CRC error.** If the CRC calculated by the receiver ECU differs from the CRC in the received CAN frame, a CRC error is detected.
- **Form error.** A form error occurs when a fixed-form bit field contains one or more illegal bits.
- **ACK error.** An ACK error is detected by a transmitter ECU if a recessive bit is found in the ACK slot.

The CAN protocol has a mechanism for fault confinement to detect faulty ECUs and switch them into passive or off state. The CAN protocol assigns the following three states:

- **Error Active.** In the Error Active state, the ECUs are assumed to function properly and can signal errors with an active error flag that consists of six dominant bits.
- **Error Passive.** In the Error Passive state, the ECUs can still transmit on the CAN bus but is restricted to signal errors with a passive error flag that consists of six recessive bits (111111).
- **Bus Off.** In the Bus Off state, the output drivers of the ECUs are switched off and cannot transmit any data on the CAN bus.

The state of an ECU changes according to the value of two counters: the Transmit Error Count (TEC) and the Receive Error Count (REC). In essence, a receiver ECU increases its REC by 1 when detecting an error during reception, whereas a transmitter

ECU increases its TEC by 8 when detecting an error during transmission. For any error-free reception and transmission, the receiver ECU and the transmitter ECU decrease REC and TEC by 1, respectively. To confine serious faults disrupting the CAN bus, the fault confinement mechanism is illustrated in Figure 15.3. The initial state of each ECU is the Error Active state. If TEC or REC is equal or greater than 128, this ECU becomes the Error Passive state. Only if both the TEC and REC go below 128, this ECU returns the Error Active state. When TEC exceeds the limit value of 255, this ECU is switched to the Bus Off state and cannot transmit or receive any data on the CAN bus. The transition from the Bus Off state back to the Error Active state is subject to the specific condition when 128 occurrences of 11 consecutive recessive bits have been detected on the CAN bus [16].

15.2.4 CAN bus network typology

CAN bus has different standardized speeds ranging from 5 to 1,000 kbit/s. Most common are 500 and 100 kbit/s, while 500 kbit/s networks have higher demands in terms of cables and CAN bus transceiver chips. Depending on the make and model, there are several CAN buses in a modern car. The most prominent reason for having more than one CAN bus is to guarantee the clear separation for safety in order to hold fault-tolerance in case one bus fails. The other benefit is that using multiple buses could reduce cost as deploying the lower speed CAN buses where high-speed CAN buses are not needed. Because of the difference in the speed, there are two types of CAN bus layout, the high-speed CAN and the low-speed CAN [16]. Notice that for both high-speed and low-speed CANs, the speed of the transition is faster when a recessive to dominant transition occurs, since the CAN wires are being actively driven.

ISO 11898-2 is the most used physical-layer standard for high-speed CANs. The data rate is defined up to 1 Mbit/s with a required bus length of 40 m at 1 Mbit/s.

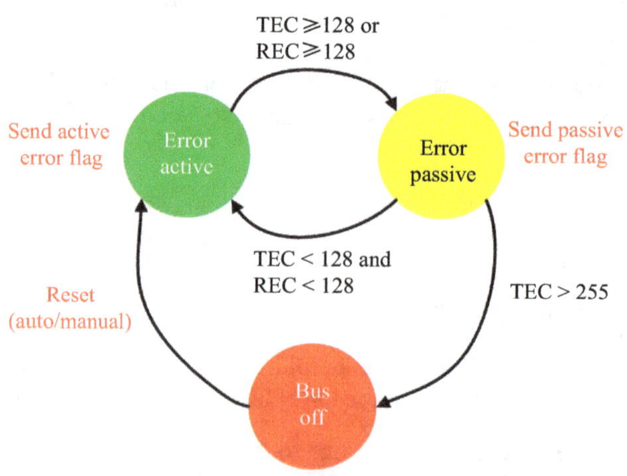

Figure 15.3 State diagram of fault confinement

The high-speed CAN uses a linear bus terminated at each end with 120 Ω resistors. The high-speed standard specifies a two-wire differential bus, whereby the number of nodes is limited by the electrical busload. The two wires are identified as CANH and CANL. The common mode voltage ranges are from −2 V on CANL to +7 V on CANH. The node can be a simple I/O device or an embedded controller with the CAN interface. The node may also be a gateway allowing a standard computer to communicate over the USB or Ethernet port to the devices on a CAN. The linear topology CAN is usually used as a high-speed CAN bus for the critical automotive applications, e.g., the engine and transmission.

The layout of low-speed CAN bus network is regulated in the ISO 11898-3 standard. It is also called fault-tolerant CAN that uses a star bus and is terminated at each node by a fraction of the overall termination resistance. Fault-tolerant CAN is often used where groups of nodes need to be connected together forming a subnetwork. Since for this specification, a short network was assumed, the problem of signal reflection is not as important as for long bus lines. This makes the use of an open bus line possible, which indicates that low-speed drivers can be used for networks with very low power consumption, and the bus topology is no longer limited to a linear structure. It is possible to transmit data asymmetrically over just one bus line in the case of an electrical failure of one of the bus lines. ISO 11898-3 defines data rates up to 125 kbit/s with the maximum bus length, and up to 32 nodes per network are specified. The common mode voltage ranges are between −2 and +7 V. The power supply is defined at 5 V. The low-speed CAN bus is usually used on some noncritical applications, such as the entertainment system or display unit of the vehicle.

In a real vehicle, both the two types of bus layout will be applied. Moreover, the length of the bus network will be extended to 3–10 m based on the scale and design of the real vehicle. The physical layout of ECUs will not be altered after manufacturing, even under cyberattacks, where the attacker could easily achieve access on the bus but have no capability in changing the physical layout of the CAN bus.

15.3 Vulnerabilities and attack interfaces

In this section, we first assess the intrinsic vulnerabilities of the CAN protocol. Then we analyze the external interfaces that can be exploited by the adversaries to compromise the in-vehicle systems.

15.3.1 Vulnerabilities

Although the CAN bus is considered as the de facto standard for IVNs, it lacks the fundamental security services [17]. By plugging in various components to determine the CAN frames associated with the component, Koscher *et al.* [9] experimentally demonstrated the vulnerabilities by monitoring the CAN bus. The critical intrinsic vulnerabilities that make the vehicles vulnerable to the adversaries include broadcast transmission, no encryption, no authentication, priority-based arbitration, limited bandwidth, and open diagnostic function.

15.3.1.1 Broadcast transmission

The CAN frames are both physically and logically broadcasted to all the connected ECUs. Every ECU receives the frames transmitted on the CAN bus and takes actions according to the frame ID [14]. Malicious ECUs can easily leverage the broadcast nature of the CAN bus to eavesdrop on the CAN frames transmitted by other ECUs.

15.3.1.2 No encryption

No encryption of the content in the CAN frame allows the adversaries to easily analyze the functions of the target ECU based on the historically recorded CAN frames.

15.3.1.3 No authentication

As shown in Figure 15.2, the CAN frame has no authentication field to indicate its source, which means a transmitter can indistinguishably transmit a CAN frame to any ECUs connected on the CAN bus. The adversaries can command a compromised ECU to take control of the target ECUs by transmitting fabricated CAN frames containing appropriate contents on the CAN bus.

15.3.1.4 Priority-based arbitration

As explained in Section 15.2, the ID field of a CAN frame determines its priority. The priority-based arbitration mechanism allows a CAN frame with smaller ID (higher priority) to be transmitted on the CAN bus while forcing all the other CAN frames to back off. If a malicious ECU asserts a dominant state on the CAN bus indefinitely, none of the legitimate ECUs can transmit any CAN frames. In this way, the adversaries can easily launch the Denial of Service (DoS) attack against IVNs.

15.3.1.5 Limited bandwidth and payload

The high-speed CAN bus has the data rate of about 500 kbit/s, and the payload of a CAN frame is up to 64 bits [16]. Limited by the bandwidth and the payload, the CAN bus cannot provide strong access control. For example, in order to protect ECUs against certain operations without authorization, ECUs are supposed to use fixed challenges (seeds) and store the corresponding responses (keys) for the challenge–response pairs [9]. Since the length of the challenges and the responses are too short, the adversaries can crack the key of an ECU within 8 days through the brute-force attack [9].

15.3.1.6 Open diagnostic function

In order to diagnose the complicated problems of the modern vehicles, the CAN bus provides the open access for diagnosis to the service technicians. They can use the external test equipment to interrogate or even manipulate the internal state of ECUs, which gives the adversaries an extended method to analyze the target ECUs [19].

15.3.2 Attack interfaces

As shown in Figure 15.4, a modern vehicle has several external interfaces that the adversaries can exploit to access the in-vehicle systems [11]. These attack interfaces can be divided into four categories according to the operational requirements on the

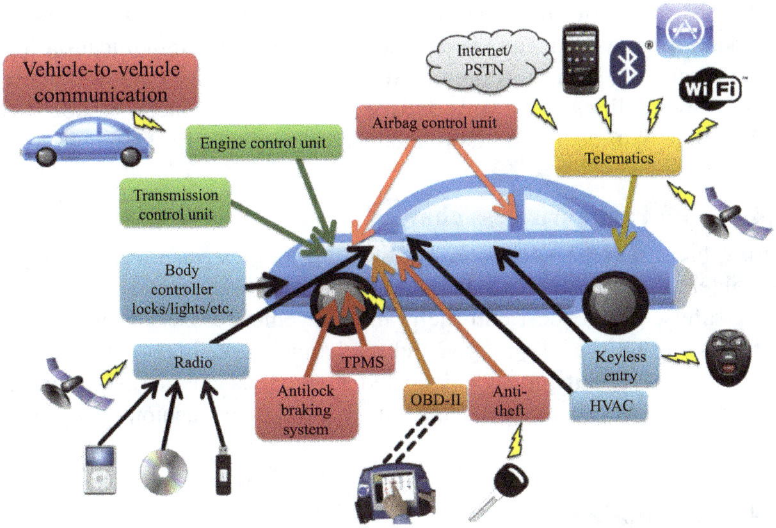

Figure 15.4 Attack interfaces in a modern vehicle [11]

adversaries: OBD-II port, entertainment system, short-range wireless channel, and long-range wireless channel.

15.3.2.1 OBD-II port

As the most significant interface for the modern vehicles, the OBD-II port provides the open and direct access to the CAN bus for routine maintenance such as diagnostic and ECUs programming [11]. Most vehicle manufacturers have developed various handheld scan tools with the PC-centric approaches. Dedicated software on the tools can interrogate or program the ECUs through the OBD-II port. If such tools are controlled by the adversaries, they can directly manipulate any ECUs in a modern vehicle.

15.3.2.2 Entertainment system

Generally, a modern vehicle is shipped with a CD player that supports a wide variety of audio formats. Also, several external digital multimedia ports such as a USB port or an iPod/iPhone docking port are provided for users to control the entertainment system on their personal devices. The entertainment system is connected on the CAN bus to directly or indirectly interact with other ECUs to support the functions, including the hands-free features, displaying messages on the console, and updating firmware for ECUs [11]. Thus, the entertainment system is another attack interface for the adversaries to compromise the vehicle components.

15.3.2.3 Short-range wireless channel

Without the access to the OBD-II port and the entertainment system, the adversaries can still compromise the ECUs through the short-range wireless channel, including

Bluetooth, Wi-Fi, the radio-frequency identification, the remote keyless entry, the tire-pressure monitoring system, and the dedicated short-range communication. Through the wireless transmitter in the range of the wireless channel, the adversaries can exploit the vulnerabilities in the software for parsing the channel messages to attack the target vehicle by transmitting fabricated CAN frames on the CAN bus.

15.3.2.4 Long-range wireless channel

A modern vehicle has ECUs for receiving the long-range signals from GPS and multifunctional telematics systems via cellular networks, which provide various features supporting safety, diagnostics, anti-theft, and convenience. Despite it is difficult to launch attacks through the long-range wireless channel due to higher technical barriers and operational complexity, the sophisticated adversaries can precisely compromise the target vehicles over arbitrary distance in an anonymous fashion [11].

15.4 Attack models

In the past decade, many researchers in both industry and academia have explored the CAN bus protocol and demonstrated a variety of attacks without direct physical access to the CAN bus. In this section, we first summarize the general procedure to remotely attack a modern vehicle. Then we explain how to manipulate the vulnerable ECUs and launch the basic attack vectors. Finally, we elaborate on four practical attack cases that successfully compromise the target vehicles.

15.4.1 Typical attack procedure

As shown in Figure 15.5, a typical attack against IVNs generally involves two main stages. In the first stage, the adversaries should compromise one or more ECUs with external interfaces. Since we assume that the adversaries have no physical access to IVNs, they have to transmit some kinds of wireless signals to inject malicious codes to the listening ECUs via the external interfaces [19]. Then the adversaries can control the compromised ECUs to eavesdrop on the CAN frames or even inject fabricated CAN frames on the CAN bus.

Since the compromised ECUs in the first stage are only responsible for receiving and processing the wireless signals, they cannot directly threaten the safety-critical features of a vehicle [19]. In order to launch a cyber-physical attack that controls various aspects of the target vehicle, it requires the adversaries to interact with other ECUs that are responsible for the functions such as steering, braking, and acceleration. In the second stage, the adversaries use the compromised ECUs from the less privileged network as a bridge to access the ECUs in charge of safety-critical features. In most cases, it is not trivial because the compromised ECUs cannot directly communicate with the safety-critical ECUs. The difficulty lies in how to manipulate the safety-critical ECUs by transmitting fabricated CAN frames on the CAN bus using the compromised ECUs.

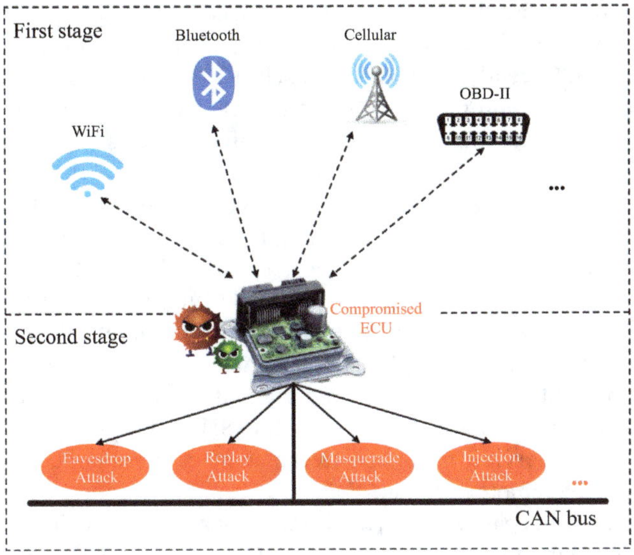

Figure 15.5 The two main stages to remotely attack a vehicle

15.4.2 Compromising ECUs

Strong attacker and weak attacker. Different ECUs with different kinds of vulnerabilities and different types of attack interfaces allow the attackers to gain different capabilities. We classify the compromised ECUs into two categories: fully and weakly compromised ECUs [20]. The attacker is able to suspend a weakly compromised ECU by transmitting certain frames or keep the ECU in the listen-only mode to eavesdrop on the CAN bus. However, a weakly compromised ECU cannot be used to inject any fabricated CAN frames. Such an attacker with limited capabilities on "a weakly compromised ECU" is called a **weak attacker**. In contrast, the attacker with full control of "a fully compromised ECU" and access to its memory data is called a **strong attacker**. A fully compromised ECU can be used to launch attack vectors by injecting fabricated frames and a strong attacker can disable the security mechanisms implemented in a fully compromised ECU [19].

In 2015, Foster *et al.* [20] demonstrated that these two types of attackers existed in the modern vehicles. They examined a popular aftermarket Telematics Control Unit (TCU) connected to a vehicle via the OBD-II port and showed that this device could be discovered, targeted, and compromised remotely. The firmware versions of TCUs determined the capabilities of the attackers. Specifically, the attacker could weakly compromise TCUs with a certain version of firmware, which was only able to receive the CAN frames, while TCUs with some other versions of firmware could be fully manipulated to both transmit and receive the CAN frames.

Generally speaking, ECUs with the most common CAN controllers can be switched to several operation modes according to the user instructions through the

Serial Peripheral Interface [21]. Thus, it is much easier to switch ECUs to the listen-only mode as a weak attacker than manipulate ECUs to inject fabricated CAN frames as a strong attacker. Recently, Miller and Valasek [12,19,22] comprehensively demonstrated how to fully compromise ECUs in a modern vehicle. Figure 15.6 illustrates the entire chain to exploit the 2014 Jeep and similar vehicles. Here, we present the four steps in the first stage.

Step 1. The first step is to get the IP address of the target vehicle. Knowing the Vehicle Identification Number (VIN) or GPS, it is simple to scan the IPs range in an area until find the target vehicle with the corresponding VIN or GPS is found.

Step 2. The second step is to exploit the Open Multimedia Application Platform (OMAP) chip of the head unit. With the IP address of the target vehicle, the attackers can run the malicious code through the appropriate Desktop Bus service, which is used for communication between multiple computer processes performing on the same machine. After uploading a Secure Shell (SSH) public key and configuration file to start the SSH service, the attackers can connect to the target vehicle and run commands from the remote terminal.

Step 3. The third step is to control the Jeep's Uconnect system. The Uconnect radio manufactured by Harman Kardon is used in all the vehicle models from Fiat Chrysler Automotive as the sole source for Wi-Fi, in-vehicle infotainment, navigation, and cellular communications [23]. The Uconnect system is also used for communication with other ECUs that are connected on the CAN bus. At this point, the attackers are able to run malicious code on the OMAP chip within the Uconnect system.

Step 4. The fourth step is to flash the v850 using the modified firmware. Since the OMAP chip cannot transmit the CAN frames without communicating with the v850 chip, the attackers can flash the v850 chip with the modified firmware that allows them to transmit the CAN frames after compromising the OMAP chip.

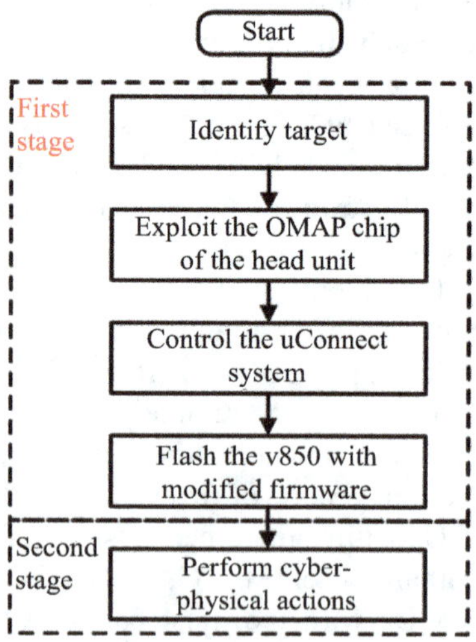

Figure 15.6 The entire chain to exploit the 2014 Jeep and similar vehicles

In this way, a strong attacker will be able to transmit fabricated CAN frames to every ECU on the CAN bus.

15.4.3 Launching attack vectors

As illustrated in Figure 15.7, the attackers can launch the basic attack vectors with the weakly and fully compromised ECUs, including the eavesdrop attack, the replay attack, the masquerade attack, and the injection attack. Since a practical attack model is a combination of one or more basic attack vectors, we first explain how to use the weak attacker and the strong attacker to launch the attack vectors. Then, we elaborate on four practical attack cases against the modern vehicles.

15.4.3.1 Eavesdrop attack

As mentioned before, the CAN frames are broadcasted to all ECUs without encryption. A weak attacker ECU-A in Figure 15.7(a) is able to eavesdrop on the CAN bus to collect and analyze the CAN frames. Through the fuzzing test [9] on the historically recorded CAN frames, the functions of the target ECUs can be figured out. Therefore, the eavesdrop attack is the foundation of all the practical attacks.

15.4.3.2 Replay attack

Without authentication and integrity for the CAN frames, a strong attack is able to launch the replay attack [24,25]. As shown in Figure 15.7(b), a fully compromised ECU-A transmits the CAN frames received from ECU-C without modifying it. As a result, the receiver ECU-B will function abnormally under the replayed control information.

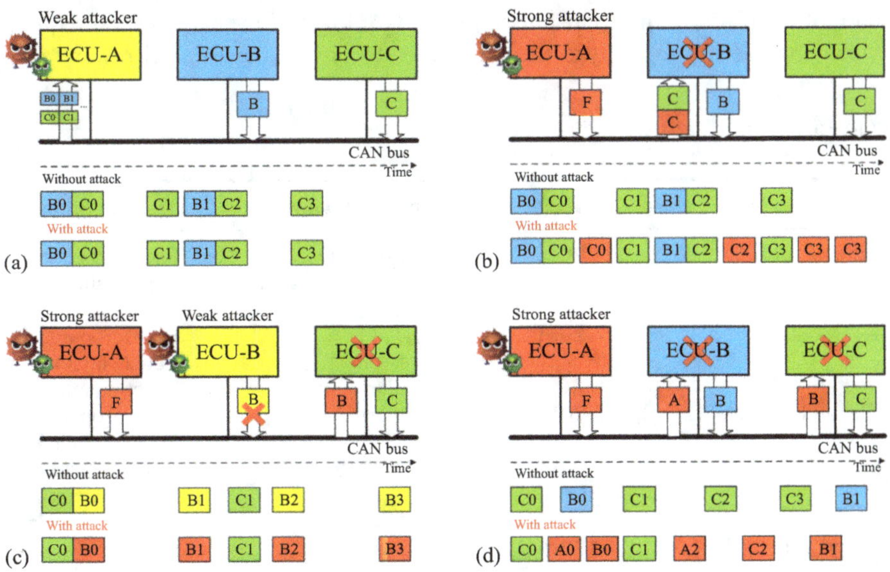

Figure 15.7 The basic attack vectors: (a) eavesdrop attack, (b) replay attack, (c) masquerade attack, and (d) injection attack

15.4.3.3 Masquerade attack

Having known the content and frequency of the CAN frames from ECU-B, the strong attack is able to launch the masquerade attack as shown in Figure 15.7(c). The weak attacker first suspends the transmission of ECU-B, then the strong attacker controls ECU-A to transmit CAN frames using ECU-B's ID to manipulate the target ECU-C.

15.4.3.4 Injection attack

As shown in Figure 15.7(d), a strong attacker ECU-A is able to inject CAN frames with arbitrary IDs and contents. On the one hand, the injected frames with the highest priority ID will always occupy the CAN bus [14]. On the other hand, an appropriate ID makes the target ECUs accept the contents in the fabricated CAN frames.

15.4.4 Representative attack case studies

Since Koscher *et al.* [9] and Checkoway *et al.* [11] experimentally assessed the vulnerabilities of the CAN protocol, many researchers in both academia and industry have dedicated to hacking a modern vehicle without physically accessing to IVNs. One of the most sensational attack case was that Miller and Valasek [12] remotely exploited an unaltered 2014 Jeep Cherokee, which forced the recall of approximately 1.4 million vehicles and the changes of the Sprint carrier networks. In Table 15.1, we briefly introduce the representative attack cases. Considering the differences in the attackers' desires and methods, we mainly analyze four representative attack cases in recent 5 years.

15.4.4.1 Remote exploitation of a 2014 Jeep Cherokee

In the second stage of Figure 15.6, the first task is to figure out the proprietary nature of the CAN frames transmitted on the CAN bus of the Jeep, which requires reverse engineering to analyze the mechanics tools and the ECU firmware. Two

Table 15.1 Attack cases against the CAN bus

Reporters	Year	Attack interfaces	Attack vectors
Koscher *et al.* [9]	2010	OBD-II	Eavesdrop attack Replay attack Injection attack
Checkoway *et al.* [11]	2011	OBD-II Bluetooth Cellular	Eavesdrop attack Replay attack Injection attack
Miller and Valasek [12]	2015	Cellular	Eavesdrop attack Injection attack
Woo *et al.* [26]	2015	OBD-II	Eavesdrop attack Replay attack Injection attack
Cho and Shin [27]	2016	OBD-II	Eavesdrop attack Masquerade attack
Keen Security Lab [28]	2017	Wi-Fi	Eavesdrop attack Injection attack

types of CAN frames, normal and diagnostic, can be used to manipulate the target vehicle. The normal CAN frames are transmitted on the CAN bus under the normal operations, while the diagnostic CAN frames are only transmitted under diagnosis or some unusual circumstances. Although the diagnostic CAN frames are more powerful than the normal CAN frames, most ECUs will ignore them when the vehicle is traveling at speed. For example, since the parking assist system will go offline when receiving conflicting CAN frames, they first switch the Parking Assist Module into the diagnostic session, in which it stops transmitting the normal CAN frames. Then they inject their own CAN frames using the compromised ECU to manipulate the steering wheel. In another example of manipulating the brakes, they first start a diagnostic session with the ECU for the Anti-Lock Braking System. Then they bleed all the brakes at maximum by injecting multiple CAN frames, which makes the Jeep lose control on the highway.

15.4.4.2 A wireless attack through malicious smartphone application

In 2015, Woo *et al.* [26] developed the malicious smartphone application to implement a long-range wireless attack against a connected vehicle. Figure 15.8 illustrates the state diagram of the attack based on the malicious self-diagnostic application. First, the victim downloads the malicious application and installs it on the smartphone. Second, the smartphone is connected to the target vehicle through Bluetooth or Wi-Fi. Third, the malicious application collects and transmits the CAN frames to the attacker's server using the smartphone. Fourth, the attacker's server analyzes the recorded CAN frames and transmits commands to control the target vehicle via the malicious application.

Although the attacker does not need compromise any ECU inside the vehicle, the main drawback of this attack model is that it requires the victim to install an OBD-II scan tool and use their malicious self-diagnostic application to monitor the status of the target vehicle. It is difficult to conceal the actions of the malicious application. Once the victims notice the malicious application is manipulating their vehicles, they will immediately delete the malicious application from their smartphones.

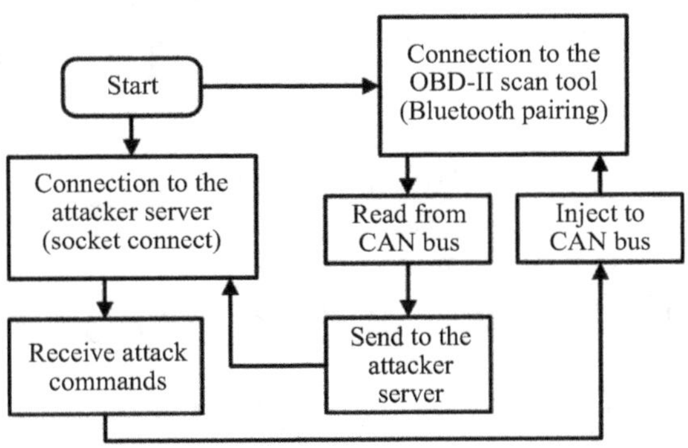

Figure 15.8 State diagram of the attack based on the malicious self-diagnostic application

15.4.4.3 The bus-off attack

In 2016, Cho and Shin [27] proposed the bus-off attack against two real vehicles. The bus-off attack is a new type of DoS attack that exploits the error-handling mechanism of the CAN protocol to disable the target ECUs. As the CAN protocol defines, if an ECU detects an error during transmission, it will increase its TEC by 8, whereas during reception, it will increase its REC by 1 [16]. An ECU will decrease its TEC or REC by 1 for an error-free transmission or reception. As explained in Figure 15.3, there are three states for each ECU according to the value of TEC and REC, i.e., the Error Active state, the Error Passive state, and the Bus Off state. If TEC or REC exceeds 127, the ECU switches from the Error Active state into the Error Passive. If TEC or REC exceeds 255, the ECU enters the Bus Off state, in which it is forced to shut down for the fault confinement. The basic idea of the bus-off attack is to increase TEC of the target ECU using a fully compromised ECU until the target ECU falls into the Bus Off state.

Figure 15.9 is an example of the bus-off attack. In the first phase when the target ECU is in the Error Active state and transmits a normal CAN frame, the strong attacker injects a fabricated CAN frame at the same time to increase TEC of the target ECU by 8. TEC of the strong attack is also increased by 8, since the target ECU transmits an active error flat that consists of 6 consecutive dominant bits. In the second phase when the target ECU enters the Error Passive state, it increases its TEC by 8 and transmits a Passive Error flag that consists of 6 consecutive recessive bits when it detects error. Thus, the strong attacker successfully transmits a fabricated CAN frame that makes its TEC decrease by 1. Then the target ECU successfully retransmits a normal CAN frame that makes its TEC decrease by 1. In an iteration of the second phase, the TEC of the target ECU is increased by 7, whereas the TEC of the strong attacker is decreased by 1. Consequently, the target ECU will fall into the Bus Off state when its TEC exceeds 255.

The bus-off attack is able to conceal the actions of the strong attacker as if there is a system error. However, the main difficulty for the strong attacker is to know

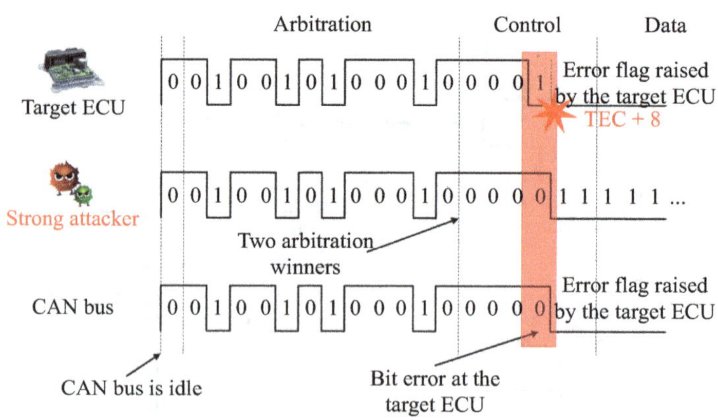

Figure 15.9 Example of the bus-off attack

the exact time to transmit the fabricated CAN frames. In other words, it requires to synchronize the transmission of the strong attacker and the target ECU within a bit resolution. If not, none of the two arbitration winners leads to the failure of the bus-off attack [27]. Therefore, the bus-off attack is a severe threat to the ECUs that transmit CAN frames at fixed intervals.

15.4.4.4 Hacking Tesla through the wireless interface

At the 2017 Black Hat Conference, the Keen Security Lab of Tencent successfully launched a remote attack against the Tesla Model S in both parking and driving modes [28]. They first dig up a vulnerability in the Tesla's browser that allows them to start running malicious code in the browser when the users visit a carefully crafted website. Then the hackers gain full privileges on the vehicle's head unit by exploiting another vulnerability in the Tesla's Linux operating system [29]. However, they cannot send commands to manipulate the safety-critical ECUs that control the steering and brakes because the head unit is separated from the CAN bus by the gateway. The gateway only allows specific commands to be transmitted from the vehicle's infotainment system to other ECUs. To defeat this safeguard, they simply overwrite the gateway's firmware with their own version. In this way, they can use the bash shell script to inject the fabricated CAN frames through the gateway to compromise the target ECUs.

15.5 Countermeasures

The countermeasures against the attack models can be classified into two categories, i.e., the IDSs and the encryption and authentication schemes for the CAN frames. IDSs focus on detecting the anomaly on the CAN bus. The encryption and authentication schemes aim to provide the necessary security services for the CAN frames. As shown in Table 15.2, each type of countermeasure is able to thwart specific attack vectors but

Table 15.2 Countermeasures

Countermeasures	References	Thwarted attack vectors
Clock-based IDSs	[21, 34–38]	Replay attack Injection attack
Voltage-based IDSs	[43–46]	Replay attack Masquerade attack Injection attack
Low-dimension-based IDSs	[49–54]	Replay attack Injection attack
MAC-based Encryption and authentication	[26, 55–61]	Eavesdrop attack Replay attack Masquerade attack Injection attack
Location-based authentication	[64, 65]	Replay attack Injection attack

has its own drawbacks. The Clock-based IDSs (CIDSs) have no encryption against eavesdrop attack and can be bypassed if the attacker gets the clock parameters. The voltage-based IDSs have no encryption against eavesdrop attack and can introduce a new attack interface. The low-dimension-based IDSs have no encryption against eavesdrop attack and cannot detect the attacks that do not change IDs. The Message Authentication Code (MAC)-based encryption and authentication incur unexpected communication delay and cannot be applied on ECUs with limited resources. The location-based authentication has no encryption against eavesdrop attack and can be bypassed by strong attackers.

15.5.1 Intrusion detection systems

The conventional IDSs employ either misuse-based or anomaly-based detection methods [30]. Since the misuse-based IDSs [31,32] only concern the patterns of the already known attacks, they cannot identify the unknown attack models effectively. The anomaly-based IDSs focus on finding the abnormal operations on the CAN bus [33]. It is possible for an appropriately designed anomaly-based IDS to detect the unknown attack models. The most essential issue to be considered while designing an anomaly-based IDS is what features the detector exploits to discriminate between normal and abnormal. Three types of anomaly-based IDSs have been proposed in recent years.

15.5.1.1 Clock-based IDSs

Song *et al.* [34] pointed out that the valid CAN frames with different IDs have unique time intervals. Several IDSs [21,34–38] have been proposed based on analyzing the time features of the CAN frames. In 2016, Cho and Shin [21] developed the CIDS that fingerprinted the transmitter ECUs according to the estimated clock skews. Since there is no timestamp carried in the CAN frame, CIDS exploits the CAN frame periodicity instead of the embedded timestamp [39,40] to extract and estimate the transmitter ECUs' clock skews. The basic idea of CIDS is illustrated in Figure 15.10, ECU-A is the transmitter ECU that transmits a CAN frame every T ms and ECU-R is the receiver ECU to periodically receive that CAN frame. We consider the clock of

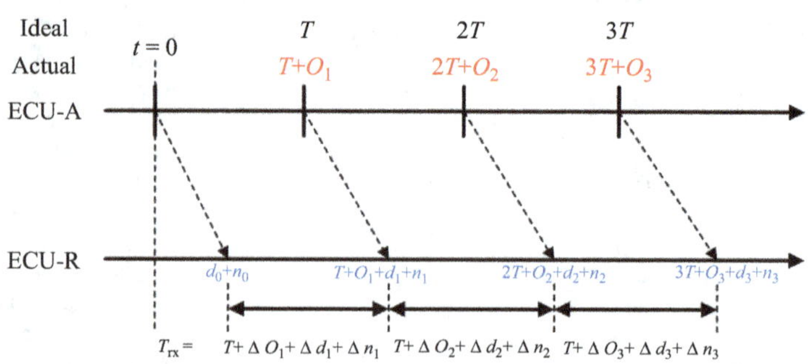

Figure 15.10 Clock skew estimation of the clock-based IDS

ECU-R as the true clock since only its timestamp is available. O_i is the clock offset of ECU-A when it transmits the ith CAN frame since $t = 0$, d_i is the network delay, and n_i denotes the noise in ECU-R's timestamp quantization. Thus, the time interval between two adjacent arrival CAN frames is $T_{rx,i} = T + \Delta O_i + \Delta d_i + \Delta n_i$, where Δ denotes the difference between ith and $(i-1)$th. Since ΔO_i is negligible, n_i is a zero-mean Gaussian noise term, and $E[\Delta d_i]$ is considered to be zero, the expected value of the time intervals, $\mu_{T_{rx}} = E[T_{rx}, i] \approx T$. Thus, the estimated arrival time of the ith CAN frame is $i\mu_{T_{rx}} + d_0 + n_0$, whereas the actual measured arrival time is $iT + O_i + d_i + n_i$. The average difference D between the estimated time and the measured time is $E[D] = E[i(T - \mu_{T_{rx}}) + O_i + \Delta d + \Delta n] \approx E[O_i]$. In this way, we can estimate the average clock offset, $E[O_i]$, as the fingerprint for the transmitter ECU.

Although CIDS is effective to thwart a variety of attack models, a sophisticated attacker can still bypass it [41,42]. In 2018, Tayyab *et al.* [41] proposed the clock phishing attack based on the estimated clock parameters. The clock phishing attack has two phases, i.e., the reconnaissance phase and the attack phase. In the reconnaissance phase, a weak attacker monitors the clock behavior of the CAN frames transmitted by the target ECU, including the period of the CAN frames, relative residual, relative clock skew, and the relative clock offset for a given CAN frame. In the attack phase, a strong attacker adjusts the period of the fabricated CAN frames by subtracting or adding the average offset according to the obtained clock behavior. Thus, CIDS cannot detect the difference between the fabricated CAN frames and the normal CAN frames.

15.5.1.2 Voltage-based IDSs

In 2017, Cho and Shin [43] proposed the Voltage-based Attacker Identification (Viden) that exploited the voltage features as the transmitter ECUs' fingerprints. In the first phase, Viden monitors the output voltages on the two dedicated CAN bus wires: CANH and CANL. Then Viden measures the voltage for each ID and learns the ACK threshold to determine whether the measured voltage signal is generated by the transmitter ECU. In the second phase, Viden uses the ACK threshold to construct the voltage instances for the transmitter ECUs as their voltage profiles via the Recursive Least Square algorithm. In the identification phase, Viden accurately identifies the attacker using the derived voltage profiles. Compared with the other voltage-based IDSs [44–47] that were designed only for intrusion detection, it is more efficient and easier to deploy Viden on any ECUs since it only relies on the voltage features and thus pinpoints the malicious ECU with a low false identification rate of 0.2%.

However, in order to acquire the voltage measurements, two additional wires are necessary to connect the microcontroller for the voltage-based IDS [48], which introduces a new attack interface to the CAN bus if the voltage-based IDS itself is compromised. In 2018, Sagong *et al.* [48] explored the potential vulnerabilities of the voltage-based IDSs and launched three voltage-based attacks through the two connected wires. In the overcurrent attack, the attacker raises the current that flows into the target ECUs to exceed the hardware limit. Consequently, the microcontrollers of the target ECUs are damaged. In the DoS attack, the attacker holds the voltage of the CAN bus to prevent other ECUs from transmitting the CAN frames. In the forced retransmission attack, the attacker intentionally generates an error during CAN frame transmission by violating the bit timing requirement so that the CAN frame has to be retransmitted.

15.5.1.3 Low-dimension-based IDSs

Considering the nature that most of the CAN frames on the CAN bus are cyclic and the format of the data field is fixed due to rigid signal allocation, the low-dimensional features of the CAN IDs can be explored to detect the anomaly on the CAN bus [38,49,50]. Wang *et al.* [51] proposed an IDS based on the entropy of the ID bits in the CAN frames. They collect the real CAN frames from a 2016 Ford Fusion and demonstrate that the injected frames can be detected because the entropy of the CAN frames is altered. References [52–54] applied the machine learning approaches to design the low-dimension-based IDSs. Kang and Kang [52] applied the Deep Neural Network (DNN) to propose an intrusion detection technique. DNN is trained with a great number of valid CAN frames to extract low-dimensional features that will be used for detecting the fabricated CAN frames, which provides a real-time response to the attacks with a high detection rate in the experiments. Weber *et al.* [53] combined the advantages of an efficient specification-based system with the machine learning measures to propose a hybrid anomaly detection system to achieve higher efficiency and effectiveness. In 2018, Seo *et al.* [54] proposed the novel Generative Adversarial Network (GAN)-based IDS (GIDS). As illustrated in Figure 15.11, GIDS only uses normal data to detect unknown attacks with high detection accuracy. After encoding each element of the CAN IDs to the one-hot vectors, the Generator (G) and the Discriminator (D) are trained simultaneously by an adversarial process. D receives the normal CAN and the fake CAN images generated by G and estimates the probability whether or not the received CAN images are normal. G and D increase

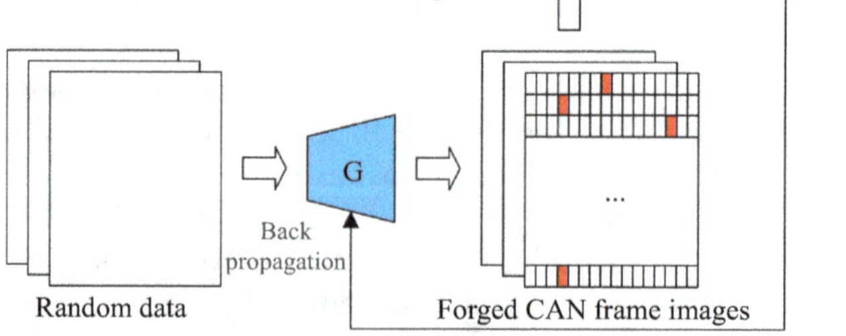

Figure 15.11 The training process of the GAN-based IDS

their performance by competing with each other. Thus, D is able to detect the fake CAN images that are very similar to the normal ones in the detection phase.

Although the low-dimension-based IDSs are effective for some unknown attacks, the eavesdrop attack and the masquerade attack can still bypass the detector, since no changes happen on the CAN IDs [51]. In summary, each type of the anomaly-based IDS only focuses on one aspect of the features on the CAN bus. Therefore, none of a single IDS is able to detect all the attack models.

15.5.2 Encryption and authentication schemes

In order to fundamentally patch the vulnerabilities of the CAN protocol, many researchers dedicated to providing encryption and authentication to the CAN frames. The MAC [26,55–61] has been widely applied for the message authentication. However, calculating the MAC for each CAN frame will definitely degrade the transmission performance, which is the main issue that the MAC-based methods should address [14]. In contrast, the location-based authentication methods [62–65] can be applied to efficiently identify ECUs on the CAN bus according to their physical location without cryptography.

15.5.2.1 MAC-based methods

MAC is a short piece of binary string used to authenticate a message [66]. The MAC value guarantees the integrity and authentication of a message. It allows the verifiers who also possess the shared secret key to detect any alteration in the content.

In 2015, Woo *et al.* [26] designed a MAC-based security protocol that consisted of an efficient key management scheme for ECUs and an encryption and authentication

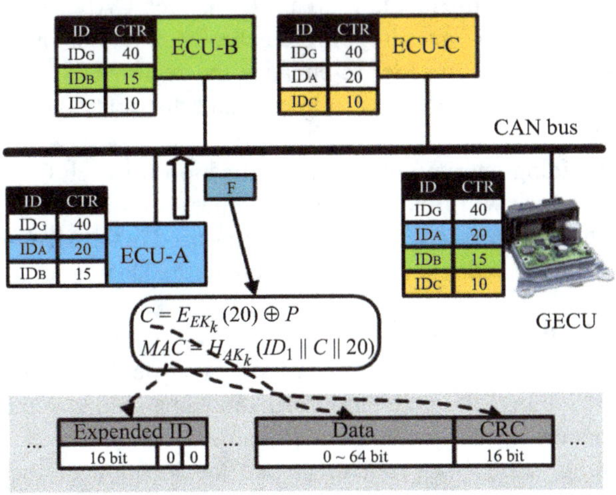

Figure 15.12 MAC-based encryption and authentication for the CAN frames

scheme for the CAN frames. The key management scheme distributes an authentication session key AK_n and an encryption session key EK_n to each ECU in the nth session, which will be used to guarantee the authentication and confidentiality of the CAN frames. In the (n+1)th session, AK_n and EK_n should be refreshed to ensure the forward and backward secrecy. As illustrated in Figure 15.12, each ECU uses AK_n and EK_n to generate MAC and encrypt the CAN frames. The Gateway ECU is responsible for the key distribution. ECU-A is the transmitter ECU and ECU-B is the receiver ECU. When transmitting a CAN frame, ECU-A generates the Ciphertext (C) using the frame Counter (CTR) and the Plaintext (P). Since the 128-bit Advanced Encryption Standard is used, the result of $E_{EK_k}(CTR)$ is 128 bits. Only the first 64 bits are used to generate C. Since the Keyed-Hash MAC is used, the result of $H_{AK_k}(ID_A||C||20)$ exceeds the length of the payload, we use a truncated 32-bit MAC [55,56] and insert the divided MAC in the extended ID field and the CRC field separately. Then, ECU-A transmits the secure CAN frame and its CTR increases by 1.

In 2019, Lu *et al.* [61] proposed a Lightweight Encryption and Authentication Protocol (LEAP) that exploited the stream cipher RC4 instead of MAC to encrypt and authenticate each CAN message. LEAP is lightweight in terms of time consumption and storage overhead. Thus, the efficiency and real-time performance of the CAN bus can be guaranteed. Moreover, LEAP only modifies the software of ECUs so that it is compatible with the existing IVNs. The key management mechanism of LEAP is of top priority to meet all the security requirements. As shown in Figure 15.13(a), the long-term symmetric keys LK_i are generated in the vehicle manufacturing stage or when an ECU is replaced. The session keys SK_{ij}^n are derived from LK_i and LK_j periodically for the message encryption and authentication mechanism. The sets of session keys are different because each ECU communicates with specific ECUs. For example, ECU_0 only communicates with ECU_1 and ECU_3. As illustrated in Figure 15.13(b), each pair of communicating ECUs with a symmetric session key perform the message encryption and authentication process based on RC4. Since the transmitter ECU generates distinct key stream K for each CAN message, it prevents a compromised ECU from eavesdropping on the CAN bus or using other ECUs' IDs to broadcast messages. In order to enhance the security of the session keys, a MAC-based method is still necessary to update and distribute the SK_{ij}^n to resist the brute-force attack.

However, providing encryption and authentication for each CAN frame incurs unexpected delay, which will hinder the normal operation of the time-critical ECUs [56]. Moreover, the costly cryptographic algorithms used in the MAC-based methods make the ECUs with limited resources vulnerable to the DoS attack.

15.5.2.2 Location-based methods

The location-based authentication systems have been deployed in the Internet of Things society, which take the ambient information as a unique symbol at a certain time and place to validate the system users [62,63].

In 2018, Wang *et al.* [64] first identified the location of the compromised ECU by monitoring the physical-based delay differences of ECUs and proposed a delay-based IDS for the CAN bus. The motivation is to fingerprint the transmitter

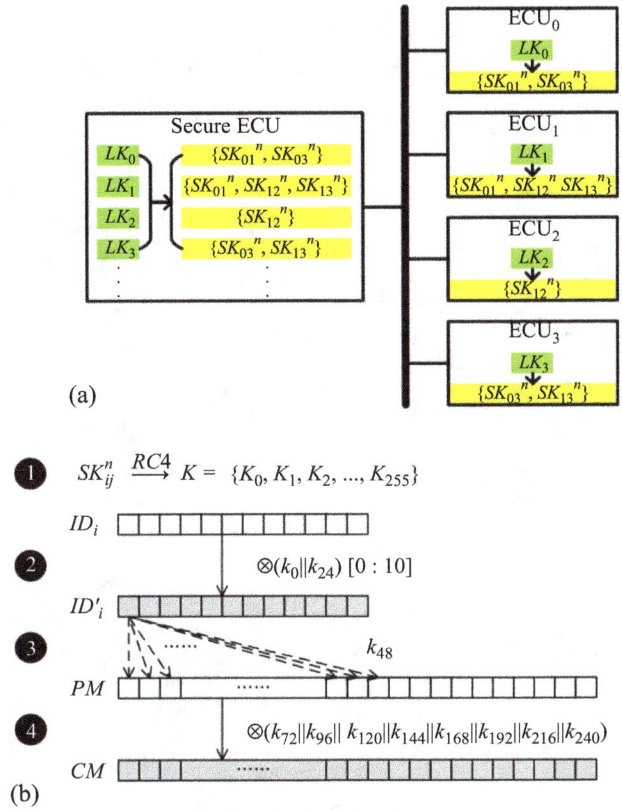

Figure 15.13 (a) Key management mechanism of LEAP and (b) RC4-based encryption and authentication for the CAN message

ECU by the delay difference between the victim ECU and the attack ECU. The two kinds of bus topology with the IDS are shown in Figure 15.14(a) for high-speed CAN and Figure 15.14(b) for low-speed CAN. The monitor is placed to connect to the CAN bus network through two wires directly attached to the bus. The monitor would watch the messages transmitted on the bus and calculate the delay difference from his two plug-in points. Thus, it is crucial to choose the monitor plug-in positions to enlarge the coverage of detection for the bus system. There are three phases for the plug-in-monitor IDS, i.e., the delay measurement, learning parameter settings, and the final intrusion detection. It is capable of an overall detection accuracy above 97% without heavy communication overhead and calculation overhead for the cryptography algorithms.

In the same year, Biham *et al.* [65] presented the Timed CAN, a location-based authentication method for the CAN bus without cryptography. As illustrated in

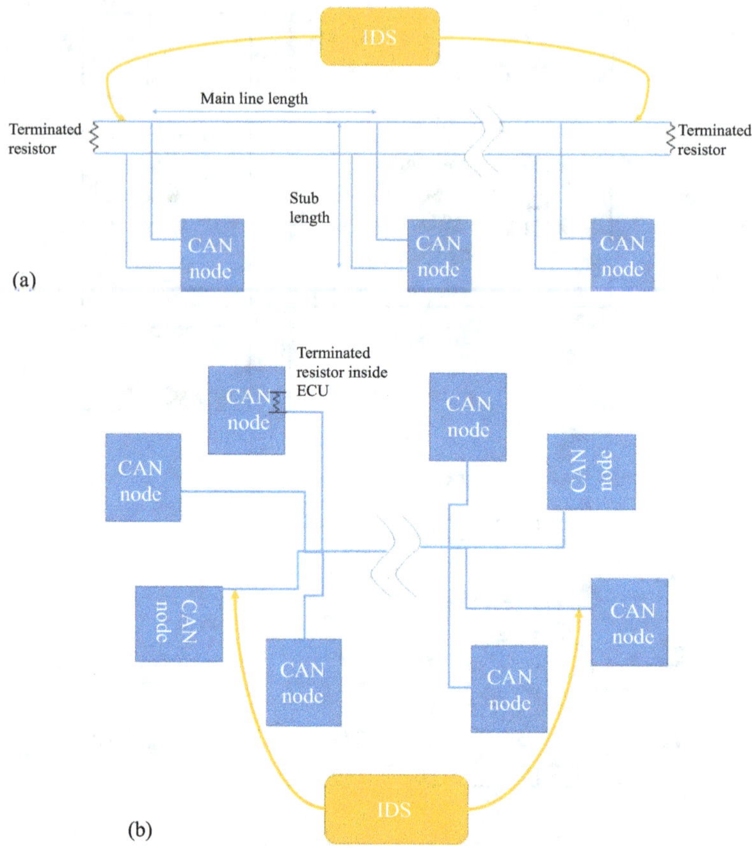

Figure 15.14 *(a) High-speed CAN ISO 11898-2 with the IDS monitor and (b) low-speed CAN ISO 11898-3 with the IDS monitor*

Figure 15.15, two dedicated ECUs (M and R) are installed on the CAN bus to measure the differences between the reception time, which are correlated to the location of the transmitter ECUs (S_1 and S_2) under the assumption of the fixed speed of propagation. The original signal transmitted from S to M and R as well as the echo signal transmitted from R to M are shown in Figure 15.13. Since the echo signal is generated immediately after the original signal reaches R, the time difference $\Delta t_S = 2a_S$; Δt_S depends on the physical location of the transmitter ECUs. In other words, the longer the distance from a transmitter ECU to R is, the longer the time difference Δt_S is. If a malicious ECU attempts to launch the masquerade attack by impersonating any other legitimate ECUs, any fabricated CAN frames transmitted by it will be immediately invalidated.

As the location-based methods do not provide encryption for the CAN frames, the attacker can still eavesdrop on the CAN bus. In addition, if a transmitter ECU

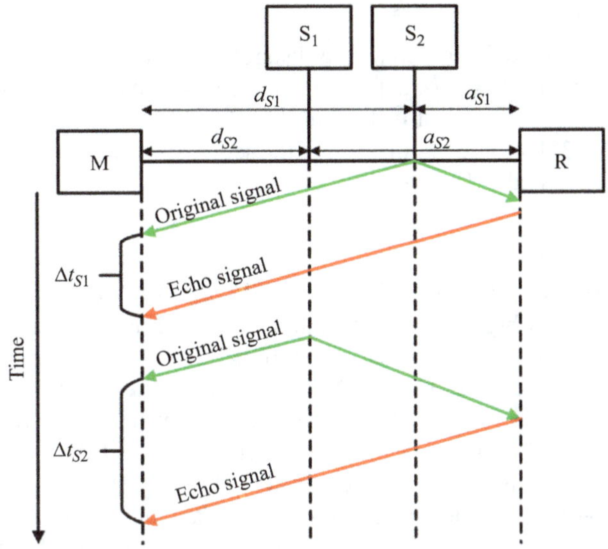

Figure 15.15 Outline of timed CAN with two transmitting ECUs

is fully compromised by the strong attacker, it cannot be stopped from transmitting fabricated CAN frames.

15.6 Future directions

With the advent of the ITSs, more advanced functions will be integrated into the future vehicles. Information security will be the first concern for the manufacturers and drivers, especially in the life-critical applications such as the automatic driving [67–69]. Even with the aforementioned countermeasures, the conventional CAN bus still cannot meet the performance and security requirements of ITSs due to the intrinsic vulnerabilities. Compatibility and cost are the two main challenges for the researchers to design the effective solutions. First, since the life cycle of a vehicle is generally more than 5 years [2], the compatibility between the existing devices and the solutions should be well addressed. Second, it is unacceptable to increase the cost significantly for the information security. Thus, the solutions in the field of Internet [70] cannot be directly applied in the automotive industry. In this section, we analyze the potential research directions to fundamentally secure IVNs.

15.6.1 Replacement of the CAN protocol

Due to the low bandwidth and the lack of security services, the CAN protocol will be inevitably replaced in the future. Thus, one of the potential research directions is to design a new protocol to guarantee better performance and security of IVNs.

15.6.1.1 FlexRay

In 2004, several vehicle manufacturers and suppliers proposed the FlexRay protocol that combined the advantages of determinism and flexibility [71]. Although FlexRay has a bandwidth of 10 Mbit/s, it is difficult for configuration because of its complicated operations [72]. Moreover, the lack of the security services as the CAN protocol makes FlexRay unsuitable for the future IVNs.

15.6.1.2 CAN-FD

In 2012, Robert Bosch GmbH [73] developed the CAN with Flexible Data Rate (CAN-FD) that had a higher communication bandwidth and a larger data payload. CAN-FD is fully compatible with the conventional CAN protocol and has no impact on the current hardware. Thus, CAN-FD is considered as the next-generation protocol for IVNs [74–76]. Figure 15.16 is the comparison between the formats of the CAN and the CAN-FD frames. The Extended Data Length (EDL), the Bit Rate Switch (BRS), and the Error State Indicator (ESI) are added to the control field. EDL indicates a frame as a CAN-FD or a standard CAN frame. The data rate is converted by BRS and an error in the ECU is identified by ESI. Moreover, the data payload is in the range of 12–64 bytes, which is defined by the Data Length Code (DLC).

CAN-FD makes it feasible to construct a secure in-vehicle communication environment. In 2016, Woo *et al.* [75] designed a practical security architecture for IVNs in accordance with the CAN-FD specification. The Second Authenticated Key Exchange Protocol [77] is used to implement a session key distribution and update mechanism. The encryption and authentication technique for the CAN-FD frames is identical to that proposed by [26]. The CAN-FD-based security architecture provides confidentiality, authentication, and access control for IVNs without impacting the communication performance. In 2018, Xie *et al.* [78] defined an AUTOSAR-complaint system model as the foundation for the Design Space Exploration of CAN-FD that integrated both the timing and the security constraints. In order to make full advantage of the large data payload of the CAN-FD protocol, they divided the Mixed-Integer Linear Programming formulation into the signal-to-message packing

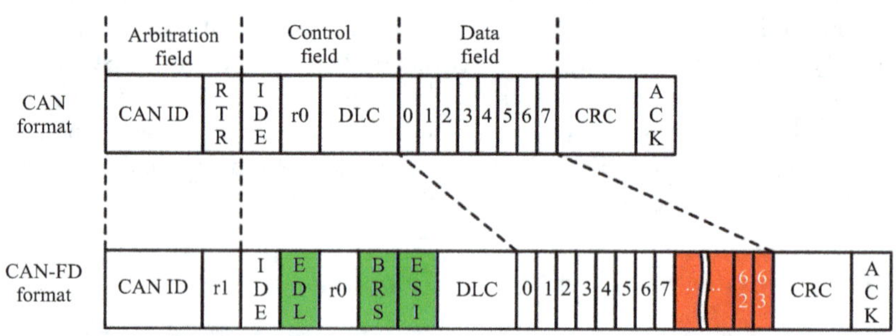

Figure 15.16 Comparison between the formats of the CAN and CAN-FD frames

with security constraint according to AUTOSAR's specification [79] and the message scheduling for the timing verification.

15.6.1.3 Automotive Ethernet

As predicted by [80], the automotive Ethernet penetration in new vehicles would grow from 1% in 2014 to 40% in 2020. The standard automotive Ethernet has a bandwidth of up to 100 Mb/s similar to the fast Ethernet in the common Local Area Networks (LANs) [81]. Unlike the previous generations of the automotive Ethernet focusing on interconnecting the domain controllers, the third generation introduces the backbone of IVNs [82–84]. Compared to other existing protocols for IVNs, the automotive Ethernet has very high payload and transmission efficiency [5]. A standard Ethernet packet format in IEEE 802.3 is shown in Figure 15.17. It indicates that the maximum payload of an Ethernet packet is 1,500 bytes, while the packet length is 1,538 bytes. Thus, the maximum transmission efficiency of an Ethernet packet equals to 97.53%. Since the automotive Ethernet is designed with the security constraint of AUTOSAR's specification, several effective security solutions [8,85] for the automotive Ethernet-based communications have recently been proposed. In addition to enhancing the security of the automotive Ethernet, another mainstream trend is to address the issue for transmitting the time-critical messages, which is critical to the automatic driving vehicles. IEEE Audio Video Bridge (AVB) [86] task group has defined the timing and synchronization, the stream reservation, and the forwarding and queuing enhancement for the audio/video bridging systems. AVB uses precisely synchronized global time to coordinate the transmission and reception for the packets in IVNs. Instead of maintaining the Quality of Service through resource reservation

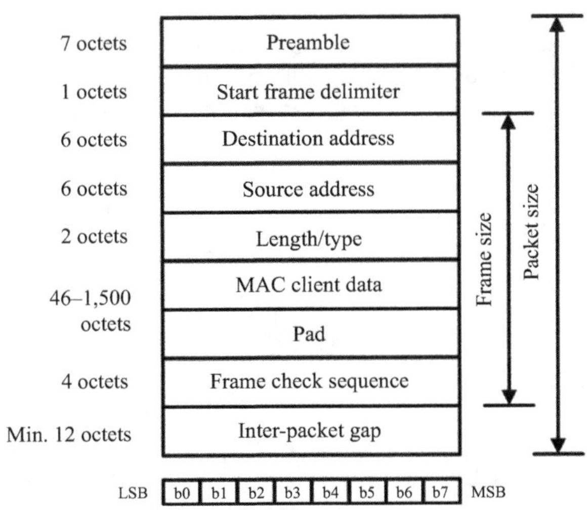

Figure 15.17 IEEE 802.3 ethernet packet format

and data prioritization, TTEthernet [87] used a preset table with global time synchronization to regulate data transmission and reception. TTEthernet defines three types of traffics: the Time Triggered (TT) traffic, the Rate Constrained (RC) traffic, and the Best Effort (BE) traffic. The latency of the TT traffic is small due to its highest priority, whereas the RC and BE traffics are event-triggered types [5]. This mechanism makes TTEthernet compatible with the conventional LAN without any protocol translation [88].

15.6.2 Next-generation gateway

In the existing vehicles, IVNs are generally split into different functional domains according to the functionality or physical location [13]. In order to avoid the difficulty to maintain the inflexible combinations caused by different protocols and topologies, Hank *et al.* [89] proposed the top-down approach as shown on the right side of Figure 15.18. As the interfaces between different domains, the automotive gateways are vital to the overall security of IVNs [90–93]. The next-generation secure gateway is responsible for three main functionalities [94,95]. First, as the bridge between different protocols, the gateway facilitates data transmission across different domains and regulates the bandwidth to avoid overloading one section of IVNs. Second, the gateway works as the firewall to isolate the unauthorized external access. Third, with high security level and abundant computational resources, the security mechanisms can be implemented on the gateway to detect anomaly or provide the security services. Thus, the secure gateway is another potential research direction to guarantee the security of the future IVNs.

In 2015, Kim *et al.* [93] proposed a gateway framework based on CAN, FlexRay, and Ethernet. It supports both the current IVNs using a central gateway and the future IVNs using the domain controllers as gateways. A security module in the gateway provides the authentication and encryption for the frames. In 2018, NXP Semiconductors [96] provided the dual-chip solution for the next-generation gateway using the MPC5748G microcontroller unit and the LS1043A microprocessor unit. As shown in Figure 15.19, MPC5748G has the Hardware Security Modules for the

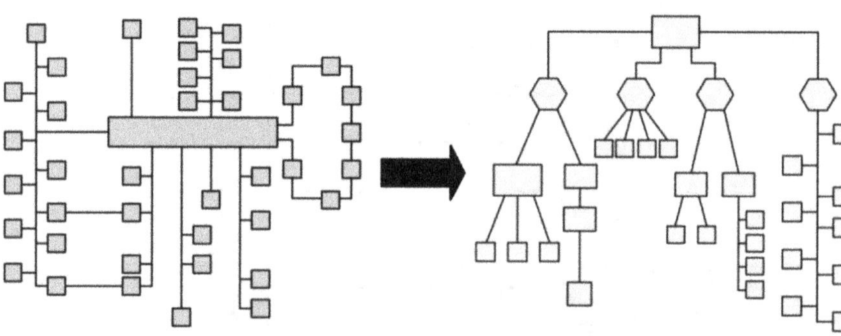

Figure 15.18 Migration from heterogeneous architecture to future top-down approach

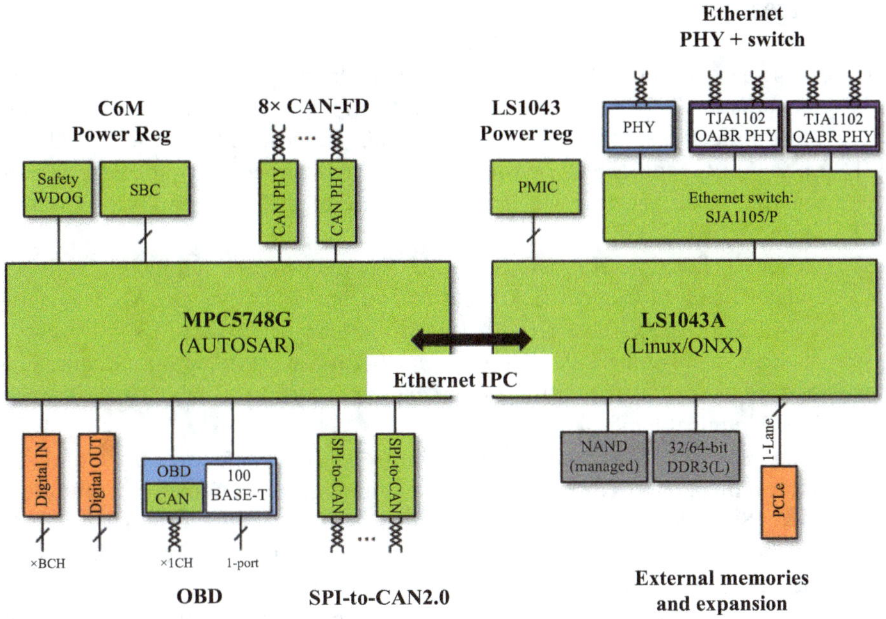

Figure 15.19 Dual-chip solution for the next-generation gateway [96]

cryptographic algorithms, and LS1043A supports high-level operating systems and hypervisors. With this dual-chip platform, the enormous amount of data from IVNs can be leveraged and processed for more complex security mechanisms [94].

15.7 Conclusions

The ECUs connected on the IVNs are responsible for processing huge volumes of information obtained from both in-vehicle and outside environment. However, the intrinsic vulnerabilities of current IVNs leave opportunities for the adversaries to remotely compromise an unaltered vehicle, which seriously threatens the safety of drivers and passengers. In this survey, we first review the basic knowledge of the most widely used CAN protocol. Then the vulnerabilities and the attack interfaces are assessed briefly. We explain the practical attack models in recent years and elaborate on how to manipulate a vehicle without physical access. Next, two types of countermeasures are analyzed but none of them can completely secure IVNs. We believe that the current in-vehicle bus protocols will fail to meet the requirements of performance and security for the ITSs. Thus, next-generation protocols and architecture of IVNs should be promising research directions. In summary, this survey fills the gaps and provides a comprehensive report about the latest advances on the security issues of IVNs.

References

[1] Kimm H and Ham Hs. Integrated fault tolerant system for automotive bus networks. In: 2010 Second International Conference on Computer Engineering and Applications. vol. 1. IEEE, Bali Island, Indonesia; 2010. p. 486–490.

[2] Lu Z, Qu G, and Liu Z. A survey on recent advances in vehicular network security, trust, and privacy. IEEE Transactions on Intelligent Transportation Systems. 2018;(99):1–17.

[3] Steger M, Dorri A, Kanhere SS, *et al.* Secure wireless automotive software updates using blockchains: A proof of concept. In: Advanced Microsystems for Automotive Applications 2017. Berlin: Springer; 2018. p. 137–149.

[4] Zheng Y, Shokouhi N, Thomsen N, *et al.* Towards developing a distraction-reduced hands-off interactive driving experience using portable smart devices. SAE Technical Paper; 2016.

[5] Zeng W, Khalid MA, and Chowdhury S. In-vehicle networks outlook: Achievements and challenges. IEEE Communications Surveys & Tutorials. 2016;18(3):1552–1571.

[6] Huang J, Zhao M, Zhou Y, *et al.* In-vehicle networking: Protocols, challenges, and solutions. IEEE Network. 2018;(99):1–7.

[7] Dvořák J and Hanzálek Z. Using two independent channels with gateway for FlexRay static segment scheduling. IEEE Transactions on Industrial Informatics. 2016;12(5):1887–1895.

[8] Carnevale B, Fanucci L, Bisase S, *et al.* MACsec-based security for automotive Ethernet backbones. Journal of Circuits, Systems and Computers. 2018;27(05):1850082.

[9] Koscher K, Czeskis A, Roesner F, *et al.* Experimental security analysis of a modern automobile. In: 2010 IEEE Symposium on Security and Privacy. IEEE, Oakland, CA, USA; 2010. p. 447–462.

[10] Ray S, Chen W, Bhadra J, *et al.* Extensibility in automotive security: Current practice and challenges. In: 2017 54th ACM/EDAC/IEEE Design Automation Conference (DAC). IEEE, Austin, TX, USA; 2017. p. 1–6.

[11] Checkoway S, McCoy D, Kantor B, *et al.* Comprehensive experimental analyses of automotive attack surfaces. In: USENIX Security Symposium. vol. 4. San Francisco, CA, USA; 2011.

[12] Miller C and Valasek C. Remote exploitation of an unaltered passenger vehicle. Black Hat USA. 2015;2015:91.

[13] Tuohy S, Glavin M, Hughes C, *et al.* Intra-vehicle networks: A review. IEEE Transactions on Intelligent Transportation Systems. 2015;16(2): 534–545.

[14] Liu J, Zhang S, Sun W, *et al.* In-vehicle network attacks and countermeasures: Challenges and future directions. IEEE Network. 2017;31(5):50–58.

[15] Wang Q, Lu Z, Gao M, *et al.* Edge computing based GPS spoofing detection methods. In: 2018 IEEE 23rd International Conference on Digital Signal Processing (DSP). IEEE, Shanghai, China; 2018. p. 1–5.

[16] Di Natale M, Zeng H, Giusto P, *et al.* Understanding and using the controller area network communication protocol: Theory and practice. New York, NY: Springer Science & Business Media; 2012.

[17] Avatefipour O and Malik H. State-of-the-art survey on in-vehicle network communication (CAN-Bus) security and vulnerabilities. arXiv preprint arXiv:180201725. 2018.

[18] Hafeez A, Malik H, Avatefipour O, *et al.* Comparative study of can-bus and flexray protocols for in-vehicle communication. SAE Technical Paper; 2017.

[19] Miller C and Valasek C. A survey of remote automotive attack surfaces. Black Hat USA. 2014;2014:94.

[20] Foster I, Prudhomme A, Koscher K, *et al.* Fast and vulnerable: A story of telematic failures. In: 9th USENIX Workshop on Offensive Technologies (WOOT 15), Washington, DC, USA; 2015.

[21] Cho KT and Shin KG. Fingerprinting electronic control units for vehicle intrusion detection. In: 25th USENIX Security Symposium (USENIX Security 16), Austin, TX, USA; 2016. p. 911–927.

[22] Miller C and Valasek C. Adventures in automotive networks and control units. Def Con. 2013;21:260–264.

[23] Automobiles FC. 2013-2017 Uconnect Voice Recognition/Control Centers with Audio Systems. Accessed January 4, 2018. https://www.allpar.com/corporate/tech/uconnect.html.

[24] Hazem A and Fahmy H. LCAP—A lightweight can authentication protocol for securing in-vehicle networks. In: 10th ESCAR Embedded Security in Cars Conference, Berlin, Germany. vol. 6; 2012.

[25] Noureldeen P, Azer MA, Refaat A, *et al.* Replay attack on lightweight CAN authentication protocol. In: 2017 12th International Conference on Computer Engineering and Systems (ICCES). IEEE, Cairo, Egypt; 2017. p. 600–606.

[26] Woo S, Jo HJ, and Lee DH. A practical wireless attack on the connected car and security protocol for in-vehicle CAN. IEEE Transactions on Intelligent Transportation Systems. 2015;16(2):993–1006.

[27] Cho KT and Shin KG. Error handling of in-vehicle networks makes them vulnerable. In: Proceedings of the 2016 ACM SIGSAC Conference on Computer and Communications Security. ACM, Vienna, Austria; 2016. p. 1044–1055.

[28] Nie S, Liu L, and Du Y. Free-fall: Hacking tesla from wireless to CAN bus. Briefing, Black Hat USA. 2017. p. 1–16.

[29] Greenberg A. Tesla Responds to Chinese Hack with a Major Security Upgrade. Accessed September 27, 2016. https://www.wired.com/2016/09/tesla-responds-chinese-hack-major-security-upgrade/.

[30] Fadlullah ZM, Nishiyama H, Kato N, *et al.* Intrusion detection system (IDS) for combating attacks against cognitive radio networks. IEEE Network. 2013;27(3):51–56.

[31] Dagan T and Wool A. Parrot, a software-only anti-spoofing defense system for the CAN bus. In: ESCAR EUROPE, Munich, Germany; 2016.

[32] Dagan T and Wool A. Testing the boundaries of the Parrot anti-spoofing defense system. In: 5th Embedded Security in Cars (ESCAR USA'17), Ypsilanti, MI, USA; 2017.

[33] Ujiie Y, Kishikawa T, Haga T, *et al.* A method for disabling malicious CAN messages by using a centralized monitoring and interceptor ECU. In: 13th Int. Conf. on Embedded Security in Cars (ESCAR 2015); 2015.

[34] Song HM, Kim HR, and Kim HK. Intrusion detection system based on the analysis of time intervals of CAN messages for in-vehicle network. In: 2016 International Conference on Information Networking (ICOIN). IEEE, Kota Kinabalu, Malaysia; 2016. p. 63–68.

[35] Kuwahara T, Baba Y, Kashima H, *et al.* Supervised and unsupervised intrusion detection based on CAN message frequencies for in-vehicle network. Journal of Information Processing. 2018;26:306–313.

[36] Moore MR, Bridges RA, Combs FL, *et al.* Modeling inter-signal arrival times for accurate detection of CAN bus signal injection attacks: A data-driven approach to in-vehicle intrusion detection. In: Proceedings of the 12th Annual Conference on Cyber and Information Security Research. ACM, Oak Ridge, TN, USA; 2017. p. 11.

[37] Taylor A, Japkowicz N, and Leblanc S. Frequency-based anomaly detection for the automotive CAN bus. In: 2015 World Congress on Industrial Control Systems Security (WCICSS). IEEE, London, UK; 2015. p. 45–49.

[38] Marchetti M and Stabili D. Anomaly detection of CAN bus messages through analysis of ID sequences. In: 2017 IEEE Intelligent Vehicles Symposium (IV). IEEE, Redondo Beach, CA, USA; 2017. p. 1577–1583.

[39] Kohno T, Broido A, and Claffy KC. Remote physical device fingerprinting. IEEE Transactions on Dependable and Secure Computing. 2005;2(2):93–108.

[40] Zander S and Murdoch SJ. An improved clock-skew measurement technique for revealing hidden services. In: USENIX Security Symposium, San Jose, CA, USA; 2008. p. 211–226.

[41] Tayyab M, Hafeez A, and Malik H. Clock phishing attack on clock based intrusion detection systems for can protocol. In: 6th Embedded Security in Cars (ESCAR USA'18), Ypsilanti, MI, USA; 2018.

[42] Sagong SU, Ying X, Clark A, *et al.* Cloaking the clock: Emulating clock skew in controller area networks. In: Proceedings of the 9th ACM/IEEE International Conference on Cyber-Physical Systems. IEEE, Porto, Portugal; 2018. p. 32–42.

[43] Cho KT and Shin KG. Viden: Attacker identification on in-vehicle networks. In: Proceedings of the 2017 ACM SIGSAC Conference on Computer and Communications Security. ACM, Dallas, TX, USA; 2017. p. 1109–1123.

[44] Murvay PS and Groza B. Source identification using signal characteristics in controller area networks. IEEE Signal Processing Letters. 2014;21(4): 395–399.

[45] Choi W, Jo HJ, Woo S, *et al.* Identifying ECUs using inimitable characteristics of signals in controller area networks. IEEE Transactions on Vehicular Technology. 2018;67(6):4757–4770.

[46] Choi W, Joo K, Jo HJ, *et al.* VoltageIDS: Low-level communication characteristics for automotive intrusion detection system. IEEE Transactions on Information Forensics and Security. 2018;13(8):2114–2129.

[47] Genge B, Haller P, and Kiss I. A framework for designing resilient distributed intrusion detection systems for critical infrastructures. International Journal of Critical Infrastructure Protection. 2016;15:3–11.

[48] Sagong SU, Ying X, Poovendran R, *et al.* Exploring attack surfaces of voltage-based intrusion detection systems in controller area networks. In: ESCAR Europe. vol. 2018, Brussels, Belgium; 2018.

[49] Groza B and Murvay PS. Efficient intrusion detection with bloom filtering in controller area networks. IEEE Transactions on Information Forensics and Security. 2019;14(4):1037–1051.

[50] Kuwahara T, Baba Y, Kashima H, *et al.* Payload-based statistical intrusion detection for in-vehicle networks. In: Pacific-Asia Conference on Knowledge Discovery and Data Mining. Cham: Springer; 2018. p. 186–192.

[51] Wang Q, Lu Z, and Qu G. An entropy analysis based intrusion detection system for controller area network in vehicles. In: 2018 31st IEEE International System-on-Chip Conference (SOCC). IEEE, Arlington, VA, USA; 2018. p. 90–95.

[52] Kang MJ and Kang JW. A novel intrusion detection method using deep neural network for in-vehicle network security. In: 2016 IEEE 83rd Vehicular Technology Conference (VTC Spring). IEEE, Nanjing, China; 2016. p. 1–5.

[53] Weber M, Klug S, Sax E, *et al.* Embedded hybrid anomaly detection for automotive CAN communication. In: 9th European Congress on Embedded Real Time Software and Systems (ERTS 2018), Toulouse, France; 2018.

[54] Seo E, Song HM, and Kim HK. GIDS: GAN based intrusion detection system for in-vehicle network. In: 2018 16th Annual Conference on Privacy, Security and Trust (PST). IEEE, Belfast, Northern Ireland, UK; 2018. p. 1–6.

[55] Van Herrewege A, Singelee D, and Verbauwhede I. CANAuth-a simple, backward compatible broadcast authentication protocol for CAN bus. In: ECRYPT Workshop on Lightweight Cryptography. vol. 2011, Louvain-la-Neuve, Belgium; 2011.

[56] Han K, Weimerskirch A, and Shin KG. A practical solution to achieve real-time performance in the automotive network by randomizing frame identifier. In: Proc. Eur. Embedded Secur. Cars (ESCAR), Cologne, Germany; 2015. p. 13–29.

[57] Kurachi R, Matsubara Y, Takada H, *et al.* CaCAN—Centralized authentication system in CAN (controller area network). In: 14th Int. Conf. on Embedded Security in Cars (ESCAR 2014), Hamburg, Germany; 2014.

[58] Radu AI and Garcia FD. LeiA: A lightweight authentication protocol for CAN. In: European Symposium on Research in Computer Security. Cham: Springer; 2016. p. 283–300.

[59] Farag WA. CANTrack: Enhancing automotive CAN bus security using intuitive encryption algorithms. In: 2017 7th International Conference on Modeling, Simulation, and Applied Optimization (ICMSAO). IEEE, Sharjah, UAE; 2017. p. 1–5.

[60] Wang E, Xu W, Sastry S, *et al.* Hardware module-based message authentication in intra-vehicle networks. In: 2017 ACM/IEEE 8th International Conference on Cyber-Physical Systems (ICCPS). IEEE, Pittsburgh, PA, USA; 2017. p. 207–216.

[61] Lu Z, Wang Q, Chen X, *et al.* LEAP: A lightweight encryption and authentication protocol for in-vehicle communications. arXiv preprint arXiv: 190910380. 2019.

[62] Xiao L, Yan Q, Lou W, *et al.* Proximity-based security techniques for mobile users in wireless networks. IEEE Transactions on Information Forensics and Security. 2013;8(12):2089–2100.

[63] Kawamoto Y, Nishiyama H, Kato N, *et al.* Effectively collecting data for the location-based authentication in Internet of Things. IEEE Systems Journal. 2017;11(3):1403–1411.

[64] Wang Q, Qian Y, Lu Z, *et al.* A delay based plug-in-monitor for intrusion detection in controller area network. In: 2018 Asian Hardware Oriented Security and Trust Symposium (AsianHOST). IEEE, Hong Kong, China; 2018. p. 86–91.

[65] Biham E, Bitan S, and Gavril E. TCAN: Authentication without cryptography on a CAN bus based on nodes location on the bus. In: 6th Embedded Security in Cars (ESCAR USA'18), Ypsilanti, MI, USA; 2018.

[66] Krawczyk H, Bellare M, and Canetti R. HMAC: Keyed-hashing for message authentication; 1997.

[67] Liu P, Liu B, Sun Y, *et al.* Mitigating DoS attacks against pseudonymous authentication through puzzle-based co-authentication in 5G-VANET. IEEE Access. 2018;6:20795–20806.

[68] Lu Z, Liu W, Wang Q, *et al.* A privacy-preserving trust model based on blockchain for VANETs. IEEE Access. 2018;6:45655–45664.

[69] Lu Z, Wang Q, Qu G, *et al.* Bars: A blockchain-based anonymous reputation system for trust management in VANETs. In: 2018 17th IEEE International Conference on Trust, Security and Privacy in Computing and Communications/12th IEEE International Conference on Big Data Science and Engineering (TrustCom/BigDataSE). IEEE, New York, NY, USA; 2018. p. 98–103.

[70] Pilz J, Mehlhose M, Wirth T, *et al.* A Tactile Internet demonstration: 1 ms ultra low delay for wireless communications towards 5G. In: 2016 IEEE Conference on Computer Communications Workshops (INFOCOM WKSHPS). IEEE, San Francisco, CA, USA; 2016. p. 862–863.

[71] Consortium F. FlexRay Communications System Protocol Specification Version 2.1. Accessed December 22, 2005. https://www.softwareresearch.net/fileadmin/src/docs/teaching/SS08/PS_VS/FlexRayCommunicationSystem.pdf.

[72] Park I and Sunwoo M. FlexRay network parameter optimization method for automotive applications. IEEE Transactions on Industrial Electronics. 2011;58(4):1449–1459.

[73] Hartwich F. CAN with flexible data-rate. In: Proc. iCC. Citeseer; 2012. p. 1–9.

[74] Bordoloi UD and Samii S. The frame packing problem for CAN-FD. In: 2014 IEEE Real-Time Systems Symposium. IEEE, Rome, Italy; 2014. p. 284–293.

[75] Woo S, Jo HJ, Kim IS, *et al.* A practical security architecture for in-vehicle CAN-FD. IEEE Transactions on Intelligent Transportation Systems. 2016;17(8):2248–2261.

[76] Nguyen TH, Cheon BM, and Jeon JW. CAN FD performance analysis for ECU re-programming using the CANoe. In: The 18th IEEE International Symposium on Consumer Electronics (ISCE 2014). IEEE, Jeju, South Korea; 2014. p. 1–4.

[77] Bellare M and Rogaway P. Entity authentication and key distribution. In: Annual International Cryptology Conference. Berlin, Heidelberg: Springer; 1993. p. 232–249.

[78] Xie Y, Zeng G, Kurachi R, *et al.* Security/timing-aware design space exploration of CAN FD for automotive cyber-physical systems. IEEE Transactions on Industrial Informatics. 2019;15(2):1094–1104.

[79] Bernardeschi C, Di Natale M, Dini G, *et al.* Modeling and generation of secure component communications in AUTOSAR. In: Proceedings of the Symposium on Applied Computing. ACM, Marrakech, Morocco; 2017. p. 1473–1480.

[80] Oyster B. Ethernet In-vehicle Networking to Feature in 40% of Vehicles Shipping Globally by 2020. Accessed January 10, 2014. https://www.abiresearch.com/press/ethernet-in-vehicle-networking-to-feature-in-40-of/.

[81] Wang R, Wu J, Qian Z, *et al.* A graph theory based energy routing algorithm in energy local area network. IEEE Transactions on Industrial Informatics. 2017;13(6):3275–3285.

[82] Schmidt K, Dannebaum U, and Zweck H. GBit Ethernet—The solution for future in-vehicle network requirements? SAE International Journal of Passenger Cars—Electronic and Electrical Systems. 2015;8(2015-01-0200): 289–295.

[83] Hank P, Müller S, Vermesan O, *et al.* Automotive Ethernet: In-vehicle networking and smart mobility. In: Proceedings of the Conference on Design, Automation and Test in Europe. EDA Consortium, Grenoble, France; 2013. p. 1735–1739.

[84] van Dijk L and Sporer G. Functional safety for automotive Ethernet networks. Journal of Traffic and Transportation Engineering. 2018;6(4):176–182.

[85] Corbett C, Schoch E, Kargl F, *et al.* Automotive Ethernet: Security opportunity or challenge? Sicherheit 2016—Sicherheit, Schutz und Zuverlässigkeit; 2016.

[86] Bello LL. Novel trends in automotive networks: A perspective on Ethernet and the IEEE Audio Video Bridging. In: Proceedings of the 2014 IEEE Emerging Technology and Factory Automation (ETFA). IEEE, Barcelona, Spain; 2014. p. 1–8.

[87] Tămaş-Selicean D, Pop P, and Steiner W. Design optimization of TTEthernet-based distributed real-time systems. Real-Time Systems. 2015;51(1):1–35.

[88] Schmidt V, Georgiev P, Horn H, *et al.* Evaluation of numerical bus systems used in rocket engine test facilities. Cell. 2013;5(U1):P5.

[89] Hank P, Suermann T, and Mueller S. Automotive Ethernet, a holistic approach for a next generation in-vehicle networking standard. In: Advanced Microsystems for Automotive Applications 2012. Berlin, Heidelberg: Springer; 2012. p. 79–89.

[90] Berg J, Pommer J, Jin C, *et al.* Secure gateway-a concept for an in-vehicle IP network bridging the infotainment and the safety critical domains. In: 13th Embedded Security in Cars (ESCAR'15); 2015.

[91] Kim MH, Lee S, and Lee KC. Performance evaluation of node-mapping-based Flexray-CAN gateway for in-vehicle networking system. Intelligent Automation & Soft Computing. 2015;21(2):251–263.

[92] Kurachi R, Takada H, Mizutani T, *et al.* SecGW secure gateway for in-vehicle networks. In: 13th Int. Conf. on Embedded Security in Cars (ESCAR 2015); 2015.

[93] Kim JH, Seo SH, Hai NT, *et al.* Gateway framework for in-vehicle networks based on CAN, FlexRay, and Ethernet. IEEE Transactions on Vehicular Technology. 2015;64(10):4472–4486.

[94] Jeong Y, Son S, Jeong E, *et al.* An integrated self-diagnosis system for an autonomous vehicle based on an IoT gateway and deep learning. Applied Sciences. 2018;8(7):1164.

[95] Sharef B, Alsaqour R, Alawi M, *et al.* Robust and trust dynamic mobile gateway selection in heterogeneous VANET-UMTS network. Vehicular Communications. 2018;12:75–87.

[96] Fortenberry R. Secure Networks & Gateways: Vehicle Network Processing with the MPC5748G MCU + LS1043A Processor. Accessed June 19, 2018. https://community.nxp.com/docs/DOC-340906.

Index

www.ingramcontent.com/pod-product-compliance
Lightning Source LLC
LaVergne TN
LVHW011559200926
848484LV00001B/333